W9-CEP-554

ANIMAL BEHAVIOR

ANIMAL

BEHAVIOR

Psychobiology, Ethology and Evolution

DAVID McFARLAND

University of Oxford

The Benjamin/Cummings Publishing Company, Inc.
MENLO PARK CALIFORNIA, READING MASSACHUSETTS

Sponsoring and Developmental Editor: Paul Elias
Copyeditor: Elliot Simon
Production Manager: Margaret Moore
Artwork: Oxford Illustrators Ltd
Book and Cover Design: Barrie Carr

Published simultaneously in Great Britain by Pitman Publishing Ltd.

Library of Congress Cataloging in Publication Data

McFarland, David.
Animal behavior.

Bibliography: p. 533
Includes Index.
1. Animal behavior. 2. Behavior evolution. I. Title.
QL751.M393 1985 591.5′1 84-28345
ISBN 0-8053-6790-X

The Benjamin/Cummings Publishing Company, Inc.
2727 Sand Hill Road
Menlo Park, California 94025

For Duncan and Kirsty

Contents

PART 2 *Mechanisms of Behavior*

Preface

The modern study of animal behavior has its roots in three different lines of scientific thought: the psychological, the physiological and the zoological. These three traditions have given rise to comparative and biological psychology, to the evolutionary analysis of behavior and to their combination in the modern field of ethology. This lattice of intersecting fields can bewilder the beginning student and often only one perspective is developed in a first course. The biology major who is never introduced to the work in comparative psychology loses as much as the psychology student who fails to learn the evolutionary approach.

This book provides a balanced and integrated introductory treatment of the entire domain of animal behavior. It is intended for both biology and psychology students enrolled in their first course in animal behavior, and recognizes the diversity of their backgrounds in the natural sciences and psychology. I have, therefore, been careful to define and develop each new concept without reliance on previous coursework.

Student Learning Aids

Every effort had been made to make this book a superior learning tool for students. Although concepts are developed to a sophisticated level, explanations are thorough and new vocabulary is defined and italicized where first introduced. I have made a special effort to explain ideas in non-mathematical terms throughout.

Each of the three parts of the book, and each three chapter section is prefaced by an introductory overview. Each section is begun with a profile of the life of a scientist whose work is central to the subject treated within the section. In addition, a set of points to remember and a list of suggested further readings appears at the end of each chapter. Original work has been cited directly in the text throughout and is fully referenced in a list at the back of the book. This reference list also serves as an index to authors by indicating which page or pages of the text contain each citation. A bibliography of films on animal behavior is available from the publisher upon written request.

Acknowledgments

I am grateful to the many colleagues who have provided permission to use photographs and illustrations. I am also indebted to the many specialists and generalists who reviewed the manuscript during various stages in its development. Although I alone bear all responsibility for any errors, they are fewer because of these helpful reviews. A list of reviewers is found below. Special thanks for editorial assistance go to Navin Sullivan, Jeremy Swinfen Green and Paul Elias. Finally, I thank especially Wendy Eadle, Sue Pusey and Audrey Wesch, who helped with the preparation of the typescript.

David McFarland, Oxford, England, January, 1985

List of Reviewers

R J Andrew (University of Sussex)
Edwin M Banks (University of Illinois)
George W Barlow (University of California, Berkeley)
C J Barnard (University of Nottingham)
D W Dickins (University of Liverpool)
Douglas D Dow (University of Queensland, Australia)
H Carl Gerhardt (University of Missouri, Columbia)
Jeremy Hatch (University of Massachusetts, Boston)
Felicity Huntingford (University of Glasgow)
P J Jarman (University of New England, Australia)
Randall Lockwood (State University of New York, Stony Brook)
Glen McBride (University of Queensland, Australia)
G A Parker (University of Liverpool)
John Staddon (Duke University)
William Timberlake (Indiana University)

Cover Photographs

The publishers are indebted to the following for the cover illustrations: (front cover) the cleaner wrasse *Labroides dimidiatus* and its host fish a coral cod (*Cephalopholis miniatus*) by courtesy of Steve Parish Photography, Paddington, Queensland, Australia; (back cover) swallow and chicks: photograph Hans Reinhard by courtesy of Bruce Coleman Ltd; hawkmoth caterpillar: photograph Nicholas Smythe; Namib lizard: photograph Gideon Louw; grasshopper *Arantia rectifolia*: photograph Malcolm Edmunds; herring gull feeding chick: photograph Jim Shaffery.

1 Introduction to the Study of Animal Behavior

The scientific study of animal behavior involves a variety of approaches. Behavior can be explained in terms of its evolutionary history, in terms of the benefits it brings to the animal, in terms of psychological mechanisms, and in terms of physiological mechanisms. Which approach you take depends upon what you want to know about animal behavior. In this book we explore the numerous ways of investigating animal behavior as well as many different aspects of behavior, from the simple responses of primitive animals to the mental life of the great apes.

1.1 Asking questions about behavior

Animals in their natural environment usually are highly adapted to their particular circumstances. For example, herring gulls make their nest upon the ground, where the eggs and chicks are vulnerable to predators. As we see later, the nest is well camouflaged, and the parent birds guard it as much as possible. The male and female each take turns foraging for food while the other remains on or near the nest. Many aspects of the behavior of herring gulls have been studied, including their behavior toward predators, their response to eggs that roll out of the nest, the cooperation between male and female, etc. In addition to asking questions about mechanisms such as those that enable the bird to recognize an egg, scientists ask questions of a different kind, such as why the eggs have a particular coloration, or why the parents sometimes engage in an elaborate nest-relief ceremony when they exchange incubation duties. Questions about the adaptedness of behavior are quite different from questions about how particular mechanisms work, and the answers require a different outlook. Suppose we wish to ask why birds sit on eggs (Figs. 1.1 and 1.2). The answer depends very much upon the way in which the scientist looks at the question. Let us consider four possibilities.

1. Why do birds sit on *eggs*?
2. Why do birds *sit on* eggs?
3. Why do *birds* sit on eggs?
4. *Why* do birds sit on eggs?

Fig. 1.1 Herring gull nest with typical clutch of three camouflaged eggs (*Photograph: Nigel Ball*)

Fig. 1.2 Herring gull sitting on its eggs (*Photograph: Jim Shaffery*)

Clearly, by emphasizing different parts of the question, different types of answers become appropriate. Thus, in the first example the implied question is why birds sit on *eggs* rather than stones or flowers. To answer this question it is necessary to show what stimulus characteristics of the egg, such as shape, color, or markings, are important in eliciting the sitting response. Many experiments have been done on the mechanisms of egg recognition in birds, and shape, size, and coloration have been shown to be important (see Chapter 12.6).

The second question emphasizes what the birds *do*: They *sit on* their eggs rather than stand on them or eat them. Here the answer must be in terms of the animals' motivation. Thus, birds sit on eggs when they are broody, but they may eat them when they are hungry. The scientist must define what he means by "broody" and by "hungry", a task that will require knowledge of the physiology of the animal.

Whereas questions 1 and 2 require answers in terms of immediate cause or mechanism, question 3 implies causality of a different order. Why do *birds* rather than cats or pigs sit on eggs? The answer is that birds have the hereditary predisposition to produce eggs and to develop the behavioral skills to respond to them appropriately. The genetic factors involved are ultimately necessary for the development of the behavior. The manner in which such traits are handed from one generation to the next is the subject of the study of behavior genetics, an important part of the study of behavior.

In answering the fourth type of question, we might say that birds sit on eggs in order to hatch them. The birds may not anticipate the consequences of their behavior but only birds that are programmed to behave in this way leave offspring. This type of argument, first put forward by Charles Darwin, implies that the survival value of any inherited trait is determined by natural selection; that is, the extent to which a trait is passed from one generation to the next in a wild population is determined by the breeding success of the parent generation and the value of the trait in enabling the animals to survive natural hazards such as food shortage, predators, and sexual rivals.

Thus, in answering questions about behavior, the scientist can take several points of view. In general, psychologists are interested in mechanisms that control behavior, and evolutionary biologists are interested in how the mechanisms came to be as they are. Ethologists believe that the distinction between mechanism and design is fundamental to the study of animal behavior. Birds sit on eggs because certain mechanisms cause them to do so. They are designed to sit on eggs (by natural selection) because the behavior fulfills a function that is important to their survival and reproduction. Questions about both design and mechanism are essential for a full understanding of animal behavior.

In this book we follow the ethological viewpoint in emphasizing the importance of both mechanistic and evolutionary explanations of behavior. Another way of appreciating the distinction is to consider the variety of ways in which animals adapt to changes in environmental conditions. First, we should distinguish between *genotypic adaptation*, in which the adjustment is genetic and takes place through evolution by natural selection, and *phenotypic adaptation*, which takes place within the individual animal on a non-genetic basis. An example of genotypic adaptation can be seen in the display which the hawkmoth caterpillar (*Leucorampha*) uses to deter predators. The caterpillar rests upside down beneath a leaf or branch. When disturbed it raises its head and inflates it. The head bears false eyes and resembles that of a snake, as shown in Figure 1.3. This display usually has the effect of frightening off small birds and other predators (Edmunds, 1974). Through evolution, the caterpillar has adapted to predation from birds by exploiting their fear of snakes. This implies no special ability on the part of individual caterpillars, but merely means that those that inherit the ability to respond appropriately are more likely to survive than those that do not.

Evolutionary adaptations among birds sitting on eggs can be seen by comparing the nest behavior of birds that nest on cliff ledges with that of

Fig. 1.3 The hawkmoth caterpillar (*Leucorampha*) mimicking the snake *Bothrops schlagelli* (*Photograph: Nicholas Smythe*) (See also back cover)

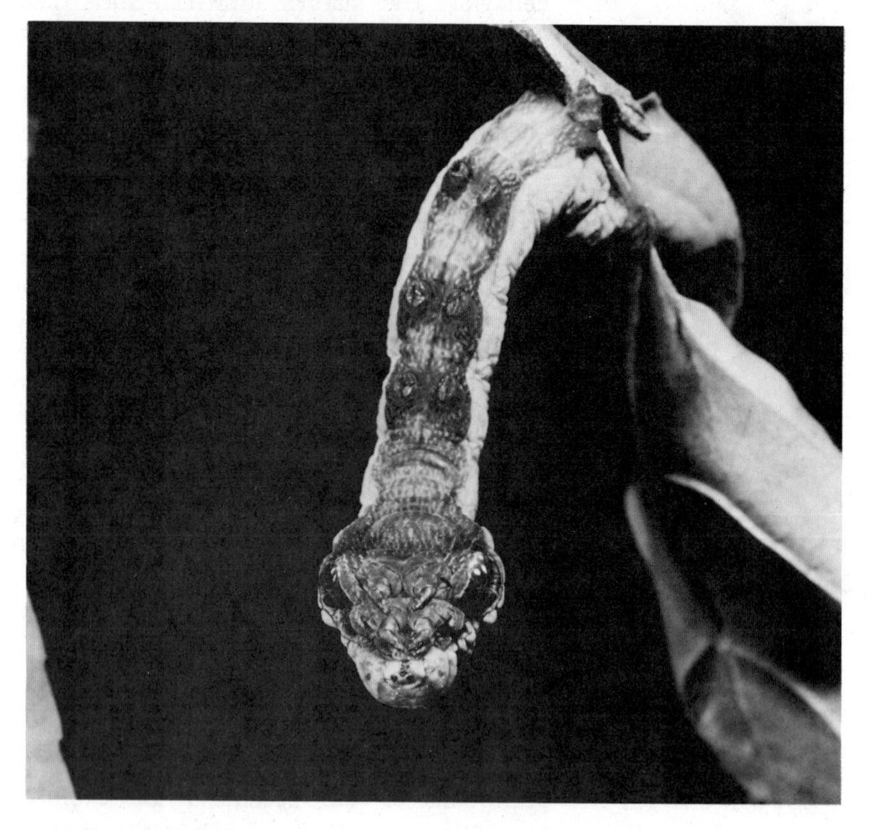

their close relatives that nest on relatively flat ground (as we do in Chapter 5.4). When we say that kittiwakes adapt to cliff nesting by adopting various behavior patterns, we mean that those individuals that employ the behavior patterns in a cliff-nesting situation have more surviving offspring than those that do not show the appropriate behavior to the same extent.

Phenotypic adaptation involves processes such as learning, maturation, and temporary physiological adjustment. For example, a gull incubating its eggs on a cold, windy day will face into the wind, sit tight on its eggs, and increase its heat production. On a hot sunny day it will cover the eggs just sufficiently to keep them shaded, and it will pant, spread its wings, and show other forms of cooling behavior. These are all short-term behavioral and physiological adaptations to the prevailing weather.

Some birds are able to learn to recognize their own eggs by their color patterning (Baerends and Drent, 1982). This form of adaptation by learning may enable them to adjust to changes in the nest situation due to predation or other forms of disturbance. The ability of individuals to respond appropriately to changing circumstances by learning or physiological adaptation is an important feature of animals that live in variable environments.

Although biologists use the term *adaptation* in many different ways, they are usually careful to distinguish between evolutionary or genotypic adaptation and individual or phenotypic adaptation. During his voyage on the Beagle, Darwin noticed many of the beautiful adaptations that are typical of particular environments, and he came to regard these as evidence for evolution. For example, the Galapagos finches diversified considerably during their evolution from a common ancestor. The differences among the various types are undoubtedly adaptive, as can be seen by comparing them with other birds. This applies particularly to their beaks, as shown in Figure 1.4, but it is also true of many related aspects of their behavior. Thus, the insectivorous tree finches move with agility among tree branches like titmice, the woodpecker finch climbs vertical trunks and probes in crevices, the warbler finches have the quick flitting movements of a true warbler, and the ground finches hop about on the ground. Originally like all finches, the Galapagos finches have become like tits, woodpeckers, or warblers, in accordance with their adopted lifestyles. The formation of a number of new species from a single-parent species as a result of adaptation to a variety of habitats is called *adaptive radiation.* The Galapagos finches provide a classic example. Another well-known example is the marsupial mammals of Australasia, which include herbivores, insectivores, and carnivores and burrowers, tree climbers, and aerial gliders.

The Galapagos finches not only illustrate adaptive radiation but also show evolutionary convergence with species in other parts of the world. Thus, we speak of "warbler finches" and "woodpecker finches" because

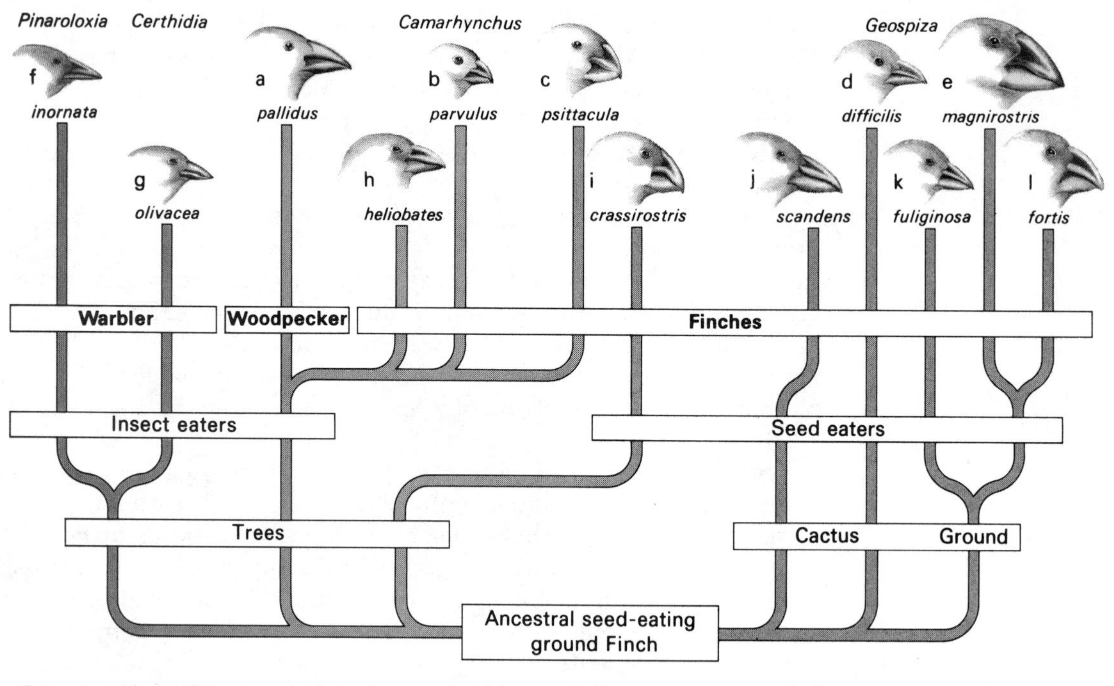

a. *Camarhynchus pallidus*
b. *Camarhynchus parvulus*
c. *Camarhynchus psittacula*
d. *Geospiza difficilis*
e. *Geospiza magnirostris*
f. *Pinarolaxia inornata*
g. *Certhidia olivacea*
h. *Camarhynchus heliobates*
i. *Camarhynchus crassirostris*
j. *Geospiza scandens*
k. *Geospiza fuliginosa*
l. *Geospiza fortis*
m. *Camarhynchus pauper*
n. *Geospiza coniostris*

we recognize the similarities between these and true warblers and woodpeckers. *Convergent evolution* occurs when different species inhabit similar environments. It sometimes results in remarkable similarities in the appearance and behavior of unrelated species, as illustrated in Figure 1.5.

1.2 Historical outline

Although the scientific study of animal behavior has its origins in the work of eighteenth-century naturalists such as Gilbert White (1720–1793) and Charles Leroy (1723–1789), it is Charles Darwin (1809–1882) who is regarded as the father of the scientific study of animal behavior. Darwin influenced the development of ethology in three main ways. First and foremost, his theory of natural selection set the stage for consideration of animal behavior in evolutionary terms, a key aspect of modern ethology. Second, Darwin's views on instinct can be regarded as a direct forerunner of those of the founders of classical ethology (see Chapter 20). Third, Darwin's observations on behavior were important, especially those that stemmed from his belief in the evolutionary continuity of

Fig. 1.4 **Adaptive radiation of Darwin's finches from an ancestral seed-eating ground finch. Isolation on the Galapagos Islands led to intense competition and to subsequent dietary specialization. This has given rise to the different types of beak found among the present-day species**

man and other animals. In his *The Descent of Man and Selection in Relation to Sex* (1871), for instance, Darwin writes: "We have seen that the senses and intuitions, the various emotions and faculties such as love, memory, attention, curiosity, imitation, reason, etc., of which man boasts, may be found in an incipient or even sometimes a well-developed condition in the lower animals." In his book *The Expression of the Emotions in Man and the Animals* (1872), Darwin elaborates on this theme: "With mankind some expressions, such as bristling of the hair under the influence of

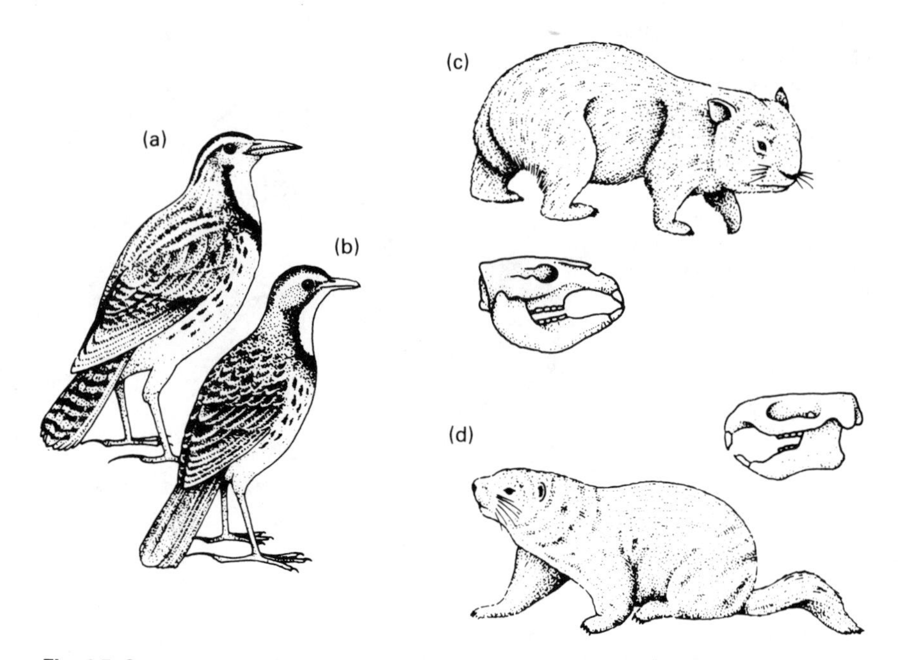

Fig. 1.5 Convergent evolution leads to similarities in appearance and behavior of unrelated species. The examples shown here are (*a*) the eastern meadow-lark (*Sturnella magna*) from America and (*b*) the yellow-throated longclaw (*Macronyx croceus*) from Africa; (*c*) an Australian wombat (*Phascolonus ursinus*), and (*d*) an American woodchuck (*Marmota monax*) (From *The Oxford Companion to Animal Behaviour*, Oxford University Press, 1981)

extreme terror, or the uncovering of the teeth under that of furious rage, can hardly be understood, except on the belief that man once existed in a much lower and animal-like condition."

Darwin's friend and disciple George Romanes energetically continued Darwin's work on animal behavior and his *Animal Intelligence* (1882) was the first general treatise on comparative psychology. However, Romanes was not very critical in his evaluation of evidence. He credited animals with mental abilities, such as reasoning, and with feelings, such as jealousy. This led to a revolt, led by Conway Lloyd Morgan. In his *Introduction to Comparative Psychology* (1894) Lloyd Morgan enunciated his famous canon: "In no case may we interpret an action as the outcome of the exercise of a higher psychical faculty, if it can be interpreted as the outcome of one which stands lower in the psychological scale." This attitude resulted in a vast improvement in the control of experiments and the assessment of evidence. This more skeptical approach to animal behavior laid the foundations for the behaviorist school of psychology initiated by John Watson (1913).

The behaviorist view is that psychology studies behavior itself, rather than mental events. In its strong form, behaviorism rejects all reference to inner processes in the explanation of behavior. This view was very influential in the development of American psychology. It led to considerable improvements in experimental technique and the interpretation of behavioral evidence. However, the behaviorist approach was criticized, particularly by the European ethologists Konrad Lorenz and Niko Tinbergen, for being too sterile and divorced from reality. The ethologists thought that animal behavior should be studied in the natural environment and not in the laboratory. Lorenz emphasized the importance of acute observation and Tinbergen showed that meaningful experiments could be carried out under natural conditions.

The modern approach to animal behavior includes many features derived from both the behaviorist and the early ethological views. In addition, modern ethology draws upon the physiological tradition, with its emphasis on the explanation of behavior in terms of the activity of the nervous system. In this book we will study all three approaches to animal behavior.

1.3 How this book is organized

This book is divided into three major parts. In the first part we look at the evolutionary approach to behavior, in the second at causal mechanisms of behavior, and in the third part we examine complex behavior using both these approaches. This structure is meant to give you a full understanding of one mode of explanation before moving on to another. Too often students remain confused about mechanistic versus design-based explanations, and are unable to follow discussion of topics in which the two are inevitably mixed. Accordingly, treatment of such topics is reserved for the third part of the book.

There are 28 chapters: This introductory chapter is followed by nine groups of three chapters, each group covering a major aspect of the study of animal behavior, preceded by a profile of an eminent scientist who played an important role in the development of the subject. The aim is to give a feeling for the continuity of the science to which you who are now reading this book may contribute in the future. The arrangement of chapters in groups makes it simple to skip over a particular group if you are already well grounded in a given field of study. Many topics are introduced at an elementary level early in the book and then taken up at a more advanced level later, an arrangement meant to encourage flexible use by students from different academic backgrounds. The book is also designed to introduce the three main ways of studying behavior.

Animal behavior can be studied in the natural environment and in the laboratory. In the natural environment the animal is free to express its full range of behavior, which is discovered primarily through observation. Direct visual observation is the traditional approach, but in recent years this has been supplemented by indirect methods, which rely on technological developments such as audio recording and radiotelemetry. The natural behavior of animals can also be investigated through experiments involving selective interference in the environment (see Chapters 6 and 23).

In the laboratory the scientist has considerable control over the animal's environment and can thus design careful tests of particular hypotheses about behavior. This approach is particularly important in investigating the sensory capabilities of animals (Chapters 12 and 13), and in the study of animal learning (Chapters 17 and 18). In addition to behavior investigation in the laboratory, physiological experiments are making an increasing contribution to our understanding of behavior. The physiological approach to behavior study has long been advocated, but only in recent years has it made a substantial contribution (see Chapters 11 and 12).

Animal behavior is quite complex, and a wide range of practical and theoretical approaches is needed to achieve a good understanding of it. It is this multidisciplinary approach that makes animal behavior such an exciting subject.

PART 1 *The Evolution of Behavior*

In the first part of this book we look at animal behavior from an evolutionary viewpoint. Since the time of Charles Darwin it has become increasingly possible to explain the role of behavior in the survival and reproduction of the animal. Just as an animal's eyes, ears, legs or wings can be viewed as mechanisms designed to enable the animal to cope with its particular mode of life, so too can the mechanisms controlling behavior. Thus we can ask why animals have particular behavior patterns, and expect to find answers in terms of the evolutionary history of the species.

This evolutionary type of explanation requires some knowledge of the genetics and development of behavior, of the theory of natural selection in relation to the ecology of the animal, and of evolutionary theory and its relevance to the social behavior of animals. The first three sections of this book are devoted to these three subjects.

1.1 Genetics and Behavior

In this group of three chapters we enter the field of behavior genetics. Chapter 2 deals with elementary Mendelian genetics and the cellular basis of heredity. Chapter 3 outlines some of the issues involved in the study of the development of behavior, including the problem of innate behavior. Chapter 4 is devoted to behavior genetics and includes discussion of the effects of mutations, polygenic inheritance and the heritability of behavior.

Gregor Mendel (1822–1884)

BBC Hulton Picture Library

Born in peasant circumstances, Gregor Mendel entered the Church and was ordained at the age of 25. He trained to be a teacher but failed examinations to qualify as a high school teacher of natural history. In 1856 he became a teacher in a school of lower rank, where he was much liked by his pupils. Mendel started his researches in 1858 in the garden of a monastery at Brunn, Moravia. His results and theory were presented in a paper read to the Brunn Society of Natural Science in 1865 and published in the proceedings of the Society in 1866. Mendel's work remained unknown to the scientific world at large until it was discovered in 1900 more or less simultaneously by three independent workers—Correns, von Tschermak, and de Vries. Its importance was quickly realized and Mendel's results were soon confirmed and extended.

The main body of Mendel's research was carried out with ordinary garden peas. He was particularly interested in characteristics of the peas that could occur in either of two contrasting forms. Thus, the seeds could be red or white, the pods could be inflated or constricted, and green or yellow, the flowers could be axial or terminal, and the stems could be long or short. Mendel sowed every seed separately. He cross-fertilized the resulting plants by hand and kept different hybrids in different plots. Instead of trying to trace individual lineages, as was typical at the time, he had large numbers of plants and simply counted the various types of offspring. On the basis of these studies, Mendel formulated his revolutionary laws of inheritance.

In 1866 Mendel wrote to the famous biologist Karl von Nägeli, who was engaged in breeding experiments with hawkweed. Mendel outlined his ideas, asked for help, and offered to work on hawkweed. However, von Nägeli was not impressed with Mendel's theories and suggested that Mendel grow more peas. Mendel had already recorded observations on some 13,000 specimens of pea plant and he tried instead to repeat von Nägeli's work on hawkweed. Hawkweed is small, difficult to fertilize by hand, and unsuitable for genetical work. Mendel persisted for a time, but made no progress. Von Nägeli offered to grow some of Mendel's pea seeds at the Botanic Gardens in Munich, and Mendel sent him 140 packets. Von Nägeli never planted the seeds and did not refer to Mendel in his major work of 1884. It is ironic that von Nägeli, who had been the first to observe and describe chromosomes under the microscope in 1842, never realized the importance of Mendel's work. The connection between chromosomes and Mendelian heredity was not established until 1914.

By 1875 the monastery at Brunn had become embroiled in a tax dispute, and Mendel, who was prelate, had no more time for research. He died in 1884.

2 Genes and Chromosomes

When Charles Darwin's theory of natural selection was published, the question of how characteristics were inherited and how variation among the offspring was maintained became of acute scientific interest. Von Nägeli, famous for his work on the cellular structure of tissues, carried out breeding experiments on hawkweed. William Bateson published his extensive studies of inheritance and variation in 1894. Dutch botanist Hugo de Vries discovered that variants of evening primrose plants arose from normal plants and seemed to breed true and he called these sudden jumps *mutations*. In 1868 Darwin published *The Variations in Animals and Plants under Domestication*, in which he put forward his theory of pangenesis—that is, each cell in the body throws off minute "gemmules" containing information about itself. "These multiply and aggregate themselves into buds and the sexual elements." However, not until Gregor Mendel's writings were uncovered in 1900 could the science of genetics be reconciled with the theory of evolution by natural selection and with the growing science of cell physiology.

In this chapter we see how Mendel's work led to our modern understanding of the cellular basis of heredity, of genetic variation, and of the nature of the gene. Genetics is of fundamental importance to the study of behavior because so much animal behavior is influenced by an animal's genetic makeup. By understanding genetics we can gain insight into the nature of this influence.

2.1 Natural selection and behavior

Natural selection operates upon the physical characteristics, or *phenotype*, of the individual, including its behavior. A mouse that does not attempt to escape from predators is less likely to survive than one that does, and a mouse that does not take advantage of opportunities to obtain food is similarly disadvantaged. Thus, we can attribute survival value to behavior patterns, just as we can to the morphological properties of animals. Moreover, the balance between different behavior patterns also may be important. For example, the mouse that pays too much attention to food may fail to notice an approaching predator, while the mouse that is too fearful of predators may miss important feeding opportunities. Such considerations lead to two important questions about the relationship between natural selection and behavior: (1) How can we determine or demonstrate the survival value of particular behavior patterns? (2) Is there a best way for an animal to divide its time among the many activities possible in a given set of circumstances? We address the first of these questions in Chapter 6.1. The second question is a little more complicated and is reserved for Chapter 24.

The effectiveness of natural selection in changing the nature of a population of animals depends upon the degree to which the phenotypic characteristics are inheritable. Although Darwin knew that there are usually variations among the offspring of particular parents and that such variations are essential for his theory, he could not really say how the variations occurred. It was in 1858 that Gregor Mendel, a Moravian monk, started experiments on plant breeding. He described his results in 1865 but in a relatively obscure publication that was not brought to the attention of the scientific world until 1900, after Darwin's death. Mendel demonstrated that heredity is not blending but particulate: offspring inherit discrete particles, which we now call *genes.* After its rediscovery, Mendel's work led to the establishment of the genetical theory of natural selection by R. A. Fisher and others in the 1930s. This theory provides the mechanism for natural selection in that it accounts for the variation within a population of reproducing animals.

In the study of animal behavior, it is important to know to what extent particular behavioral characteristics are inheritable. Not only will this knowledge enable us to estimate the extent to which behavior traits are subject to selection, but also it highlights the distinction between innate and acquired behavior, a currently controversial topic with far-reaching implications for human philosophy and politics.

Chapter 4 introduces the subject of behavior genetics. For our present purposes it is sufficient to understand that the effectiveness of natural selection in influencing the evolution of behavior depends on the extent to which the behavior is under genetic control. The evolution of behavior is complicated by the fact that natural selection is not always the only important mechanism. In certain animals, especially humans, behavior characteristics can evolve via cultural means. Individuals can learn from each other in a variety of ways (see Chapter 27), thus allowing information to be passed from one generation to the next. The behavior of birds and mammals is the result of a complex interaction of genetics and experience. To understand this interaction, ethologists must be familiar with the elements of genetics.

2.2 Mendel's laws

Mendel carried out most of his experiments with garden pea plants. In one typical experiment, Mendel crossed plants that display red flowers with plants displaying white flowers: all the offspring had red flowers. He then allowed these offspring to breed freely among themselves and obtained 705 plants with red flowers and 224 plants with white. Two aspects of these results are important. First, the first generation of offspring has no representatives of the white flowered parent, but the second generation does. Secondly, the red flowers and the white flowers of the second generation appear in a 3:1 ratio. Darwin also experimented with peas and noticed the 3:1 ratio. Mendel, however, established this and other important ratios in numerous repeated experiments. Moreover, he realized the significance of the disappearance of one parental type (white) in the first generation of offspring. Mendel proposed that each pea plant possessed two hereditary factors for flower color as well as two hereditary factors for each of the other characters. One factor dominates the other so that the flowers are of only one color when two different factors are present. Thus, white-flowered plants possess two white factors, while red-flowered plants possess two red factors or a red factor and a white factor. This type of reasoning enabled Mendel to account for all the observed ratios of offspring types and to formulate his revolutionary laws of inheritance.

On the basis of his breeding experiments with pea plants, Mendel came to a number of conclusions:

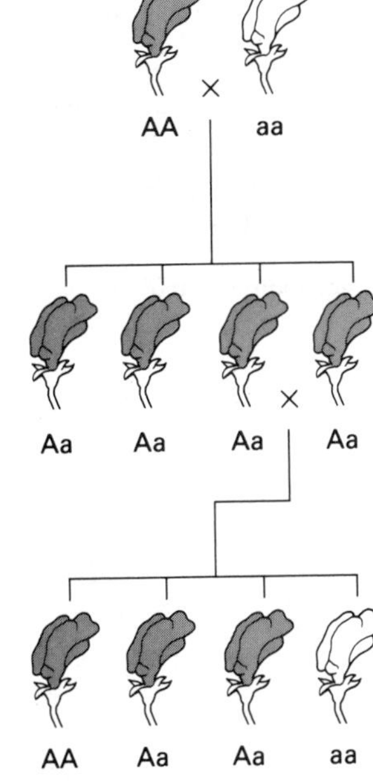

Fig. 2.1 Mendelian table showing crosses between red and white flowered peas. Red (A) is dominant over white (a).

1. Inheritance is particulate and the genetic contributions from the parents are equal. The genetic material from one parent cannot mix with, or contaminate, that from the other. Mendel's particulate factors are now called *genes.*
2. Each external character contributing to the structure and appearance of the individual is controlled by a pair of genes, one inherited from each parent. These external characters are called the *phenotype.*
3. Each gene may exist in two or more alternative forms, now called *alleles.* The total collection of genes within the individual is called the *genotype.* Thus, three kinds of combinations are possible when there are two alleles. For a gene that takes two forms, A and a, the three possible genotypes that an individual might have are AA, aa, and Aa. When an individual carries two copies of the same allele, as in the first two cases, the individual is said to be *homozygous* for the character in question. When an individual carries one copy of each allele, as in the third case, the organism is said to be *heterozygous.*
4. One allele may be dominant over another, in which case the non-dominant allele is called "recessive." *Dominant genes* control the nature of the phenotype when they occur in homozygous or heterozygous combination. *Recessive genes* control the phenotype only when they occur in homozygous combination, as illustrated in Figure 2.1.
5. Of the genes making up a pair in a parent only one will be copied and inherited by an offspring. The offspring receives another copy from its other parent. Thus, there may be one phenotype in the first filial (F_1)

generation (the first generation following a specific cross) but more than one in the second filial (F_2) generation, as illustrated in Figure 2.1. *The law of segregation* (sometimes called *Mendel's first law*) results in typical Mendelian ratios of phenotypes, like the 3:1 ratio depicted in Figure 2.1.
6. When two or more pairs of genes segregate simultaneously, the distribution of any one gene is independent of the distribution of the others. This is called the *law of independent assortment* (or *Mendel's second law*). The result is that phenotypic characters paired in one generation need not be paired in subsequent generations, as shown in Figure 2.2.

Mendel's conclusions apply to the inheritance of characters in many

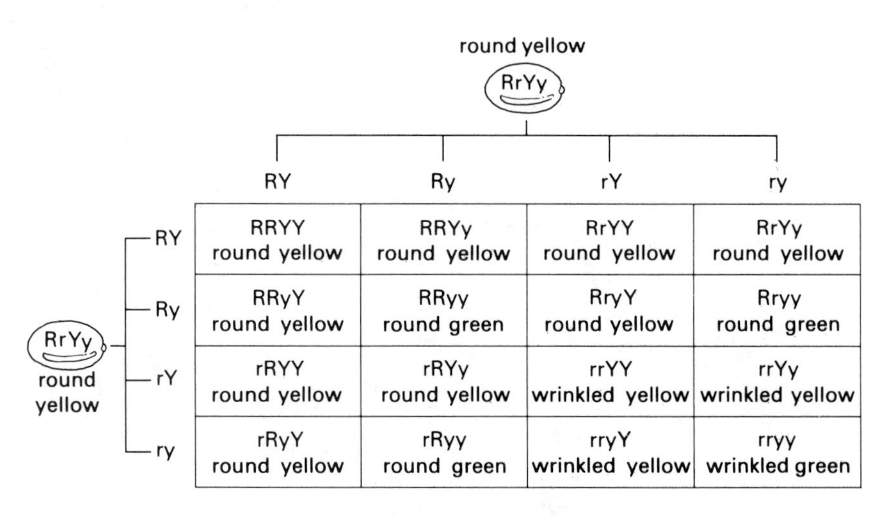

	RY	Ry	rY	ry
RY	RRYY round yellow	RRYy round yellow	RrYY round yellow	RrYy round yellow
Ry	RRyY round yellow	RRyy round green	RryY round yellow	Rryy round green
rY	rRYY round yellow	rRYy round yellow	rrYY wrinkled yellow	rrYy wrinkled yellow
ry	rRyY round yellow	rRyy round green	rryY wrinkled yellow	rryy wrinkled green

Fig. 2.2 Mating of two F_1 individuals in one of Mendel's hybrid pea crosses. Round (R) is dominant over wrinkled (r), and yellow (Y) is dominant over green (y).

Fig. 2.3 Graphical representation of four different types of gene expression (After McClearn and DeFries, 1973)

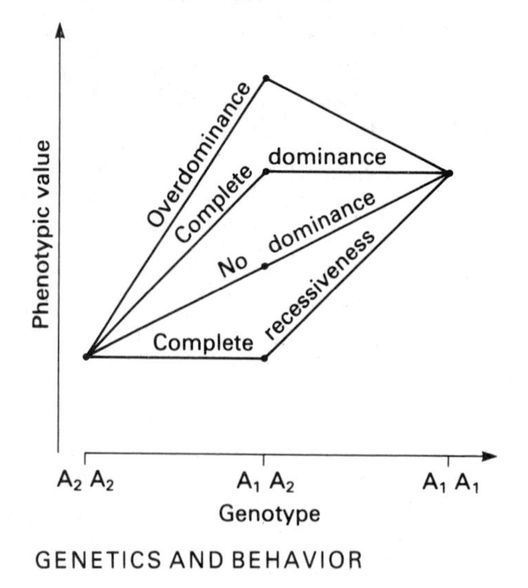

species of plant and animal, but they have not been found to be universally applicable. One departure from the classical Mendelian picture stems from variability in the relationship between the hereditary constitution, or genotype, and the observed character, or phenotype. For most of the characters studied by Mendel, the dominance of a gene over its allele was complete. However, we now know that there is a range of possible gene expression, as indicated in Figure 2.3, and that dominance is not always complete.

(a)

(b)

(c)

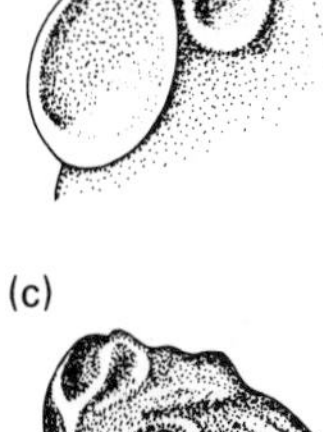

(d)

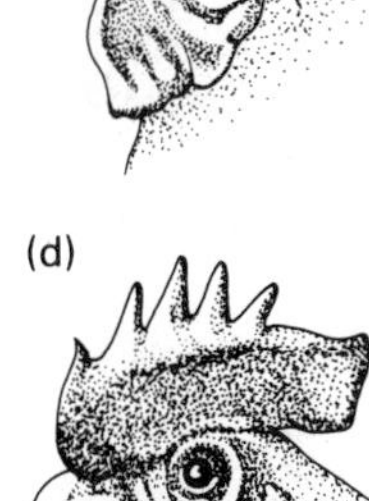

Fig. 2.4 Comb types in chickens (*a*) pea comb, (*b*) rose comb, (*c*) walnut comb, (*d*) single comb

Another complication of the simple Mendelian picture is that a given gene can influence many different characters. This phenomenon is known as *pleiotropism.* The first demonstration of behavioral pleiotropism was by Sturtevant in 1915. Working with the fruit fly *Drosophila*, he found that male flies appeared to mate preferentially with mutant females showing either white-eye or yellow-body phenotypes. He found conversely that the females of various types preferentially mated with non-yellow-body and non-white-eyes males. These results are due to a pleiotropic effect whereby the white-eye and yellow-body mutations both have the side effects of lowering the mutant fly's overall activity level. Because the normal, active females tend to flee the advances of males, the mutant females are more commonly inseminated. Conversely, because the males must approach females to initiate mating, the inactive, mutant males had lower success than their normal male counterparts. In other words, the mutant genes were pleiotropic in that they influenced more than one character. In a situation where mating was random, the mutant flies would be selected against on account of their lower sexual activity.

Gene interactions can also produce deviations from simple Mendelian principles. When two or more different genes interact with each other in determining a single character, the ratios obtained from matings may differ from the classical Mendelian ratios. For example, the form of the comb in domestic fowl is determined by the two genes R and P. The dominant gene R produces the rose comb (see Fig. 2.4), and its recessive allele r produces the single comb, where homozygous. The dominant gene P produces the pea comb, while the recessive p produces the single comb. When R and P occur together, they interact to produce the walnut comb, which neither gene could produce on its own.

Another form of gene interaction is the case of *complementary genes*, which are mutually dependent: Neither gene can produce its phenotypic effect without the other gene. Sometimes one gene masks the effect of another, a condition known as *epistasis.* Epistasis is similar to dominance, but whereas dominance involves different alleles, epistasis involves different genes.

A *modifier gene* may alter the expression of another gene. Thus, for instance, human eye color is controlled largely by a single pair of genes. B is responsible for brown eyes and b for blue eyes. The genotypes BB and Bb both produce brown eyes, and bb gives rise to blue eyes.

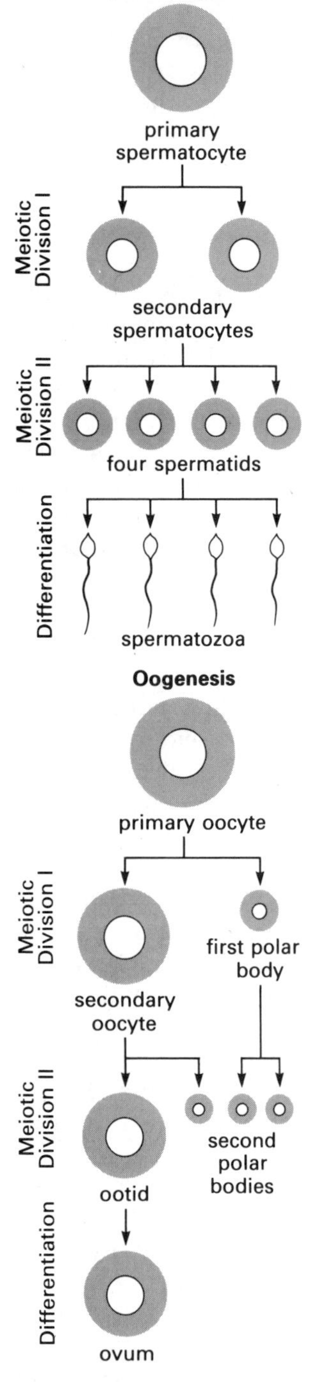

Fig. 2.5 Schematic representation of spermatogenesis and oogenesis in an animal

However, due to the action of modifier genes, many variations exist on these two basic eye colors. Modifier genes can switch on the phenotypic effects of other genes.

Some phenotypic characters, like height in humans, are influenced by a large number of genes, and thus such characters vary continuously across different members of a population. With height, for instance, if only a few genes were involved, we would expect to find certain heights to be much more prevalent than others.

2.3 The cellular basis of heredity

The idea that plants and animals are composed of cells developed gradually and was first explicitly stated in 1839 by Schwann, who pointed out the correspondence between cellular structure and the growth of plants and animals. Albrecht von Koelliker was the first to recognize, in 1840, that sperm and ova were cells; it soon became established that all living tissue was made up of cells.

The first recorded observations that cell division included the nucleus and its chromosomes were made in 1842 by von Nägeli, but a detailed account of how the cell nucleus divides awaited improvements in the techniques of microscopy. The first correct account of the behavior of chromosomes during cell division was made in 1882 by Walther Flemming, who observed the process in the cells of salamander larvae. By 1885 Eduard van Beneden had shown that chromosomes remain unaltered from one cell division to the next. He also discovered that the number of chromosomes is fixed for a given species. In 1856 Pringsheim was the first to see a sperm enter a female cell, but it was Oscar Hertwig (1876) who realized that, where two nuclei were seen within a fertilized egg, one nucleus must have come from the sperm. Thus, he proposed that it is the chromosomes that carry the genetic material, which is provided by both parents at the time of fertilization. A clear account of cell division and fertilization thus became possible, setting the stage for a rapid assimilation of Mendel's work on heredity once it was unearthed in 1900.

The somatic cells of most animals are *diploid*—that is, each cell has two copies of each type of chromosome. Human somatic cells have 46 chromosomes, or 23 different pairs, and the fruit fly *Drosophila* has eight chromosomes forming four different pairs. Normal cell division involves *mitosis* and produces diploid cells. Mitosis is characteristic of the formation of all kinds of cells except the germ cells, which give rise to the male and female gametes, the sperm and the egg. Since sperm and egg combine to form a single cell with two chromosome sets (the diploid zygote), each gamete cell carries only one set of chromosomes, and is said to be *haploid*. In the formation of the germ cells, therefore, the diploid number of chromosomes is reduced to the haploid number by a process called *meiosis*.

Meiosis involves two separate cell divisions. The first division achieves a reduction in the number of chromosomes, while the second division is similar to normal mitosis, except that the daughter cells are haploid. Complete meiosis thus results in the division of a single diploid cell into four haploid cells. In male animals, all four haploid cells differentiate into spermatozoa, as illustrated in Figure 2.5. In females, the cells produced by the first cell division are usually of unequal size. The smaller cell, called the *first polar body*, may or may not undergo the second division. The larger cell divides into a *second polar body* and a large cell that differentiates into the *ovum* (Fig. 2.5). The polar bodies do not survive. Thus, meiosis in males usually produces four spermatozoa of equal size, while in females it produces a single large ovum and two or three non-functional polar bodies.

2.4 Genetic variation

The American biologist Thomas Hunt Morgan realized that the phenomenon of mutation, discovered by de Vries, provided a means of investigating genetics. After pilot studies on mice, pigeons, and rats, he settled on the fruit fly *Drosophila* as a suitable experimental animal. Since the fruit fly reaches maturity in 12 days and breeds rapidly, 30 generations can be bred in 12 months, thus providing an abundant source of experimental material. Morgan attempted to induce mutations by subjecting the flies to extremes of temperature, radioactivity, etc. At first he had no success, but in 1910 he discovered a male fly with white eyes, in contrast to the red eyes of normal flies. He bred this solitary specimen with normal females and obtained only red-eyed flies in the first generation. In the second generation, however, both red-eyed and white-eyed flies appeared.

Morgan soon identified many more mutants. He discovered that white eyes occurred only with yellow wings, never with grey wings. Similarly, a black body appeared only with vestigial wings and an ebony body only with pink eyes. This could be explained if the three pairs of characters each occur on a different chromosome. Since *Drosophila* has three large chromosome pairs, this interpretation seemed reasonable. Furthermore, in 1914 bent wings appeared, followed by eyelessness and shaven appearance, all linked together but unlinked to other mutations. Morgan postulated that these forms were mutations on the fourth small chromosome of *Drosophila* (see Fig. 2.6).

Some mutations, like white eyes, appeared only in males. Morgan showed that one chromosome in the male is slightly larger than the corresponding chromosome in the female and that this additional material carries the sex-determining factor plus some other factors found only in the male. The chromosomal basis of heredity was now established.

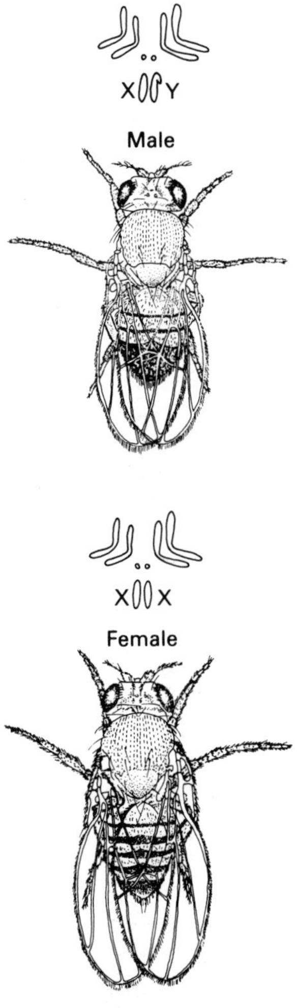

X Y

Male

X X

Female

Fig. 2.6 *Drosophila melanogaster* and its chromosomes

One day, yellow wings and white eyes were discovered unlinked. Morgan postulated that a chromosome had broken during meiosis and that the pairs had rejoined in a different arrangement. He proved to be correct, and the phenomenon is now called *crossing over* (see Fig. 2.7). Crossing over is an example of a chromosome mutation, many other types of which are now known. The phenomenon of crossing over proved to be particularly useful in experimentally determining which parts of a chromosome normally carry particular genes. Thus, by studying numerous cases of crossing over, it is possible to calculate the relative positions of different factors and to establish the order of genes on the chromosome (see Fig. 2.7).

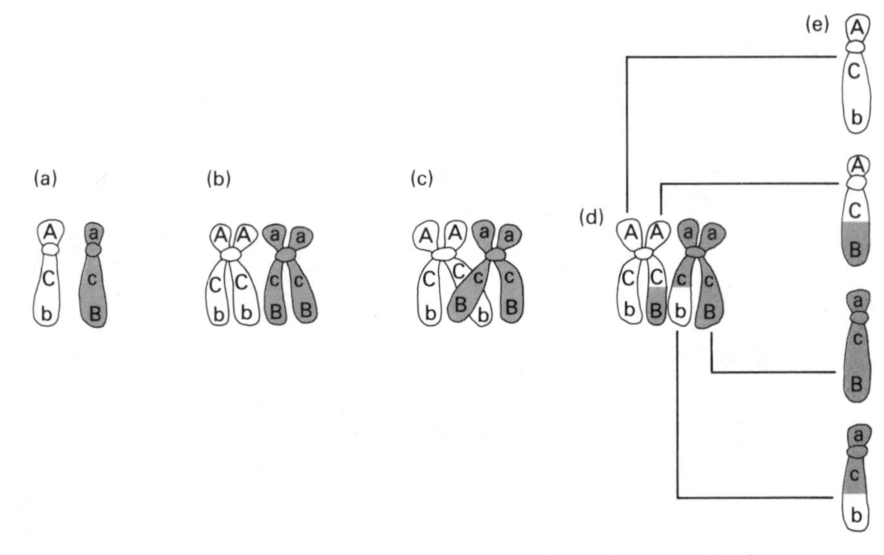

Fig. 2.7 Schematic representation of crossing over (After McClearn, 1963)

Mutations also can occur within a single gene. The chemical structure of the gene is normally very stable and resistant to changes induced by normal environmental variables, such as temperature extremes and toxic chemicals. However, Morgan's student H. Muller discovered that the rate of mutation could be enhanced greatly by strong radioactivity, which is not normally present in the natural environment.

Gene mutations occur at random, in the sense that there is usually no way of knowing which gene in which individual will mutate. Most mutations are deleterious and usually lethal in the homozygous form. For example, there is a mutant form of domestic fowl, called *creeper*, that has very short crooked legs. When two creeper fowl are crossed, the offspring include normals and creepers in a 1:2 ratio. This is not a straightforward Mendelian ratio, but a quarter of the fertilized eggs fail to hatch and these are homozygous for the creeper gene, a lethal combination. Thus, the true ratio is 1:2:1 normal, creeper, lethal—a classical Mendelian ratio.

We now have three known sources for the genetic variation that Darwin recognized as so important for his theory of evolution by natural selection: (1) Mendelian variation, due to new combinations of characters as described by Mendel's laws, (2) chromosome mutation, and (3) gene mutation.

2.5 The genetic material

Mendel did not use the term *gene.* He wrote about the observable (phenotypic) characters of living organisms and postulated the existence of heredity elements, or factors, that behaved in a particulate manner. When Mendel's work was brought to the attention of the scientific world in 1900, scientists already suspected that the hereditary material was carried by the chromosomes. Intensive research confirmed the generality of Mendel's law, although exceptions were recognized, as mentioned earlier. In 1905, William Bateson gave the name *genetics* to this field of research; the term *gene* was applied to the Mendelian elements by Danish botanist Wilhelm Johannsen in 1909. Johannsen also made the important distinction between genotype and phenotype.

Morgan regarded the genes as the smallest units of recombination, arranged on the chromosome like beads on a necklace. Linked genes could recombine by crossing over if there were a break in the chromosome, between the two genes. The gene was also regarded as a functional unit of control over the phenotype and as the unit of mutation. Thus, a gene mutation was understood as the smallest genetic change that could change the phenotype. For the first part of the century, the concept of the gene was a unified one. However, this was to be undermined by advances in understanding the chemical nature of the gene: Through the work of Oswald Avery (1944) and James Watson and Francis Crick (1953) it was discovered that the genetic code is carried by molecules of DNA (deoxyribonucleic acid).

DNA molecules consist of two strands (Fig. 2.8), each made up of a string of smaller molecules called *nucleotides.* These nucleotides are the elements of the genetic code. DNA does two things. It reproduces itself, providing the hereditary connection between generations, and it provides the information necessary for the production of thousands of different *proteins.* Just as the DNA is composed of a string of smaller nucleotides, so a protein is composed of a string of smaller *amino acids.* Each possible combination of three nucleotides from the four nucleotide types specifies one of the 20 amino acid types (there are also nucleotide triplets, or *codons,* which indicate the end of the protein and which transmit other messages). The string of DNA itself is immobile within the nucleus, so that it must first be copied into a matching *messenger*

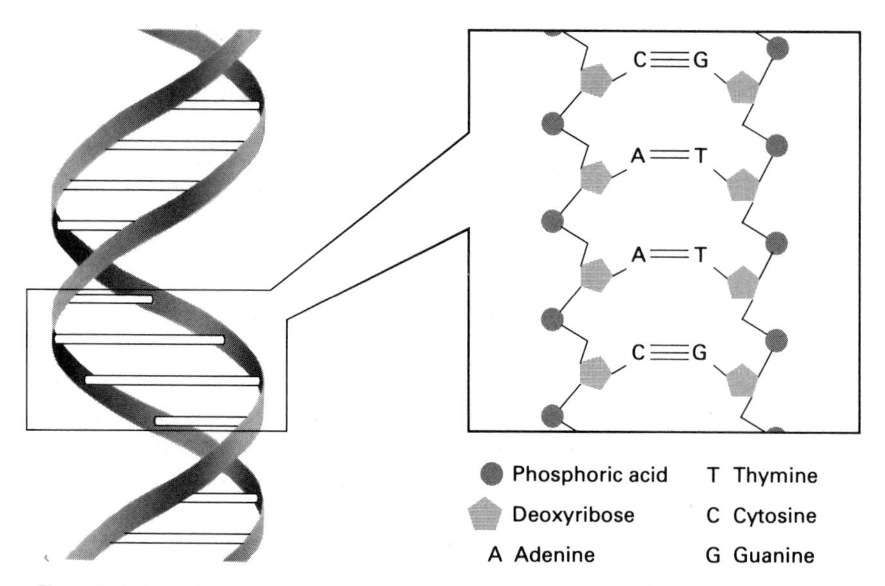

Fig. 2.8 Schematic representation of the DNA double helix. Each helical chain is made up of alternating deoxyribose sugar and phosphate. The two chains are joined by weak links between the nucleotide bases adenine (A), thymine (T), cytosine (C) and guanine (G).

RNA (ribonucleic acid), which can move out of the nucleus into the cytoplasm where the amino acids are stored. In the cytoplasm a string of amino acids is matched to the string of nucleotide triplets in the RNA. When the amino acid chain, or *polypeptide*, is complete it separates from the RNA and takes up its role as a functional protein. The gene, then, is simply that stretch of DNA which includes the entire nucleotide sequence for one protein (or for one RNA active in gene regulation without translation into protein). In the chromosomes of any complex animal there are tens of thousands of genes as well as long pieces of DNA that do not code for any known product.

Gene mutations can induce various types of disruption of the DNA molecule. In particular, it has been shown that *deletions* or *substitutions* in particular nucleotides can occur. If such a change happens near the beginning of a gene, then the triplets are misread all along the line and a completely different non-functional protein results. If the addition or deletion occurs near the end of a gene, then the first part of the protein is coded correctly, the rest being altered in some way. If the alterations do not affect the active sites of the molecule, then some function might remain. Another type of mutation is base substitution—that is, the exchange of one nucleotide for another—which may create a new triplet coding for a different peptide, or it may produce a spacer triplet which separates active parts of the chain. Mutations can also occur in regulator genes that control the activity of other genes.

Chromosome mutations produce recombinations of genes such that the genes appear in a different order on the chromosome, or appear on a different chromosome altogether. This type of mutation does not alter any single gene but can affect the way in which proteins are produced, since the order of genes on the chromosome is an important aspect of gene expression.

The rate at which any particular gene undergoes mutation is quite low. However, there is a large number of different genes within an individual, and the number in the gene pool is vast. Thus, mutations frequently occur within a population, pure chance determining which individual is affected. The vast majority of mutations are deleterious and cause death at an early stage of development. Only a few mutations allow the progeny to remain viable. These viable mutations increase the genetic variability upon which natural selection depends, but they do not determine the direction of evolution.

Points to remember

● Natural selection depends upon genetic variation in the population. Those variants that are best suited to the environment tend to have more offspring. The variation comes about as a result of meiosis, chromosome mutation and gene mutation.

● During meiosis there is a reassortment of the genetic material in accordance with Mendel's laws. This can result in changes in the genetic combinations from one generation to the next.

● Chromosome mutations also occur during meiosis, but their occurrence is not systematic. They are errors in meiosis, which occur at random and result in novel genetic combinations.

● Gene mutations are changes in the chemical makeup of the genes. They occur at random and are generally deleterious. However, a few result in beneficial innovations which endow the individual with an evolutionary advantage.

Further reading

Ayala, F. J. and Kiger, J. A. (1984) *Modern Genetics*, 2nd edition, Benjamin/Cummings, Menlo Park, California.

3 Development of Behavior

Behavior patterns result from the complex interactions of external stimuli and internal conditions. However, any behavior pattern is constrained by the way in which information is processed by the animal. The internal information-processing systems are established during the course of development from the fertilized egg, to the embryo, to the adult animal, a process called *ontogeny*. Through the study of ontogeny we can discover the ways in which genetic and environmental (or learned) information interact to give rise to the behavior of the animal.

3.1 Ontogeny

Almost all animals are composed of the same basic materials, the differences between species resulting from differences in the way in which these basic materials are put together. Although this regulatory function is poorly understood, it is known that genes control development by producing proteins that regulate the complex organization of embryological processes.

The regulatory proteins, however, are only effective insofar as the developing organism is sensitive to the messages they convey. A regulatory molecule does not act by constructing a system according to a map; rather, it causes a specific reaction in cells, whose surfaces contain receptors for the regulatory molecule. The response of a developing embryo to a regulatory gene product thus depends upon the embryo's internal organization at the time the regulator is released. In addition, the regulator is only functional under a certain range of environmental conditions. As an example we first look at some aspects of development in a species with a relatively simple behavioral repertoire.

Aplysia is a shell-less marine snail that can attain a body weight of up to 10 pounds. As is typical of snails, *Aplysia* is a true hermaphrodite. Fertilization is internal and mating chains commonly form in which up to a dozen individuals mate with one another, each animal inseminating one animal as it is being inseminated by another. Following fertilization, *Aplysia* lays over a million eggs, connected in a long strip. The egg strand is expelled from the reproductive duct initially through contraction of duct muscles. When the strand is long enough, the snail picks up the end in its mouth and performs a series of head-waving motions that help to pull the eggs free of the reproductive duct (Fig. 3.1). A special mucous gland in the animal's mouth secretes adhesive onto the egg strand, which is then wrapped over itself to make a more compact egg mass. Finally, the snail uses its head to press the egg mass firmly onto a solid substrate (Fig. 3.1), where the eggs adhere until hatching. The entire series of egg-processing actions is a tightly stereotyped pattern of behavior. During the sequence the animal's heartbeat and respiratory rates rise and it stops all feeding and locomotion.

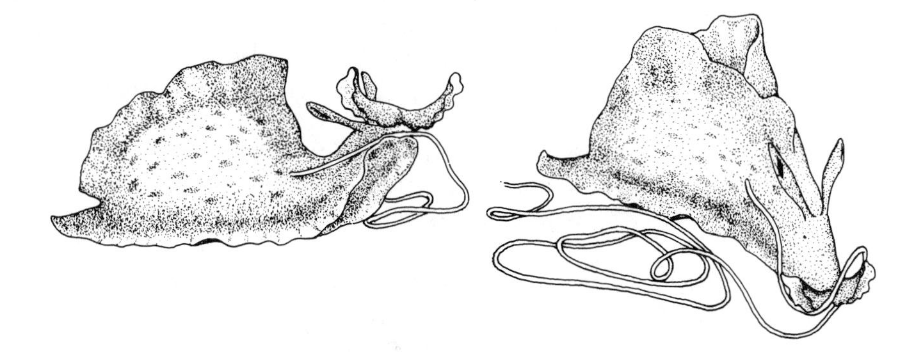

Fig. 3.1 The egg-laying behavior of *Aplysia*. The string of egg cases is expelled from the reproductive duct in the side of the body. The animal grasps the egg string in its mouth (left) and fixes it to the substrate (right). (After Scheller and Axel, 1981)

Richard Scheller and Richard Axel (1981) found the behavior sequence to be the result of gene action at a number of levels, both during development and from minute to minute. The cells that give rise to *Aplysia*'s entire nervous system are descendants of a few cells in the embryo's body wall. Early in development a subset of these pre-nerve cells produces a specific protein called *ELH* (*egg-laying hormone*). It is thought that these pre-nerve cells may all be daughters of a single cell in the very early embryo that has the ability to use the appropriate gene to make ELH. All other cells in the embryo possess this gene as well, but in these it is never functional. The ELH-producing cells divide further and migrate to their final locations in the developing organism, where they differentiate into adult nerve cells. In the adult snail, therefore, certain nerve cells have inherited the ability to produce ELH, while others have not.

By the time of maturation of the snail, all of these systems are in place. At a specific moment another hormone is released that causes the ELH-producing cells to go into full production. ELH then causes neighboring nerves to fire (that is, it acts as a neurotransmitter). It also circulates in the blood, causing specific muscle fibers to contract (thus acting as a hormone), a response that results in the coordinated behavior pattern observed as egg extraction and deposition. In this sequence we see the importance of embryogenesis, genes, neurotransmitters, nerves, and muscles all acting in coordination to generate a behavior pattern critical to the reproduction of the snail.

In *Aplysia*, the action of the genes is of direct importance in guiding the production of a stereotyped behavior pattern. While genetic influences underlie all behavior at some level (for instance, the embryogenesis through which the body structure is laid down), there are also varying degrees of environmental influence that can modify the behavioral results of development. The same genes may have different phenotypic effects when the animal is subjected to different environmental influences during development. For example, the crustacean *Gammarus* normally has a red eye color. E. B. Ford and Julian Huxley (1927) discovered a single mutation that affects the rate at which eye pigment is deposited during a certain stage of ontogeny. If the mutant's temperature is raised to a certain level, the eyes become red; at higher temperatures they become a chocolate brown, and at intermediate temperatures there will be intermediate colors. Given the vast complexity of the biochemical processes involved in gene expression and in the growth and differentiation of cells (see Ham and Veomett, 1980), it is not at all surprising that the medium in which these processes occur affects the course of these cells' development. To a certain extent, built-in stabilizing, or regulatory, mechanisms serve to correct deviations and to control the speed and direction of the developmental processes. It is even possible for a structure or behavior pattern to develop via different routes, a phenomenon known as *equifinality*. Nevertheless, deviations from the general pattern of development are inevitable. While in some cases these deviations may be lethal, in others they may give rise to differences among individuals having the same genetic makeup.

Some early ethologists tended to think of behavior as being determined directly by genes. Thus, Konrad Lorenz (1965) proposes that genetic determination is analogous to the architectural blueprint of a building. The genetic blueprint represents a plan for the construction of the adult animal. Recognizing that bricks, mortar, and a work force are necessary for the construction of a building, Lorenz draws a sharp distinction between the conditions necessary for the translation of the blueprint into a building and the information contained in the blueprint upon which the characteristics of the finished building depend. Critics of this view (e.g., Lehrman, 1970) point out that, whereas an architectural blueprint is isomorphic with the structure it represents, this cannot be the case with a genetic blueprint. While there is a one-to-one correspondence between the measurements marked on an architectural blueprint and those of the finished building, there is no such correspondence in biological ontogeny. Even if it is correct that the genes code for certain key enzymes, this is far removed from the idea of a blueprint. The idea that information provided by the genes can be separated from that provided by the environment during development was criticized by

Hebb (1953), who argued that it is as meaningless to ask how much a given piece of behavior depends upon genetic factors and how much upon environmental as it is to ask how much the area of a field depends upon its length and how much upon its width.

The major emphasis of research into the ontogeny of behavior has been to demonstrate the variety of ways in which patterns of behavior develop. There appears to be a spectrum of processes, ranging from those that seem relatively uninfluenced by environmental factors to those that are heavily dependent upon experience.

The genetic determinism of the early ethologists has given way to the recognition that genetic and environmental influences are inextricably bound together in ontogeny—not, as Lorenz (1965) has suggested, like the relationship between a blueprint and the materials necessary for the construction of a building but more along the lines of the process of *epigenesis*, by which each developmental event sets the stage for, but does not dictate, the next. As suggested by Brown (1975), epigenesis can be summarized as follows. Starting with the fertilized egg or zygote (P_1), its phenotype at the next stage of development (P_2) will be determined jointly by the genes that are active in guiding its growth and differentiation during the intervening interval (G_1) and by the environment in which the development takes place (E_1); that is, zygote plus genes plus environment leads to the next-stage phenotype, or:

$$P_1 + G_1 + E_1 \rightarrow P_2$$

The phenotype at the following stage of development (P_3) will be determined by the way in which P_2 has been changed by the genes (G_2) and environmental influences (E_2) affecting development between P_2 and P_3:

$$P_2 + G_2 + E_2 \rightarrow P_3$$

This notation points up that all three elements—genes, environmental conditions, and starting phenotype—are aspects of normal development. Development can only proceed to the next stage if the phenotype of the developing animal is appropriate, if the correct gene products are available, and if environmental conditions fall within certain ranges.

During the early stages of development ($P_1 + G_1 + E_1 \rightarrow P_2$), the environmental component E_1 consists mainly of biochemical factors that surround the early embryo. At a later stage of development ($P_2 + G_2 + E_2 \rightarrow P_3$), the environmental component E_2 may consist of the post-hatching environment. After hatching or birth an animal's environment includes information taken in by its sense organs as well as biochemical conditions based upon feeding level, and the like. In this situation, learning can take place and can permanently influence the animal's behavioral phenotype P_3.

It is interesting to explore the relationship between learned and innate, or non-learned, behavior by comparing the baby bird of a *precocial species* (a species in which hatching occurs late in development) to the baby of an *altricial species* (a species in which hatching occurs early). If two such birds are compared at the same stage of physical development, such as when the first feathers appear, we find that the precocial bird is still inside the egg, while the altricial bird has already hatched. By the time the precocial bird hatches, the altricial juvenile has had the opportunity to learn a great deal, while the precocial bird has not yet begun to learn. What we find, interestingly, is that the precocial hatchling may have innately developed capabilities that the altricial chick has had to learn. For example, the altricial white-crowned sparrow learns the song of its species while a nestling. By contrast, chickens are precocial—passing early infancy in the egg—yet they have the innate ability to produce normal vocalizations, even though never previously exposed to them.

We may be inclined to think of what happens inside an egg as purely maturational, in the sense that the relevant environmental factors are predetermined. We should not forget, however, that the normal environment of the white-crowned sparrow also is predetermined, in the sense that the nestlings normally hear the song of their own species during the critical period of song learning.

Later still in development, $P_3 + G_3 + E_3 \rightarrow P_4$. By this stage, the effects of postnatal experience will be observable in both altricial and precocial species. Behavior that clearly is affected by experience is not likely to be called "innate". However, there is little difference in principle here from the situation at earlier stages of development. The extent to which the environmental factors are predetermined depends largely upon the variability of the environment. Inside the egg such variability is usually small, and we therefore are inclined to regard such behavior as innate. Outside the egg, however, the variability also may be small, depending upon the ecological circumstances typical of the species. Thus, the postnatal behavior of a precocial animal may be just as predetermined as the prenatal behavior of an altricial animal. If it is evolutionarily important that the development result in a particular end product—as might well be the case with the song of a passerine bird, then we might expect the environmental influences to be so contrived as to ensure that outcome.

3.2 Environmental influences upon behavior

Some forms of behavior do not appear until a particular stage of development is reached. Some of these behaviors seem to develop without any obvious practice. For example, pigeons start to flap their wings and fly erratically at a particular age, and their flying ability appears to improve with practice. However, Grohmann (1939), in a classical experiment, reared a group of pigeons in tubes so that they

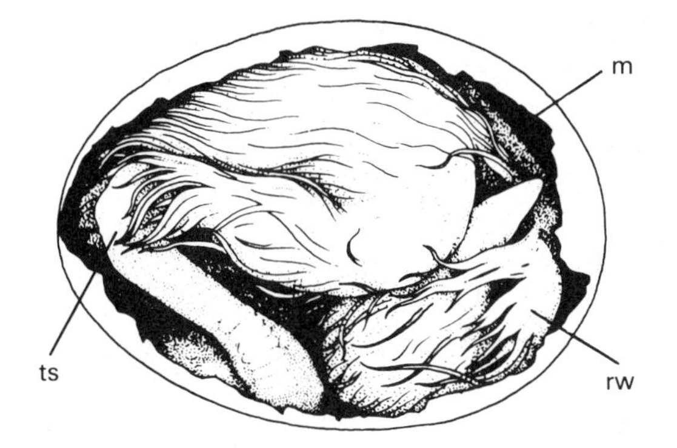

Fig. 3.2 The hatching position of the chick embryo approximately 1–2 days before emergence from the shell. m membrane; ts tarsal joint of leg; rw right wing (From *The Oxford Companion to Animal Behaviour*, 1981)

could not move their wings. Another group of the same age was allowed to develop without restraint. When the unrestrained pigeons had reached the stage at which they could fly satisfactorily, those that had been restrained were freed. Grohmann discovered that they also were able to fly immediately upon being released. Chickens that are featherless as a result of mutation develop normal wingflapping vestibular reflexes, even though these are completely ineffective (Provine, 1981). At first sight it would appear that the behavior is independent of environmental factors, but we must not assume that the ability to fly develops irrespective of any ontogenetic eventuality. Rather, flying ability is contingent upon some essential conditions, although our knowledge of these is sketchy (Ham and Veomett, 1980). A similar example is the vocalizations of pigeons and domestic fowl, which are highly stereotyped and appear at certain stages of development. They are not dependent upon auditory experience but upon certain hormonal conditions that are a normal part of overall maturation.

Many movement patterns appear to develop in the absence of practice or example (Hinde, 1970). However, study of the developmental history of some behavior patterns has suggested that they involve fragmentary and incomplete movements that might influence the course of development (e.g., Kuo, 1932; 1967; Kruijt, 1964; Anthoney, 1968). As an example, we can consider Kuo's (1932) work on the development of behavior in chick embryos. Kuo placed windows in the eggshell and was able to directly observe the behavior of the embryo (see Fig. 3.2). He found that the embryo is constantly subject to stimulation both from its own activity and from outside the egg. Various movements can be observed, including initial passive movements of the head caused by the beating heart. Active head movements appear within a few days, and these may be accompanied by opening and closing of the beak. By the seventeenth day, movements similar to pecking are apparent. Some scientists believe that the chick learns to peck in this way (e.g.,

Lehrman, 1953; Schneirla, 1965). Others (e.g., Thorpe, 1963; Lorenz, 1965) ridicule this idea. Whether or not true learning occurs, there is little doubt that the developing embryo responds to stimuli from both inside and outside the egg and that these stimuli may influence development. Thus, Margaret Vince (1969) showed that in some species, stimuli produced by the embryo can influence the development of other eggs in the clutch and can lead to synchronization of hatching. Gilbert Gottlieb (1971) shows that duck embryos can respond to maternal calls five days before hatching and that such experience may influence the ducklings' subsequent behavior. However, other workers (e.g., Hamburger, 1963) provide evidence that movement in the chick embryo develops in the absence of sensory stimulation and that the movement patterns that occur later in ontogeny are not necessarily influenced by earlier movements (for a review, see Gottlieb, 1970).

Readers should remember that the young of different species develop in very different ways. While ducklings and goat kids are mobile as soon as they are hatched or born, blackbird nestlings and kittens are helpless at this stage. We are not surprised to discover that the postnatal development of the latter is influenced by experience. Should we be surprised at prenatal influences in the case of animals that are still inside the egg or womb at the same overall stage of development? There seems little difference in principle between the protection the duckling receives while in its egg and that the young blackbird enjoys in the nest. Similarly, there would seem to be little difference in principle between the effects of experience in the two cases. For each species, certain types of experience are essential for normal development. For example, adult cat vision is abnormal if there has been any disruption in the coordination of the vision from its two eyes during the first few months of life—by inducing an artificial squint or by covering each eye on alternate days so that the two eyes never work together (Hubel and Wiesel, 1965). The exact nature of the experience needed for normal development varies considerably from one species to another.

3.3 Sensitive periods during development

Some animals appear to be preprogrammed to learn about certain aspects of the environment during particular periods of their development, for example, language learning in humans. In contrast to other types of learning, language learning is "preprogrammed" in the sense that it occurs in the absence of any obvious reward or punishment. For example, the young of many precocial species show a fairly indiscriminate attachment to moving objects. Thus, newly hatched mallard ducklings, separated from their mother, will follow a crude model duck, a person, or even a simple box moved slowly away from them (see Fig. 3.3). Some stimuli are more effective than others in eliciting this

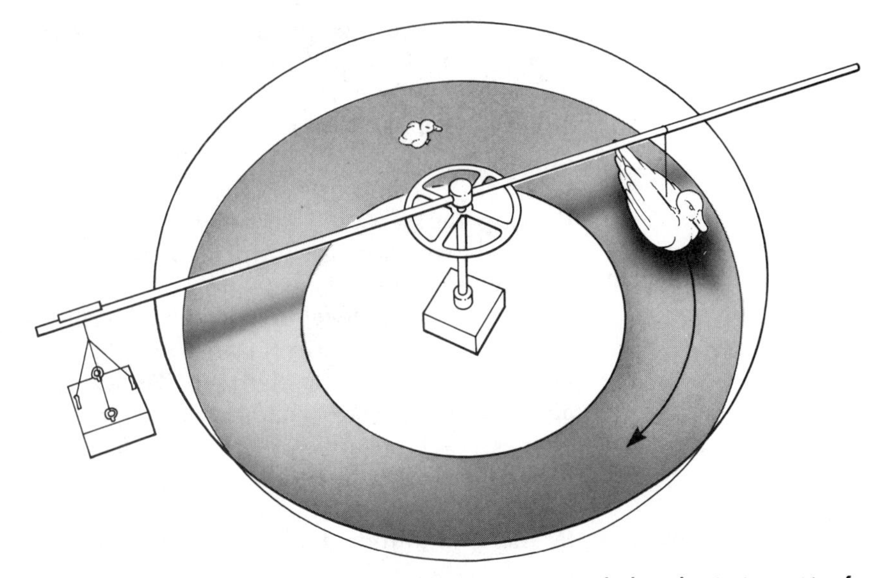

Fig. 3.3 A duckling following its 'mother' in an apparatus designed to test aspects of imprinting (After Gottlieb, 1971)

following response (see Chapter 20.3). In the natural environment the most effective stimuli normally are provided by the mother, and approaches to the mother often are rewarded by body contact and warmth or by food that the mother uncovers. In the laboratory the attachment to a model can be enhanced by food rewards.

The more an animal develops an attachment to one object, the less interested it is in others. This process of learning, through which attachment to the mother normally develops, is called *imprinting*. It occurs during a particular, sensitive period of development, which varies according to the species and the circumstances. Imprinting may have long-term effects, beyond the attachment to a parent or foster parent. In many mammals such early experience affects subsequent social adjustment. In a number of bird species, imprinting has been shown to affect subsequent sexual behavior (see Chapter 20.5). In short, imprinting is a process of learning that occurs at a particular stage of development and that affects subsequent behavior toward parents, peers, or sexual partners. If the object of attachment is a cardboard box, then the duckling will become attached to the box as to a parent. If male zebra finches (*Taeniopygia guttata*) are raised by Bengalese finches (*Lonchura striata*), then they court Bengalese finch females when adult.

Although this type of learning may be influenced by rewards, it is not dependent upon them or upon any particular consequences of the behavior. The learning is preprogrammed to take place as part of the normal process of development and in whatever circumstances pertain at the time.

A similar arrangement applies to the song learning of some passerine birds. The first attempts at song, called *subsong*, usually occur in the young bird's first spring or first autumn, some months after hatching. The subsong resembles the adult song in length, pitch, and tonal quality, but it is lacking elements and motifs typical of the adult song and is usually rather variable and imprecise (see Fig. 3.4). If white-crowned sparrows (*Zonotrichia leucophrys*) are reared in isolation, they develop the subsong but fail to develop the normal adult song. If exposed to the normal song of adult males when about 10 to 90 days old, male white-crowned sparrows subsequently (that is, when about 8 months old) will develop the normal adult song. However, if exposed to a normal song before they are 8 days old, and not thereafter, they will not develop the adult pattern. Similarly, if exposed only after the age of about 100 days, male white-crowned sparrows will not develop normal song. Thus, there appears to be a critical period, between the ages of 10 to 90 days, when it is necessary for young male white-crowned sparrows to hear adult male song if they are eventually to learn to sing that song (Marler and Mundinger, 1971). Similar critical periods have been found for other species.

If white-crowned sparrows are surgically deafened before being

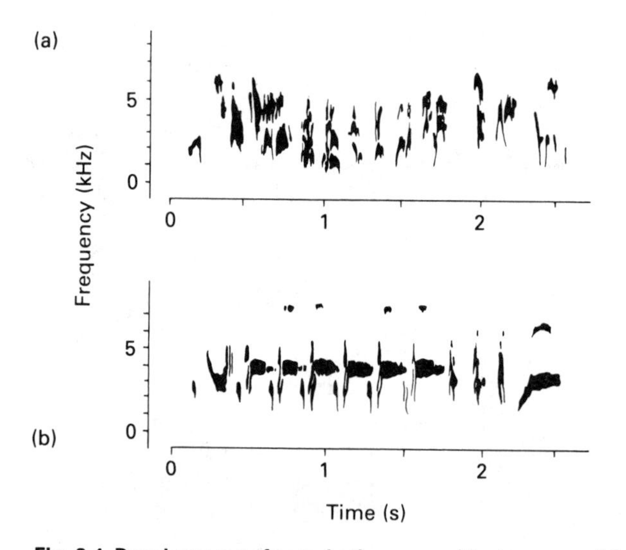

Fig. 3.4 Development of song in the canary (*Serinus canaria*). (*a*) Subsong: The sound spectrogram shows the ill-defined phrasing and lack of tonal purity. (*b*) Full song, by contrast, shows regular phrasing and notes which are relatively pure and free of harmonics (From *The Oxford Companion to Animal Behaviour*, 1981)

exposed to normal song during the critical period, they do not develop normal song. If they are isolated after having heard normal song during the critical period, then they do produce normal song. Thus, the birds that hear the song during the critical period remember it even though they do not sing until they are about 8 months old. However, if they are deafened after hearing normal song during the critical period but before they sing, then they do not produce normal adult song (Konishi, 1965). They may be able to hear themselves sing and to make use of their auditory memory in developing a song that matches the song they heard during the critical period. If white-crowned sparrows are deafened after their song has been rehearsed and has crystallized, then their song is not affected. This means that during song rehearsal and perfection, the song is encoded in another kind of memory, perhaps as a set of singing instructions.

3.4 Juvenile behavior

The animal develops throughout the whole of its life, and must be well adapted to its environment throughout its life history. Thus behavioral development is not simply a matter of constructing adult behavior patterns. The juvenile animal will often show behavior that never occurs

Fig. 3.5 An example of specialized juvenile behavior. Swallow nestlings begging for food (*Photograph: Hans Reinhard: courtesy of Bruce Coleman, Ltd*) (See also back cover)

during its later life (Fig. 3.5). This behavior is usually tailored to suit the young animal's needs. For example, the alarm behavior of herring gull chicks is quite different from that of adults. When alarmed, the chicks move a short distance from the nest and crouch motionless and silent amongst the vegetation. The adults, on the other hand, fly away from the nest, uttering alarm calls. Chicks that run around during an alarm are very vulnerable to predation from crows and other gulls. By remaining motionless they increase their camouflage and reduce the risk of predation.

The extent to which the behavior of a juvenile resembles that of the adult varies considerably from species to species. In some, such as the wildebeest (*Connochaetes*), where the young animal runs with the herd within a few minutes of being born, the juvenile life style is very similar to that of its parents. In others, such as the larval forms of butterflies and frogs, the life of the juvenile is quite different from that of the adult.

An animal's life history is subject to natural selection, and different species have different life-history strategies. While some genes influence behavior from the outset, others have effects that are considerably delayed. Indeed, it has been suggested that the delayed effects of some genes are responsible for the process of aging (Dawkins, 1976).

3.5 Innate behavior

The question of what it means to say that behavior is "innate" has long been a subject of controversy. The term *innate* has come to have different meanings, which should be distinguished. The early ethologists considered innate behavior to be behavior that is determined by heredity and that is part of the animal's original makeup and therefore that is independent of the experience of the individual. Thus, Lorenz (1939) writes about characteristics of behavior as being hereditary, individually fixed, and thus open to evolutionary analysis. Tinbergen (1942) refers similarly to instinctive acts as being highly stereotyped coordinated movements, the neuromotor apparatus of which belongs, in its complete form, to the hereditary constitution of the animal. These views were criticized soundly by Lehrman (1953) and others as being too rigid, as incorporating a naive idea of genetic determinism, and as being semantically confusing. A rigid distinction between innate and learned behavior is unsatisfactory because many aspects of behavior are influenced both by genetic factors and by the experience of the individual.

The idea that genes determine behavior is naive, since genes cannot possibly contain detailed instructions for particular aspects of behavior. The genes may influence the processes of development in various ways, but these processes are affected also by environmental factors. As we have seen, the genes can never dictate the course of ontogeny without reference to the environmental medium in which the development occurs. For the student of behavior it is convenient to use the term *innate behavior* as shorthand for "behavior that develops without obvious environmental influence". This is the way it is used in this book.

Points to remember

- The development of the individual from embryo to adult involves a continual interaction between the animal's genetic makeup and its environment. The interaction is one in which each phase of development sets the stage for the next, a process called epigenesis.

- Environmental influences upon development are most important just after birth, or hatching, but may occur at any stage of the developmental process. The circumstances in which the parents raise their young are usually designed to protect the young from unfavorable environmental influences.

- Many species have periods of development during which they are sensitive to particular kinds of environmental influence. What the animal learns during these sensitive periods usually affects it for the rest of its life.

- Juvenile animals often have characteristic behavior which enables them to respond in an appropriate way to environmental occurrences such as the appearance of a predator or the provision of food by a parent. This typical juvenile behavior is lost in adulthood.

- It is convenient to use the term innate for behavior that occurs without obvious environmental influence, provided it is recognized that environmental factors influence the development of all behavior to some extent.

Further reading

Bateson, P. P. G. and Klopfer, H. (eds) (1982) *Perspectives in Ethology*, Volume 5, *Ontogeny*, Plenum, New York.

4 Behavior Genetics

Although Charles Darwin certainly was interested in hereditary factors in relation to behavior, the scientific pioneer in this field was another grandson of Erasmus Darwin, Francis Galton. *On the Origin of Species* inspired Galton to devote the rest of his life to the study of the inheritance of mental characteristics. In 1869, Galton published *Hereditary Genius: An Inquiry into Its Laws and Consequences*. He argued that people of outstanding mental ability are to be found more often among the relatives of similarly outstanding people than among the population at large. Not having any satisfactory way of measuring mental ability, Galton relied upon an index of reputation, "the reputation of a leader of opinion, of an originator, of a man to whom the world deliberately acknowledges itself largely indebted" (1869). He examined the pedigrees of some 300 families containing eminent judges, statesmen, military commanders, literary men, scientists, poets, musicians, painters, etc. His results showed that eminent social status was most likely to appear in close relatives, and as the degree of relationship became more remote, so the likelihood of eminence decreased.

Galton was aware of the fact that eminent people would share social, educational, and financial advantages. He argued that the reputations of eminent people are an indication of their natural ability and are not due to environmental factors. To support his argument, Galton pointed out that many eminent men had humble family backgrounds. He also judged that the adopted kinsmen of Roman Catholic popes, who enjoyed great social advantages, were a less distinguished group than the sons of eminent men.

Galton devoted considerable effort to improving the means of assessing mental characteristics. He developed procedures for measuring smell, touch, and visual acuity; judgments of length, of weight, and of the vertical, and reaction time and memory span. In statistics, he pioneered the concepts of correlation, median, and percentile. In 1883, he introduced the twin-study method of distinguishing between the effects of nature and of nurture. He studied 35 pairs of twins who were very similar at birth and who were raised in similar conditions. He noted that their behavioral similarities persisted even after they had developed separate adult lives. He also studied 20 pairs of twins who were dissimilar at birth and raised in similar environments:

There is no escape for the conclusion that nature prevails enormously over nurture when the differences of nurture do not exceed what is commonly to be found among persons of the same rank of society and in the same country. My fear is that my evidence may seem to prove too much, and be discredited on that account, as it appears contrary to all experience that nurture should go for so little (1883).

The study of the genetic inheritance of behavioral traits developed rapidly following Galton's lead, but it has remained controversial to this day. In the early part of this century, the main challenge came from the rise of behaviorism. John Watson (1930), in particular, persuaded many psychologists that genetics was irrelevant to behavior development, and the strong environmentalist view adopted by the behaviorists remained influential until the 1960s. The field of behavior genetics may be said to have become firmly established by 1960 when Fuller and Thompson published their *Behavior Genetics*. This book relates the history of psychological studies of human behavior and intelligence from the beginning of the century and reviews the evidence for genetic influences upon behavior. Despite the overwhelming weight of evidence, many sociologists and psychologists remained opposed to the idea of genetic influences in behavior and the debate continued (see Hirsch, 1963; 1967). Even today some theories that are based upon genetic arguments remain controversial. In this chapter we review the evidence concerning the influence of genetic factors upon behavior.

4.1 Single genes and behavior

Behavioral traits under the control of single genes can provide a clear demonstration of genetic analysis of behavior. A classic example is Margaret Bastock's (1956) study of mating success in the fruit fly *Drosophila melanogaster*. She crossed a sex-linked yellow mutant with wild stock for seven generations. This ensured that the wild stock was genetically similar to the yellow stock, except in the region of the yellow gene. Bastock found that males with the yellow mutation were less successful in mating with wild-type females than were wild-type males. The yellow males had an altered courtship pattern that reduced their mating success. They were less stimulating to the females because their courtship contained a smaller proportion of wing vibration.

Rothenbuhler (1964) carried out an elegant genetic analysis of the nest-cleaning behavior of honeybees (*Apis mellifera*). The larvae sometimes are killed by a disease called "American foulbrood." To maintain a hygienic environment within the hive, the worker bees normally uncap, or open, the comb cells that contain diseased larvae and remove them. Some strains of bees, called "unhygienic," do not follow this procedure. When these are crossed with normal hygienic bees, all the offspring are unhygienic, indicating that this is a dominant character. When these hybrids were back crossed with the parental hygienic strain, Rothenbuhler obtained the following results. Out of a total of 29 back-crossed colonies he found that nine uncapped the infected cells but did not remove the diseased larvae, six did not uncap the cells but would remove the larvae if the cells were uncapped by the experimenter, and eight would not uncap cells and remove larvae.

These results show that different genes must control the uncapping and the removal behaviors. The results can be explained on the basis of two pairs of alleles, of which the unhygienic alleles are dominant. Thus, worker bees with Uu or UU will not uncap the infected cells, and those with Rr or RR will not remove larvae.

The male bees, called "drones," are haploid, having only one set of chromosomes. Among his 29 colonies, Rothenbuhler must have had four types of drones (UR, Ur, uR, ur) that he crossed with hygienic queens (uu, rr), as shown in Figure 4.1. According to this simple Mendelian scheme, the back cross should produce four genotypes in equal proportions, which does not differ significantly from the results Rothenbuhler obtained. No physical or physiological differences have been discovered among totally hygienic, partially hygienic, or unhygienic workers, although there is some evidence that unhygienic workers do perform hygienic activities at a very low fequency and require a stimulus that is more powerful than normal. This suggests that the alleles U and u

Fig. 4.1 Mendelian scheme proposed to account for resistance to American foulbrood in honeybees (After Rothenbuhler, 1964)

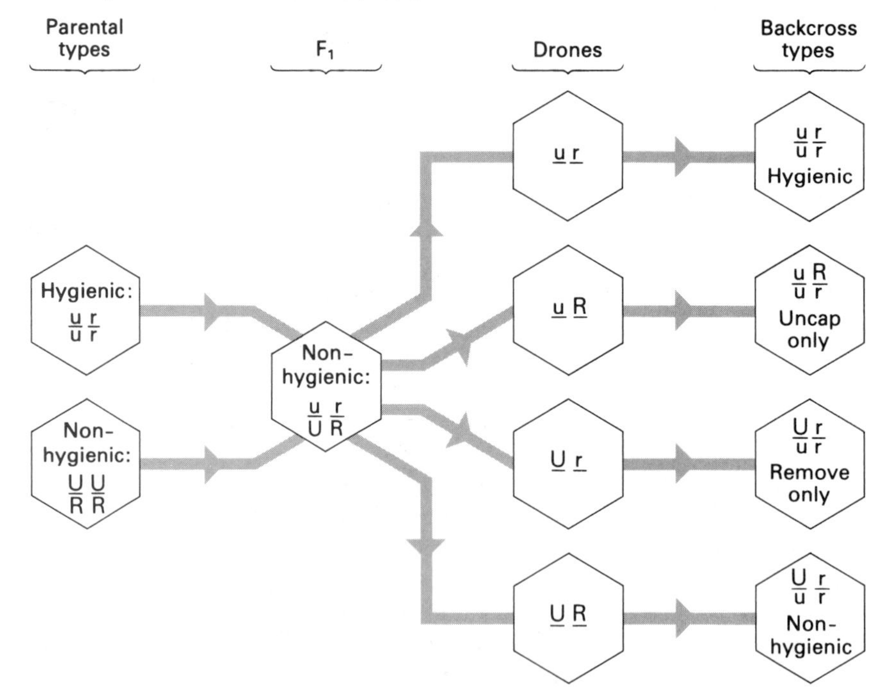

act as switches that release the uncapping behavior, provided there is a certain threshold of stimulation.

Genes that act as switches, activating a group of other genes, are known in a variety of circumstances. For example, some butterflies have elaborate wing patterning that mimics that of other species that are distasteful to predators. The development of the mimic patterns is under the control of many genes, but it appears that a single switch gene determines whether or not the pattern appears (Sheppard, 1961). Single-gene effects are also known in humans. An example is the lactase deficiency found in certain racial groups and discussed in Chapter 5.2. Lactase production seems to be controlled by a gene with three alleles—L, l_1, and l_2. Both l_1 and l_2 are recessive to L, and l_2 is recessive to l_1. Individuals with LL, Ll_1, or Ll_2 genotypes produce lactase both as adults and as children. Those with l_1l_1 or l_1l_2 do not produce lactase as adults, and those with l_2l_2 cannot produce lactase even in infancy. Adults with l_1l_1 or l_1l_2 genotypes can digest milk that has been soured or turned into yogurt or cheese.

4.2 Chromosome mutations

The arrangement and number of chromosomes can often be observed directly under the microscope. There are various known types of chromosome mutations, and some of these have identifiable effects upon the phenotype. The study of behavioral correlates of chromosome structure thus provides a useful way of investigating genetic influences upon behavior.

A favorite animal for this type of study has been the fruit fly *Drosophila*. *Drosophila* larvae have giant chromosomes in the salivary glands that can be prepared and observed relatively easily. The application of chromosomal analysis in *Drosophila* to the study of behavior was pioneered by Jerry Hirsch and his co-workers. They investigated the tendency of *Drosophila melanogaster* to move toward or away from the direction of gravity (positive and negative geotaxis). In their experiments the behavioral response is tested in a vertical plastic maze (Fig. 4.2), through which the flies are attracted by the odor of food. Backward movement is discouraged by cone-shaped funnels at the junctions in the maze. The flies are introduced, in large numbers, in a vial on the left-hand side of the maze and collected from a series of vials on the right. In this way thousands of flies can be tested without being handled by humans. The flies recovered in the various right-hand vials are segregated into those that are strongly positively geotaxic, those that are strongly negatively geotaxic, and those in between. By breeding from those flies that are collected from the extreme upper and lower vials, it is possible to produce strains selected from positive or negative geotaxis, as shown in Figure 4.3.

Fig. 4.2 A geotactic maze for *Drosophila*. The flies are introduced in the vial on the left and collected from the array of vials on the right (*Photograph: Jerry Hirsch*)

In one experiment, three populations of *Drosophila* were compared (Hirsch and Erlenmeyer-Kimling, 1962). One was selectively bred for positive geotaxis, one for negative geotaxis, and there was an unselected control population. These were crossed with a special stock that carried various chromosomal inversions and marker genes. *Drosophila melanogaster* have four pairs of chromosomes, three large and one small. The marker genes were used to identify the three large chromosomes. They were dominant genes controlling phenotypic features by which their presence in the genotype is made visible. By means of a special mating design, females were produced that were either homozygous or heterozygous for the chromosomes to be investigated.

The three chromosomes were identified by marker genes as follows: chromosome X by bar eyes (B), chromosome II by curly wings (Cy), and chromosome III by stubble bristles (Sb). Tester females carrying these markers were mated with males from one of the stocks to be investigated. Of the progeny, only those carrying all three marker genes were used in the subsequent part of the experiment. These were back crossed with the original male population, producing eight possible genotypes. Each of the three major chromosomes is therefore heterozygous or homozygous for an S chromosome obtained from the sample line (s) to which the father belonged. From the eight classes of genotype, the individual effects of the chromosomes and their interactions can be studied.

Jerry Hirsch and Linda Erlenmeyer-Kimling (1962) found that, in an unselected population, chromosomes X and II contributed to positive geotaxis and chromosome III contributed to negative geotaxis. In a strain selected for positive geotaxis, there was little change on chromosomes X and II, but chromosome III now contributed to positive geotaxis. In a strain selected for negative geotaxis, the negative contribution of chromosome III was increased, while the positive effects of chromosomes X and II were reduced. When the effects on the three chromosomes were taken together, the magnitude of the effect was greater for negative geotaxis. This is not surprising because the total response to selection (see Fig. 4.3) is greater for negative geotaxis. These results demonstrate that geotaxis behavior is controlled by a number of genes, which are distributed over all three of the major chromosomes.

Chromosome analysis has been applied extensively to various types of *Drosophila* behavior, including mating speed and other aspects of courtship (Ehrman and Parsons, 1976). Chromosome inversions are common, and it has been found that in *D. pseudo-obscura* that inversion heterozygotes have greater fitness than inversion homozygotes, due to the effects upon courtship behavior. Chromosome anomalies in humans have been the subject of considerable study and are thought to be

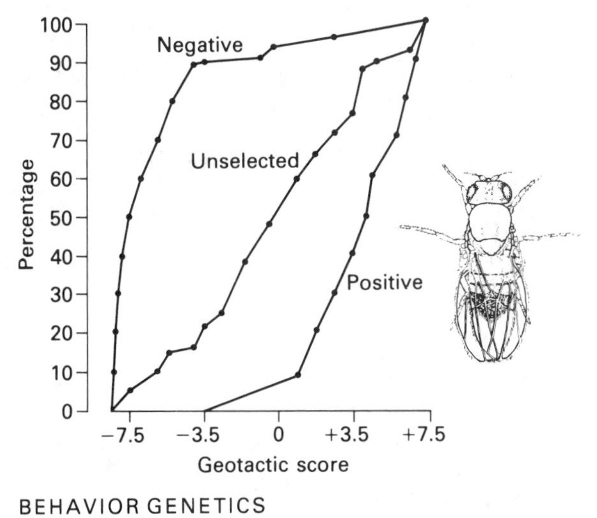

Fig. 4.3 Geotactic scores obtained for three strains of *Drosphila* in the maze shown in Fig. 4.2 (After Hirsch, 1963)

involved in a number of genetically induced behavioral disorders, including epilepsy, manic depression, mental retardation, and schizophrenia (McClearn and DeFries, 1973; Ehrman and Parsons, 1976).

4.3 Polygenic inheritance of behavior

Many behavioral traits are influenced by a large number of genes as well as environmental factors. A variety of methods can be used in the genetic analysis of such complex situations. Their effectiveness depends largely upon the degree of heritability of the behavioral trait in question. However, some aspects of behavior have high heritability, and with these we begin our discussion of polygenic inheritance.

Male crickets attract females over long distances by a calling song. The sound is produced by rhythmic opening and closing of specialized forewings that carry friction mechanisms. Each closing stroke of the wings produces a sound pulse, while the opening stroke is silent. The songs are remarkably stereotyped among members of a local population, but they differ considerably from one species to another. The differences occur primarily in the temporal patterning of the pulses, as illustrated in Figure 4.4. Although hybrids are not common in nature (Hill et al., 1972), they can be produced in a laboratory. For example, Leroy (1964) obtained hybrids between the Australian field crickets *Teleogryllus commodus* and *Teleogryllus oceanicus*. David Bentley and Ronald Hoy (1972) found that the songs of the F_1 hybrids are distinctly different from either parental song. In particular, the intrachirp and intratrill intervals of the hybrids are intermediate between those of the parents. Figure 4.5 shows how the parental songs compare with the hybrids. Bentley and Hoy found that reciprocal hybrids differed from each other. The hybrid from female *T. oceanicus* by male *T. commodus* was similar to that of *T. oceanicus* in having a well-defined intertrill interval.The hybrid from female *T. commodus* by male *T. oceanicus* lacked a well-defined intertrill

Fig. 4.4 Phrase structure of the calling song of the cricket *Teleogryllus*. Each phrase is composed of two types of pulse: A-pulses contained in the chirp portion of the phrase, and B-pulses contained in the trill. The song of *T. oceanicus* is shown above and that of *T. commodus* below (After Bentley and Hoy, 1972).

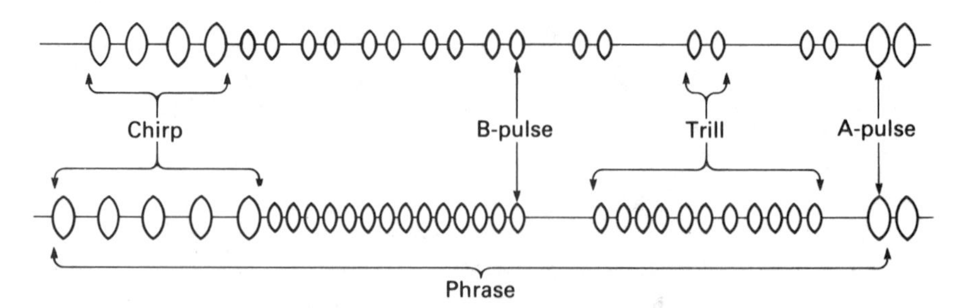

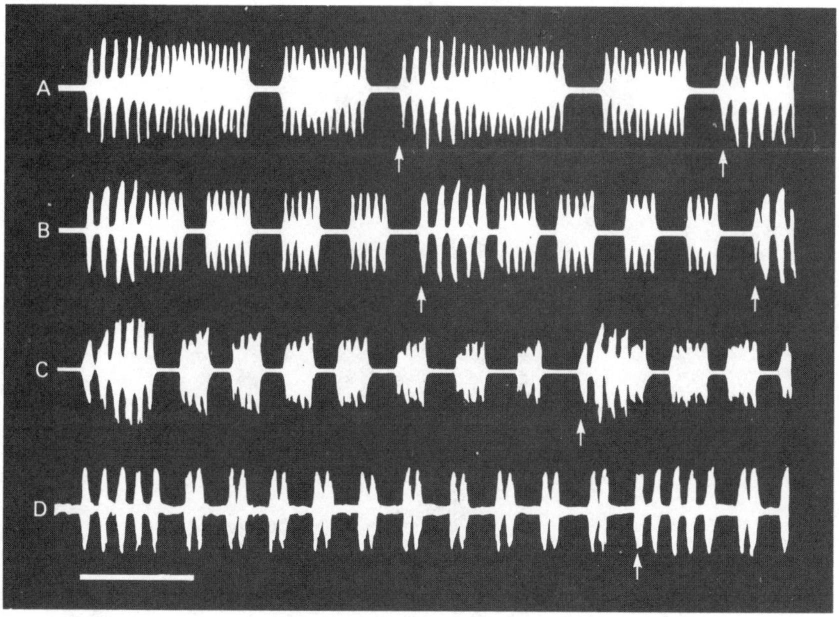

Fig. 4.5 Oscillograms of calling songs of *T. oceanicus* (A), *T. commodus* (D), and their hybrids *T. oceanicus* female × *T. commodus* male (B) and *T. commodus* female × *T. oceanicus* male (C) (arrows mark the beginning of phrases) (After Bentley and Hoy, 1972)

interval, and its song was similar to that of the *T. commodus* parent. These differences are best seen in the interpulse-interval frequency histograms in Figure 4.6.

These results suggest that the inheritance of the song pattern is polygenic. There is no evidence of any simple dominance of any particular song feature, and the hybrid songs are intermediate between those of the parents. The hypothesis is supported by the fact that back-crossed hybrids also show intermediate inheritance (Bentley, 1971). There is also some indication that sex-linked factors are involved. The songs of the reciprocal hybrids differ in a way that suggests that the intertrill interval characteristic of *T. oceanicus* is present in one hybrid but not the other. Sex determination in crickets is XO, meaning that there is no Y chromosome. The male receives the X chromosome from his mother but nothing to match it from his father. Thus, it appears that the intertrill interval characteristic of the female *T. oceanicus* by male *T. commodus* hybrid is determined by the X chromosome of the *T. oceanicus* parent. This trait is present neither in the female *T. commodus* by male *T. oceanicus* hybrids nor in their mothers.

The male songs characteristic of different species presumably are responded to by the corresponding female. It is, therefore, of interest to discover how hybrid females respond to the songs of their parents and

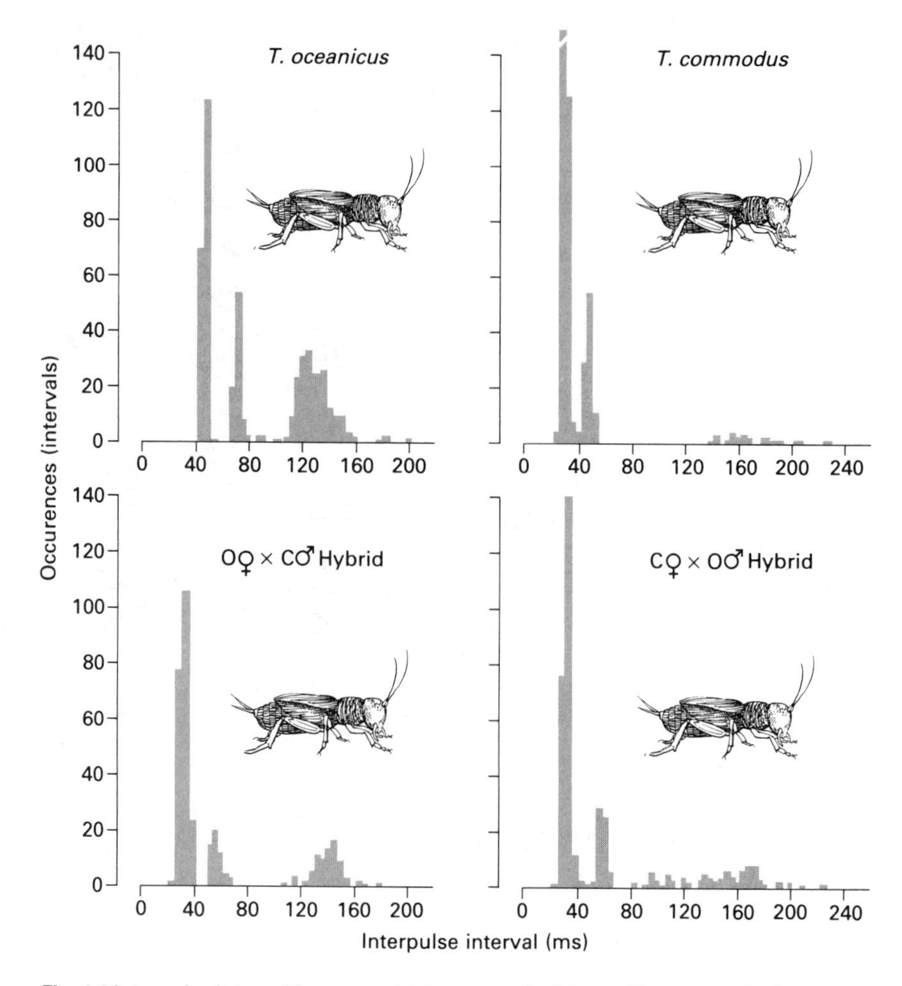

Fig. 4.6 Interpulse interval frequency histograms of cricket calling songs. Each histogram represents the analysis of one individual song. Intratrill, intrachirp and intertrill intervals can be distinguished from these histograms (After Bentley and Hoy, 1972)

siblings. Female crickets are mute and respond to the male song by walking toward the source of the sound. Hill et al. (1972) showed that free-walking *T. oceanicus* and *T. commodus* females can discriminate between the song of their own species and that of another species. To obtain a more quantitative measure of female preference, Hoy and Paul (1973) employed the Y maze illustrated in Figure 4.7. They presented recorded songs from loudspeakers placed on the right and left of the maze. In this way they were able to measure the relative attractiveness of any test song. Hoy and Paul found that *T. oceanicus* and *T. commodus* females make species-specific discriminations, just as Hill et al. had found in free-walking females. They also found that females that were female *T. oceanicus* by male *T. commodus* hybrid preferred the song of

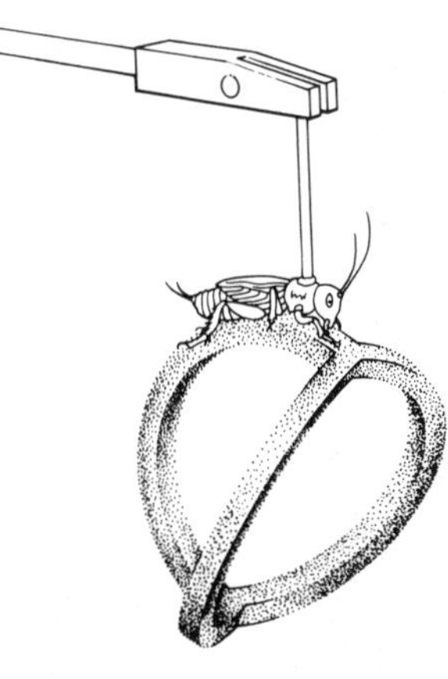

Fig. 4.7 Female field cricket on a Y-maze. Her movements on the maze indicate her preferences among male calling songs (After Hoy, 1974)

sibling hybrid males to that typical of either parent species. As can be seen from Figure 4.6, the song of the male *T. oceeanicus* by female *T. commodus* hybrid is easily distinguishable from that of either parent. (Note that this is not so clear-cut in the case of the alternative hybrid.) The fact that the hybrid song is preferred by the female implies that the production of song in the male and its reception and recognition by the female somehow are coupled genetically.

It is well known that different breeds of dogs have different behavioral characteristics. In 1965, John Scott and John Fuller published some of the results of their extensive research into the genetic basis of such differences. In one study they compared the behaviors of the cocker spaniel, the basenji, and their hybrids. Spaniels have long been bred as sporting dogs, obedient and devoted to their master (see Fig. 4.8). Spaniels originally came from Spain, where they were used in hawking and in net-hunting of birds. The dog was trained to crouch close to the ground when he detected partridge or other birds. A net then was thrown across both the dog and the birds. When shotguns were invented, nets were no longer used and land spaniels were developed to stop and point whey they located birds. Springer spaniels were developed to spring and flush birds. The original tendency to crouch remains only in the cocker spaniel.

Basenjis used to be widespread in Africa and were used for hunting by Pygmies and several other African tribes (see Fig. 4.9). They are a general-purpose hunting dog with a somewhat wary temperament. Basenjis bark little, if at all, although they do sometimes indulge in bouts of prolonged howling.

Scott and Fuller (1965) carried out extensive breeding programs with these two breeds (see Fig. 4.10), including reciprocal hybridization and various types of back cross. Table 4.1 summarizes the characteristics of the two breeds, together with the most likely modes of inheritance as judged on the basis of the results of the experiments. Basenji puppies tend to be wild in contrast to the tameness and friendliness of spaniel puppies. They avoid being handled and struggle against restraint. Handling tests showed that the F_1 behavior of basenjis is similar to parental basenji behavior, suggesting that dominant genes are involved. The results of back-cross experiments pointed to a single dominant gene controlling wildness, normally present in basenjis. The tameness of spaniels is controlled by a single recessive gene. To take another example, basenji females come into estrus once per year, at the time of the autumnal equinox. Spaniels, like most European breeds, may show heat once every six months, at any time of year. The basenji type of estrus cycle appears to be controlled by a single recessive gene. Overall, Scott and Fuller come to the conclusion that the behavioral traits they investigated were controlled by one or two genes. The situation appears to be intermediate between simple Mendelian inheritance and polygenic inheritance.

Fig. 4.8 Use of spaniels in hawking. The dogs crouch in response to a handsignal given by their master (From Blome's *Gentleman's Recreation*, 1686) (*Courtesy of J P Scott*)

Fig. 4.9 Pygmies returning from a hunting trip with a basenji. The nets are used to trap game which the dogs flush out of the brush (*Courtesy of J P Scott*)

It may seem surprising that such complex behavioral traits are controlled by so few genes. However, the two breeds of dogs have been isolated for a very long period of time and have been subjected to intense artificial selection. This seems to have led to genotypes that are homozygous for particular traits, so that there is little segregation within breeds. The genetics of dog behavior is particularly interesting because the ability to learn particular types of behavior is often the inherited trait.

Fig. 4.10 Parental cocker spaniels (above) and basenjis (below) used in the genetic studies of Scott and Fuller, 1965 (*Photographs: J P Scott*)

Table 4.1 Characteristics of basenjis and cocker spaniels

Characteristic	*Basenji*	*Cocker spaniel*	*Most likely mode of inheritance*
Wildness and tameness			
Avoidance and vocalization in reaction to handling	High	Low	One dominant gene for wildness
Struggle against restraint	High	Low	One gene with no dominance
Playful aggressiveness at 13 to 15 weeks of age	High	Low	Two genes with no dominance
Barking at 11 weeks			
Threshold of stimulation	High	Low	Two dominant genes for low threshold
Tendency to bark a small number of times	High	Low	One gene with no dominance
Sexual behavior (time of estrus)	Annual	Semi-annual	Basenji type as a recessive gene
Tendency to be quiet while weighed	Low	High	Two recessive genes for high tendency

After Scott and Fuller, 1965.

4.4 Heritability of behavior

The question whether the nature or the nurture, the genotype or the environment, is more important in shaping man's physique and his personality is simply fallacious and misleading. The genotype and the environment are equally important, because both are indispensable. . . . The question about the roles of the geotype and the environment in human development must be posed thus: To what extent are the differences observed among people conditioned by the differences of their genotypes and by the differences between the environments in which people were born, grew and were brought up? (Dobzhansky, 1964).

Much of the work of Francis Galton was based on the idea that phenotypic similarities among relatives are due partly to similarities of genotype. However, Galton was not aware of the work of Mendel, and he assumed that inheritance was blending. He thought the hereditary material was continuous, so that offspring were genetically intermediate between their parents. On the basis of such assumptions, Galton pioneered the technique of using correlations among relatives as a means of estimating heritability. In 1918, Ronald Fisher published a paper that showed that it was possible to predict the correlations expected among relatives, on the basis of Mendelian theory. Although Mendel's work had been brought to public attention in 1900, it had been applied only to discontinuous traits. Fisher showed how Mendelian theory could cope with continuous traits such as height and weight. It is interesting that his conclusions were similar to those of Galton, with many predictions of the two theories being almost identical.

The concept of *heritability* was first defined as the fraction of the observed variance caused by differences in heredity (Lush, 1940). In other words, when we examine a particular trait, such as bodyweight, in a population of animals, we are looking at a phenotypic value made up of both genotypic and environmental components. By looking at related populations in different circumstances, it is sometimes possible to estimate how much of the variability between individuals is due to environmental factors, and how much to genetic factors.

There are various statistical methods for estimating the heritability of behavior. Some are based upon examining phenotypic variation in genetically identical individuals. The total phenotypic variability among genetically identical individuals is compared to the total phenotypic variability in a natural, genetically variable population. The comparison generates a ratio that contrasts the genetic versus environmental components of phenotypic variability. This technique has certain technical statistical problems, but it also makes clear the fallacy of calling one behavior "genetic" and another "environmental." If a new gene variant appears, or if the population is exposed to new types of environmental variation, the ratio may well be altered. A behavior pattern that is "genetically determined" in one environment (identical twins reared apart are found to have identical phenotypes) may be shown to have a strong environmental component in a different climate (identical twins reared apart are found to have divergent phenotypes). Conversely, a behavior pattern that is "environmentally determined" (genetically variable individuals all have identical phenotypes when reared in the same environment) can come to have an apparently strong genetic component in an environment in which the genetic variability is expressed. Because of these interactions, the determination of behavioral heritability, just as with the distinction between innate and learned behavior, is always determinable only within strict limits. We can say that a given behavior is 80 per cent heritable across the range of environments and genotypes in a specific population, but we cannot generalize to other environments or populations.

Estimates of heritability may be subject to error from various sources, the most important of which are genotype-environment interaction and genotype-environment correlation. Interaction between genotype and environment introduces a variability that is not taken account of in normal heritability calculations. For example, Tryon (1942) was able to selectively breed rats for maze dullness and maze brightness. These rats were bred and tested in ordinary laboratory conditions. Cooper and Zubek (1958), however, raised some rats in ordinary conditions and others in impoverished conditions with bare surroundings or in enriched conditions with complex mazelike surroundings. When tested in a standard maze, the maze-bright and maze-dull strains of rats bred in ordinary conditions showed large differences in performance, as other workers had found previously. However, both strains of rats raised in a restricted environment performed equally badly in the test maze, while both strains raised in the enriched environment performed equally well, as illustrated in Figure 4.11. Although a definite genotypic difference exists between the two strains, this is only evident phenotypically in rats raised in particular environmental conditions.

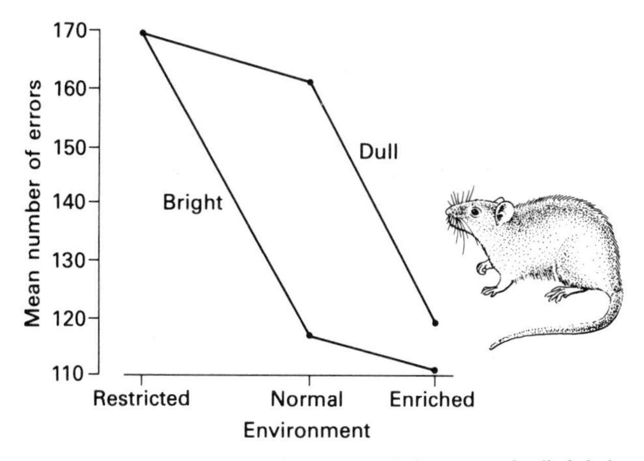

Fig. 4.11 Mean number of errors made in a maze by 'bright' and 'dull' rats reared in enriched, normal and impoverished environments (After Cooper and Zubek, 1958).

Correlation between genotype and environment, another potential source of error in estimating heritability, can arise when individuals select particular environments or develop particular habits to compensate for genetic defects. Despite these difficulties, however, heritability measures have proved to be particularly useful in separating the effects of nature and nurture.

Estimates of heritability based on similarity among relatives are much used in human behavior genetics. There are, however, a number of complicating factors. The first problem is that genetic dominance lowers the correlation among relatives. Under genetic dominance the parental contributions to a trait are not equal, and an extra source of variation is introduced. This is called the *dominance variance*. However, dominance variance has differential effects on parent-offspring correlations and sibling-sibling correlations. In fact, the difference between these two correlation coefficients should be exactly one-quarter of the proportion of the total variance due to dominance (Bodmer and Cavalli-Sforza, 1976).

The second problem is that, for certain traits, mating is likely to be non-random in humans. While we can assume that random mating occurs with respect to unobservable traits like enzyme levels, this is not the case with traits such as height and intelligence. Correlations between husband and wife are usually positive for such traits, indicating non-random, or assortive, mating such that people tend to take partners of similar height and I.Q. The husband-wife correlation for height is about +0.3, while for I.Q. it is about +0.4. These correlations may be influenced by social effects because positive correlations also exist between these variables and socioeconomic status. The tendency to choose a mate from the same social group, therefore, would lead to positive correlations between husband and wife. Assortative mating has the effect of increasing the frequency of homozygotes.

The third problem lies in estimating environmental variance in human studies, in which environmental conditions cannot be controlled as they can in some animal studies. An approach adopted by Cavalli-Sforza and Bodmer (1971) is to divide the environmental variance into separate components. Initially, we have the variance among individuals within the family. This may result from age differences, birth order, sex differences, or the like. Families of different sizes usually will experience different environmental effects, such as nutritional level and degree of crowding.

Next we have variance among families within socioeconomic groups. Correlations between adopted children and their foster parents can give some estimate of the importance of such effects, but the results may be biased by the selection of parents by adoption agencies.

We then have the variance among socioeconomic groups. Cultural inheritance may give rise to correlations among relatives that are difficult to distinguish from those due to genetic factors. Cultural transmission from parent to child is confounded with biological inheritance to a considerable extent (Cavalli-Sforza and Feldman, 1974). This remains one of the outstanding problems of biology.

We also may have to consider the variance in environmental conditions accompanying racial differences. For example, the environmental conditions typical of black and white Americans show considerable variation. In part, these may be due to cultural differences and in part to socioeconomic factors.

Finally, we have the variance due to interaction between the genotype and the environment. In humans, such interaction may occur, for example, between age of maturity and the type of schooling available to different age groups.

Overall, we can specify the variance due to environmental factors by adding the component variances outlined above.

Cavalli-Sforza and Bodmer (1971) concluded that estimates of heritability are not adequate yardsticks of the relative importance of nature and nurture, except in particular circumstances. Changes in environmental conditions can invalidate even straightforward measures of heritability. For example, although the heritability of human stature is high, there have been large changes in the average stature of some human populations, due to environmental factors like improved nutrition. While the heritability measures provide an indication of the ratio of prevailing genetic differences to prevailing environmental differences, they cannot be extrapolated to other populations, environments, or time periods.

Points to remember

- In this chapter we review some methods of investigating genetic influences upon animal behavior. We see that in a few cases it is possible to demonstrate the effect of single genes; usually however, many genes are involved in even simple aspects of behavior.

- Polygenic inheritance can be investigated by breeding experiments designed to study the effects of chromosome mutations upon behavior or to study quantitative differences in the behavior of different genetic strains.

- The heritability of behavior is measured in terms of the variance in phenotypic traits within a particular population. The heritability of a trait is the fraction of the observed variance which is due to differences in heredity. It can be studied by looking at differences in closely related animals raised in different environments, and at differences among animals raised in the same environment.

Further reading

Bodmer, W. F. and Cavalli-Sforza, L. L. (1976) *Genetics, Evolution, and Man*, Freeman, New York.
Ehrman, L. and Parsons, P. A. (1976) *The Genetics of Behavior*, Sinauer Associates, Sunderland, Massachusetts.

1.2 Natural Selection

In this section we discuss the elements of evolutionary theory as they affect behavior. The aim is to provide a basic understanding of some key concepts and to show how these apply to the study of behavior. Chapter 5 introduces the theory of natural selection and some assessment of the evidence. It also outlines the physical and biotic pressures of natural selection that shape the evolution of animal behavior. Chapter 6 gives an account of some of the classical experiments demonstrating the survival value of behavior. It also provides an introduction to the various concepts of fitness that are so important in evolutionary biology. Chapter 7 introduces the concept of evolutionary strategy, including evolutionarily stable strategies (ESS).

Charles Darwin (1809–1882)

Mary Evans Picture Library

Charles Robert Darwin was born at Shrewsbury, England, in 1809. His father, Robert Darwin, was a physician, and his grandfather, Erasmus Darwin, was a noted biologist. Although interested in natural history, Darwin showed no early academic promise. He studied medicine briefly at the University of Edinburgh, and then transferred to the University of Cambridge to study for the ministry. In 1831 he was offered a place as a naturalist on H.M.S. Beagle. The Beagle carried Darwin to Brazil, Argentina, the Falkland Islands, Tierra del Fuego, Chile, the Galapagos Islands, Tahiti, New Zealand, Tasmania, and Australia. He discovered both the riches of the tropical forest and the fossilized bones of long-extinct reptiles. He experienced earthquake and insurrection. Darwin kept extensive notes on animals, plants, fossils, geological formations, and coral reefs. He thought a great deal about the adaptiveness, geographical variation, and competitiveness that characterizes living things. Upon returning to England, Darwin initially lived in London but then settled in Kent, where he lived for the rest of his life. He died there in 1882.

Darwin's thinking was greatly influenced by Lyell's *The Principles of Geology*, which he took with him on the Beagle. This book presented the view that the history of the world is the result of natural laws and forces that are active and observable in the present-day world. It argued against the prevailing view that the world was shaped by a series of catastrophes, either natural or divinely inspired. Darwin's voyage on the Beagle opened his eyes to the variability of species and to their remarkable adaptations to the environment. In 1838 Darwin read Malthus's *Essay on Population*, first published in 1798. In this essay Malthus pointed out that populations tend to increase geometrically, and maintained that food supplies could increase only in arithmetic ratio. He argued that population increase will inevitably be checked by starvation or some other form of "vice and misery." Darwin realized that competition for scarce resources would inevitably lead to the survival of those individuals with attributes that gave them an advantage in the struggle for life. He knew that domestic animals and plants could be changed over generations by artificial selection, and he began to form the idea that the characteristics of wild species were established through natural selection.

Darwin wrote a first draft of his theory in 1842, and another in 1844. He intended to marshal the evidence necessary for a much larger work, but in 1858 he received a letter from Alfred Russel Wallace outlining a theory of evolution by natural selection, which Wallace had conceived independently. Darwin and Wallace jointly published a preliminary sketch of their theory in 1858 and in 1859 Darwin published his book entitled *On the Origin of Species by Means of Natural Selection, or the Preservation of Favoured Races in the Struggle for Life.*

This book created a revolution in scientific thought and provided the foundation of modern biology. Darwin wrote a number of other books that have implications for the study of animal behavior and that greatly influenced the development of ethology, the most important of which were *The Descent of Man and Selection in Relation to Sex* (1871) and *The Expression of the Emotions in Man and the Animals* (1872).

5 Natural Selection, Ecology, and Behavior

The elements of Charles Darwin's theory of natural selection are as follows: (1) There is considerable variation among individuals belonging to any population of animals of the same species. (2) Much of this variation is genetically inherited. (3) There are many more individuals born in each generation than can survive to maturity. From these facts it follows that different individuals will not have the same likelihood of survival. Those whose characteristics are best suited to the environment will be more likely to survive and to pass on their beneficial characteristics to the next generation. Thus, certain characteristics will tend to be perpetuated within a population. In other words, certain features of animals are selected.

It is important to distinguish between the logic of the theory of natural selection and the evidence that natural selection is actually effective in the natural environment. Logically, provided it is true that animals reproduce in larger numbers than can survive, and that there are inherited variations among the offspring, then it is inevitable that the variants that survive will tend to be those best fitted to, or selected by, the environment.

Darwin had no direct example of natural selection at work, but since his time the industrial revolution has provided a natural example that has been intensively studied. Since the mid-nineteenth century many moth species have become darker in industrial areas of England. For example, the first black specimens of the peppered moth *Biston betularia* were caught near Manchester in 1848 and by 1895 they formed about 98 per cent of the total population of the area. Normally, this species is light in color, with small dark markings (see Fig. 5.1). In unpolluted parts of the country they commonly alight on light-colored tree trunks, where they are well camouflaged. Bernard Kettlewell (1956) carried out a number of experiments in rural areas of the English county of Dorset, where the trees were covered with light-colored lichens. He released both pale and dark moths in large numbers and observed that moths resting on tree trunks were preyed upon by birds and that out of 190 moths taken, 164 were dark and only 26 were light. Similar experiments conducted in industrially polluted countryside near Birmingham, England, gave opposite results: Birds were observed to capture three times as many light moths as dark moths. Kettlewell (1955) found that the light moths

Fig. 5.1 Typical and melanic forms of the peppered moth *Biston betularia* on a lichen-covered tree trunk (left) and a soot-blackened tree trunk (right) (*Photographs: M W F Tweedie*)

prefer to rest on pale backgrounds, while the dark form rests more often on a dark background. Thus, it appears likely that the survival of the moths in different habitats depends not only on their color, but also on their behavior.

These and other experiments (Kettlewell, 1973) show that changes in the environment due to industrial pollution change the consequences of natural selection. In both rural and industrial areas it appears that birds are the main selective agent. Their ability to detect resting moths is the main factor in determining which moths survive in a particular environment. From the moths' point of view, the predatory birds are the main feature of the environment to which they have to adjust. Those which happen to have inherited dark coloration will have an increased chance of survival in an industrial area and a decreased chance in a rural area. Those that survive to reproduce pass on their traits to the next generation, so that the characteristics of the population change from one generation to the next. We have here an example of evolutionary adaptation due to natural selection.

5.1 The evidence for evolution by natural selection

Although Darwin could not point to natural selection in action, he noted that artificial selection was practiced by man in the domestication of animals. The rapid changes that occur in animals under domestication show that evolution can result from selection, in this case from selective breeding. He also put forward many other types of evidence to support his theory.

In Darwin's day most people were resistant to the idea of evolutionary change, preferring to imagine that the world has always remained much the same. However, during the late eighteenth and early nineteenth century, geologists discovered numerous fossils and realized that many of them were the remains of species no longer living. Initially it was thought that climatic catastrophes could account for the extinction of species, and separate creation for the appearance of others. As more and more fossils were discovered, however, it became possible to trace changes in some species that had occurred over many thousands of years. Much of the fossil evidence was known to Darwin and was used by him as evidence that some kind of evolution had occurred (Romer, 1958).

Darwin also regarded the resemblances between living species as important evidence for evolution. For example, most vertebrates have the same basic bone structure in their forelimbs. This would not be expected if species were independently created, but it would be expected among species that had evolved from a common ancestor. Darwin pointed out that parts that were functional in some species were vestigial in others. Thus, pigs walk on two toes and have two other vestigial toes that protrude from the leg well above the ground. In his *The Descent of Man* (1871) Darwin wrote, "It is notorious that man is constructed on the same general type or model with the other mammals. All the bones in his skeleton can be compared with the corresponding bones in a monkey, bat, or seal. So it is with his muscles, nerves, blood vessels, and internal viscera. The brain, the most important of all the organs, follows the same law". Darwin realized that similarities between species were sometimes (though not always) due to common ancestry, while differences between closely related species were due to evolutionary adaptation to different environments. By comparing species one can often make important deductions about their biology, and this comparative approach is an important part of ethology.

During his voyage on the Beagle, Darwin discovered that the geographical distribution of animals was hard to account for in terms of separate creation. The presence of subspecies or variants of a species on neighboring islands, for example, is easy to account for in terms of evolution, but it is hard to see why a creator should choose to make such minor variations. During his voyage on the Beagle, Darwin visited the Galapagos archipelago, situated some 600 miles off the west coast of

South America. In the *Voyage of the Beagle* (1841) Darwin wrote,

I never dreamed that islands about 50 or 60 miles apart, and most of them in sight of each other, formed of precisely the same rocks, placed under a quite similar climate, rising to nearly equal height, would have been differently tenanted. . . . It is the circumstance, that several of the islands possess their own species of the tortoise, mocking thrush, finches, and numerous plants, these species having the same general habits, occupying analogous situations, and obviously filling the same place in the natural economy of the archipelago, that strikes me with wonder.

In *On the Origin of Species* (1859) he wrote,

How has it happened in the several [Galapagos] islands . . . that many of the immigrants should have been differently modified, though only in a small degree. This long appeared to me a great difficulty: but it arises in chief part from the deeply seated error of considering the physical conditions of a country as the most important for its inhabitants: whereas it cannot be disputed that the nature of the other inhabitants with which each has to compete, is at least as important, and generally a far more important element of success. . . . When in former times an immigrant settled on any one or more of the islands, or when it subsequently spread from one island to another, it would undoubtedly be exposed to different conditions of life in the different islands, for it would have to compete with different sets of organisms. . . . If then it varies, natural selection would probably favour different varieties in the different islands.

The tendency for closely related species to diverge in characteristics that reduce competition between them is nowadays called *character displacement*. Darwin's finches, the Geospizini that inhabit the Galapagos Islands, are a well-studied example of this phenomenon (see Fig. 1.4).

5.2 Frequency distribution of phenotypes

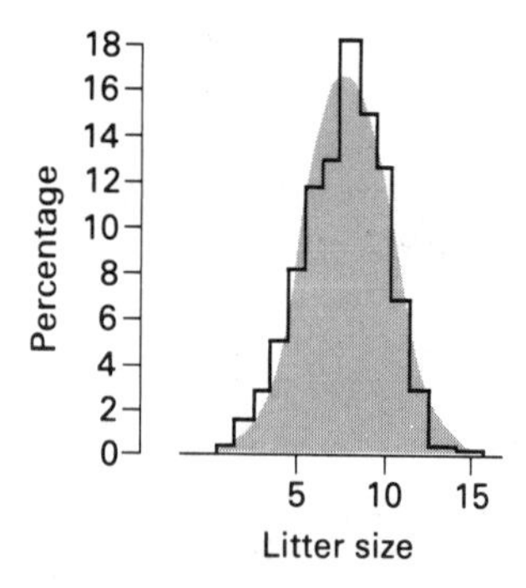

Fig. 5.2 Frequency distribution of litter size in mice (After Falconer, 1960)

Within a given population, each characteristic of the species varies between individuals, and one can plot the frequency distribution of the various values. For example, in a population of mice, the size of the litter might range from 2 to 16. If we were to plot the percentage of litter of each possible size on a graph like that illustrated in Figure 5.2, then we would have established the frequency distribution. If there were no variation among offspring and if individuals of all phenotypes had identical numbers of surviving offspring, then the form of the frequency distribution would remain unchanged from one generation to the next. Where a change in the frequency distribution persists for more than one generation, evolution may be said to have occurred. Changes can occur if some phenotypes leave fewer surviving offspring than others. Mice with a small litter size may or may not have the same number of surviving offspring as mice with a large litter size. The former may have

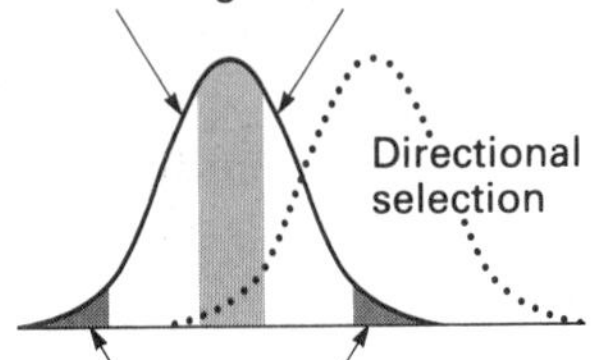

Fig. 5.3 Sections of a normal population distribution favored under disruptive, directional and stabilizing selection

more litters than the latter, or the survival rate may be lower in large litters than in small ones. However, a change in circumstances may favor litters of a particular size so that more mice from such litters survive and reproduce. Evolution of litter size will then begin.

Natural selection can have a variety of effects upon the frequency distribution of phenotypes within a population. As a result of mutation, recombination, and genetic drift, the variance of a character tends to increase from one generation to the next. This tendency usually is opposed by *stabilizing selection*, which acts relatively more severely at the extremes of a frequency distribution, as illustrated in Figure 5.3. When there is a differential selective pressure along a phenotypic gradient, *directional selection* occurs. This may result in a shift in the mean of the frequency distribution or in a skewed distribution without alteration of the mean. Sometimes there may be strong selective pressures against characters typical of the population, so that selection is stronger against animals near the mean than against extreme individuals, as illustrated in Figure 5.3. Such *disruptive selection* is rare, and it tends to result in a bimodal frequency distribution. In some cases it may lead to division of a single species into two separate species.

The relative abundance of different genes can change at random from generation to generation, by accident and without the action of natural selection. This effect, known as *random genetic drift* (Sewell Wright, 1921) is greatest in small populations. Figure 5.4 illustrates how a particular genotype can be eliminated from a small population purely by chance. Variation due to chance fluctuation sometimes gives rise to unusual

Fig. 5.4 Schematic representation of genetic drift. In each generation each individual of the grey and white genotypes produces two identical offspring. Half the juveniles die and the population remains constant in size. However, which juveniles die is entirely random and not dependent on genotype. The frequency of each fluctuates at random until by chance grey becomes extinct and white is fixed in the population (After Futuyma, 1979)

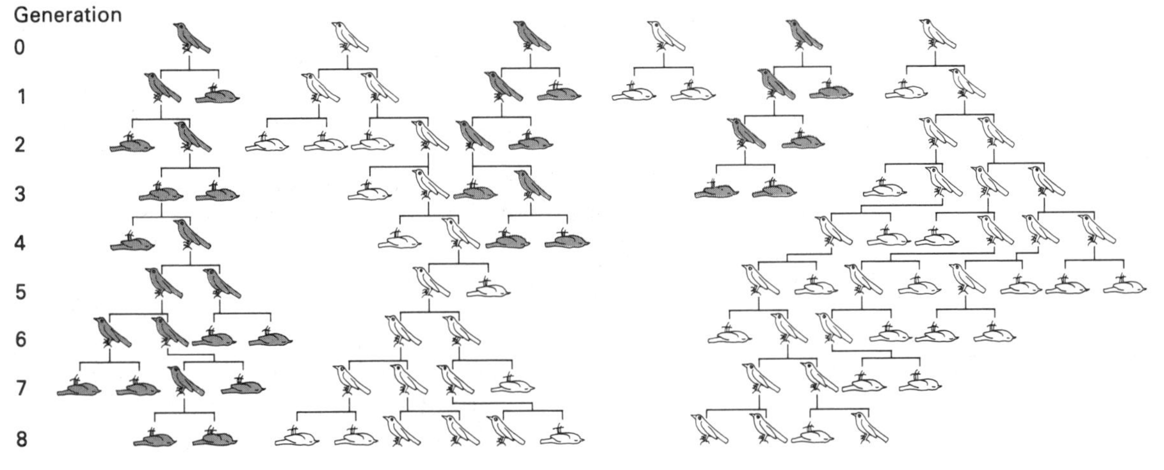

proportions of genotypes, especially in small, isolated populations (see Fig. 5.5). For instance, in some isolated alpine villages the frequency of albinism is 10 times higher than normal. Other isolated human communities are known to have abnormal incidences of color blindness, deaf-muteness, and certain types of mental deficiencies.

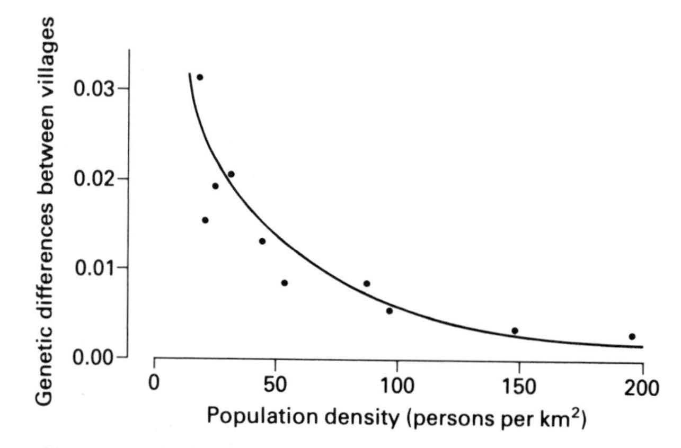

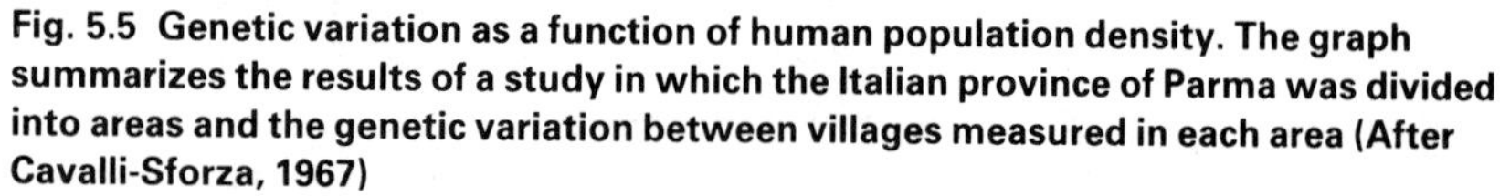
Fig. 5.5 Genetic variation as a function of human population density. The graph summarizes the results of a study in which the Italian province of Parma was divided into areas and the genetic variation between villages measured in each area (After Cavalli-Sforza, 1967)

The rate of change of gene frequency induced by natural selection depends upon the relative fitness of the various genotypes. The *relative fitness* of a genotype reflects the difference in fitness between that genotype and the others represented in the same population. If the fitness of the genotype with the highest rate of increase is designated 1.00, the relative fitness of another genotype with, say, 85 per cent as great a rate of increase would then be 0.85. The difference, 0.15, is called the *coefficient of selection* against the inferior genotype.

Selection coefficients can be used to calculate the rate of change of phenotype in a population. An interesting example arises from the domestication of animals some 10,000 years ago. Before the domestication of animals, the only source of milk for human babies was the human mother's milk given during the first years of life. The child's stomach produces the enzyme lactase, which is necessary for the digestion of the lactose sugar present in milk. In later life the enzyme disappears, and among many present-day human populations it is completely absent in adults.

People who lack lactase are intolerant to lactose and experience nausea, vomiting, and abdominal pains if they drink milk. The retention of lactase in late life is an inherited trait, which is thought to be the result of a mutation that became advantageous in populations for which milk became a regular part of the diet. Presumably, the mutant was rare before the domestication of animals, when humans lived entirely by hunting animals and gathering plants. Thus, in some populations the lactose-tolerant genotype must have increased from an initially low frequency to the present high frequency among northern Europeans and other milk drinkers.

Let us assume that the human generation time is 30 years, on average. On this basis, 300 generations have passed since the beginning of domestication. During this period, the proportion of adults that can tolerate lactose has risen from almost zero to 75 per cent among northern Europeans. If we assume a single gene is responsible for the trait, the gene frequency must have increased from about 0.001 per cent to the 50 per cent required to account for the present genotype frequency of 75 per cent. It is possible to calculate that a selection coefficient of 0.04 would be necessary to achieve this change. If the initial frequency of the lactose-tolerant gene was 1.0 per cent, then a selection coefficient of 0.015 would be sufficient (see Fig. 5.6). This means that a relatively rapid evolutionary change can take place even if the difference in fitness between two alleles is less than 5 per cent.

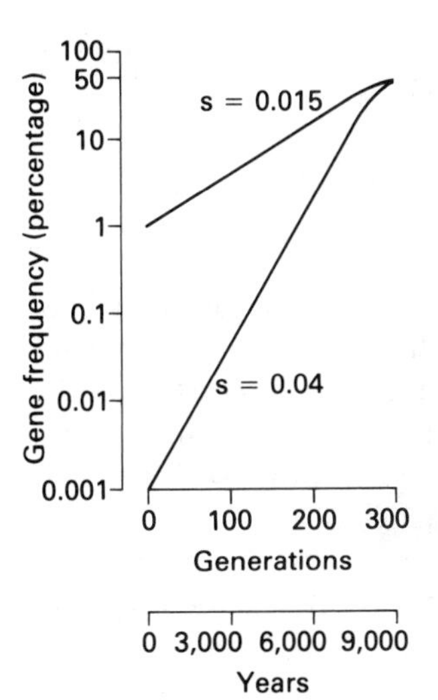

Fig. 5.6 Selection for lactose tolerance of adults. Assuming 300 generations have elapsed since cattle were first domesticated and it became advantageous to be able to digest milk, then the present gene frequency of 50 per cent can be explained in terms of a particular coefficient of selection (s) acting upon the assumed initial frequency of the gene (After Bodmer and Cavalli-Sforza, 1976)

It is possible to calculate (e.g., Futuyma, 1979, pp. 314–316) that an advantageous allele increases in frequency at a rate proportional to the selection coefficient and to the frequencies of both the advantageous allele and the disadvantageous allele. The rate of genetic change therefore is only high when both genes are common in the population. New mutations initially will increase in frequency very slowly. Moreover, complete replacement of one allele by another will take a very long time because the process is slow initially, when the favorable allele is rare, and terminally, when the disadvantageous allele becomes rare. Deleterious recessive alleles are seldom completely eliminated by natural selection because they show themselves phenotypically only when homozygous. Natural populations, therefore, have many deleterious recessive genes that occur at low frequency and that create a kind of genetic reservoir.

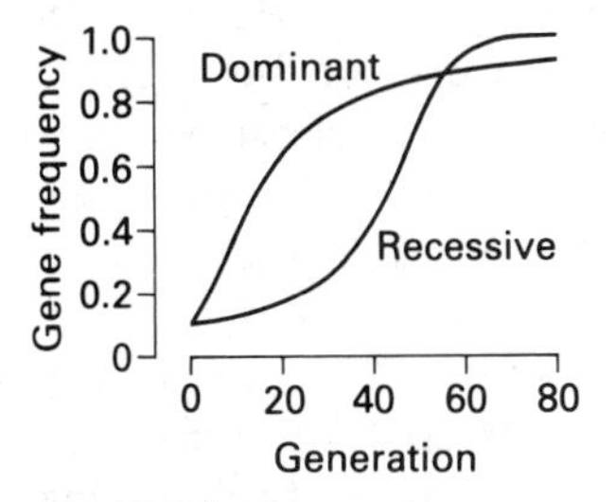

Fig. 5.7 A rare advantageous dominant gene increases more rapidly than a rare advantageous recessive, but it achieves fixation more slowly because the last few recessive genes are protected in the heterozygous form

The rate of evolution of a gene depends upon the phenotypic expression of the trait as well as the gene frequencies and the coefficient of selection. For example, a rare advantageous dominant gene increases in frequency much more rapidly than a rare advantageous recessive gene, as illustrated in Figure 5.7. However, a dominant gene will achieve complete fixation (replacement of the competing allele) more slowly than the recessive because the last few recessive genes occur in heterozygous form and are thus shielded from natural selection.

No animal species exists in the absence of other life, and the processes of adaptation of one species may change the conditions under which another species lives. This may induce compensatory adaptation in the affected species. Thus, if a bird species adapts to the scarcity of its principal prey species by broadening its diet to include new insects, then a new selective agent is introduced into the ecology of the newly preyed-upon insect population. Because such interconnections exist in every ecosystem, we must always imagine the behavior of an animal as embedded in a complex of ecological relationships.

5.3 Ecology and behavior

In considering the behavior of animals in the natural environment, it is important to have some understanding of the effects of the consequences of the behavior on the survival of the animal. The consequences of a particular activity depend largely upon the animal's immediate environment. If the animal is in a situation to which it is well adapted, the consequences of the activity may be beneficial. The same activity performed in a different environment may be harmful. In order to appreciate how the behavior of animals has been shaped during evolution, we need to understand the adaptive relationship between the animal and its environment.

Ecology, or scientific natural history, is the study of the relationships among animals and plants in the natural environment. It is concerned with all aspects of these relationships, including the flow of energy in the environment, the physiology of the animals and plants, animal populations, animal behavior, etc. In addition to specific knowledge of particular animals, the ecologist seeks to understand the general principles of ecological organization, and we are concerned with some of these here.

During the course of evolution, animals become adapted to a particular environment, or *habitat*. Habitats usually are described in terms of the physical and chemical features of the environment. Plant communities depend upon the physical features of the environment, such as the type of soil and the climate. The plant communities provide a variety of possible habitats that animals are able to exploit. The association of plants and animals, together with the physical features of the habitat, constitute an *ecosystem*. There are about 10 general types of ecosystems in the world, and these are known as *biomes*. Figure 5.8 shows the distribution of the major terrestrial biomes of the world. There are also marine and freshwater biomes. To take one example, the savannah biome occupies large areas of Africa, South America, and Australia, and consists of a combination of grassland and scattered trees found in tropical or subtropical parts of the world. Savannah is typified by seasonal rainfall. At the higher end of the rainfall range, the savannah grades into tropical forest, and at the lower end it falls into desert. The dominant trees of the African savannah are the acacias, in South America, the palms, and in Australia, the eucalyptus. The characteristic feature of the African savannah is the large variety of grazing ungulate species, which supports a variety of carnivores. In South America and Australia the same niches, or roles, are occupied by other species.

The association of animals and plants living in a particular habitat is

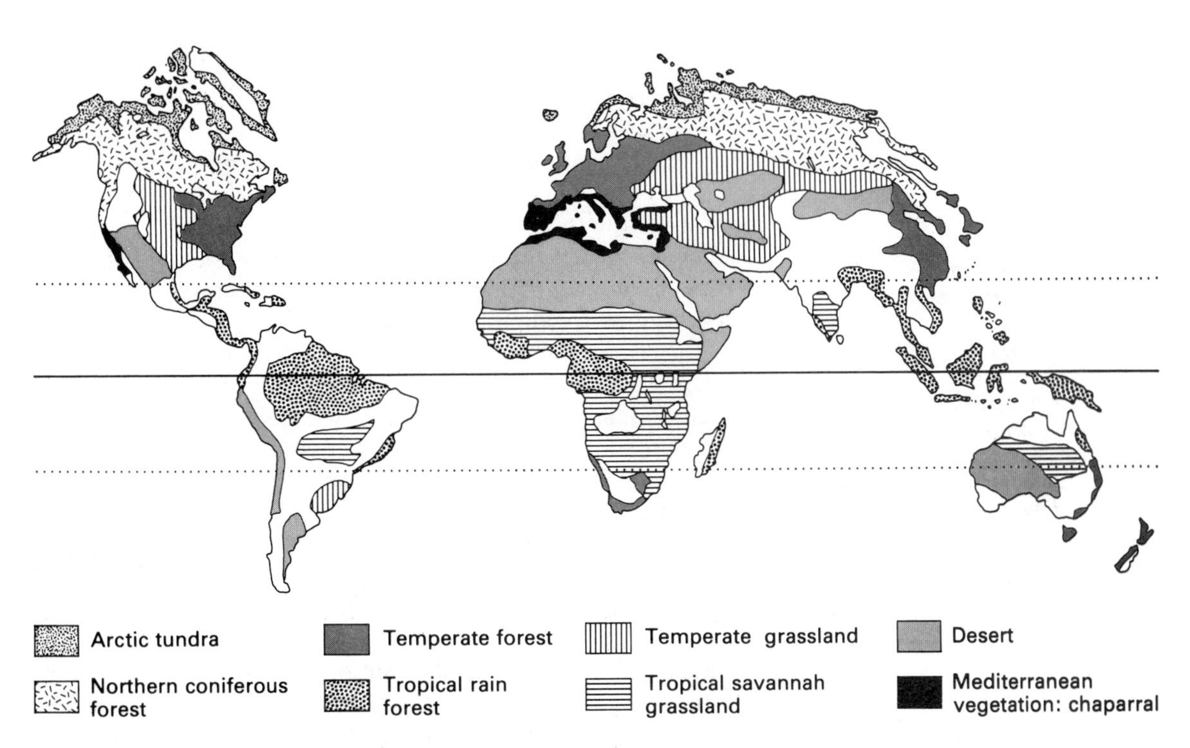

Fig. 5.8 The distribution of the main terrestrial biomes of the world

called a *community*. The species that make up a community may be classified into producers, consumers, and decomposers. The *producers* are the green plants that trap solar energy and convert it into chemical energy. The *consumers* are the animals that eat the plants or that eat herbivorous animals and thus depend indirectly upon plants for energy. The *decomposers* are usually fungi and bacteria that break down animal and plant material into a form that can be reused by plants.

The *niche* is the role the animal plays in the community in terms of its relationship both to other organisms and to the physical environment. Thus, a herbivore eats plant material and usually is preyed upon by carnivores. The species occupying a given niche varies from one part of the world to another. For example, a small herbivore niche is occupied by rabbits and hares in northern temperate regions, by the agouti and viscacha in South America, by the hyrax and mouse deer in Africa, and by wallabies in Australia.

In 1917 the American ecologist Joseph Grinnell pioneered the study of niches in his research on the California thrasher (*Toxostoma redivivum*), a bird that nests in dense masses of foliage one or two meters above the ground. This nesting habit is an aspect of the animal's niche. In mountain areas, the necessary vegetation is available only in that ecological community called *the chaparral*. The habitat of the thrasher, in terms of the physical characteristics of the environment, is determined

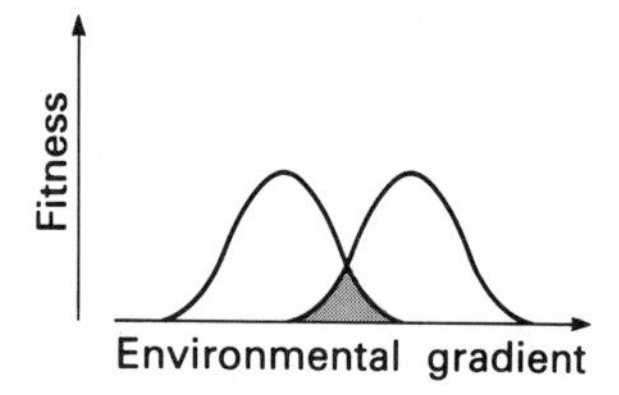

Fig. 5.9 Niche overlap. Along an environmental gradient, such as temperature, the fitness of an animal can often be represented as a bell-shaped curve. Niche overlap (shaded area) occurs along that part of the gradient shared by representatives of different species.

partly by the response of the thrasher population to the niche situation. This, in turn, depends upon the status of the thrasher in the ecological community. Thus, if height of the nest above ground were a critical factor in the avoidance of predators, there would be competition within the population for nest sites at the optimal height. If this factor were less critical, more individuals could nest in suboptimal sites. The niche can also be affected by competition from other species, for nesting sites, food, etc. The habitat of the California thrasher is determined partly by the niche situation, by the distribution of the various shrub species that characterize the chaparral, and by the population density of the thrasher. Clearly, if the population were small, only the best nest sites would be occupied, and this would affect the habitat of the species. Thus, the thrasher's total relationship to its environment, for which the term *ecotope* is sometimes used, results from a complex interaction of niche, habitat, and population factors.

When animals of different species use the same resources or have certain preference or tolerance ranges in common, *niche overlap* occurs (see Fig. 5.9). Niche overlap leads to competition, especially when resources are in short supply. The *competitive exclusion principle* states that two species with identical niches cannot live together in the same place at the same time when resources are limited. The corollary is that, if two species coexist, there must be ecological differences between them.

As an example, let us consider the niche relationships of a group of bird species that comprises a "foliage-gleaning guild," feeding in oak trees on coastal mountains in central California (Root, 1967). A *guild* is a group of species that exploits the same environmental resources in a similar manner. Such species overlap significantly in their niche requirements, and thus, they compete with each other. One advantage of the guild concept is that it focuses attention on all competing species in a given geographic area, without regard to their taxonomic relationship. To be considered as a member of the foliage-gleaning guild in the oak woodland, the major portion of the animals diet must consist of arthropods obtained from the foliage. This is an arbitrary classification, since a species may be a member of more than one guild. For example, the plain titmouse (*Parus inornatus*) belongs to the foliage-gleaning guild on account of its foraging habits, and it is also a member of the hole-nesting guild by virtue of its nesting requirements.

Although there are five species feeding on insects in the same area, each tends to eat insects of different characteristic sizes and taxonomy. An overlap exists between the taxonomic categories of insects eaten by the five bird species but each species tends to specialize on certain taxa. Similarly, Figure 5.10 shows complete overlap in size, but different mean values and variances, at least in some cases. Root also found that the bird species used three types of foraging technique: (1) Gleaning—taking prey from leaf surfaces while the bird walks on a solid substrate;

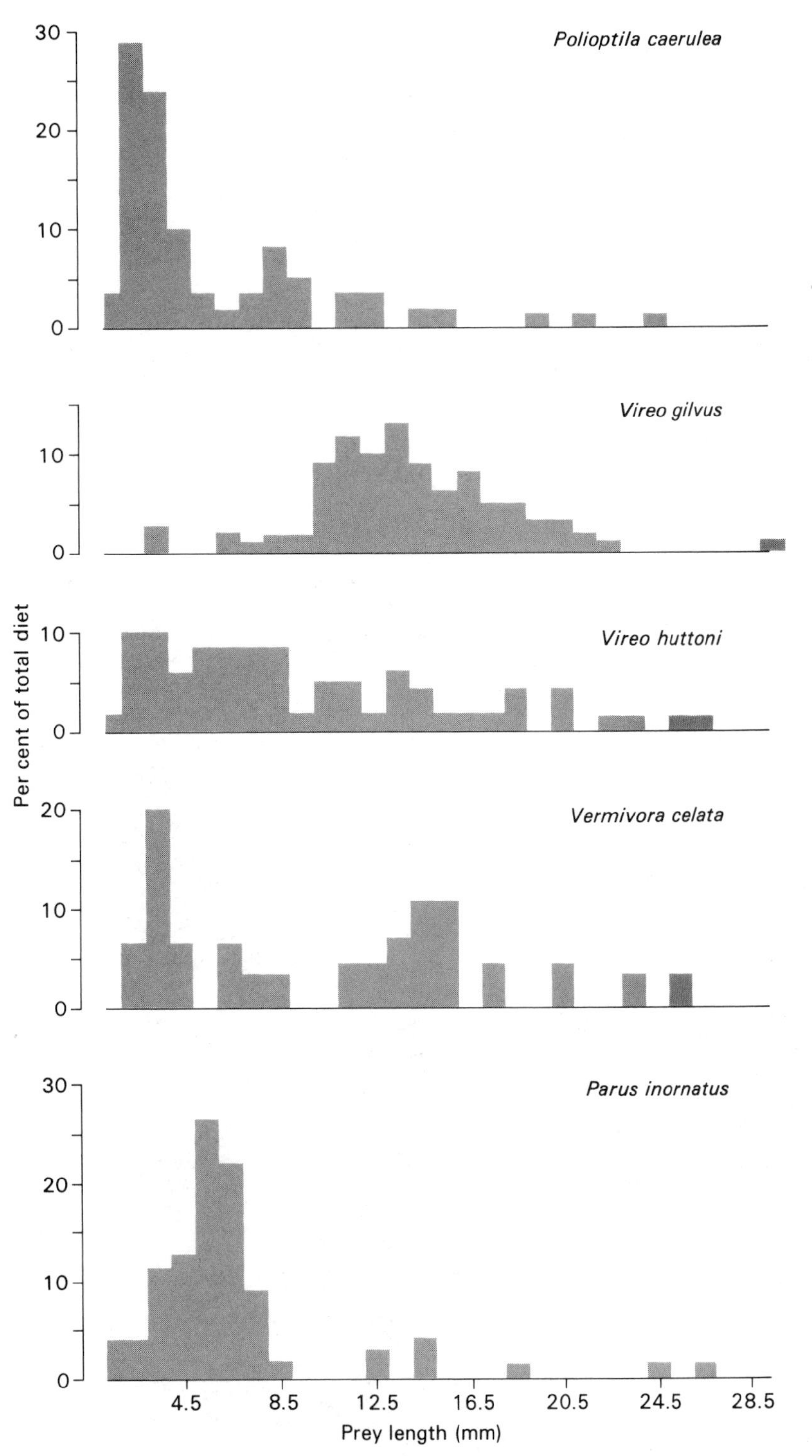

Fig. 5.10 Size distribution of intact prey in the stomachs of foliage-gleaning birds (After Root, 1967)

(2) hovering—taking prey from leaf surfaces while in flight; and (3) hawking—taking insects on the wing. The proportions of time the five bird species devote to these methods of foraging are illustrated in Figure 5.11. In this example we see the operation of ecological specialization in behavior. The behavior of each species influences that of the others such that the species of the guild are spread out over the available foraging sites and prey types.

Competition often results in dominance of one species over another, in that the dominant species has priority in use of resources such as food, space, and shelter (Miller, 1967; Morse, 1971). On theoretical grounds, we would expect that a species that becomes subordinate to another species should shift its use of resources in a way that decreases overlap with the dominant species. This typically involves the subordinate's reducing its use of certain resources, thus reducing its niche breadth. In some circumstances the subordinate species may be able to expand its niche to include previously unutilized resources, implying that either the subordinate species can dominate another species in an

Fig. 5.11 Three types of foraging tactics by members of a foliage-gleaning guild are represented as the three sides of a triangle. The length of a line perpendicular to a side of the triangle is proportional to the percentage of time spent in that foraging tactic. The sum of all three lines for each species equals 100 per cent (After Root, 1967)

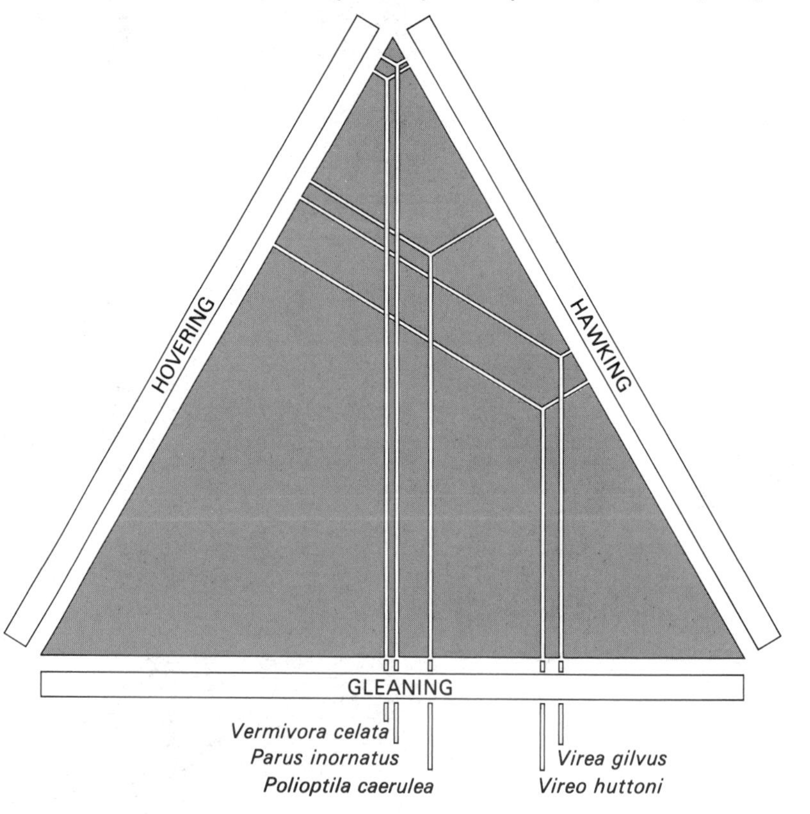

adjoining niche or it can make fuller use of its own fundamental niche.

If a subordinate species is to survive in competition with a dominant species, it must have a fundamental niche broader than that of the dominant species. Such cases have been documented for bees and new-world blackbirds (Orians and Willson, 1964). Since dominant species have priority in use of resources, subordinate species may be excluded from areas of the niche when resources are limited in availability, are unpredictable, or require considerable effort to find, which may considerably reduce the fitness of subordinate species in regions of overlap. In such cases we would expect subordinate species to be under considerable selective pressure to modify their fundamental niche by becoming specialists or by evolving a tolerance for a wider range of physical environmental conditions.

5.4 The adaptedness of behavior

Naturalists and ethologists have discovered numerous examples of the marvelous ways in which animals appear to be perfectly adapted to their environment. The problem with such accounts of animal behavior is that they seem convincing precisely because the various details and observations fit together so well; that is, a good story may seem convincing simply because it is a good story. This is not to say there is anything wrong with a good story. Indeed, in any correct account of behavioral adaptation, the variety of details and observations *should* fit together. The problem is that biologists, as scientists, should evaluate the *evidence*, and a good story is not necessarily good evidence. As in a court of law, the evidence must be more than circumstantial and must have some element of independent corroboration.

One way to obtain evidence about the adaptedness of behavior is to compare related species that occupy different habitats. A classic example of this approach is Ester Cullen's (1957) comparison of the nesting habits of cliff-nesting kittiwakes (*Rissa tridactyla*) with those of ground-nesting gulls, such as the black-headed gull (*Larus ridibundus*) and the herring gull (*Larus argentatus*). The kittiwake nests on cliff ledges inaccessible to predators and is thought to have evolved from ground-nesting gulls as a result of predation pressure. Kittiwakes retain some features of ground-nesting gulls, like the partial cryptic coloration of their eggs. The eggs of ground-nesting birds usually are well camouflaged as a protection from predators, but this cannot be the function of egg coloration in kittiwakes because each nest is marked conspicuously by white droppings. The adults and young of ground-nesting gulls are careful not to defecate near the nest and thus advertise its whereabouts. Thus, it seems most likely that the camouflage of kittiwake eggs is a vestigial indication of their ground-nesting ancestry.

Cullen (1957) studied a breeding colony of kittiwakes in the Farne

Islands, off the east coast of the United Kingdom, where they nest on very narrow cliff ledges. She observed that the eggs were free from ground predators like rats and from aerial predators like herring gulls, which frequently prey on the eggs of ground-nesting birds. Kittiwakes feed mainly on fish and plankton and do not cannibalize the eggs and chicks of their neighbors as do many ground-nesting gulls. The kittiwakes seem to have lost many of the antipredator adaptations of other gulls. For example, as well as failing to camouflage the nest, they rarely give alarm calls and they do not mob predators.

Kittiwakes have many special adaptations to cliff nesting. They are lightly built and have strong toes and claws that enable them to cling to ledges too small for other gulls to manage. Compared with ground-nesting gulls, kittiwake adults appear to have a number of behavioral adaptations to their cliff habitat. Their fighting behavior is limited and stereotyped compared with that of ground-nesting relatives (Fig. 5.12). They build fairly elaborate cup-shaped nests using both sticks and mud, whereas ground-nesting gulls build rudimentary nests of grass or seaweed, without using mud as cement. Young kittiwakes are different in many ways from the young of ground-nesting gulls. For example, they stay in the nest for a longer period and face the cliff wall much of the time. They take regurgitated food directly from the throat of the parent, whereas most gulls pick it up from the ground where it has been deposited by the adult. The young of ground-nesting gulls run and hide when alarmed, but kittiwake young remain in the nest. The young of

Fig. 5.12 Red-legged kittiwakes (*Rissa brevirostris*) nesting on cliff ledges off the Pribilof Islands, Bering Sea (*Photograph: Ronald Squibb*). Note the restricted form of fighting. Compare with the herring gull nest (Fig. 1.2)

ground-nesting gulls are cryptic in both appearance and behavior; kittiwake young are not.

Comparison of species can provide evidence about the function of behavior patterns in the following ways: When a behavior pattern is observed in one species but not another, there may be correlated differences in the way that natural selection has affected the two species. For example, herring gulls remove broken eggshells from the region of the nest, a possible function of which is to help maintain the nest camouflage, since the white interior of the eggshell is so conspicuous. Evidence in support of this hypothesis comes from the observation that kittiwakes do not remove broken eggshells. As we have seen, kittiwakes are not subject to much nest predation, and their nests and chicks are not camouflaged. If eggshell removal functions primarily to maintain nest camouflage, then we would not expect to find it among kittiwakes. If, however, it serves some other function such as prevention of disease, then we might expect to observe it in kittiwakes. Kittiwakes usually maintain very clean nests and throw away any foreign objects. Herring gulls do not normally bother to do this.

The type of evidence discussed here is strengthened if we can show that other related species, faced with similar selective pressures, show similar adaptations. An example is provided by Jack Hailman's (1965) study of the cliff-nesting swallow-tail gull (*Larus furcatus*) of the Galapagos. Hailman investigated various aspects of behavior that appeared relevant to the danger of falling off the cliff. Swallow-tail gulls nest neither on such steep cliffs as kittiwakes nor so high above the ground. Thus, we might expect the relevant adaptations of the swallow-tail gull to be intermediate between that of the kittiwakes and typical ground-nesting gulls. Swallow-tail gulls are subject to greater predation than kittiwakes, and Hailman observed several aspects of behavior that appeared to be correlated with this difference. For example, as mentioned above, kittiwake chicks defecate on the nesting ledge, thus rendering it conspicuous. The chicks of swallow-tail gulls, however, defecate over the edge of the ledge. He found that in a number of characteristics, likely related to the amount of predation, swallow-tail gulls are intermediate between kittiwakes and other gulls. Hailman similarly evaluated those behavioral characteristics of swallow-tail gulls that appear to be adaptations to the amount of available room for nesting and the availability of nesting sites and nesting materials. He then set out to evaluate the evidence relevant to Cullen's (1957) hypothesis that the peculiarities shown by kittiwakes are the result of selective pressures accompanying cliff-nesting habits. He took 30 characteristics of swallow-tail gulls and divided them into three groups according to the degree of similarity to kittiwake behavior. Taken as a whole, the comparison supports Cullen's hypothesis that the special features of the kittiwake are the result of selective pressures that accompany cliff nesting.

John Crook's (1964) work on some 90 species of weaver bird (Ploceinae) is another example of the comparative approach. These small birds live throughout Asia and Africa. Although similar in appearance, the different weaver bird species vary markedly in their social organization. Some defend large territories in which they build camouflaged nests, while others build their nests in conspicuous colonies. Crook found that the species living in forests tend to be solitary and insectivorous and to have camouflaged nests in large defended territories. They are monogamous and have little sexual dimorphism. Species living in the savannah are usually seed eaters that live in flocks and nest colonially. They are polygamous, males being brightly colored and females dull.

Crook argued that since food is difficult to find in the forest, necessitating that both parents feed the young, the parents have to stay together as a pair throughout the breeding season. The insects that the forest birds feed on are widely dispersed, so a pair of birds must defend a large territory to ensure an adequate supply of food. The nests are camouflaged and the adult birds are dull colored so their visits to the nest do not so easily alert predators to its whereabouts.

Fig. 5.13 A colony of the village weaver (*Ploceus cucullatus*). Note the large number of nests positioned relatively free from predation (*Photograph: Nicholas Collias*)

In the savannah, seeds may be abundant in some places and absent in others, an example of patchy food distribution. Foraging in such conditions is more efficient if the birds form flocks in order to search over a wide area. Nesting sites that offer protection from predators are scarce in the savannah, and many birds nest in a single tree. The nests are bulky to insulate against the heat of the sun, and the colonies tend to be conspicuous. To gain some protection from predators, the nests are usually built high up in a spiny acacia or similar tree (Fig. 5.13). The female can feed the young by herself because food is relatively abundant. The male invests little in the young and is free to court other females. Males compete for nest sites within the colony, and the successful males may each attract several females, while other males fail to breed. In the colonial village weaver (*Textor cucullatus*), for example, the males steal nest material from each other. They therefore have to remain near the nest to protect it. To attract females the male performs an elaborate display while hanging from the nest. If the male is successful in courting, the female enters the nest. Such nest displays are typical of colonial weaver birds and contrast with the courtship of the forest-living species, in which the male chases the female, courts her distant from the nest, and then leads her to it.

The comparative approach has proved to be a fruitful method of studying the relationship between behavior and ecology. There have been several studies of birds (Lack, 1968) and also extensive studies of ungulates (Jarman, 1974) and primates (Crook and Gartlan, 1966; Clutton-Brock and Harvey, 1977). Though not without its critics (e.g., Clutton-Brock and Harvey, 1977; Krebs and Davies, 1981), the comparative approach can provide satisfactory evidence concerning evolutionary aspects of behavior, provided that sufficient care is taken to avoid confounding variables and circular arguments. Hailman (1965) regards the comparative method as satisfactory only in cases where comparison of two populations of animals leads to predictions about a third population that has not been studied already at the time the predictions are formulated. In this way a hypothesis resulting from a comparative study can be tested independent of the data used in the study. Simply to discover that correlated differences exist in behavior and ecology between two populations is not sufficient evidence to conclude that the characters reflect selective pressures arising from environmental differences between the populations. Differences due to confounding variables or to comparison at inappropriate taxonomic levels may be eliminated by careful statistical analysis (Clutton-Brock and Harvey, 1979; Krebs and Davies, 1981).

Points to remember

● Evidence for evolution by natural selection can be found by comparing living species with each other and with their fossil ancestors, and by studying the effects of geographical isolation. There are also some present-day examples of natural selection in action.

● The frequency distribution of phenotypes may change as a result of selective pressure, and the rate of change depends partly upon the genetic makeup of the population and partly upon the phenotypic expression of the trait.

● The main types of interaction among animals of different species are predation and competition. In obtaining food, each species occupies a particular niche. Species with similar niches in the same habitat compete for food and may also compete for other resources, such as nest sites. No two species can occupy the same niche in the same habitat.

● Comparison of closely related species living in different habitats can often reveal those aspects of behavior which are particularly important in adapting the animal to its environment.

Further reading

Futuyma, D. J. (1979) *Evolutionary Biology*, Sinauer Associates, Sunderland, Massachusetts.

Dawkins, R. (1976) *The Selfish Gene*, Oxford University Press, Oxford.

6 Survival Value and Fitness

In Chapter 5 we saw that the survival value of a hereditary trait within a population depends upon the extent to which the trait contributes to reproductive success, which depends partly upon the selective pressures inherent in the environment. A number of features of the environment could jeopardize reproductive success by leading to the death of the parent by starvation, predation, etc.; failure to breed as a result of competition for mates or nesting sites; or failure of the young to thrive through lack of parental care, food, or protection from predators.

In studying the ways in which structural or behavioral traits contribute to survival and reproductive success, we can think of the animal as designed, via natural selection, to fulfill certain functions. In everyday language, "function" refers to the job that something is designed to do. Thus, for instance, the function of a bicycle is to transport a person from place to place. The biologist uses the word function in a slightly different way. Strictly speaking, as we shall see, the function of a trait is to increase, via natural selection, the genetic contribution to future generations.

However, biologists also use the word function in a less rigorous sense, to indicate the trait's role in the survival and reproductive success of the individual animal. For example, a biologist might say that the function of incubation behavior in birds is to keep the eggs warm up to the point of hatching. Implicit in this statement is the understanding that birds that do not incubate adequately have low reproductive success. That is, that incubation behavior has evolved through natural selection, to ensure that the eggs hatch. We use the same type of logic in ordinary language when we say that the function of a bicycle is to transport people, or that a bicycle is designed to transport people.

The idea that the function of incubation behavior is to keep the bird's eggs warm is only one among a number of possibilities. In some cases, incubation behavior serves to shield the eggs from the hot sun or to protect the eggs from predation (Drent, 1970). Moreover, there is also the question of the costs and benefits of incubation behavior: While the benefits include maintaining the eggs at correct temperatures and protecting them from predators, the incubating bird may suffer costs through its own vulnerability to predators and its lost opportunities for feeding. In addition, there are other consequences of incubation that can affect future survival and reproductive success: The eggs are kept dry and the incubating bird's presence on the territory is maintained. For the time being, however, we confine ourselves to the more straightforward ways of investigating the survival value of behavior.

6.1 Experimental studies of survival value

We have seen that natural selection acts upon the phenotype of an individual and that how effective natural selection is in changing the nature of a population depends upon the degree to which the phenotypic characteristics are controlled genetically. That is, the effectiveness of natural selection depends upon the genetic influence that an individual can exert upon the population as a whole. Obviously, an individual that never has offspring will exert no genetic influence, however great its ability to survive. Evolutionary biologists distinguish between two aspects of survival value. First, the survival value of the traits of a particular individual can be estimated, as we see in Chapter 24. Thus, individuals with high-survival traits are said to be well adapted to the environment in that they efficiently obtain food, avoid predators, etc. Second, the survival value of a trait within a population depends upon how much the trait contributes to reproductive success.

The term *survival value* is akin to the concept of *fitness*. Fitness is a measure of the ability of genetic material to perpetuate itself in the course of evolution. It depends not only upon the individual's ability to survive but also upon its rate of reproduction and the viability of its offspring. In this book we use the term survival value in reference to the survival of the individual and the term fitness in reference to long-term reproductive success.

Estimating the survival value of behavior is partly a matter of conjecture, since we have no direct way of measuring the selective pressures of the past. As we have seen, one of the problems in interpreting the results of comparative studies is that differences among species are the result of events that occurred long ago. However, we can ask questions about present-day selective pressures, and one way of doing that is to conduct experiments.

A classic example of the experimental approach is the study by Niko Tinbergen and his co-workers (1962) on eggshell removal in the black-headed gull (*Larus ridibundus*). Many birds dispose of the empty eggshell after a chick or nestling has been hatched. This can be done in a variety of ways, but usually the eggshell is either trampled into the nest, eaten by the parent, or picked up and carried away (Nethersole Thompson, 1942). The black-headed gull removes the eggshell by picking it up in its bill (Fig. 6.1) and flying some distance away before dropping it, invariably within a few hours of hatching. Although eggshell removal takes only a few minutes, by leaving the nest unguarded the parent

Fig. 6.1 Black-headed gull (*Larus ridibundus*) removing an egg shell from its nest (*Photograph: Niko Tinbergen*)

exposes its chicks to predation. It would seem, therefore, that there must be considerable survival value to eggshell removal for the parent to take such a risk.

Tinbergen and his co-workers considered various ways in which disposal of the eggshells might benefit the gulls. For example, the sharp edges of the shell might injure the chicks; the shell might slip over an unhatched egg, thus trapping the chick in a double shell; the shells might interfere with brooding; the inside of the shell might become a breeding ground for infectious organisms; or the white inside of the shell might attract the attention of predators.

The results of their previous work gave the researchers some clues. First, black-headed gulls remove not only eggshells but also many other objects of equivalent size, even those that do not resemble eggshells in any obvious respect—almost any foreign object. Second, this tendency occurs not only at the time of hatching but also for a number of weeks before and after hatching. Thus, it seems unlikely that the response is connected with injury or disease in newly hatched chicks. Third, as we have seen, the kittiwake never disposes of its broken eggshells. These observations combine to suggest that the prime function of eggshell removal in black-headed gulls is protection from predators through maintenance of the nest's camouflage. The kittiwake is not subject to nest predation, but presumably it would dispose of eggshells if they were injurious to the chicks. The black-headed gull maintains the response long before and long after hatching, which makes sense for antipredator behavior but not for behavior directly related to the health of the chicks.

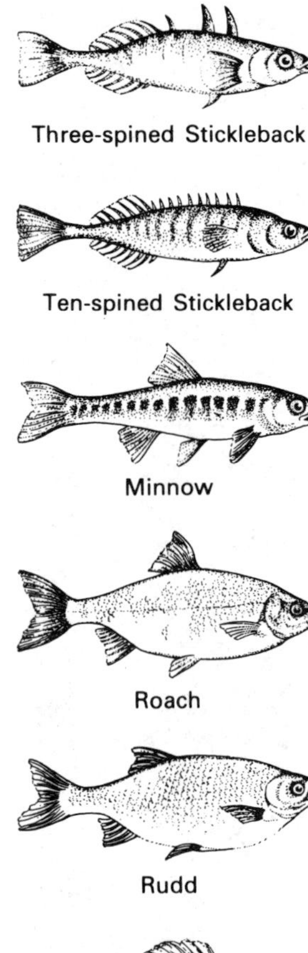

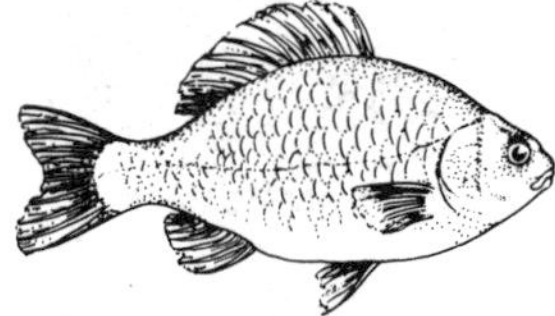

Fig. 6.2 Fish used in experiments on prey capture by pike (After Hoogland et al, 1957)

In order to test the hypothesis that eggshell removal serves to maintain nest camouflage, Tinbergen and his team conducted a series of experiments. They laid out eggs in a widely scattered pattern (about 20 meters apart) in an area of land outside the gull colony. This site was chosen to resemble the gullery, without provoking interference from the gulls' defensive behavior. In some experiments, hen eggs were laid out, some of them painted to look like gull eggs. In others, gull eggs were laid out, camouflaged with bits of vegetation (some with a broken eggshell nearby) or painted white. After the eggs were laid out, a watch was kept from a hide, or blind, and a count was kept of the number of eggs discovered and taken by typical nest predators, such as carrion crows (*Corvus corone*) and herring gulls (*Larus argentatus*). The results showed that normal hen eggs were taken much more quickly than artificially camouflaged hen eggs. Similarly, black-headed gull eggs painted white were more vulnerable than unpainted black-headed gull eggs. Eggs with a broken eggshell placed nearby also were much more vulnerable to predation than those not so marked.

Tinbergen and his co-workers also discovered that carrion crows soon learned to associate broken eggshells with the presence of eggs. First, they showed that the crows ignored empty eggshells put out alone. Then they paired broken eggshells with real eggs, some of which were taken by crows. When empty eggshells subsequently were put out alone, the crows paid particular attention to them and searched in their vicinity. Overall, the results of the experiment strongly suggest that the presence of a broken eggshell near the nest of a black-headed gull makes the nest more vulnerable to predators because it makes the nest easy to spot from the air. Moreover, if nests are distinguishable by the presence of broken eggshells, then crows and other predators soon learn to use them as a cue to the presence of eggs or chicks. It is interesting, however, that the parent black-headed gull does not remove the eggshell immediately but waits about an hour before doing so. Tinbergen suggested that this delay allows time for the chicks to dry. They are then more robust, better camouflaged, and less easily swallowed by a predator.

The experimental approach to the study of survival value may be combined with the *comparative approach*, for example, the study by Tinbergen and co-workers (1957) of the function of the spines of sticklebacks. Two species of sticklebacks occur in European fresh water—the three-spined stickleback (*Gasterosteus aculeatus*) and the ten-spined stickleback (*Pygosteus pungitius*)—and they are preyed upon by perch (*Perca fluviatilis*) and pike (*Esox lucius*). The spines of sticklebacks are thought to give some protection against predators, and to test this possibility Tinbergen and his co-workers carried out experiments in which they compared sticklebacks with fish of similar size but no spines, such as minnow (*Phoxinus phoxinus*), roach (*Rutilus rutilus*), rudd (*Scardinius erythrophthalmus*), and crucian carp (*Carassius carassius*), as illustrated in Figure 6.2.

The experiments were conducted in large aquariums and involved observations of the predatory behavior of perch and pike and the reactions of the prey fish. The researchers observed that pike and perch usually attempt to take their prey head first but that they are not always successful. They can swallow minnows tail first without difficulty, but they cannot cope with a stickleback in this way and usually spit it out and try to catch it again immediately. When attacked, sticklebacks raise their spines and attempt to keep their head away from the predator. Once caught, some sticklebacks managed to escape, especially if the predator attempted to maneuver its prey in order to swallow it head first (Fig. 6.3).

When presented with mixed schools of minnows and sticklebacks,

Fig. 6.3 Drawing made from film showing the variety of positions of a stickleback in the mouth of a pike (After Hoogland et al, 1957)

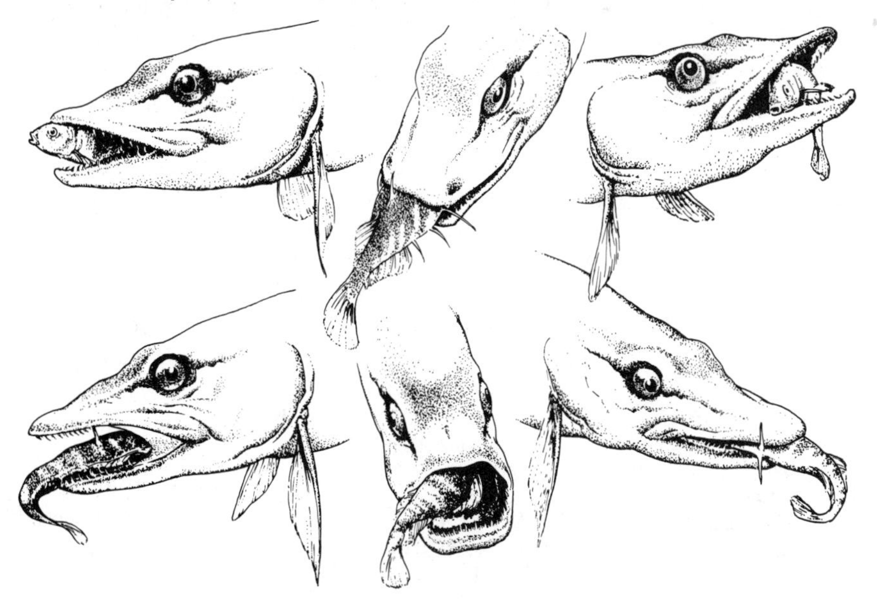

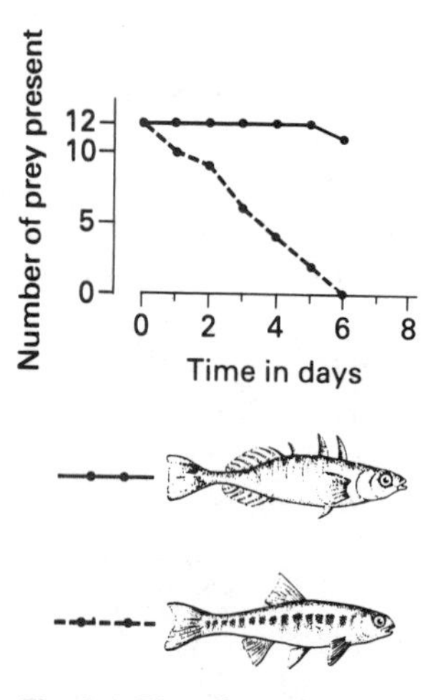

Fig. 6.4 The effect of predation by pike on a mixed shoal of 20 three-spined sticklebacks, (solid line) and 20 minnows (dashed line) (After Hoogland et al, 1957)

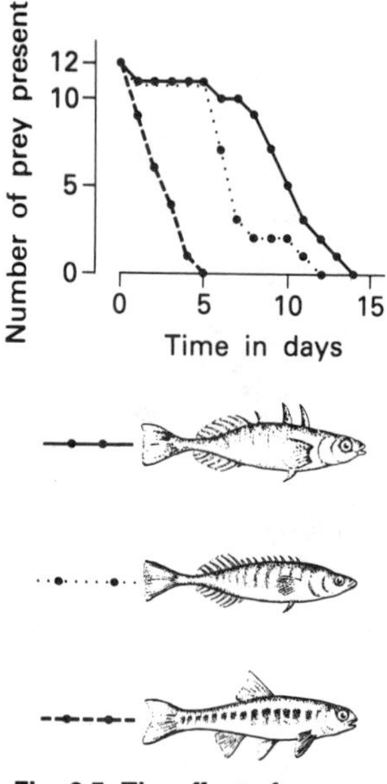

Fig. 6.5 The effect of predation by pike on a mixed shoal of 20 three-spined sticklebacks (solid line), 20 ten-spined sticklebacks (dotted line) and 20 minnows (dashed line) (After Hoogland et al, 1957)

both pike and perch ate the minnows more quickly than the sticklebacks (Fig. 6.4), partly a result of the predators' preferring to attack prey without spines and partly a result of sticklebacks' being caught and then rejected or being caught and then escaping. Three-spined sticklebacks survived longer than ten-spined sticklebacks (Fig. 6.5), presumably because their spines are much larger.

In some tests, sticklebacks were offered to perch and pike after their spines had been cut off. Such fish survived much less well than intact sticklebacks. This supports the main conclusion gained from these experiments, which is that the spines of sticklebacks do indeed provide some protection against predators. The differences between the three-spined and ten-spined sticklebacks are large, and the tests showed that the three-spined sticklebacks survived longer than the ten-spined sticklebacks. How, then, are the ten-spined sticklebacks—or any fish without protective spines—able to survive in nature? Comparison of the habitats and behavior of three-spined and ten-spined sticklebacks gives some clues. The male three-spined stickleback builds its nests on a substratum relatively free of weeds, and it has bright red nuptial coloration. The male ten-spined stickleback builds its nest among thick weeds and has black coloration during the breeding season. Thus, the ten-spined stickleback apparently compensates for its relative lack of protection by its cryptic habitat, coloration, and behavior.

6.2 Assessment of mortality

The comparative and experimental studies of survival value are indirect ways of gaining evidence about the action of natural selection. A more direct approach is to measure the mortality that results from a particular selective pressure. For example, Hans Kruuk (1964) studied the predators and antipredator behavior of the black-headed gull (*Larus ridibundus*).

As discussed earlier, these gulls nest on the ground in large colonies, where they are vulnerable to attack from a variety of predators. Some predators, such as other gulls, crows, and hedgehogs, prey only on eggs and chicks. Others, including hawks, stoats, foxes, and humans are a danger to both the adults and their brood. The peregrine falcon attacks only the adult birds. Kruuk (1964) found that the reactions of the gulls depended upon the type of predator. They would flee from the peregrine falcon, but in reaction to humans, foxes, stoats and hawks, their behavior was ambivalent. The gulls fleed the nest but launched aerial attacks at the predators. They did not flee from other gulls, hedgehogs, or crows, but attacked them if they came near the nest.

These differences in reactions to predators are highly adaptive. If the gull leaves the nest, it is leaving its eggs or chicks protected only by their camouflage, which does not matter when the predator is a peregrine falcon that preys only on the adult birds. The gulls can remain at the nest and protect their brood from hedgehogs and ferrets, which are a threat only to the eggs and chicks. By vigorously attacking these predators, the gulls can not only deter them on a particular occasion but can also possibly keep them from visiting the colony in the future. Kruuk (1964) found that crows tended to avoid the colony, though hedgehogs remained persistent, especially at night. In reacting to foxes and humans, the gulls are faced with a problem: To flee the nest endangers the brood, but not to flee endangers the adult. A fine balance of costs and benefits exists in this type of situation, and the gulls seem to make some good compromises.

In addition to direct responses to predators, gulls can protect themselves in various indirect ways. Ian Patterson (1965) showed that choice of nest site and of egg-laying date were important factors affecting reproductive success. Patterson noted that the majority of black-headed gulls in his study area nested in the densely packed colony, where the nests are about one meter apart. Some, however, nested more than a hundred meters from the colony. He showed there was a marked difference in breeding success between birds that nested within the colony and those that nested outside it. Obviously, there must be some advantages in nesting within the colony. To investigate these, Patterson followed the case histories of some 800 marked nests and found that nests in the center of the colony had a higher success rate than those near the edge, in terms of the number of young fledged. Patterson thus used naturally occurring variation to see whether differences in behavior were correlated with differences in breeding success.

Black-headed gulls tend to synchronize their egg laying, but again there is natural variation. Some birds lay much later than the majority (Fig. 6.6). Patterson found that breeding success was correlated closely with laying date. The most successful birds were those that laid in synchrony with the majority, while those that laid later had much

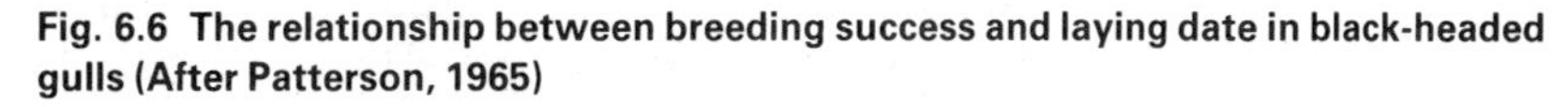

Fig. 6.6 The relationship between breeding success and laying date in black-headed gulls (After Patterson, 1965)

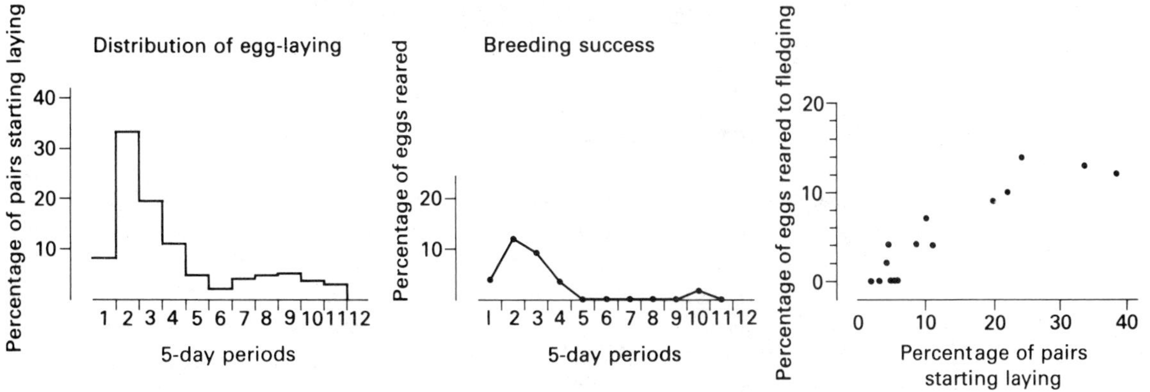

greater mortality among their eggs and chicks. Predation was by far the most common cause of chick and egg losses (Patterson, 1965). The greater success of birds that synchronize their egg laying probably is due to the fact that this leads to an overabundance of food for the predators (Kruuk, 1964). It is difficult for predators to exploit a sudden and short-lived increase in potential food when the predator population is adjusted to a low level of food availability. Thus, gulls that lay in synchrony have a lower probability of being attacked by a predator than those that are providing a larger proportion of the available food by laying at a different date.

Patterson (1965) found that breeding success was not correlated with nest density, which means that the greater success of birds nesting within the colony was due not to closer packing but to the total number of neighbors alerted by a predator. This conclusion is supported by Kruuk's (1964) observation that the gulls often joined together in combined attacks on predators, often deterring the predators from entering the highly populated parts of the colony.

This work shows that the black-headed gulls that synchronize their egg laying and that can establish themselves in the center of the colony benefit in reproductive success, and therefore, have more offspring than those birds that nest on the edge of, or away from, the colony or that lay their eggs later than the majority. Since these offspring presumably inherit the traits of colonial behavior, these traits are maintained in the population through the action of natural selection. To confirm this hypothesis, it would be necessary, ideally, to show that the offspring of gulls exhibiting the typical colonial behavior are more likely to survive to maturity. This is difficult to do, however, because black-headed gulls do not reach sexual maturity until they are three to four years old.

Sometimes it is possible to measure the number of offspring surviving to maturity. The females of the side-blotched lizard (*Uta stansburiana*) hold territories, and in one study (Tinkle, 1969) it was found that those that held larger territories had a greater number of mature progeny than those with small territories. The territories were similar in quality, so the main reason for the correlation was probably the amount of food in each territory.

6.3 Darwinian fitness

We have seen that the extent to which a genetic trait is passed from one generation to the next, in a wild population, is determined by the breeding success of individuals of the parent generation and the value of the trait in enabling the animals to overcome natural hazards such as food shortage, predators, and sexual rivals. Such environmental pressures can be looked upon as selecting those traits that fit the animal to the environment.

The concept of fitness has proved to be something of a problem, for the reason elucidated by the historian William Dampier (1929): "Herbert Spencer's phrase for natural selection, the survival of the fittest standing alone, begs the question what is the fittest? The answer is: The fittest is that which best fits the existing environment. That which is fit survives, and that which survives is fit."

Darwin discussed the survival of the fittest without rigorously defining fitness. "Darwinian fitness," however, is now widely recognized by biologists as a measure of the capacity to produce offspring. The apparent circular reasoning in the phrase can be broken if we recognize that the fittest is not that which best fits the existing environment but that the fit are those who fit their existing environments and whose descendants will fit future environments (Thoday, 1953). Thus, in measuring fitness, we are looking for a quantity that will reflect the probability that, after a given lapse of time, the animal will have left descendants.

In any population of animals there is variation among individuals, and consequently, some have greater reproductive success than others. Individuals that produce a higher number of viable offspring are said to have greater "Darwinian fitness." An individual's fitness depends upon its ability to survive to reproductive age, its success in mating, the fecundity of the mated pair, and the probability of survival to reproductive age of the resulting offspring.

The fitness of a genotype in a Darwinian sense can be measured by means of numbers of its progeny, different generations being counted at the same stage of the life cycle. The fitness of an individual must take its age into account, since the potential for reproduction changes with age. In this respect, Ronald Fisher (1930) was the first to draw attention to the importance of the concept of *age-specific reproductive value*—an index of the extent to which the members of a given age group contribute to the next generation between now and when they die. In the example of Australian women illustrated in Figure 6.7, we can see that reproductive value increases up to the age of 20 and then declines. The initial increase is probably due to childhood mortality, which means that the average 10-year-old Australian girl is less likely to have children than the average 20-year-old woman (because some 10-year-olds will die before childbearing age). Conversely, the average 20-year-old woman has more childbearing potential left than the average woman of 30.

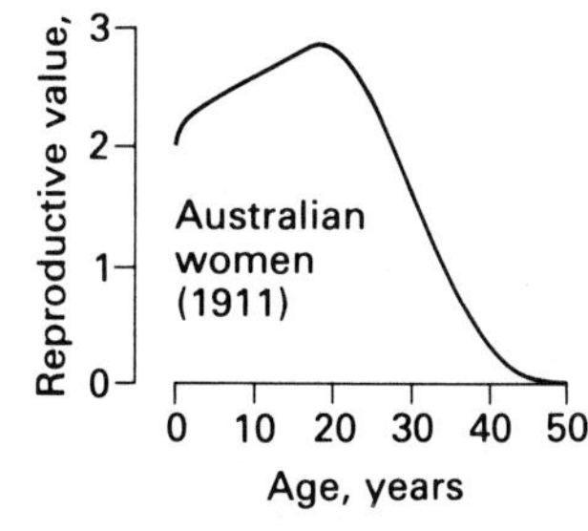

Fig. 6.7 Reproductive value plotted against age for Australian women about 1911 (After Fisher, 1958)

The age at which an animal should ideally become sexually mature and capable of reproduction is a matter of evolutionary life-history strategy. In unpredictable environments, natural selection usually favors early maturity and large numbers of offspring left to fend for themselves. In more stable environments it is a better strategy to mature late and have few offspring well cared for.

Reproduction may expose an animal to risks, thus reducing the chances of subsequent survival and reproduction. In Figure 6.8, the data

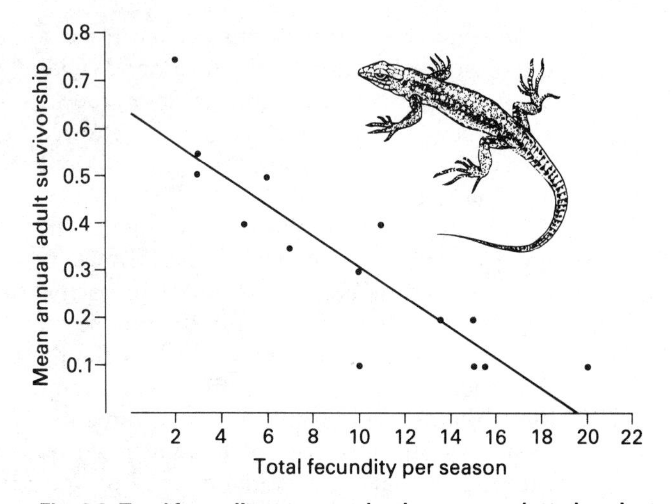

Fig. 6.8 Total fecundity per reproductive season plotted against probability of surviving to a subsequent reproductive year for 14 lizard populations (After Tinkle, 1969)

do not provide conclusive evidence of the risks involved in reproduction because they compare differences among rather than within species. However, Loschiavo's (1968) data on the length of life and number of eggs laid by the beetle *Trogodenna parabile* (Fig. 6.9) do provide more direct evidence of risks to the individual. We must distinguish here between the evolutionary strategy represented by a particular species and the factors affecting individuals within a species. On the one hand, high reproductive output may be such a drain on the energy resources of the individual that its subsequent survival is endangered—for example, many birds lose a considerable amount of weight during incubation. On the other hand, species with short life expectancy, perhaps due to predation, can be expected to have high fecundity to compensate, while longer-lived species may find that relatively low fecundity increases

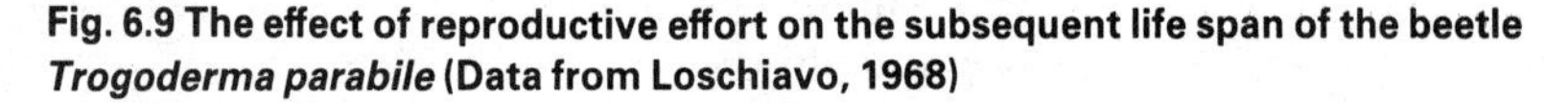

Fig. 6.9 The effect of reproductive effort on the subsequent life span of the beetle *Trogoderma parabile* (Data from Loschiavo, 1968)

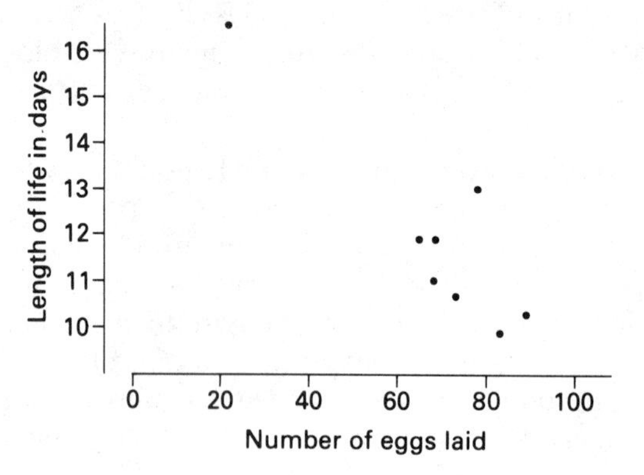

overall fitness, especially if the resources available to the offspring are limited.

In general, the more time and energy a parent expends upon a particular offspring, the fitter that offspring will be. There is often an inverse relationship between the total number of offspring produced and their average fitness. As an example, let us briefly consider the work of David Lack (1954; 1966) on the great tit (*Parus major*) in Marley Wood near Oxford. These studies show that annual fluctuations in the breeding population are due primarily to variations in juvenile mortality before the winter. Survival of the young was greater for broods reared early in the season, largely due to a decline in food availability toward the end of the season. Chris Perrins (1965) found that the average weight of the nestlings in the great tit broods declined with increasing brood size (Fig. 6.10), an effect more marked in seasons when food for the female at egg-production time was in short supply. Perrins also found that the larger nestlings had a greater chance of survival. To be able to rear a brood early in the season, the parents must be successful in establishing a territory during the spring. Behavioral factors like aggressiveness are of importance here. In general, it appears that parental effectiveness—including the ability to establish a territory, the tendency to produce a clutch of the optimal size, and experience in rearing the nestlings—is an important factor determining fitness in this species. To maximize fitness, the reproductive strategy must be a compromise between having a large number of progeny and attaining a high individual fitness for each offspring.

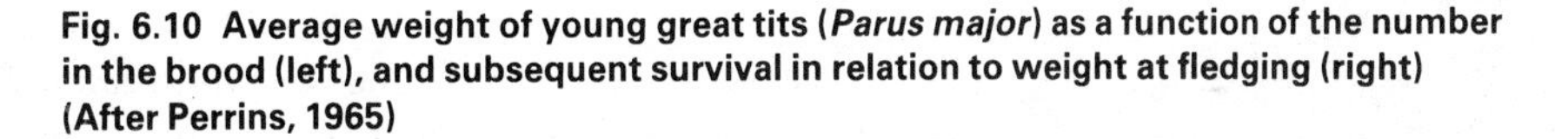

Fig. 6.10 Average weight of young great tits (*Parus major*) as a function of the number in the brood (left), and subsequent survival in relation to weight at fledging (right) (After Perrins, 1965)

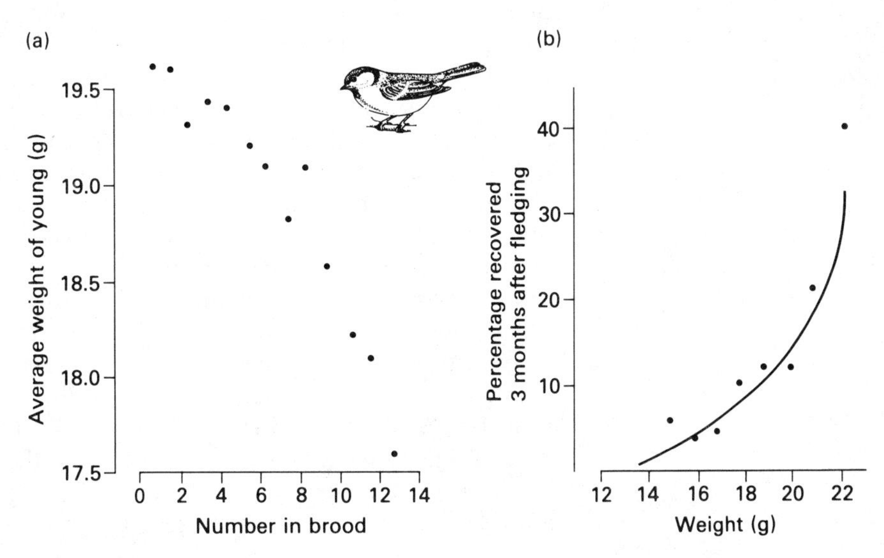

The studies of Lack and his co-workers show that an optimal clutch size exists from which the greater number of offspring survive to breed in the next generation. The smaller clutches produce fewer offspring, while young birds from the larger clutches leave the nest at a lighter weight and have a smaller chance to survive. Here we have an example of stabilizing selection in which the intermediates in a population leave more descendants than the extremes. In a stable environment, genetic recombination increases the variability among the individuals of a population during each generation, but stabilizing selection reduces the variability back to the level of the previous generation. In a changing environment, on the other hand, the average individual may not be the fittest member of the population. Under such conditions directional selection can occur, such that the population average shifts towards a new phenotype better adapted to the altered environment.

The components of fitness, such as viability, fecundity, and fertility, can be described in various ways. For example, Prout (1971) distinguishes between larval and adult components of fitness. The larval component consists of the probability of survival to adulthood, while the adult component is divided into a male and a female subcomponent. The female subcomponent includes viability and fecundity, and the male subcomponent includes viability and virility. Mating ability is the most important aspect of virility, and therefore is the aspect of fitness connected with sexual selection (see Chapter 8).

6.4 Inclusive fitness

We have seen that an animal's individual fitness is a measure of its ability to leave viable offspring. The process of natural selection determines which characteristics of the animal confer greater fitness. However, the effectiveness of natural selection depends upon the mixture of genotypes in the population. Thus, the relative fitness of a genotype depends upon the other genotypes present in the population, as well as upon other environmental conditions.

The concept of fitness can be applied to individual genes by considering the survival of particular genes in the gene pool from one generation to another. A gene that can enhance the reproductive success of the animal carrying it thereby will increase its representation in the gene pool. It could do this by influencing the animal's morphology or physiology, making it more likely to survive climatic and other hazards, or by influencing its behavior, making the animal more successful in courtship or raising young. A gene that influences parental behavior will probably be represented in the offspring so that by facilitating parental care, the gene itself is likely to appear in other individuals. Indeed, a situation could arise in which the gene has a deleterious effect upon the animal carrying it but increases its probability of survival in the offspring. An obvious example is a gene that leads the parent to endanger its own life in attempts to preserve the lives of its progeny. As we shall see, this is a form of *altruism*.

Both Fisher (1930) and Haldane (1955) realized that the fitness of an individual gene could be increased as a result of altruistic behavior on the part of animals carrying the gene. However, Bill Hamilton (1964) was the first to enunciate the general principle that natural selection tends to maximize not individual fitness but *inclusive fitness*; that is, an animal's fitness depends upon not only its own reproductive success but also that of its kin. The inclusive fitness of an individual depends upon the survival of its descendants and of its collateral relatives. Thus, even if an animal has no offspring, its inclusive fitness may not be zero, because its genes will be passed on by nieces, nephews, and cousins.

In normal diploid animals, each parent contributes a copy of one of its two sets of genes to each of its offspring (see Chapter 2). The chance that any given gene in a parent will appear in one of the progeny is therefore one-half. By the same logic the chance that a single gene in a parent will be inherited in common by two of the offspring is one-half. The chance that one gene in a grandparent will be transmitted to a given grandchild is one-quarter, and the chance that two first cousins both carry copies of a gene present in a grandparent is one-eighth.

We can now define the *coefficient of genetic relatedness* (r) as a measure of the probability that a gene in one individual will be identical by descent to a gene in a particular relative. An equivalent (Dawkins, 1979) and alternative measure is the proportion of an individual's genome that is identical by descent with the relative's genome. Note that the genes must be identical by descent and not simply genes shared by the population as a whole. This means that r is the probability that a gene common to two individuals is descended from the same ancestral gene in a recent, common relative.

Inclusive fitness is sometimes equated with a simple weighted sum based on the animal's various coefficients of relationship. Thus, it is sometimes seen as the sum of the animal's individual fitness and that of the relatives, discounted in proportion to the coefficient of relationship This measure counts all the animal's offspring, and although this may be a useful measure for practical purposes, it does not accord with Hamilton's (1964) original formulation. Thus inclusive fitness has often been misdefined (Grafen, 1982). It should exclude those of the animal's own offspring that exist because of help received from others, and include those offspring of relatives whose existence is a result of the animal's help being offered to the relative (Grafen, 1984).

6.5 Fitness in the natural environment

Implicit in our discussion of inclusive fitness are two alternate ways of describing natural selection. In population genetics, the unit of selection is the gene, and replication of the gene is the quantity maximized by natural selection. While this approach has an appealing logic that sometimes can be used to illuminate aspects of animal behavior (e.g., Dawkins, 1976), it is not really convenient for those interested in the behavior of individual animals. An equivalent approach is to regard the individual animal as the unit of selection and inclusive fitness as the quantity maximized by natural selection (Dawkins, 1978). For practical purposes, the reproductive success of an individual is a good guide to its inclusive fitness (Grafen, 1982).

Fig. 6.11 Pair of toads mating. The male is repelling a rival (Drawing from Davies and Halliday, 1977. *Photograph: Tim Halliday*)

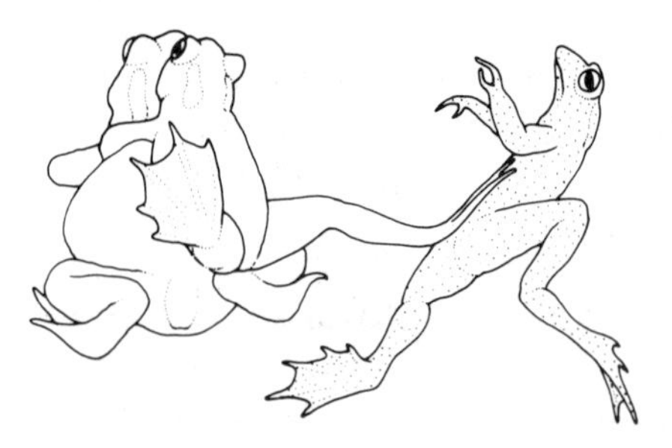

The fitness derived from behavior in the natural environment depends upon the selective pressures that are operating. In theory, every aspect of an animal's behavior will make a difference to its fitness, either because of the direct consequences of the behavior or simply because one type of behavior precludes another that might be more, or less, beneficial. In practice, the natural environment is so complex that it is extremely difficult to measure the changes in fitness that derive from behavior. There are two main approaches to this problem. One approach is to attempt to specify in detail the various costs and benefits associated with behavior in the natural environment. This approach is discussed in Chapter 24. The other approach is to measure directly the changes in reproductive success that result from particular aspects of behavior.

In Chapter 6.1 we saw that the survival value of particular aspects of behavior, such as eggshell removal by the black-headed gull, could be estimated from experiments carried out in the field. While such experiments are valuable in providing an understanding of the function of particular features of morphology and behavior, they do not bear directly upon the question of fitness. The concept of survival value is usually applied to relatively short-term questions like the survival of the young in a particular breeding season. Ideally, estimates of fitness should relate to the long-term survival of particular genetic traits or the long-term consequences, in terms of reproductive success, of particular behavior patterns.

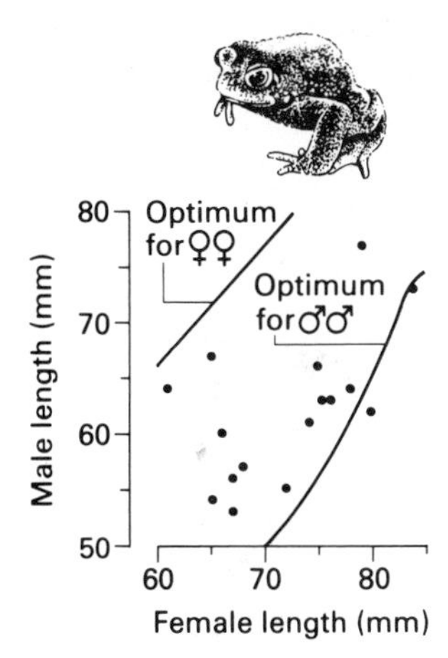

Fig. 6.12 Optimal and observed pairings of toads (*Bufo bufo*) of various sizes. Optima are calculated on the assumption that the number of fertilized eggs produced in a season is maximized. The female optimum curve represents a minimum, because no experimental pairings with males larger than females could be obtained. The observed pairings are represented by the solid dots (After Davies and Halliday, 1977)

Sometimes differences in reproductive success can be attributed to fairly simple aspects of behavior. For example, Nick Davies and Tim Halliday (1977) found that the reproductive success of a pair of toads (*Bufo bufo*) was influenced by both the size of the female and the degree to which the mating pair (Fig. 6.11) are matched for size. Thus, a larger female produces more eggs, and more eggs are fertilized when the mating pair are well matched for size. A male that mates with a female just slightly larger than himself should have the greatest reproductive success, while a female should prefer larger males that can ward off rivals. Thus, the male and female should each prefer a mate larger than itself. Observed pairings lie between these two theoretical extremes, with the females usually being larger than the males (Fig. 6.12).

The lifetime mating success of a natural population of the damselfly *Enallagma hageni* has been studied in northern Michigan (Fincke, 1982). These flies breed in a shallow pond separated from a lake by a sand beach 10 to 20 meters wide. Feeding occurs in a marshy open field adjacent to the breeding site. On sunny days male flies move from the feeding area to the pond at about 9:30 A.M. Up to 1:00 P.M. most of the males present at the pond are unmated. The number of mated pairs (clasped together in a tandem formation) increases until 3.00 P.M. and remains high until about 5:30 P.M. On a particular day, nearly all females are found in tandem pairs, but usually more than half the males fail to mate. Since there are more males than females in the breeding area, competition among males is intense. The males intercept females on their way to the pond, and nearly all successful males are paired before arriving at the water. Competition among males takes the form of interference with tandem pairs. Such harassment sometimes results in the uncoupling of a tandem pair and displacement of one male by another.

Surprisingly, Fincke found no correlation between the body size of males and their mating success. Apparently, a small male is as capable as a large male at flying in tandem with a female. Although large males may be more successful at takeovers than small males, this advantage is probably offset by the fact that harassment of tandem pairs exposes single males to predation near the water. Fincke found that the larger males tended to spend a greater amount of time near the water, which suggests that interference with mated pairs was their preferred strategy. However, the alternative strategy of searching for unmated females is probably more successful overall.

Mating success among males varied considerably from day to day, 39 per cent of males mating on average. Lifetime mating success was also highly variable. Only 59 per cent of males in the population mated at least once, and this measure was highly correlated with longevity. The major cause of death at the breeding site was predation from spiders, dragonflies, or frogs. Thus, it seems that the most important component of lifetime breeding success in this species is to avoid predators and live to mate another day.

Many animals defend territories during the breeding season. In some cases, this is a form of spacing out that has to do with the acquisition of mates (Davies, 1978). Territories sometimes differ in quality, and males

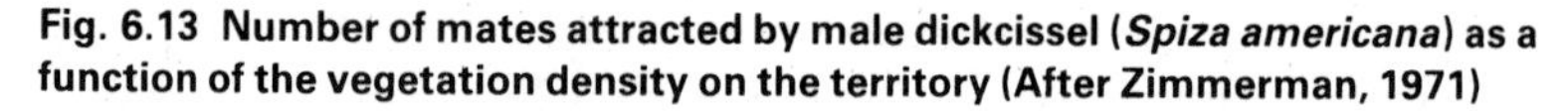

Fig. 6.13 Number of mates attracted by male dickcissel (*Spiza americana*) as a function of the vegetation density on the territory (After Zimmerman, 1971)

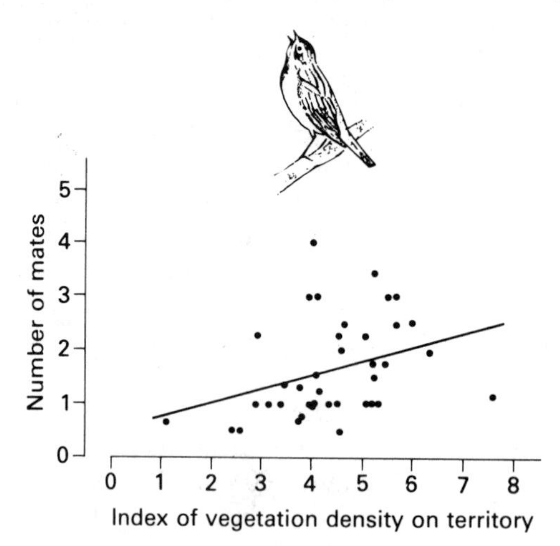

that defend the better territories can be expected to have greater reproductive success. In the three-spined stickleback, for example, males with large territories are more successful in attracting females (Assem, 1967). Females that deposit their eggs in large territories have greater reproductive success because fewer eggs are lost to predation by other males. Rival males from neighboring territories try to interfere with the nest, but this is harder to do when the territory is large (Black, 1971; Wootton, 1976). Territory quality may depend not only upon size but also upon food or nest-site availability, as shown in Figure 6.13. In cases where animals compete for mates, territory, or habitat (Partridge, 1978), the successful may be fitter in various respects. They may have greater reproductive success not only because they acquire better mates or territories but also because they are fitter in other respects. This is a perennial problem with the assessment of fitness. In the natural environment there are so many interacting factors that it is virtually impossible to account for all of them.

Points to remember

- The survival value of particular aspects of behavior can be investigated by experiments performed under natural conditions.

- Investigation of the causes of mortality within a population can often provide useful evidence about the selective pressures.

- Fitness in the Darwinian sense is measured by the number of progeny, counted at a particular stage of the life cycle.

- The inclusive fitness of an individual depends upon the survival of its descendants and of its collateral relatives. Even if an animal has no offspring, its inclusive fitness may not be zero because its genes will be passed on by nieces, nephews and cousins.

- In studying fitness in the natural environment, it is usually necessary to use an index of mating success or reproductive success. Such indices only approximate fitness because they take no account of the viability of the offspring.

Further reading

Dawkins, R. (1982) *The Extended Phenotype*, Freeman, New York.
Krebs, J. R. and Davies, N. B. (1981) *An Introduction to Behavioural Ecology*, Blackwell Scientific Publications, Oxford.

7 The Evolution of Adaptive Strategies

Evolutionary biologists are interested in explaining how a state of affairs observed today is likely to have come about as a result of evolution by natural selection. To account for the establishment of a particular genetic trait, they imagine a time before the trait existed. Then they postulate that a rare gene arises in an individual, or arrives with an immigrant, and that individuals carrying the gene exhibit the trait. They then ask what circumstances will favor the spread of the gene through the population. If the gene is favored by natural selection, then individuals with genotypes incorporating the gene will have increased fitness. The gene may be said to have invaded the population. This chapter considers the implications of this kind of argument for studies of animal behavior.

7.1 Evolutionary strategies

To become established, a gene not only must compete with the existing members of the gene pool but also must resist invasion from other mutant genes. Indeed, evolutionary biologists, when speculating about a particular evolutionary situation, often impose upon themselves the test of postulating an invasion by hypothetical mutants in order to see whether a particular theory will stand up to the competition. Before going into the detail of such procedures, let us look at a few examples. A high degree of similarity between an animal and its visual background may result from selection for features that enable animals to elude predators or predators to lurk undetected and ambush a suitable prey. This type of strategy sometimes takes the form of an astonishing resemblance between an animal and the plant with which it normally associates. For example, the leaflike grasshopper (*Arantia rectifolia*) shown in Figure 7.1 can rest with its wings held close together so it resembles a complete leaf, or it can rest on a leaf and hold its wings flat so it appears to be part of the leaf (Edmunds, 1974). It takes up these camouflaged postures as a protection against predation. Conversely, the mantis (*Phyllocrania paradoxa*) may be disguised similarly (Fig. 7.2), but its camouflage is partly defensive and partly aggressive, since it sits motionless waiting to seize its prey.

Fig. 7.1 The leaf-like grasshopper *Arantia rectifolia* (see back cover) (*Photograph: Malcolm Edmunds*)

Fig. 7.2 The praying mantis (*Phyllocrania paradoxa*) mimicking a dead leaf (*Photograph: Malcolm Edmunds*)

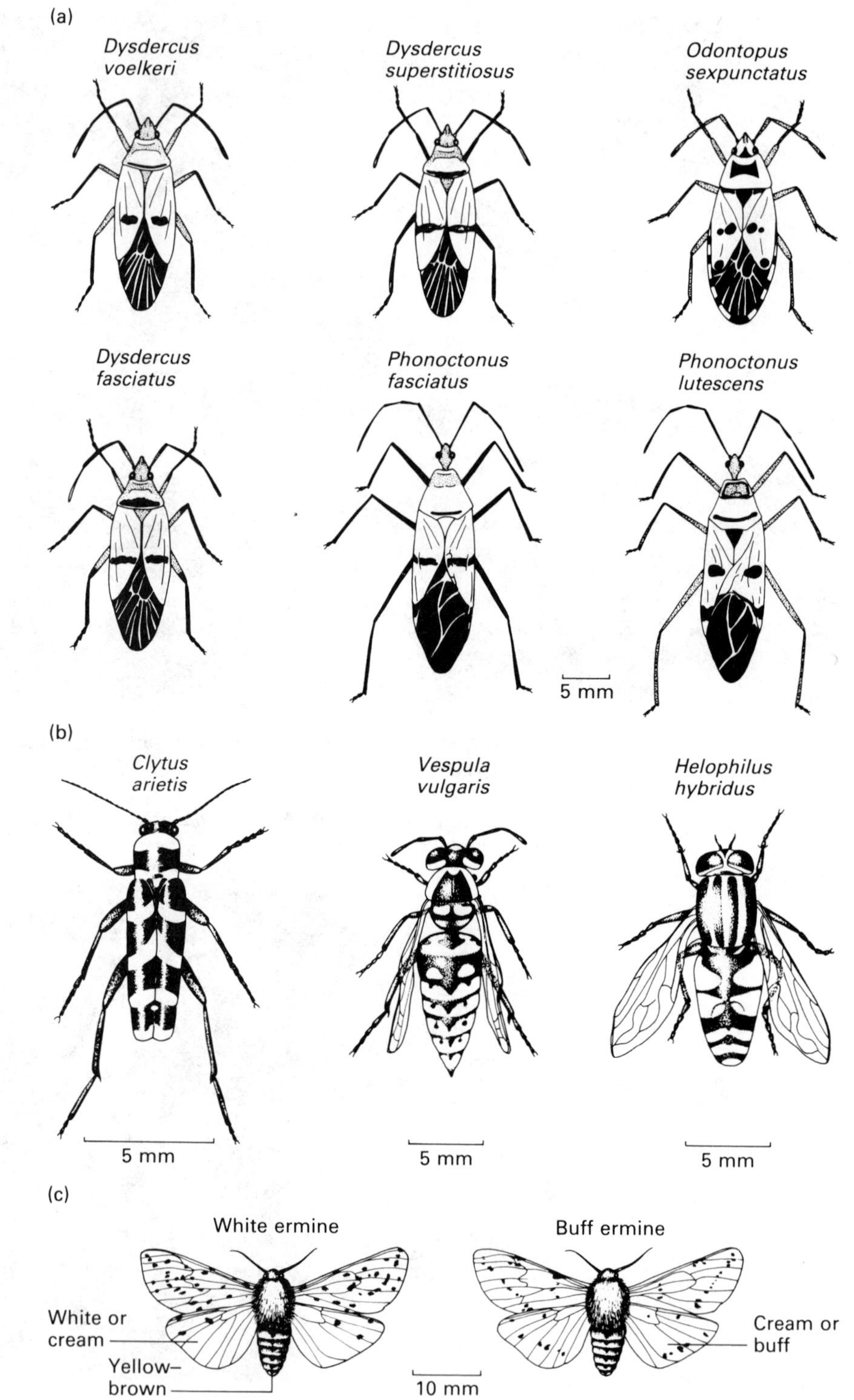
(a)
Dysdercus voelkeri
Dysdercus superstitiosus
Odontopus sexpunctatus
Dysdercus fasciatus
Phonoctonus fasciatus
Phonoctonus lutescens
5 mm
(b)
Clytus arietis
Vespula vulgaris
Helophilus hybridus
5 mm
5 mm
5 mm
(c)
White ermine
Buff ermine
White or cream
Yellow–brown
10 mm
Cream or buff

◁ **Fig. 7.3 Some examples of Mullerian and Batesian mimicry. (*a*) Mullerian mimicry amongst West African *Heteroptera*. The reduviid *Phonoctonus fasciatus* normally lives in colonies of one or more of the pyrrhocorids *Dysdercus fasciatus, D. superstitiosus* and *D. voelkeri. Phonoctonus lutescens* normally lives in colonies of *Odontopus*, sometimes with *D. voelkeri* as well. All of these bugs are black, red (shown stippled) and greyish orange (shown white). (*b*) Batesian mimicry: The wasp *Vespula vulgaris* and two mimics, the beetle *Clytus arietis* and the hoverfly *helophilus hybridus*. In all three the body is dark brown or black with yellow bands. (*c*) The distasteful white ermine moth (*Spilosoma lubricipeda*) and its Batesian mimic the buff ermine (*S. lutea*). The abdomen of both is yellowish with darker markings, while the wings are white or cream with black spots (After Edmunds, 1974)**

In addition to resembling the background or vegetation, animals also may resemble parts of other animals. This is often a feature of the deimatic (Edmunds, 1974) displays by which animals attempt to scare off their enemies, for example, the caterpillar of the hawkmoth (*Leucorampha*). This animal normally rests upside down beneath a branch or leaf. When disturbed, it raises and inflates its head, the ventral surface of which has conspicuous eyelike marks and whose general patterning resembles the head of a snake, as illustrated in Figure 1.3. Many moths and butterflies have eye spots on their wings that they reveal suddenly when disturbed, with the possible effect of frightening the predator.

True mimicry is the resemblance of one animal, called the *mimic*, to another animal, called the *model*, so that the two are confused by a third animal, usually a predator. Various types of mimicry exist, of which the most exemplary is *Batesian mimicry*. In this type, the mimic resembles a model that is noxious or distasteful to predators. There is no advantage to the model, but experiments have shown that mimics definitely can benefit as a result of predators' becoming confused between model and mimic. For example, birds soon learn to avoid the salamander *Notophthalmus viridescens*, which is unpalatable. They readily eat the similarly colored salamander *Pseudotriton ruber*, provided they have not previously experienced the unpalatable *Notophthalmus*. Thus, *Pseudotriton* gains some protection from its resemblance to *Notophthalmus*, provided that the birds are already familiar with *Notophthalmus* and that they are not so experienced as to be able to discriminate between the two salamanders (Howard and Brodie, 1971). Some other examples of Batesian mimicry are illustrated in Figure 7.3.

Sometimes a number of noxious species share the same warning coloration. This is called *Mullerian mimicry* and is of advantage to all participant species because once a predator has learned to avoid one species, it will avoid all the mimics. Figure 7.3 illustrates some examples of Mullerian mimicry.

A remarkable case of *aggressive mimicry* is found among fish. The black-and-white-striped cleaner wrasse (*Labroides dimidiatus*) lives among the Pacific coral reefs. Other fish recognize it because of its conspicuous coloration and its advertising dance or display. Large fish permit the cleaner wrasse to approach and to remove parasites from the body surface and the inside of the mouth. The relationship is symbiotic because the wrasse benefits by obtaining food and the host fish benefits by having its parasites removed. However, the cleaner wrasse has a mimic that resembles it closely, the sabre-toothed blenny (*Aspidontus taeniatus*) (see Fig. 7.4). Other fish often mistake it for a cleaner wrasse and allow it to approach them. The sabre-toothed blenny then quickly bites a piece from the fin of the host fish and escapes.

In the case of Batesian mimicry the antipredator markings on one species, the model, are imitated by another species, the mimic. This will be of advantage to the mimic, provided that predators are unable to

(*a*)

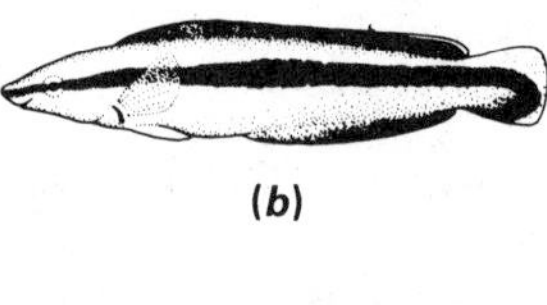

(*b*)

Fig. 7.4 **(*a*) The cleaner wrasse (*Labroides dimidiatus*) with its host fish a coral cod (*Cephalopholis miniatus*) and (*b*) its mimic the sabre-toothed blenny (*Aspidontus taeniatus*) (From *The Oxford Companion to Animal Behaviour.* 1981) (See also front cover)**

discriminate mimic from model and provided that the mimics do not become too common in relation to the models. If the model is distasteful to predators, the mimic will benefit, because predators are likely to avoid both model and mimic, while the mimic does not have to pay the cost of bearing the deterrent chemical. In some situations, a model may be mimicked by a number of species, and there is a danger predators will take more models than they might otherwise. For example, the large and conspicuous monarch butterfly (*Danaus plexippus*) of the eastern United States is distasteful to birds, which learn to avoid it after one or two experiences (Brower, 1958). As a caterpillar, the monarch feeds on species of milkweed, which are poisonous to many vertebrates. The caterpillars accumulate toxins, which are retained in the pupae and which are thus present in the butterfly. Predators that eat the butterflies usually vomit (Fig. 7.5). The emetic monarch butterfly is mimicked by the viceroy (*Limenitis archippus*), which resembles it closely (Fig. 7.6). The viceroy is edible, and, as Brower (1958) has shown, jays will eat them readily if they have no experience of the monarch butterfly. Once having eaten a monarch, however, the jays avoid both monarchs and viceroys. The emetic monarch also has other mimics, so there is a possibility that, in a particular area, the number of mimics may be large in relation to the number of models. In such a situation, predators will not so quickly learn to avoid the monarch and its mimics, since their first experiences may be with edible butterflies, and thus, the level of predation on both model and mimic is likely to be higher.

In attempting to trace the evolution of such a situation, the biologist must try to calculate the consequences of the possible combinations of frequencies of different genotypes in the populations of models and mimics. For example, what is likely to happen if some monarchs evolve

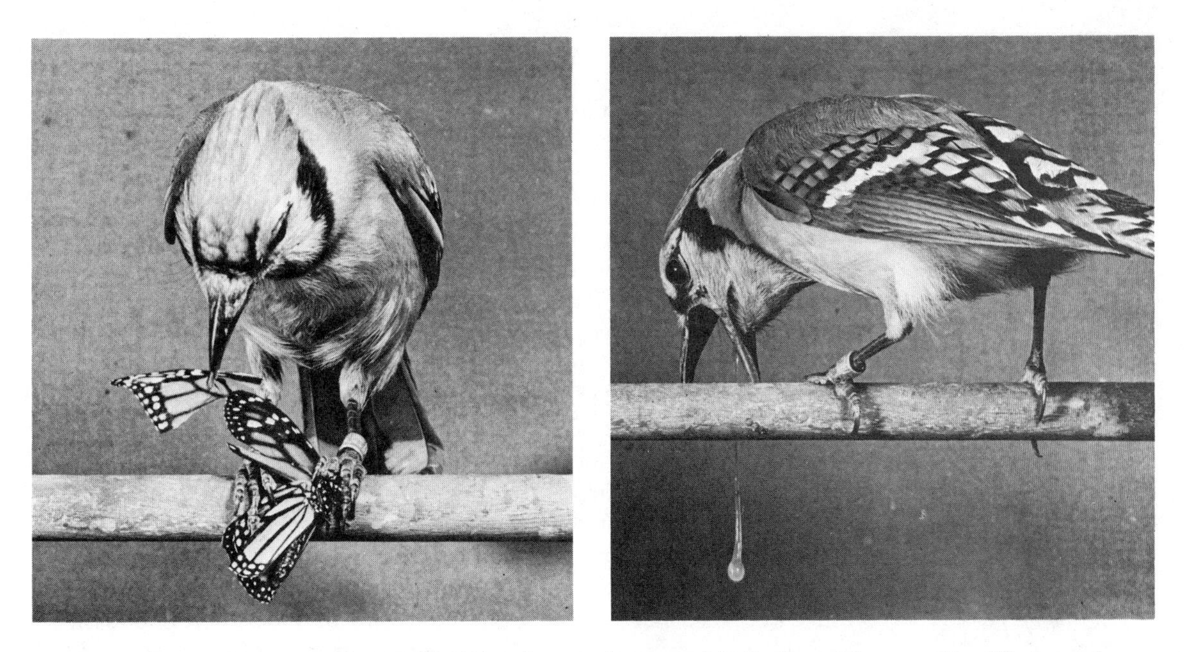

Fig. 7.5 Blue jay eating a toxic monarch butterfly and then vomiting (*Photographs: Lincoln Brower*)

Danaus plexippus

Limenitis archippus

20 mm

Fig. 7.6 The monarch butterfly *Danaus plexippus* and its mimic the viceroy *Limenitis archippus*. The stippled areas are orange (After Edmunds, 1974)

the habit of eating non-toxic plants and are no longer distasteful to predators? On the one hand, we might expect such individuals to gain an advantage by being perfect mimics of the unpalatable members of their species, without having to pay the price of being restricted to eating toxic milkweed, on which basis the genotype should increase in frequency. On the other hand, if the edible monarchs become too common, predators presumably would start to eat them. It is possible to calculate, on the basis of certain assumptions, the chances of survival of the edible monarchs as a function of the proportions of emetic and edible types in the population. The assumptions concern the degree of predation and the number of trials it takes a predator to avoid a monarch after having eaten an emetic one. In fact, it has been discovered that there is considerable variation in the toxicity of milkweeds and in the palatability of the monarchs raised on them (Brower, 1969). Thus, the natural populations do contain some monarchs that eat particular species of milkweed and that are palatable to avian predators, a form of mimicry called *automimicry* (Brower, 1969). In some localities the automimics outnumber the models 3:1, and yet predators are deterred from eating them.

If the change to a non-toxic diet is genetically based and not merely fortuitous, then such a change can be regarded as an *evolutionary strategy*. Thus, evolutionary biologists sometimes find it useful to think of genes as employing a strategy to increase their numbers at the expense of other genes, even though, in reality, the whole exercise is conducted passively by natural selection.

In considering evolutionary strategies that influence behavior, we have to visualize a situation in which changes in genotype lead to changes in behavior. It is not necessary to suppose that the behavior is genetically determined in a direct manner, since there are many routes by which genetic changes can influence behavior. It is convenient, however, to refer to genes for behavioral traits, provided we realize this is merely shorthand. Thus, by "gene for sibling care," we mean that genetic differences exist in the population such that some individuals aid their siblings while others do not. Similarly, by "dove strategy," we mean that animals exist in the population that do not engage in fights and that this trait is passed from one generation to the next.

At first sight, it might seem that the most successful evolutionary strategy will spread throughout the population and eventually supplant all others. While this may sometimes be the case, it is not always so. In many situations it may not be possible to say what is the best strategy, because the effectiveness of a strategy may depend upon the behavior of other animals; that is, competing strategies may be interdependent in that the success of one depends upon the existence of the other and the frequency with which the other is represented in the population. For example, the strategy of mimicry has no value if the warning strategy of the models is not effective.

Some species have communal mating territories, called *leks*, in which the dominant males defend small territories. In the ruff (*Philomachus pugnax*), for instance, there are two types of male birds: *territorial males*, which are aggressive and compete with each other for females, and *satellite males*, which are not aggressive and which attach themselves to territorial males and steal copulations while the territorial male is busy elsewhere. The satellite males have a conspicuous white ruff, while the territorial males have darker coloration, as shown in Figure 7.7. It has been suggested that the territorial males tolerate the presence of the satellite males because their conspicuous plumage attracts females to the lek. There are two possible explanations for the existence of satellite males. First, they may be males that have lost out in the competition for territories or young and inexperienced males that adopt the satellite strategy to make the most out of their predicament and to wait for a territory to become vacant. Something like this is thought to be the case in the leks of the black grouse (*Lyrurus tetrix*). The second possibility is that the satellite males are employing an alternative evolutionary strategy and have reproductive success equivalent to that of the territorial males. This is likely to be the case with the ruff. As in the case of automimicry, discussed earlier, such a strategy is essentially a form of exploitation of a pre-existing strategy.

Although it has long been realized that alternative strategies could exist, it is only recently that it has been possible to give a satisfactory

Fig. 7.7 Ruffs displaying at their lek. This is the species with the most marked individual variations in male plumage. Males with dark ruffs and tufts are territorial, while the white one is a satellite male. The bird without a ruff is a female (From *The Oxford Companion to Animal Behaviour*, 1981)

explanation in terms of evolutionary theory. The key to this type of problem is the concept of "the evolutionarily stable strategy," developed by John Maynard Smith, Geoff Parker, and others (see Maynard Smith, 1982).

7.2 Evolutionarily stable strategies

An *evolutionarily stable strategy* (*ESS*) is a strategy that cannot be bettered by any feasible alternative strategy, provided sufficient members of a population adopt it. This is another way of saying that the best strategy for an individual depends upon the strategies adopted by other members of the population. Since the same applies to all individuals in the population, a true ESS cannot be invaded successfully by a mutant gene. As an example, let us consider the question of fighting and assessment of rivals. Animals that employ strategies to avoid unnecessary fighting usually will be at an evolutionary advantage. This idea can be formalized in the following way: Suppose we imagine that two types of strategies are represented in a population: the *hawk strategy*—which is to fight to kill or injure an opponent, even though there is a risk of injury to oneself—and the *dove strategy*—which is to threaten and display but to avoid serious fighting. These two strategies are extreme examples of those that might occur in real life. Suppose we now assign fitness-increment payoffs to the consequences of an encounter between the two animals. Let the winner of the contest score +50 and the loser 0. Let the cost of wasting time in a display be −10 and the cost of injury be −100. We have four possible types of encounters in a population containing both hawks and doves, the average payoffs from which are illustrated in Table 7.1. We can see that in an encounter between hawk and hawk, each stands to lose because while there is a 50 per cent chance of

Table 7.1 The game between Hawk and Dove (after Maynard Smith, 1976)

The payoffs are as follows: Winner +50, Injury −100, Loser 0, Display −10. The matrix of average payoffs to the attacker appears below.

Attacker	*Opponent*	
	Hawk	Dove
Hawk	$\frac{1}{2}(50) + \frac{1}{2}(-100)$ $= -25$	+50
Dove	0	$\frac{1}{2}(50\text{–}10) + \frac{1}{2}(-10)$ $= +15$

Note that when a Hawk meets a Hawk we assume that on half of the occasions it wins and on half of the occasions it suffers injury. Hawks always beat Doves. Doves always immediately retreat against Hawks. When a Dove meets a Dove we assume that there is always a display and it wins on half of the occasions.

winning a given encounter (a gain of 50), there is also a 50 per cent chance of losing (a loss of 100 through injury). On average, each hawk will gain a payoff of −25 as a result of fighting another hawk. When a hawk meets a dove, the hawk always wins 50 while the dove gains 0, because by avoiding a fight, it avoids injury. When a dove meets a dove, each threatens the other and wins half the contests without fighting. The average payoff is therefore +15.

It is easy to see that a hawk could invade a population made up entirely of doves. While the doves would average +15 in contests against each other, an invading hawk would get +50 in every contest with a dove and, thus, would be at an advantage. In a population with only hawks, the average payoff is −25, so that a dove could easily invade, since a dove gains 0 in a contest with a hawk, a better score than −25. Thus, we see that neither all-dove nor all-hawk populations are proof against invasion; neither one is an ESS. However, it is possible that a mixture of hawks and doves could provide a stable situation when their numbers reach a certain proportion of the total population. Let the proportion of hawks be h and the proportion of doves be $(1-h)$. The average payoff for a dove (D) can be calculated from the probability of meeting a hawk or another dove and the payoff from each type of encounter. Thus,

$$D = 0h + 15(1-h)$$

Similarly, the average payoff for a hawk (H) is:

$$H = -25h + 50(1-h)$$

When D is equal to H there will be a stable equilibrium. If $D = H$, then $h = 7/12$ and $(1-h) = 5/12$. Therefore, an ESS will exist in which 7/12 of the population are hawks and 5/12 are doves. An alternative possibility is that individuals behave as hawks for 7/12 of their encounters and as doves for 5/12, the strategy being chosen at random on each occasion.

This is called a *mixed strategy*. In either case, equilibrium is achieved only when the dove strategy is deployed on 5/12 of all contests and the hawk strategy on 7/12. Each strategy is at an advantage when it is relatively rare. Its representation in the population then increases until the equilibrium point is reached again. No invasion of hawks or doves can upset the ESS.

Another strategy is conceivable in this case, a so-called *bourgeois strategy*, where the individual behaves like a hawk when it is the owner of a territory and like a dove when it is an intruder into the territory of another. If we assume that a bourgeois is an owner for half its encounters and an intruder for the other half, then the payoffs will be as illustrated in Table 7.2. If the population were made up entirely of bourgeois, then the average payoff per contest would be +25, more than could be gained by invading doves or hawks, who would get +7.5 and +12.5, respectively, in contests against bourgeois. Thus, the bourgeois strategy is an ESS. It seems that the stable strategy is the conditional strategy—fight hard if the territory owner, but be prepared to retreat if an intruder.

Table 7.2 The Hawk, Dove, Bourgeois game (after Maynard Smith, 1976)

The payoffs, as in Table 7.1, are: Winner +50, Injury −100, Loser 0, Display −10. The matrix of average payoffs to the attacker appears below.

Attacker	*Opponent*		
	Hawk	Dove	Bourgeois
Hawk	−25	+50	+12.5
Dove	0	+15	+ 7.5
Bourgeois	−12.5	+32.5	+25

Note that when Bourgeois meets either Hawk or Dove we assume it is winner half the time and therefore plays Hawk, and intruder half the time and therefore plays Dove. Its payoffs are therefore the average of the two cells above it in the matrix. When Bourgeois meets Bourgeois on half the occasions it is winner and wins while on half the occasions it is intruder and retreats. There is never any cost of display or injury.

These examples, first formulated by Maynard Smith (1976), are gross simplifications of real-life situations. Nevertheless, they represent a considerable advance in evolutionary theory because they show that this coherent approach to a complex situation can in principle provide answers to long-standing problems. The basic principles of ESS may have to be elaborated to account for particular natural situations, but this is true of any scientific law that has reasonable generality.

In fact, bourgeois, dove, and hawk, strategies can be observed in nature. Male speckled wood butterflies (*Pararge aegeria*), for example, compete for mating territories (Fig. 7.8). Contests are brief and the owner always wins (Davies, 1978). This is an example in which the bourgeois strategy is followed by all members of the population. Dove

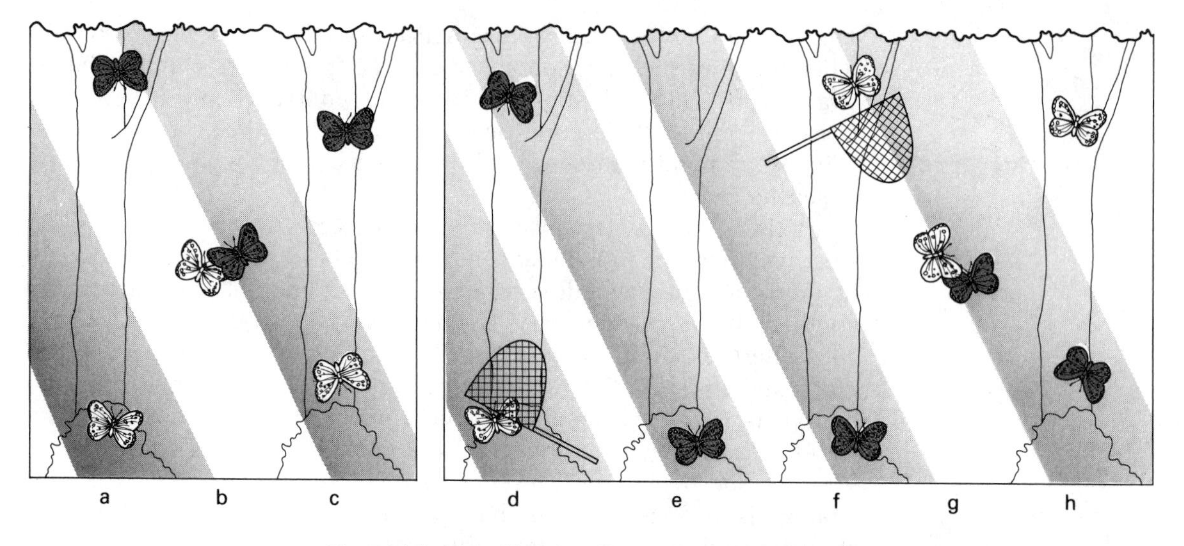

Fig. 7.8 (*a*) **A speckled wood butterfly basking in a patch of sunlight is challenged by an intruder, and a contest results (*b*), which the resident always wins (*c*). If the resident is removed (*d*), another butterfly becomes resident (*e*). If the original resident is then released (*f*), a contest results (*g*), but the new resident wins (*h*) (After Davies, 1978)**

Fig. 7.9 Ritualized technique of fighting in the oryx (*Oryx gazella*) (From *The Oxford Companion to Animal Behaviour*, 1981)

Fig. 7.10 Technique of fighting in the rattlesnake (*Crotalus*). The snakes do not bite each other, but each attempts to pin the other to the ground (From *The Oxford Companion to Animal Behaviour*, 1981)

strategy is often shown by animals that carry potentially dangerous weapons. For example, the oryx (*Oryx gazella*) (Fig. 7.9) has sharp pointed horns that could inflict mortal wounds. These may be used in defense against predators, but in contests among each other, the horns are used in a purely ritualized manner. It is against the rules to stab a rival in the side. Similarly, rattlesnakes settle their contests with ritualized trials of strength in which one attempts to pin the other to the ground (Fig. 7.10). They do not use their poisonous bite against rivals. The hawk strategy is not observed commonly among animals, but it may occur in contests over a valuable resource like the opportunity to mate. In species in which access to females is difficult to attain, or is short-lived, we might expect to see hawklike fighting because the payoff may be the one chance of a lifetime to contribute genetically to future generations. For example, male fig wasps (*Idarnes*) engage in lethal combat for the opportunity to mate with females inside the fig. The males have large mandibles and can bite another wasp in half. Bill Hamilton (1979) found one fig that contained 15 females, 42 males that were dead or dying from injury, and 12 uninjured males. In the musk ox (*Ovibos*) up to 10 per cent of adult males may die per year as a result of fights over females (Wilkinson and Shank, 1977). Serious injury may occur among red deer, though only after fairly prolonged assessment routines (see Chapter 8).

Fig. 7.11 A female digger wasp (*Sphex ichneumoneus*) at the entrance to its burrow (*Photograph: Jane Brockmann*)

7.3 Digger-wasp strategies

Female great golden digger wasps (*Sphex ichneumoneus*) (Fig. 7.11) lay their eggs in underground burrows that they have provisioned with katydids (long-horned grasshoppers) as food for the larvae. Jane Brockmann studied the female wasp's behavior in detail. She maintained almost continuous records of the nest-related activities of 68 individually color-marked females at three different fields sites over a total of six breeding seasons (Brockmann and Dawkins, 1979). Brockmann discovered that the females obtain a burrow either by digging one for themselves or by entering an already dug burrow. It takes a female an average of 100 minutes to dig a burrow. She then provisions it with stung and paralyzed katydids (a process that may take a few days), lays a single egg, and seals up the burrow prior to starting the cycle again. There is a 5 to 15 per cent chance that her burrow will be entered by another female wasp, who also provisions the burrow. Both wasps will be engaged fully in provisioning the same burrow and will not have another burrow open at the same time. Because both wasps spend most of their time hunting, it may be some time before they meet. When they do meet they fight, and one wasp is usually driven away. Only one wasp eventually lays an egg in the brood cell.

A female wasp has two strategies open to her. She can undertake to dig her own burrow and run a small risk of being invaded by another wasp, or she can enter an already dug burrow, saving herself the work

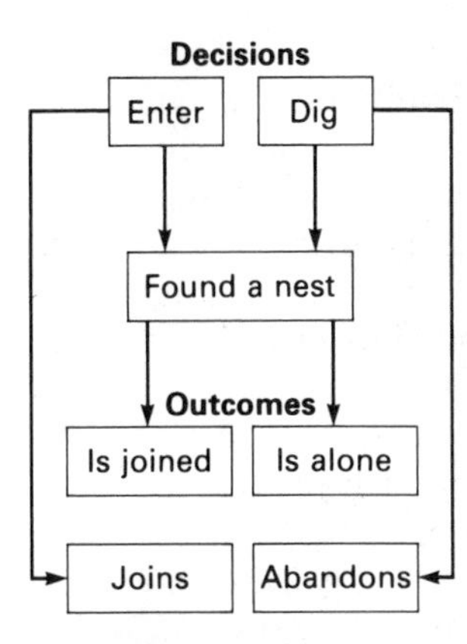

Fig. 7.12 The mixed strategy of the female digger wasp. The consequences of the decisions to dig a new burrow or enter an existing one are shown in this figure. Digging may result in abandoning the burrow, or in founding a nest where the wasp may remain alone or be joined by another. Entering an existing burrow may result in obtaining a nest, in which she may or may not be joined by another wasp, or in joining an already occupied nest (After Brockmann *et al*, 1979)

of digging but running the risk that the burrow is being used by the owner. The best strategy will depend upon that adopted by the majority of the other female wasps in the vicinity. On the one hand, if nearly all other females dig their own nests, then it is better to adopt the entering strategy, since there would be plenty of empty burrows and few other wasps exploiting them. On the other hand, if the majority of other wasps is employing the entering strategy, competition will be fierce and it will be better to dig one's own burrow. Thus, it seems that ESS lies between these two extremes.

It is possible to measure the success of the two strategies in terms of the number of eggs laid in a given period of time (Brockmann et al., 1979). Digger wasps employ a mixed strategy, so instead of comparing the success of individuals, the results of the decisions to enter or to dig have to be compared. On the basis of the hypothesis that the alternative decisions constitute a mixed ESS, we would expect the number of eggs laid per unit time to be the same whether the wasp decides to enter an existing burrow or to dig her own. On the basis of Brockmann's observations, Brockmann et al. (1979) calculated the success attributable to all possible outcomes of the two decisions, as illustrated in Figure 7.12. They discovered no significant difference between the two strategies on the basis of number of eggs laid. This conclusion supports the hypothesis that entering and digging are components of a mixed ESS.

As we have seen, when two females share a burrow, they usually end up fighting. They rear up, lunge with open mandibles, and wrestle with each other. The duration of fights varies between 2 and 16 minutes. A fight ends when one wasp, the loser, leaves the area. Out of 23 fights observed, in 18 cases, the loser never returned to the nest, while in the remaining 5 cases it returned many hours later (Dawkins and Brockmann, 1980). The winner gains the use of the burrow, but the value of the prize depends upon the number of katydids it contains. A burrow containing four katydids is ready for egg laying, a prize worth fighting for. A burrow containing no katydids still would be worth some effort because the winner is saved the trouble of having to dig a new burrow. On this basis we would expect each wasp to fight to an extent that is related to the payoffs. It would not be worth fighting very hard for an empty burrow because the effort and risk involved could amount to more than that required to dig a new burrow.

The problem is that the situation is the same for both wasps. If both wasps know how many katydids the burrow contains, then we can expect them to fight equally hard. Would such a situation be evolutionarily stable? Dawkins and Brockmann (1980) argue that it would not. If the burrow is valuable, both wasps would fight for a long time but would surrender at about the same time, the winner being determined by chance. A less valuable burrow would result in a less-prolonged fight, but both participants still would bear a substantial cost. Such a

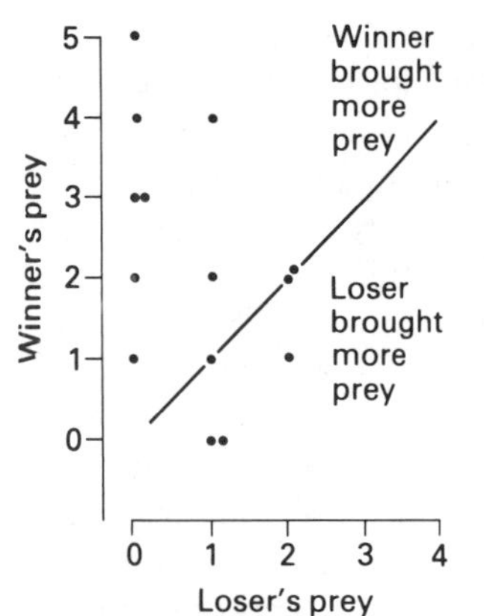

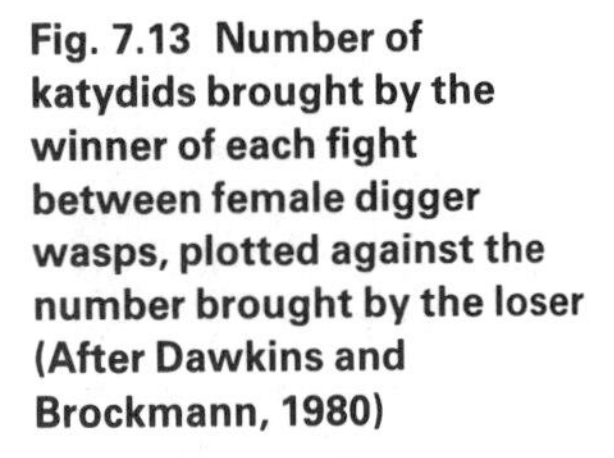
Fig. 7.13 Number of katydids brought by the winner of each fight between female digger wasps, plotted against the number brought by the loser (After Dawkins and Brockmann, 1980)

situation could be invaded by a gambler strategy: "On a random half of your encounters give up immediately without a fight; on the remaining half of your encounters, persist indefinitely until your rival surrenders" (Dawkins and Brockmann, 1980). In a population dominated by rational wasps that fight according to the total value of the burrow, the gambler would be at an advantage because she would win half her fights on average but would pay nothing for the fights lost. However, the gambler strategy alone probably would not be an ESS.

Dawkins and Brockmann analyzed the circumstances surrounding observed fights between two female wasps. For each of 23 fights they ascertained: the duration, which wasp dug the burrow, how long each wasp had been associated with the nest, which wasp was larger, the number of katydids each wasp had placed in the nest, and which wasp won. They found that the winner was not more likely to be the larger wasp or the one that dug the burrow or the one that most recently visited the burrow. Eleven wasps fought more than once, but there was no significant individual effect on the distribution of wins and losses. However, Dawkins and Brockmann did discover that the wasp that had placed the most katydids in the nest was more likely to be the winner (Fig. 7.13). This is a surprising result because it appears to be an example of the so-called "Concorde effect" (see Dawkins and Carlisle, 1976); that is, future behavior is decided on the basis of past investment instead of on future prospects. Decisions about the Concorde airliner reputedly were based upon such considerations, but we can hardly expect natural selection to act in such a way (Dawkins, 1976).

The duration of a fight was highly correlated also with the number of katydids the loser had supplied. A fight ends when the loser breaks off, so the loser effectively determines the duration of the fight. Dawkins and Brockmann show convincingly that the loser breaks off the fight at a time determined by the extent of her investment in the nest. This also is an example of the Concorde effect. The best strategy is always to base decisions upon the expected future payoff, but the female digger wasps do not seem to do this.

It appears that the wasps are not able to evaluate the contents of the burrow, but presumably the ability to do so could evolve. Apparently, they can tell how many katydids they have caught, though it is not known on what basis. As we have seen, the best policy for the individual is to fight in proportion to the total number of katydids in the nest, although this is probably not an ESS. However, it is not readily apparent that the Concorde strategy is necessarily the one most likely to be an ESS. Dawkins and Brockmann merely suggest that it might be. However, they do point to an important principle in evolutionary theory: The strategy that is best for the individual may not necessarily be the one that evolves, since it may not be an ESS.

Points to remember

- Evolutionary strategies are the passive result of natural selection which give the appearance of a ploy employed by genes to increase their numbers at the expense of other genes.

- An evolutionarily stable strategy (ESS) is an evolutionary strategy that cannot be bettered by any feasible alternative strategy, provided sufficient members of the population adopt it. Where the best strategy for an individual depends upon the strategies adopted by other members of the population, the resulting ESS may be a mixture of strategies.

Further reading

Edmunds, M. (1974) *Defence in Animals*, Longman, Harlow.

Parker, G. A. (1984) 'Evolutionarily stable strategies.' In Krebs, J. R. and Davies, N. B. (eds) *Behavioural Ecology*, 2nd edn, Blackwell Scientific Publications, Oxford.

1.3 Evolution and Social Behavior

In this section we are concerned with the ways in which evolutionary theory can illuminate the social behavior of animals. Chapter 8 is an introduction to the theory of sexual selection as conceived by Darwin. It provides a necessary background for understanding sexual strategy and its effects on social behavior. Chapter 9 is concerned with explanations of altruism in terms of kin selection and with reciprocal altruism and cooperation among animals. Chapter 10 discusses sexual strategies and their influence upon the evolution of mating systems and upon social organization in general. The aim is to provide a good grounding in the modern applications of evolutionary theory and a prelude to further reading in sociobiology.

Sir Ronald Fisher (1890–1962)

Ronald Aylmer Fisher was born in London to a prosperous business family. As a child he was academically precocious, especially in mathematics. In 1909, after attending Harrow School, Fisher entered Cambridge University, where he studied mathematics. He was, however, deeply interested in biology, particularly genetics. He was a prominent member of Cambridge University Eugenics Society, and retained that interest throughout his life.

After graduating, Fisher took a job as a statistician in London. When war broke out in 1914 he volunteered for active service, but was rejected because of his poor eyesight. He became a schoolmaster, teaching physics and mathematics in various schools. It was during this period that he began to make important contributions to statistics and to evolutionary theory. For example, he tackled the problem of reconciling the particulate nature of Mendelian genetics with the continuously varying characteristics (such as height in humans) measured by biometricians. The prevailing view was that such characters were the result of blending, and not Mendelian, inheritance. In a paper published in 1918, Fisher showed that the biometrical results follow logically from Mendelian principles in cases where a number of gene pairs contribute to the character in question.

After the war, Fisher took up a post as statistician at Rothamstead Experimental Station, where he remained for 14 years and where he made many very important contributions to statistics, biometry, and experimental design. In 1929 Fisher was elected a Fellow of the Royal Society in recognition of his work in statistics.

While at Rothamstead, Fisher worked at home on various projects, including genetics. He made a number of important contributions to genetics, eugenics, and evolutionary theory, culminating in his major work *The genetical theory of natural*

Reproduced by courtesy of the Trustees of the National Portrait Gallery

selection, published in 1930. In this work Fisher brought together the subjects of genetics and evolution by natural selection, and developed his fundamental theorem: The rate of increase in fitness of any organism at any time is equal to its genetic variance in fitness at that time. Fisher argued that since variation in populations is maintained by mutations, the rate of mutation occurrence determines the speed of evolution, while natural selection determines its direction. Many of Fisher's formulations were incompletely explained and some were not well understood by his fellow geneticists. Some were independently discovered by subsequent workers. It was said of Fisher that he was a geneticist of such prescience that the genius of his conclusions is still unfolding today. As we shall see, Fisher anticipated some of the fundamental ideas of modern sociobiology.

In 1933 Fisher was elected to the Chair of Eugenics at University College, London. In 1949 he became Balfour Professor of Genetics at Cambridge. He was knighted in 1952. Fisher retired in 1957 and died in Adelaide, Australia, in 1962.

8 Sexual Selection

In 1871, Darwin published *The Descent of Man*, in which he considered the subject of sexual selection (to which he had referred in his *On the Origin of Species* (1859)) as a form of natural selection. According to Darwin (1871), sexual selection depends on the advantage which certain individuals have over others of the same sex and species solely in respect of reproduction. Darwin reasoned that females make a definite choice of sexual partner and that males have acquired particular adornments and courtship behavior "not from being better fitted to survive in the struggle for existence but from having gained an advantage over other males, and from having transmitted this advantage to their male offspring alone". Darwin realized that there are two ways in which a male could gain advantage over other males. First, they can compete directly with one another by fighting or by some form of ritualized combat, now called *intrasexual selection* (selection within a sex). Second, males can compete indirectly in attracting females by special displays and adornments, called *intersexual selection* (selection between sexes). The two types of selection can occur at the same time, but in this chapter we consider them separately.

8.1 Intrasexual selection (male rivalry)

Direct competition between males can be seen among stags of the red deer *Cervus elaphus*, a species native to Europe and common in Scotland. Stags grow antlers each year, and in the autumn rutting season, they directly challenge each other for ownership of females, which have no antlers and which are herded into harems by the successful males. The females have little or no choice of sexual partner because the males defend their harems against possible rivals. Tim Clutton-Brock and his co-workers have made an intensive study of this species on the Isle of Rhum. Males challenge each other initially by roaring (Clutton-Brock and Albon, 1979), starting slowly and then speeding up. The challenger usually retreats if the harem owner can roar faster. These roaring matches are thought to help the stags assess each other, since a stag has to be in good condition to be able to roar well. If the challenger is able to match the roaring of the harem owner, then the two approach and walk parallel to each other, which enables the rivals to assess each other more closely, particularly with respect to body size. Many contests end at this stage, but some escalate into fighting proper. The stags interlock antlers and push against each other. The fighting is quite dangerous, and during its life there is about a 25 per cent chance that a stag will be injured permanently from fighting. Usually the larger stag wins, although one that is handicapped by a previous injury or exhausted by the strenuous rutting season may be supplanted by a younger stag.

The older, larger stags normally have larger antlers, and these confer an advantage in fighting. Apart from this advantage, however, the antlers contribute little to survival. Since stags do not possess antlers outside the breeding season, and since females do not grow them at all, it is unlikely that antlers are an important protection against predators. Moreover, for their growth, each year the antlers require large dietary quantities of materials such as phosphorous and calcium salts, so growing them must constitute a burden on the stag's metabolism. It seems, therefore, that with respect to antlers, at least, the advantages of intrasexual selection outweigh the pressures of natural selection.

Competition between two stags involves both fighting and assessment. Fighting is risky because it can lead to injury and may distract the animal from other important aspects of behavior. For example, while a harem-owning stag is engaged in a fight, other males may attempt to steal some of his unguarded females. Although winning a fight can bring considerable benefits, these may be offset by the risk of loss of assets or of future fighting potential. It is not surprising, therefore, that natural selection has led to the evolution of modes of assessment of the fighting potential of rivals. There is not much point in engaging in a fight that is certain to be lost, and animals that avoid unnecessary fighting will be at an evolutionary advantage. The question of assessment is directly relevant to sexual selection and merits some discussion here.

Fig. 8.1 Body size is exaggerated in these courtship displays of the Siamese fighting fish (*Betta splendens*) and frigate bird (*Fregata magnificans*) (From *The Oxford Companion to Animal Behaviour*, 1981)

One way to assess the likelihood that a rival will win a fight is to look for direct indications of fighting potential, such as body size and weapons. This may seem simple enough. However, natural selection may favor cheaters. If an animal exaggerates its body size, for instance, it may deter rivals from fighting. Figure 8.1 shows some examples of this. In a similar manner, natural selection may favor weapons exaggerated beyond the point of usefulness. For example, fossils of the extinct giant deer (*Megaloceros giganteus*) (which lived in Europe during the Ice Age) have been found with antlers measuring more than 3 meters in span (Fig. 8.2). Some scientists (e.g., S. J. Gould, 1978) think these animals became extinct because the intrasexual advantage of their large antlers was offset by their disadvantage, in terms of natural selection, when the climate became colder, since the antlers probably had to be regrown every year, a huge cost when food resources became scarce.

One way for a male to circumvent possible cheating is for him to base

Fig. 8.2 The extinct giant deer *Megaloceros giganteus*

his assessment of his competitor on some factor well correlated with real body size and physical fitness, like the roaring of red deer mentioned earlier. Similarly, two males may take part in ritualized trials of strength that do not involve real fighting and injury. Such contests are found in many species. Thus, some beetles engage in pushing matches, which the larger individual usually wins. Buffalos (Sinclair, 1977) and bighorn sheep (Geist, 1971) charge each other and clash head on in ritualized trials of strength (Fig. 8.3). Darwin thought that the evolution of

Fig. 8.3 Two male bighorn sheep in a contest of rivalry over females (*Photograph: L Lee Rue: courtesy of Frank Lane Picture Agency*)

weapons in species where they appear in the males but not in the females was due to their usefulness in fighting sexual rivals. However, some biologists take the view that the effectiveness of weapons is primarily psychological, serving to threaten and intimidate the rival.

The way to assess the likelihood that a rival will win a fight is to gauge his aggressive motivation or tendency to attack. For example, many animals undergo changes of posture that reflect their motivational state. Figure 8.4 shows that the posture of the greylag goose (*Anser anser*) in an aggressive state is almost the opposite of that in a fearful state. Research has shown that other geese of the same species recognize these signals and respond to them appropriately (Fischer, 1965). As we shall see in Chapter 22, the occurrence of overt displays of emotion is commonplace among animals.

Fig. 8.4 Threat postures of the greylag goose (*Anser anser*). From the 'at ease' posture at the bottom left picture, components of fear become stronger in the upwards direction, and aggression increases towards the right-hand side (From the *Oxford Companion to Animal Behaviour*, 1981)

However, such displays pose something of a problem for evolutionary biologists because it is difficult to see why natural selection does not favor cheating. It would seem that a goose that adopts a posture indicating a high level of aggression would be at an advantage in a sexual contest, even if it was a poor fighter. Such deceitful behavior should be passed to the next generation at a higher rate than the honest behavior so that the proportion of liars in the population would rise. At some point, however, the signal would become devalued because it no

Fig. 8.5 Two male elephant seals fighting (*Photograph: Burney Le Boeuf*)

longer provided useful information, there being so many liars in the population. This is a problem to which we return in Chapter 22, when we discuss animal communication.

Although intrasexual selection is primarily a matter of male rivalry, this rivalry is not always influenced by females. Among elephant seals (*Mirounga angustirostris*), for example, male rivalry is intense and often involves fighting (Fig. 8.5). When a male attempts to copulate with a female, she protests loudly, thus attracting the attentions of other nearby males, who attempt to interfere (Cox and Le Boeuf, 1977). A male is likely to be successful in copulating only if he is dominant and can ward off his rivals. The female intensifies the competition among males by her protests and ensures herself a dominant male. Thus, while the female does not directly exercise a choice, her behavior indirectly has that effect. If the female does not protest, the copulation is less likely to be interrupted and a low-ranking male will have a greater chance of success.

8.2 Intersexual selection (female choice)

Darwin maintained that the advantage of behavior evolved through sexual selection lies primarily in the satisfaction of female choice. He did not say, however, why such female preference might arise or be maintained within a population. Many examples exist of male adornment and courtship behavior that has evolved as a result of sexual selection (see Fig. 8.6). Males that succeed in attracting females by virtue of their special features are likely to father more offspring than less attractive males and thus to pass on their features to the next generation. The problem is that, although it is clear how males benefit by acquiring attractive features, it is not so clear why females should benefit by choosing males with features that may be irrelevant to survival and that may even be a disadvantage in the face of natural selection. For example, the enormous tail of the peacock (*Pavo cristatus*) is costly to produce, in terms of food and metabolic load. It is unwieldy and likely to hinder escape from predators—it is a handicap. Why, then, do pea hens not prefer to mate with males that carry less of a burden on their chances of survival?

Amos Zahavi (1975) suggested that females prefer males precisely because they carry a handicap and therefore must be robust individuals. The handicap is an advertisement for male quality. If a peacock can survive despite the encumbrance of his large tail, he must be a worthy male, since his good qualities will be passed on to the next generation. This suggestion has been criticized (e.g., Halliday, 1978; Maynard Smith, 1976) in that even a modest handicap, if inherited, places a burden on the next generation that outweighs any possible correlated advantages. A female might do well to choose a male that had survived despite an injury or other noninheritable handicap, although this could not explain the evolution of male adornments. The handicap principle may operate in cases where the male's attractive features are not fixed genetically but are a direct indication of male quality (Zahavi, 1977; Halliday, 1978). However, this possibility cannot account for more than a few of the many examples of sexual selection.

Ronald Fisher (1930) provided the most popular explanation of intersexual selection. He pointed out that females that mate with attractive males will tend to have attractive sons, provided the attractive characteristics are inherited. These sons, in turn, will be successful in attracting females and in reproducing themselves. Therefore, a female that chooses to mate with a male on the basis of his sexual attractiveness is likely to have more grandchildren than a female that mates with a less attractive male. It does not really matter what male feature is attractive to females, provided that it is not too strongly opposed by the forces of natural selection. Fisher suggested that initially, females are attracted to male characteristics that have survival value and that these characteristics became exaggerated during evolution through the action of intersexual selection. For example, female birds might show a preference for males with well-maintained plumage, since this would indicate not only a direct survival value in flying efficiency but also that the male could afford the time and trouble to maintain his plumage in good order. Once such a female preference was established in a population, almost any arbitrary feature of the plumage could become exaggerated and evolve into a superplumage irrelevant to the survival of the individual male bird. Fisher recognized that the interaction of male attractiveness and female preference should lead to an escalating evolution of a particular fashion that eventually would be checked by natural selection.

Darwin assumed that the female selects her sexual partner while the males compete for her attentions. Though this is typically true throughout the animal kingdom, there are exceptions. Usually the female pays the greatest cost in the process of reproduction. She provides the developing embryos with nutrition and may devote much of her time to their care. Before making such a large investment, the female can be expected to exercise a certain amount of caution. In particular, it is important that she should mate with a male of the correct species that is fully capable of carrying out his functions and likely to endow her offspring with high survival value.

In some species, however, the male invests considerably in the reproductive process. In the three-spined stickleback (*Gasterosteus aculeatus*), for example, the male establishes a territory that he defends against other males. He builds a nest and courts any ripe females that enter his territory. Once the female has laid her eggs in the nest, the male fertilizes them and then vigorously drives the female out of his territory. Care of the eggs and young is entirely the responsibility of the male. Although there is competition among males in establishing territories and attracting females, the males are discriminating as to who is allowed into the territory. Only females evidently ready to spawn are permitted to approach the nest. Thus, choice of mate is partly the prerogative of the male. It is most important for the male to guard against raids by other males seeking to steal eggs and destroy the nest (Wootton, 1976). Thus, any mistakes made in identifying intruders into his territory could have serious consequences. The female stickleback is also able to exercise choice, and some researchers have shown that females prefer to spawn with males whose nests already contain eggs (Ridley and Rechten, 1981).

8.3 Sexual dimorphism

Darwin (1871) was convinced that many of the differences between the sexes (sexual dimorphism) were the result of sexual selection:

> There are many other structures and instincts [of males] which must have been developed through sexual selection—such as the weapons of offence and

the means of defence of the males for fighting with and driving away their rivals—their courage and pugnacity—their various ornaments—their contrivances for producing vocal or instrumental music—their glands for emitting odours, most of these latter structures serving only to allure or excite the female. It is clear that these characters are the result of sexual and not of ordinary selection, since unarmed, unornamented, or unattractive males would succeed equally well in the battle for life and in leaving a numerous progeny, but for the presence of better-endowed males. We may infer that this would be the case, because the females, which are unarmed and unornamented, are able to survive and procreate their kind.

Fig. 8.6 Male and female African Paradise birds (*Vidua paradisea*). Outside the breeding season the male looks similar to the female (After Halliday, 1980)

In the case of intrasexual selection, it is obvious that we should expect the sexes to differ in morphology and behavior. As we have seen, male rivalry results in selection for features such as large body size, weapons, and aggressiveness that assist their possessors in the competition for females, which tend to lack weapons and to be smaller and less aggressive than the males.

In the case of intersexual selection, Darwin's argument is that males evolve particular adornments and behavior that attract females and that are therefore maintained through sexual selection. As long as males compete for the attentions of females, the females are in a position to choose among alternative males, and there is no selective pressure for them to become sexually attractive. Although some species provide good examples of this argument (see Fig. 8.6) a number of questions remain to be answered. Why is sexual dimorphism common in monogamous species, such as the majority of birds, in which opportunities for females to select mates are severely limited compared with polygamous species? To what extent can sexual dimorphism be attributed to sexual selection and to what extent to natural selection?

There have been various attempts to answer these questions since the time of Darwin. Recently, however, there has been a change in emphasis. Rather than attempting to account for differences between the sexes in terms of sexual or natural selection, we can assume that the sexes are fundamentally asymmetrical, and then ask what evolutionary consequences follow from this basic fact.

The fundamental feature of sexual reproduction is that it involves fusion of two gametes to form a zygote that develops into a new individual. The gametes come from the male and female parents and usually take the form of a small sperm and a large egg, respectively. Originally, we may imagine, the gametes were of the same size, as with some protozoa, such as *Paramecium*. Early in evolution there was probably some genetic variation in gamete size, and then the larger gametes produced larger zygotes, which had a better chance of survival because of their greater food reserves. Therefore, we might expect the larger gametes to have been favored by natural selection. If this were the case, then immediately there would have been selection for small gametes that could find a large partner to fuse with them, effectively parasitizing on its food reserves. Thus, small gametes could survive only if they became highly mobile and able to discriminate between large and small mating partners. This theory of the evolutionary origin of eggs and sperm, first suggested by Geoff Parker et al. (1972), is accepted widely as an explanation of the fact that virtually all present-day sexually reproducing multicelled animals produce eggs and sperm (Krebs and Davies, 1981).

Males produce small sperm in large numbers and females produce large eggs in relatively small numbers. Although these may cost roughly the same amount of energy to produce, the disparity in numbers of gametes means that sperm must compete for the chance to fertilize eggs. With relatively little cost per female, a male can increase his reproductive success by fertilizing many females. The reproductive success of a female, however, depends upon the number and viability of the eggs she produces. A female can follow the strategy of producing many eggs, investing little in each, and releasing them in conditions where at least some have a chance of fertilization and ultimate survival. Alternatively, she may produce fewer eggs and devote more resources to each one, thereby increasing each one's chance of survival, in which case we would expect the female to be highly discriminating in her choice of mate, since she has much to lose from an unsatisfactory partnership. Examples of both extreme types of strategies can be found in the animal kingdom. Most species follow some kind of intermediate position, and the best strategy for a species is determined largely by its ecological circumstances. Robert Trivers (1972) was the first to relate sexual competition to investment in offspring. He used the term *parental investment* to indicate the effort, in terms of time and resources, put into rearing an individual offspring. Trivers argued that "where one sex invests considerably more than the other, members of the latter will compete among themselves to mate with members of the former." The degree of competition depends upon a number of factors, however. Obviously, much depends upon the relative numbers of males and available females, which in turn depend upon the ratio of males to females in the population (the sex ratio) and the type of mating system typical of the species. Chapter 10 shows that types of mating systems and parental care are intimately related. The sex ratio within a population is usually 1:1 for evolutionary reasons that were first enunciated by Fisher (1930).

We have seen that sexual dimorphism can arise as a result of selection for male ability to attract females and to compete with rival males. There are, however, a number of other ways in which differences between the sexes can evolve. In those species that have few offspring, to which they devote considerable parental care, the female may have special adaptations connected with care of the young. The female mammal provides an obvious example, for not only does she possess special adaptations that enable her to retain embryos within her body during their development and to feed the newborn on milk, but she also displays highly developed maternal behavior. The male mammal is incapable of the type of parental care offered by the female. Therefore, there inevitably are differences between the parental roles played by each sex. Such differences tend to promote sexual dimorphism, since males and females become adapted for their respective roles through the action of natural selection.

Most birds are monogamous and opportunities for sexual choice are few. Sexual selection should theoretically be of little importance in monogamous birds, yet many such species are sexually dimorphic (Selander, 1972). Feeding ecology is thought to be an important factor responsible for such differences. In omnivorous species like the house sparrow (*Passer domesticus*), the sexes tend to be similar in bill size and skull dimensions because there is bound to be overlap in the types of food taken by the two sexes. Among food specialists, however, competition between the sexes can be reduced if they take different foods. This would be advantageous for a breeding pair with a limited territory. Thus, morphological differences between the sexes can arise as a result of differing adaptations to feeding. In most monogamous birds, males are slightly larger than females. It has been suggested that this difference has to do with social dominance, but it may also be connected with feeding ecology. For example, in Britain a typical male herring gull weighs about 1,500 grams, while the female weighs about 1,200 grams. The male also has a larger bill. During incubation the male becomes involved in territorial disputes more often than the female. He also tends to feed in places where his body size is an advantage. Thus, the males often forage on garbage dumps, where the food is concentrated and where much fighting and squabbling occur among birds. The females favor mussel beds and pastureland, where the food items are spread out and where little aggressive competition exists for food. Herring gulls share incubation duties, but in many other birds the female is solely responsible for incubation. In such species the female is often highly camouflaged, while the male is more brightly colored. Among monogamous hole-nesting ducks, the females tend to be much smaller than the males. The females alone incubate the eggs, and there is a clear advantage in choosing a site with a small entrance hole as a protection against predators (Bergman, 1965). Thus, natural selection may act to promote sexual dimorphism in many ways.

8.4 Sexual selection in humans

Darwin (1871) recognized that there are important similarities and differences among the various human races. "Although the existing races of man differ in many respects . . . yet if their whole organization be taken into consideration they are found to resemble each other closely on a number of points." However, Darwin found the differences to be something of a problem and felt "baffled in all our attempts to account for the differences between the races of man". He was inclined to think that natural selection was of little importance because "we are at once met by the objection that beneficial variations alone can thus be preserved" whereas, "not one of the external differences between the races of man are of any direct or special services to him. . . . For my own part I conclude that of all the causes which have led to the differences in external appearance between the races of man, and to a certain extent between man and the lower animals, sexual selection has been by far the most efficient". Thus, Darwin appears to have thought that sexual selection was more important than natural selection in the evolution of the differences between the human races.

Darwin probably underestimated the importance of natural selection, particularly in respect of human adaptation to different climates. People who inhabit hot sunny regions are usually darker than those who live in cold conditions, and they tend to have longer, more slender limbs. Nasal passages tend to be narrower among inhabitants of cold regions, helping to reduce loss of heat from the lungs. Eskimos and Alaskan Indians, both racial mongoloids, show a marked increase in blood flow into the hand when it is held in cold water. This phenomenon does not occur in people of caucasoid (European) race, including even the Lapp reindeer herders, who have been living in similar cold conditions since prehistoric times (Krog et al., 1960; Baker and Weiner, 1966). Among mongoloids living in the Far East there is a gradation in response. The Orochons, a nomadic hunting and reindeer-herding tribe from northern Manchuria, have a response similar to that of Eskimos. The northern Chinese have a lesser response and the Japanese have least of all (Coon, 1962). These studies indicate that genetic differences exist among human races in adaptation to cold. Similarly, differences have been found in respect to heat adaptation (Coon, 1962; Riggs and Sargent, 1964).

The significance of differences in skin color has proved to be more difficult to investigate. Although dark-skinned people tend to inhabit hot climates, there are numerous exceptions, such as the Indians of South America (Dobzhansky, 1972). Dark pigmentation may have some protective value, but also it has been suggested that it serves to absorb solar radiation and thus to save on the energy that must be expended to maintain body temperature at dawn and dusk in otherwise hot climates.

Many scientists are of the opinion that the genetic differences among the human races are insignificant compared with the differences between humans and related primates. Some of these, such as large brain size and upright posture, can be accounted for without difficulty in terms of natural selection. Others, however, like lack of body hair, are more difficult to account for.

It may be, as Darwin thought, that sexual selection has been important in the evolution of mankind. In order to evaluate this possibility, it is helpful to consider how sexual selection might operate today. Before looking at sexual strategies in humans, we consider a theoretical analysis carried out by Maynard Smith (1958).

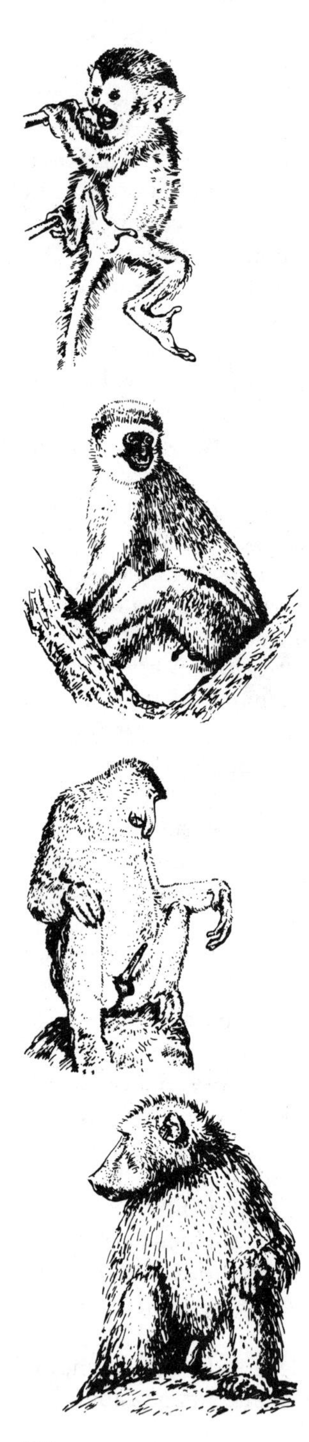

We have to imagine a society in which women prefer redheaded husbands and in which this preference is determined genetically. The red hair color of men is inherited. Redheaded men will have a choice in the selection of mates and are likely to marry early and to produce more offspring. If the society is not strictly monogamous, there will be many oportunities for sexual selection to exert its effect. In a strictly monogamous society, sexual selection will have a weak effect unless the fertility of redheaded men is greater than normal or redheaded men marry women who have more children for some other reason.

Maynard Smith's (1958) analysis suggests that degree of polygamy is an important factor in determining the effectiveness of sexual selection in human society. This is a difficult matter to investigate, but a number of studies have been done. Among the Xavantes and Yanomamas, two of the surviving primitive tribes of South American Indians, the men vary considerably in their reproductive performance (Salzano et al., 1967; Chagnon et al., 1970). In one of the villages studied, a quarter of the total population were the offspring of two head men. Statistical studies of urban man show evidence that active mating preferences (assortative mating) occur for physical traits like stature and for psychological traits such as intelligence and musical ability (Parsons, 1967).

Another way to evaluate the role of sexual selection in the past evolution of humankind is to compare humans with the other primates. Primate sexes may differ in body weight, hair color, skeletal characteristics, and secondary sexual characteristics. Such sexual dimorphism is more marked among the apes and humans than among the monkeys (Crook, 1972). Some of the differences between the sexes can be attributed to natural selection and to the differing roles of males and females. For example, the differences between the skeletons of male and female humans are primarily due to the greater muscularity of the man and to the fact that the female pelvis is designed to enable her to give birth to an infant with a large head. Among our remote ancestors, women with too small a birth passage or men too weak for the rigors of the hunt would soon be eliminated by natural selection.

Although it may be true that strong men may be sexually attractive to women and that female hips are attractive to men, this does not necessarily mean that sexual selection is responsible for these features. Even in the total absence of sexual selection, men and women must have some features that enable them to recognize each other as members of the opposite sex. In looking for evidence of sexual selection, we must

Fig. 8.7 Genital displays of male primates. Top to bottom: squirrel monkey (*Saimiri*), vervet monkey (*Cercopithecus*), proboscis monkey (*Nasalis*), and baboon (*Papio*) (After Wickler, 1966)

look for those different male and female features that appear to have no role in survival or reproduction. The secondary sexual characteristics of humans include the beard (in some races) and other body hair of the man, the change in the male voice that occurs at puberty, and the protruding and rounded breasts of the woman. A number of authors (e.g., Goodhart, 1964; Morris, 1967; Wickler, 1967) have suggested that the latter are the result of sexual selection because they are attractive to men and, by comparison with other primates, appear to be unnecessarily large for their role in providing milk for infants. Other authors, however, point out that the attractiveness of female breasts may be a cultural phenomenon, since it is apparently not a feature of sexual behavior in all human societies (Crook, 1972).

We discuss cultural phenomena in Chapter 27. At this point, the student should be aware that differences among populations that appear to be culturally based are difficult to interpret. We cannot assume that, where such cultural variants exist, natural selection and sexual selection can be dismissed as important factors in the evolution of the features in question. In the case of sexual selection in humans, there are a number of complicating factors.

Present-day human mating patterns range from strict monogamy to various forms of polygamy. In some societies, mating follows strict culturally based rules and taboos. In others, there is relative freedom from social control. Some authors take the view that sexual selection has been of the utmost importance in the evolution of early hominid society (Fox, 1972), but others are more cautious (Caspari, 1972). It seems likely that hominids have become progressively more monogamous in response to an increasing requirement of parental care. Among the primates, humans have by far the longest period of development from birth to sexual maturity. It is difficult to see how infant humans could be successfully brought to maturity without close cooperation between the parents. If our primitive ancestors were polygamous, then sexual selection probably would have been more effective than it is now, and this may account for some of the differences, like lack of body hair, between humans and the other primates (Crook, 1972). How, then, are we to account for present-day polygamous societies? Societies are regarded as primitive if they do not use metal tools and practice little or no agriculture, so that their subsistence is derived mainly from hunting and gathering. They are often adapted to life in specialized habitats, such as arctic, desert, or forest conditions. The question arises, therefore, whether societies that we label as primitive in their biological and cultural characteristics are truly primitive or whether these characteristics are special adaptations to particular living conditions.

The behavior of such people is often difficult to interpret. For

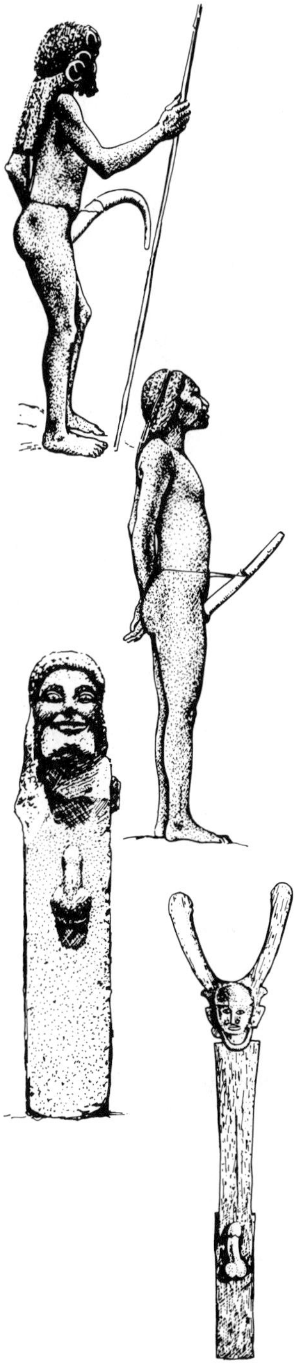

Fig. 8.8 Genital displays in man. Top to bottom: two Papuans from the Konca river, Herme of Siphnos (490 BC) from the Athens National Museum; house guardian of the natives of the island of Nias (After Wickler, 1966)

example, Wolfgang Wickler (1966) discusses the significance of penis display in some primate species (Fig. 8.7) and compares it with penis rituals and artifacts in humans (Fig. 8.8). In Papua, men of certain tribes enhance the penis with a sheath that is kept in an erect position by means of a waist string. By analogy with other primates, this may appear to be a sign of male dominance and, therefore, an important indicator of sexual selection. However, some anthropologists (Heider, 1969) maintain that the penis sheath is not associated with social position or with erotic practices. In the interpretation of cultural practices, a certain amount of skepticism is advisable. After all, so many parts of the body are decorated by people living in different parts of the world that it is not surprising to find some places where the penis is currently in fashion.

Points to remember

- Males which have a (genetically based) mating advantage over other males have more offspring to which the advantage is transmitted. This is called sexual selection.

- Males may have an advantage over other males as a result of male rivalry or of female choice. Male rivalry involves ritualized combat, bluff, assessment of the rival's fighting potential and, sometimes, actual fighting.

- Females that choose to mate with males as a result of their sexual attractiveness will tend to have attractive sons, provided the attractive features are inherited. Therefore, a female that chooses a mate on the basis of sexual attractiveness is likely to have more grandchildren than a female that mates with a less attractive male.

- Some of the differences between the sexes (sexual dimorphism) are thought to be due to sexual selection. Others are due to natural selection.

- Fundamental differences between the sexes result from the fact that males produce many small sperm while females produce few large eggs. This means that the female invests more in the outcome of each fertilization. This difference in initial parental investment means that male and female are likely to pursue different evolutionary strategies.

- Sexual selection is thought to have been important in human evolution and may be partly responsible for some of the differences between the sexes and between the races of mankind.

Further reading

Maynard Smith, J. (1978) *The Evolution of Sex*, Cambridge University Press, Cambridge.

Campbell, B. G. (ed.) (1972) *Sexual Selection and the Descent of Man*, Heinemann, London.

9 Altruism

Altruistic behavior benefits other animals at some cost to the donor. In evolutionary biology, altruism is defined by its effects on survival prospects, without reference to any motivation or intention involved. The possibility that animals have altruistic or selfish intentions is, of course, of interest (see Chapter 26.7), but it is not relevant to consideration of altruism from an evolutionary point of view. This distinction sometimes is forgotten.

In the strict sense, altruism is an evolutionary possibility only if defined in terms of individual fitness. An altruistic act increases the individual fitness of the recipient while it decreases the individual fitness of the donor. In certain circumstances, as we see later, natural selection may favor such behavior. In terms of inclusive fitness, however, this is not the case. Thus, natural selection would not favor behavior that benefited another animal at the expense of the donor's inclusive fitness because inclusive fitness is maximized by natural selection (see Chapter 6.4).

Natural selection will favor altruistic behavior under two main circumstances: (1) if the benefit in fitness to the recipient of an altruistic act exceeds the cost (decrement in fitness) to the donor by more than their coefficient of relationship, and (2) if one individual benefits another at little cost and this situation is later reciprocated. This second aspect of natural selection was proposed first by Trivers (1971) as a form of altruism that could occur between unrelated individuals.

There have been various other explanations of apparently altruistic behavior. The most notorious of these, called *group selection*, claims that natural selection will favor behavior that reduces the fitness of the donor if it benefits the group or species as a whole. However, evolutionary biologists have been unable to demonstrate how such a situation could arise during evolution, and many consider it to be impossible (e.g., Maynard Smith, 1964; Dawkins, 1976).

The main problem with the group-selection explanation is the possibility of cheating. For example, let us imagine a population of rabbits in which the members do not warn each other of approaching danger. Suppose that a gene is introduced that promotes thumping on the ground when a rabbit senses danger. The thumping serves to alert other rabbits, but it also attracts the attention of the predator. Thus, it would seem that neighboring rabbits benefit from the warning, while the thumper endangers itself. If we assume that the thumper manages to pass on this thumping gene before he is eaten by a predator, then we have a subpopulation of thumping rabbits within a population of non-thumpers. Because thumping is of benefit to the group, the group-selection argument maintains that the thumping group will evade predation better than the non-thumping group. Although the thumpers endanger themselves by attracting the attention of the predator, the nearby rabbits escape. Thus, the predator gets only one rabbit, whereas a stealthy predator in a non-thumping population might be able to pick off the prey one at a time and end up with more than one rabbit per visit.

However, this situation is not evolutionarily stable, since a rabbit in the thumping group that did not possess the thumping gene would not endanger itself when it detected a predator, yet it would benefit from the warnings given by other members of the group. This rabbit would be a cheat and would have an advantage over other members of the group. The greater reproductive success of the cheating rabbits would mean that the thumping gene would gradually be eliminated from the thumping population. There have been various attempts to circumvent this type of theoretical argument, but none has won much support among evolutionary biologists (see Grafen, 1984).

9.1 Kin selection

Natural selection favors genes that promote altruistic behavior toward individuals that are genetically related to the altruist. John Maynard Smith (1964) coined the term *kin selection* to distinguish this type of selection from group selection. Some authors (e.g., Wilson, 1975) erroneously define kin selection as a special case of group selection. Kin selection is a special consequence of gene selection (Dawkins, 1976).

The degree to which altruistic behavior should be extended toward other individuals depends upon the probability of the gene's being represented in those other individuals (that is, upon the coefficient of relationship, r, between the two animals involved). This does not mean that the individual altruist has to calculate its relatedness to each possible recipient. J. B. S. Haldane joked that "on the two occasions when I have pulled possibly drowning people out of the water (at an infinitesimal risk to myself) I had no time to make such calculations." He realized, of course, that the animal behaves as if it had made the calculations. Nevertheless, the animal somehow has to direct its altruistic behavior toward its kin rather than toward other animals. There are two main ways by which this can be achieved. The first is by kin recognition, which we discuss below. The second is simply a result of living near one's relatives. For example, in the case of the rabbits discussed previously, the thumping gene may spread if the thumping rabbit tends to be surrounded by kin; that is, if the gene for thumping primarily benefits rabbits who also carry some of the thumper's genes, then the inclusive fitness of the thumper will be increased by thumping behavior. The cheat, who does not carry the thumping gene, may benefit initially, but once surrounded by its own non-thumping relatives, it will be at a disadvantage. Thus, it will be difficult for a population of thumping rabbits to be invaded by cheats, provided that related individuals stay close to each other.

Studies of communal animals reveal that they often closely interrelate (Brown, 1975). This is particularly true in birds in which the young are aided by non-parents (see Chapter 9.4) (Harrison, 1969; Brown, 1974). It is also true of rabbits. In such species there is relatively little dispersal of the young. In migratory species of birds there is little communal breeding, and altruistic behavior is less evident (Brown, 1975). Cooperative breeding occurs among communal birds and some mammals, and this is probably the most prevalent form of altruism apart from parental care. (We discuss cooperative breeding later in this chapter.)

The alternative strategy is for altruists to recognize their kin and to confine their altruistic behavior to them. Such kin recognition is known to occur in some species and is often due to early experience. Perhaps as important as individual recognition is the detection and eviction of outsiders. In communal species, where the individuals are closely related, outsiders are quickly recognized and repulsed, while in colonial species, where the individuals are not closely related, outsiders are ignored.

9.2 Parental care

Parental care is a form of altruism. In spending time and energy to aid its offspring, the parent is increasing their fitness, to the detriment of its individual fitness, in that it is favoring current offspring at the expense of possible future offspring. The degree of parental care varies considerably from species to species, and depends upon the number of offspring produced, the types of mating system involved, and the aid given to the offspring by animals other than the parents.

The female usually spends more time and energy in caring for the young than does the male. The unequal contribution of the parents in caring for the young can be seen as an evolutionary contest in which each sex, while having an interest in the offspring in common with the other, also has an interest in minimizing the cost to itself of the necessary parental care. Thus, if a gene in a male can so influence the behavior of the animal that the burden of parental care is shifted toward the female, then the gene is likely to become more frequent in the population. Conversely, if a gene in a female has influenced her behavior so that the male is required to raise his contribution, without any extra risk to the offspring, then that gene is likely to spread.

To facilitate discussion of this topic in nongenetic terms, Robert Trivers (1971) introduced the concept of *parental investment*, defined as any investment by the parent in an individual offspring that increases the offspring's chances of surviving (and hence its reproductive success) at the cost of the parent's ability to invest in other offspring (see Chapter 8). Although undoubtedly useful, this definition suffers from the implication that past investment can influence future behavior. For example, suppose a mother has a choice of saving the lives (by investment) of two of her offspring of different ages. The one she neglects is bound to die, but if she can save only one, which should it be? In her lifetime, the mother has limited resources to invest. She stands to lose a higher proportion of her life's investment if she neglects the older offspring; thus, it would seem that she should save the older and let the younger die (Dawkins, 1976). However, we would expect natural selection to evolve animals that behave so as to maximize their future reproductive success irrespective of past investment. Richard Dawkins and Tamsin Carlisle (1976) pointed out this fallacy in some applications of the notion of parental investment. They likened it to the folly of governments who argue that they should continue to spend money on a project like the Concorde airliner because they have already spent so much on it.

John Maynard Smith (1978b) points out that the optimal behavior of one parent depends upon what the other parent is likely to be doing. Therefore, we should look for a pair of strategies, one for the male and one for the female, that together form an evolutionarily stable combination. For instance, if a female had brought an offspring near to the stage of independence, only to abandon it to start on a new offspring, would be bad policy, not because of the past investment but because the chances of survival of a new offspring would be less than that for the existing one. However, if she could rely upon the male to care for the offspring, the best strategy for the female would be to start on a new offspring. From the male's point of view, it might be better not to take care of the nearly weaned offspring, thus forcing the female to do so, since this would leave the male free to pursue other reproductive opportunities.

Maynard Smith (1977) analyzes the situation in which reproductive success is governed primarily by parental care. A male bird who deserts his mate after she has laid her eggs has a chance of mating with a second female. Thus, desertion is favored by natural selection if there is a good chance of the male's finding a second female and if one parent is almost as good as two in caring for the young. However, the female can adopt a strategy that makes it less likely that the male will desert. In a population of females that mate only after the male has courted and paired with them for a fairly long time, males that desert have little chance of finding a second mate, especially if there is a 1:1 sex ratio and if the breeding is synchronous. Maynard Smith shows that such a strategy can be an evolutionarily stable strategy (ESS), and he argues that once an ESS has evolved in which the male has nothing to gain by desertion, he would increase his fitness by investing before copulation in activities like nest building.

In situations in which the success of a breeding pair depends not only upon parental care but also upon the extent of investment by the female prior to egg laying, a number of ESSs are possible, involving guarding or deserting the eggs on the part of the male or female. If one parent is nearly as effective as two in caring for the young, and if there are good prospects of mating again following desertion, then it will pay one parent to desert. It is not entirely clear why evolution has taken the path of male desertion in some cases and female desertion in others (Maynard Smith, 1977), but perhaps we can learn something by looking for comparative evidence.

In reviewing the patterns of parental behavior observed in birds and fishes, Maynard Smith (1978b) makes the following points. Among fishes the most common pattern is no parental care. If there is parental care, it is usually provided by one parent only, and consists of protecting the eggs and fry from predators and removing parasites and waste products of respiration. It does not include the provision of food, so one parent is almost as effective as two. In species where internal fertilization occurs, parental care is usually provided by the female, probably because the male is not present when the eggs are laid and because, even if present, he has no guarantee that the eggs have been fertilized by his sperm, making it not in his interests to help care for the offspring. If external fertilization occurs, usually the male cares for the young. He may construct a nest, as in sticklebacks, or brood the eggs, as in sea horses and pipe fish, or he may protect a particular egg-laying site. In the absence of parental care, female fish exhaust their food reserves in the egg laying, so for parental care to evolve from such a situation, the female would have to reduce the number of eggs laid, or the care would have to be provided by the male. This is thought to be the evolutionary origin of male parental care in fish.

The most common pattern among birds is for parental care to be provided by both sexes, which usually involves bringing food for the young, and may also involve incubation. In herring gulls, for example, the incubation duties are shared equally. This is necessary because the eggs are quickly eaten by other gulls if left unguarded, and thus, one parent must sit on the eggs while the other forages for food. Similarly, once the young are hatched, they can easily fall prey to predators and so usually are guarded by one of the parents. During the first few weeks of the chick's life, the parents are kept very busy providing food and they usually lose weight.

In species in which only one parent incubates, it is nearly always the female. Whereas a male can desert a female immediately after mating, a female cannot seek a second mate until she has laid the first batch of eggs. In some waders, such as the sanderling (*Calidris alba*), the female lays one clutch of four eggs that is incubated by the male, and then lays another clutch of four eggs for which she cares herself. This strategy allows parents to take maximum advantage of the short productive arctic summer. Once males have evolved the habit of caring for a clutch of eggs after the female has laid them, females might compete with each other to acquire more males. This is what may have happened in the polyandrous jacana mentioned earlier (Maynard Smith, 1978b). Compared with fish, birds have fewer offspring, which are more carefully tended, and their reproductive success depends not upon the number of eggs laid but upon the number of young that can be successfully raised.

In mammals, the main evolutionary trend has been for males to mate with many females. In some species this has led to intense rivalry among males (see Chapter 8). Female mammals feed their young on milk, and female desertion is thus unlikely to be an ESS. In some species, the males obtain food for the young and protect them from predation. Some primates and carnivores are monogamous, and it is surprising that male lactation has not evolved among those species in which the male shares in the care of the young (Maynard Smith, 1977).

Maynard Smith (1978b) summarizes the main factors that determine whether or not parental care will take place and, if so, which parent will provide it. He cites four variables:

1. The effectiveness of the parental care provided by one versus two parents
2. The chance that a male who deserts a female after mating will be able to mate again
3. How able a female is to guard her offspring after producing them
4. The confidence with which a male can assume that a particular batch of eggs was fertilized by him

The first and third factors depend mainly upon the extent to which the female exhausts her reserves in producing eggs. In some species, greater fitness is achieved by producing as many eggs as possible, while in other species, fewer eggs are required, leaving some resources available for maternal care. The second factor depends partly upon the length of the breeding season and partly upon the availability of females. The fourth factor depends partly upon the mode of fertilization and partly upon the ability of the male to guard the female against rival males. Males that protect their females, either directly or by maintaining a territory, have greater certainty of paternity but less time to pursue other females.

In summary, parental care is a form of altruism. Parents invest in their offspring at the expense of their own survival and chances of future reproduction. The role of the sexes in parental care varies from species to species in accordance with ecological factors that are only partly understood.

9.3 Reciprocal altruism

Altruism towards kin can be regarded as selfishness on the part of the genes responsible, because copies of these genes are likely to be present in relatives. Altruism could also be regarded as a form of gene selfishness if by being altruistic an individual ensured that it was a recipient of altruism at a later date. The problem with the evolution of this kind of altruism is that individuals that cheated, by receiving but never giving, would be at an advantage.

It is possible that cheating could be countered if individuals were altruistic only toward other individuals likely to reciprocate. For example, Craig Packer (1977) observed that when a female olive baboon (*Papio anubis*) somes into estrus, a male forms a consort relationship with her. He follows her around, waiting for an opportunity to mate, and guards the female from the attentions of other males. However, a rival male

Fig. 9.1 When !kung women go gathering fruit and vegetables, they have to carry their children with them (*Photograph: Gerald Cubitt: courtesy of Bruce Coleman Ltd*)

Fig. 9.2 Bushmen going hunting (*Photograph: Jen and Des Bartlett: courtesy of Bruce Coleman Ltd*)

sometimes may solicit the help of a third male in an attempt to gain access to the female. While the solicited male challenges the consort male to a fight, the rival male gains access to the female. Packer showed that the altruism shown by the solicited male often is reciprocated. Those males that most often gave aid were those that most frequently received aid.

This type of situation obviously provides scope for cheating. An individual that receives aid may refuse to reciprocate at a later date. However, if opportunities for reciprocal altruism arise sufficiently often, and if the individuals involved are known to each other, then a non-cooperative individual can be identified easily and discriminated against. Thus, for natural selection to favor reciprocal altruism, the individuals must have sufficient opportunities for reciprocation, they must be able to recognize each other individually and remember their obligations, and they must be motivated to reciprocate. These conditions are found in primitive human societies, and reciprocal altruism has played an important role in human evolution (Trivers, 1971).

The !kung bushmen of the Kalahari desert provide a good example of the importance of reciprocal altruism in hunter-gatherer societies. The women (Fig. 9.1) bring in about 60 per cent of the protein and carbohydrate by gathering vegetables and fruit. The men spend many hours hunting for game, which provide essential amino acids and minerals (Fig. 9.2). The food supply is highly variable and times of plenty may be followed by periods of hardship. The men usually hunt in pairs and if one pair is successful, the meat is shared among all members of the group. The successful provide food for the unsuccessful on the understanding that the situation may be reversed. In this way the bushmen maximize their chances of having some meat available, and minimize waste.

9.4 Cooperation

Cooperation among animals usually involves some form of altruism. In cooperation among members of different species, called *symbiosis*, the relationship is reciprocal. For example, many aphids gain protection by associating with ants, and the ants benefit by obtaining food from the aphids. Thus, when a garden ant (*Lasius niger*) encounters a bean aphid (*Aphis fabae*), it strokes the aphid with its antennae, which induces the aphid to exude honeydew, a sugary liquid by-product of digestion that the ant then consumes.

Anemone fish, *Amphiprion*, are able to gain protection from predators by swimming unharmed among the tentacles of sea anemones. Each

Fig. 9.3 Anemone fish (*Amphiprion*) are not stung by anemones to which they have become acclimatized. They can then gain protection from predators by hiding among the tentacles (*Photograph: Tim Halliday*)

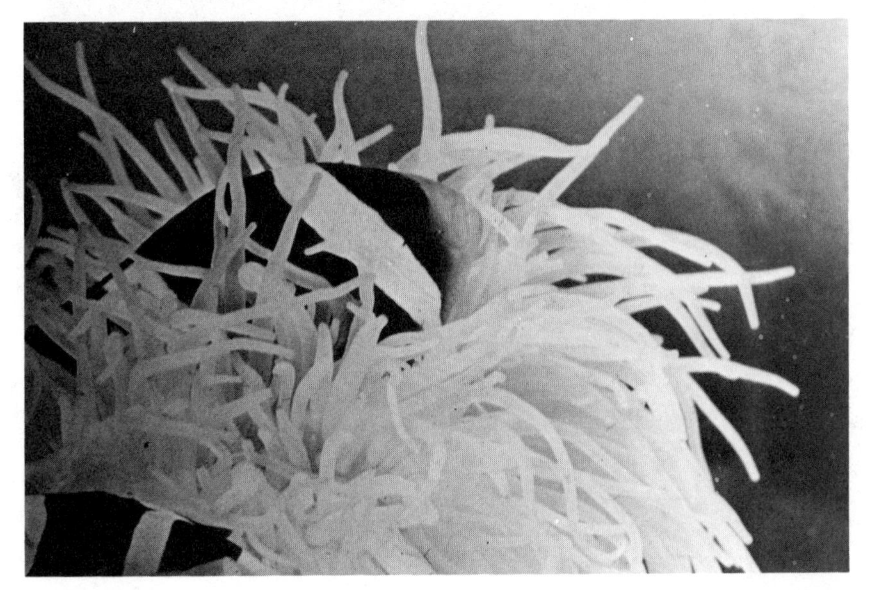

fish has to develop an immunity to the anemone sting, and it is highly probable that the anemones could evolve an effective deterrent to anemone fish. However, the fish do not harm the anemones, and some fish defend their anemone against predators that bite the ends of anemone tentacles, such as the butterflyfish (*Chaetodon*). The anemones also benefit from the anemone fish's habit of bringing its food into the anemone, some of which is eaten by the anemone. Thus, the relationship between the two animals is reciprocal: The anemone benefits by obtaining food and by gaining some protection from predators; the anemone fish gains protection for itself and for its eggs and fry, which are able to develop among the anemone tentacles undisturbed (see Fig. 9.3).

Cooperation among members of the same species often involves some form of altruism. The cooperative hunting of wild dogs, lions, and hyenas usually occurs among relatives. Wild dogs (*Lycaon pictus*) prey upon animals much larger than themselves, such as zebra and wildebeest (Fig. 9.4). They select a single quarry and chase it over a long distance. The hunt involves cooperation in the choice of prey and in bringing the animal down. During the chase, there may be exchange of leaders, and thus a sharing of the burden over long distances. A dog in

Fig. 9.4 Wild dogs pulling down a wildebeest after running it down (*Photograph: Hans Kruuk*)

the rear sometimes will cut corners in an attempt to head off prey. The food is shared among the members of the pack, and upon arrival home, adults often regurgitate food for the young. Some adults may stay behind to guard the juveniles, and these animals are also fed by the returning pack. Thus, individual dogs do not act purely in their own interest but are altruistic toward other members of the pack.

The altruistic behavior involved in cooperative hunting is fairly straightforward and not difficult to account for in terms of kin selection. Cooperative breeding, however, poses more difficult problems for the evolutionary theorist. For example, the Mexican jay is a communally breeding species that lives in flocks of 4 to 15 individuals. It is a non-migratory species, and each flock communally defends a territory in oak or pine forest. A mated pair builds a nest, and the female lays her eggs only in her own nest. Each flock may have between one and four nests. When the eggs hatch, the nestlings are fed not only by their parents but also by other members of the flock. About 50 per cent of the food an infant receives is provided by birds other than its parents. These helpers at the nest are behaving in an apparently altruistic manner, for at considerable cost to themselves, they are helping the offspring of another individual.

It is necessary to explain why a mutation that led to failure to aid another individual's offspring should not spread through the population and eventually destroy the basis for the cooperative behavior. One explanation is that the helpers are related sufficiently closely to the progeny of other members of the flock for the altruistic trait to be maintained in the population by kin selection. Another possibility is that helpers at the nest obtain benefit from their apparently altruistic behavior. They might obtain protection from the parents, or they might gain valuable experience. A third possibility is that some form of reciprocal altruism is involved. Perhaps the parents tolerate the other birds on the territory at some cost in depletion of resources in exchange for the aid received in rearing the young. Before evaluating these possibilities, we should see if the various species of cooperatively breeding birds have any features in common.

A review of the literature (Emlen, 1978) suggests that most species of cooperatively breeding birds have several features in common. For example, most are non-migratory and inhabit the tropics or subtropics. The environment is relatively stable and seasonal changes are small, conditions that lead to competition for suitable habitats and territory; nest sites often are in limited supply. Robert Selander (1964) has suggested that such conditions led to the evolution of communal territoriality and cooperative breeding, a view supported by a number of ethologists working on a variety of species (Emlen, 1978). Cooperatively breeding birds usually have the characteristics typical of populations that live in an environment near its carrying capacity, including low fecundity, long life span, late sexual maturity, and lower dispersal

(Brown, 1970). Typically, the young are retained on the parental territory and recruited as helpers. It may seem that the juvenile that remains on the parental territory increases its probability of survival and gains valuable experience at the same time. However, this does not explain why the juvenile should incur the extra cost involved in helping its parents to raise another brood. It is also not obvious that the breeding birds actually benefit from such a helper at the nest. Emlen summarizes the advantages and disadvantages associated with cooperative breeding as follows.

There are several potential advantages for the breeding birds. They may obtain valuable assistance in the care and rearing of their current offspring. Comparison of the reproductive success of parents with and without helpers, in 12 different species, shows that helpers usually do help. However, among five species the breeding pairs with helpers occupied larger or better territories than those without helpers. Thus, their greatest success might be due to the quality of the territory. Alternatively, the larger groups (i.e., those with helpers) might be better able to defend a larger territory.

The helpers may provide insurance. That is, they could raise the young in the event of the death of one of the parents.

The experience gained by the helpers improves their chances of being successful parents in the future. If the helpers are related closely to the breeders, then their future success will increase the inclusive fitness of the original breeders.

Retention or recruitment of helpers is an important way of increasing the size of the group. Competition among groups might mean that the larger groups are better able to obtain and defend the better territories. There may be other advantages of large group size, like improved detection of predators, that will increase the probability of survival of all members of the group.

There are also several disadvantages incurred in tolerating helpers in the group. The additional birds may deplete the food supply to the extent that the reproductive success of the breeding pair is diminished.

The extra activity around the nest might attract the attention of predators.

The inexperience of novice helpers might be detrimental to the offspring. This is true if experience is important in parental care. Studies of the Tasmanian native hen (*Tribonyx mortierii*) and the Florida scrub jay (*Aphelocoma coerulescens*) show that experienced parents have greater reproductive success than inexperienced parents (Emlen, 1978).

As part of a strategy for taking over the territory, the helpers could sabotage the efforts of the breeding pair. There is evidence that this does occur in some species. Many workers have described serious disputes between breeders and helpers; and Arabian babblers (*Turdoides squamiceps*) (Zahavi, 1974) and Florida scrub jays (Woolfenden, 1973) have been observed to destroy eggs in nests of their own group.

Non-breeding birds may benefit in the following ways by becoming helpers: (1) They may gain experience in caring for young; (2) they may benefit from belonging to a group; (3) they may increase their inclusive fitness by aiding relatives; or (4) they may inherit part of the parental territory.

The main disadvantage for the non-breeder that joins a breeding group is that the helper may lose or postpone the opportunity to breed itself. However, the chances of successfuly breeding independently are offset by the difficulty of finding a mate, establishing an independent territory, and breeding successfully as a novice.

Altruistic behavior can be distinguished from other types of social interactions on theoretical grounds in terms of the apportionment of benefit between the two participants. Bill Hamilton (1964) suggested the terminology shown in Figure 9.5 to describe the four main types of interactions. This classification can be usefully applied to the problem of helpers at the nest. If the relationship between a breeding bird and its helper is truly cooperative, then we would expect both to benefit in terms of individual fitness. This could happen in situations where the consequences of living in a group are of benefit to both the breeding pair and the helpers. Thus, a group might be able to hold a better territory than a pair. There might also be benefits in combating predation. The help given by non-breeding birds could be a payment for joining a group (Gaston, 1976), or it could be a form of apprenticeship (Emlen, 1978), the helper eventually taking over the role of the female.

Fig. 9.5 Types of behavioral interaction (After Hamilton, 1964)

Fitness change to donor	Fitness change to recipient: Gain	Fitness change to recipient: Loss
Gain	Cooperative	Selfish
Loss	Altruistic	Spiteful

In cases in which the helper is altruistic, the breeding birds should benefit, while the helper loses individual fitness. The inclusive fitness of the helper is increased in cases in which it is aiding its relatives, a strategy that could become established by kin selection. The alternative strategy, for the juvenile bird to mate and set up independently, might not be viable at the outset. Thus, it might pay the helper to wait for a vacant territory.

The selfish helper is one that gains from associating with a breeding pair but that may decrease the pair's reproductive success. Among some woodpeckers (Skutch, 1969) and the white-winged chough (Rowler, 1965), novice helpers tend to be inefficient. The extra birds on the territory could deplete the food resources, and the extra activity around the nest may attract predators. Why should the parents tolerate the presence of helpers if they do not benefit from it? A possible reason might be that they have a genetic interest in the welfare of their kin and that their tolerance of helpers is an extended form of parental care.

If neither the breeders nor the helpers benefit, then the relationship is spiteful. The helpers could be tolerated by the parents, even though their presence was detrimental, for the reason discussed in the last paragraph. The helper may gain nothing in the short term, but it may be able to sabotage the efforts of the breeding pair and eventually take over the territory (Zahavi, 1974; 1976).

Points to remember

- Natural selection will favor altruistic behavior if the benefit in fitness to the recipient of an altruistic act exceeds the cost (decrement in fitness) to the donor by more than their coefficient of relationship. It may also favor altruism in which one individual benefits another at little cost and this situation is later reciprocated.

- Natural selection will favor genes that promote altruistic behavior toward individuals that are genetically related to the altruist. This form of selection is known as kin selection.

- Parental care is a form of altruism because the parent diminishes its own fitness by investing time and energy in the care of its current offspring. This increases their fitness at the expense of possible future offspring because the parent has a limit to the resources that it can invest during its reproductive life.

- The best strategy for one parent depends upon what the other parent is likely to be doing. Males and females may follow different strategies, and some of these may be evolutionarily stable strategies. The role of the sexes in parental care varies from species to species in accordance with their ecological circumstances.

- Cooperation among animals usually involves some form of altruism. The (symbiotic) cooperative relationships between members of different species are usually reciprocal. Cooperation among members of the same species may also be based upon kin selection.

Further reading

Grafen, A. (1984) 'Natural selection, kin selection and group selection.' In Krebs, J. R. and Davies, N. B. (eds) *Behavioural Ecology*, 2nd edn, Blackwell Scientific Publications, Oxford.

10 Sexual Strategy and Social Organization

Social organization varies considerably from one species to another. It ranges from simple cooperation between male and female to the complex societies of some primates. One of the most fundamental influences on the social organization of a species is the species' mating system. To understand mating systems we need to appreciate the alternative evolutionary strategies open to males and females.

10.1 Sexual strategy

In Chapter 8 we saw an inevitable difference between the initial investment of male and female in their offspring. Males produce many small sperm and females produce relatively few large eggs. Sperm compete for the chance to fertilize eggs, and a male can increase his reproductive success best by fertilizing many eggs at relatively little cost per egg. Initially, the female invests more in each offspring, and it is normally the case that females continue to invest more than males in bringing each to a viable, independent stage of life. However, a female would increase her reproductive success if she could persuade the male to invest more in her offspring, leaving her free to divert more of her resources to the production of other progeny. Since, conversely, it is often in the male's interest to reduce his investment in parental care, an evolutionary conflict of interest exists between the sexes.

The main evolutionary problem for males is to attract females in the face of competition from other males. The main evolutionary problem for females is to select, as sexual partners, males that will endow the offspring with the greatest chances of survival and reproduction. Initially, it would seem, the female has the advantage because she can refuse to cooperate. As we saw in Chapter 8, the female usually exercises the choice in any sexual encounter. A male can do little if the female is unreceptive. It is possible that the male could enforce copulation, but this is usually rather difficult, since in most species the female must take up a specific posture before copulation can take place. Forced mating, however, has been observed in certain animals, including mallard ducks and scorpion flies (*Panorpa*). The male scorpion fly usually presents the female with a nuptial gift during courtship, often a dead insect obtained from a spider's web. Copulation occurs while the female is eating the gift. Sometimes the male forces mating without presenting a gift (Thornhill, 1980). The male benefits from a successful forced mating because he does not have to run the risk of finding a nuptial gift. Apparently, 65 per cent of adult male scorpion flies die as a result of being caught in a spider web. The female loses because she does not receive a gift, which would normally provide energy for her eggs. However, the success rate from forced mating appears to be small, so it may be a strategy of last resort.

Males usually benefit from a given sexual encounter if they mate, whereas females do not always benefit and may even suffer a reduction in fitness (Parker, 1979). Therefore, it is a good strategy for females to be coy. George Williams (1966) has described courtship as a contest between male salesmanship and female sales resistance. A female

Fig. 10.1 Isolating mechanisms in gulls. The principal visual features thought to be important in species recognition are shown on the left. The results of altering eye color in mated female Kumlien's gulls are shown on the right. The numbers on the right show the extent to which pairs were broken as a result of the experimental manipulations. When the purple eye was changed to white, 21 out of 34 pairs broke up, whereas the other experiments and controls had very little effect on pair break-up (After Smith, 1966)

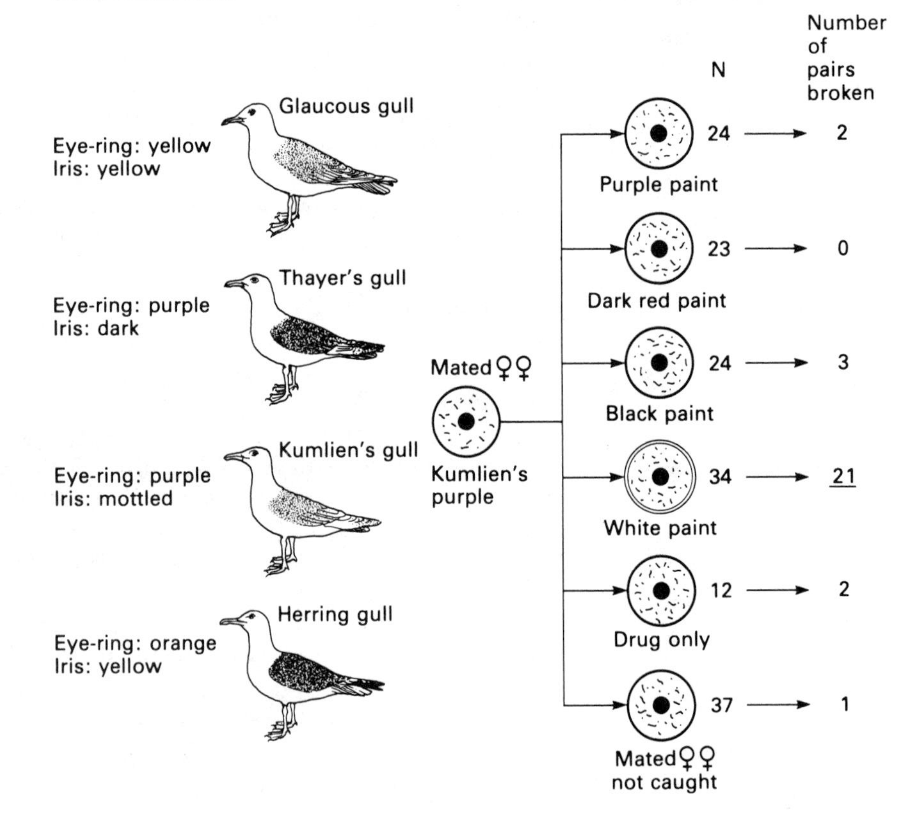

Fig. 10.2 A brown leghorn cock with white feathers attached to alter neck color contour (From *The Oxford Companion to Animal Behaviour*, 1981)

would not need to be coy if she could directly gauge the male's fitness and likely future behavior. All the female has to go on, however, is the male's appearance and present behavior. The female has to induce the male to reveal his true nature, or she has to arrange matters so his interests coincide with hers. In both cases, the female's best policy is to be cautious and to prolong the courtship. The first task for the female is to ensure that her potential mate is of the correct sex and species. This is also important for the male, but as we have seen, the female has much more to lose by making a mistake. Natural selection favors individuals that have distinctive markings and behavior that can identify them as members of a particular species. These features play a prominent role in courtship, and they are called *isolating mechanisms* because they promote reproductive isolation among species by reducing the possibility of hybridization. We now briefly discuss three examples of isolating mechanism.

There are four closely related species of gull that breed in the Canadian Arctic. In some localities, all four species nest in the same colony, without interbreeding. The birds look quite similar, differing primarily in markings on wing tips and the region around the eye. By experimentally altering these color patterns, it is possible to induce cross-mating between two species, as illustrated in Figure 10.1 (Smith, 1966).

Second, extensive studies of interbreeding among different breeds of domestic fowl have shown that appearances are very important (Lill, 1966; 1968). For example, during courtship the cock displays to the hen with a waltzing motion during which his plumage is displayed. Receptive hens crouch in a typical solicitation posture. They are then mounted by the cock. Since hens judge their mates primarily by appearance rather than by differences in behavior, experimental interference with the color patternings of cocks (Fig. 10.2) has decreased their effect on solicitation by hens.

Third, in the courtship of guppies, it appears that males are fairly indiscriminate in their choice of mate, while females exercise a fairly rigid choice based upon the male's color markings and his courtship behavior. In experiments in aquariums, Robin Liley (1966) found that males of the species *Poecilia picta* and *Poecilia reticulata* (Fig. 10.3) frequently court and attempt to mate with females of either species. Receptive females would respond to courtship of males of their own species but not of other species. However, if the females had no choice, they eventually would breed with males of the other species.

A female must try to ensure that she mates with a male that is fully mature and sexually competent. By prolonging the courtship she is more likely to mate with males that are fully motivated and sexually vigorous. In the European smooth newt (*Triturus vulgaris*) (Fig. 10.4), the intensity of the male's display is correlated with the number of spermatophores (sperm capsules) he produces in the sexual encounter

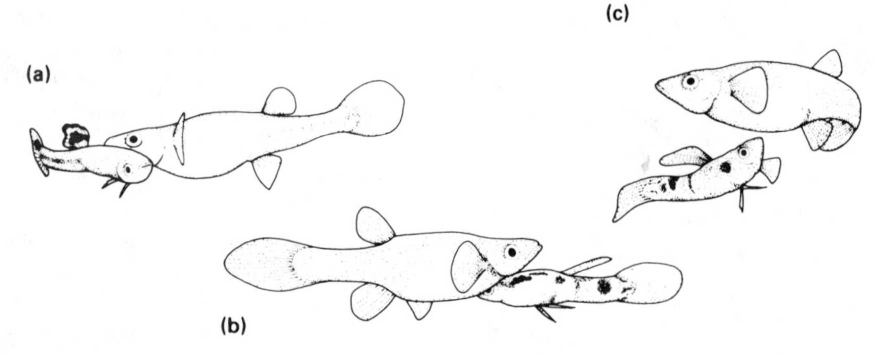

Fig. 10.3 (*a*) the courtship dance of the guppy (*Poecilia picta*). The male circles repeatedly around the head of the female displaying the distinctive markings on his caudal fins. (*b*) The sigmoid display of *Poecilia reticulata*. The male holds a position in front of the female and quivers for several seconds. The males have conspicuous though variable markings on the body. (*c*) A mating attempt by a male *P. reticulata*. After displaying, the male swings round and attempts to insert his gonopodium (a modified anal fin) into the genital pore of the female. Fertilization is internal (From *The Oxford Companion to Animal behaviour*, 1981, after photographs by Robin Liley)

(Halliday, 1976). During courtship, the male deposits the spermatophore on the substratum of the pond; the probability that the female will pick it up successfully increases during the course of encounter (Halliday, 1974). To be able to deposit three spermatophores, the male must remain underwater without breathing for much longer than normal (Halliday, 1977a). Thus, it is possible that in prolonging courtship, the female is providing a test of male fitness.

The assessment of the male by the female is facilitated if the male can indicate that he is a good choice. Some researchers have suggested that courtship feeding may be important in this respect. Courtship feeding, in which the male presents food or foodlike objects to the female, is observed in many species. In the herring gull (*Larus argentatus*) the female begs for food as a juvenile would, and the male regurgitates food, which the female then eats. In most species the quantity of food transferred from male to female is small in relation to the female's normal daily intake. In the pied flycatcher (*Ficedula hypoleuca*), however, it is about half what a nestling of equivalent weight would receive. In some species the courtship feeding is highly ritualized (see Chapter 22.1), and there is no exchange of real food. Courtship feeding influences clutch weight in the common tern (*Sterna hirundo*) and it may be a predictor of the male's future performance in providing food for the young (Nisbet, 1973; 1977). Virginia Niebuhr (1981) found that courtship feeding was a reliable predictor of chick feeding by the male herring gull. It also provides an indication of the male's tendency to incubate and to protect the chicks after hatching. For these reasons courtship feeding would appear to provide a good indication of a male's parental abilities.

Fig. 10.4 A male European smooth newt (*Triturus vulgaris*) (*Photograph: Tim Halliday*)

As in the case of the assessment of fighting ability, it is difficult for the female to find a criterion of male value that is not open to cheating. Her best strategy is simply to prolong or elaborate the courtship, since this makes it more difficult for the male to mate with other females, particularly if the breeding season is short or if the females are scarce; that is, males that invest heavily in precopulatory behavior such as courtship, nest building, or territorial defense will have fewer resources in time and energy to pursue other females. Having already invested so much in one female, a male will be less likely to desert his mate, thus forcing her to care for the offspring alone. Moreover, if the male loses the opportunity to mate with other females, he can improve his fitness only by investing the remainder of his resources in his offspring. This has been called the "domestic bliss situation" by Richard Dawkins (1976).

There is no doubt that domestic bliss can be observed in some monogamous species. However, it is not an evolutionarily stable situation under all circumstances because it is open to invasion by females operating under a different strategy. Dawkins considers a model in which there are two female strategies—*coy* and *fast*, and two male strategies—*faithful* and *philanderer*. Coy females insist on a prolonged courtship before they permit copulation; fast females will copulate without prior courtship. Faithful males are willing to undertake a prolonged courtship, and after mating, they help the female to raise the young. Philanderer males are not willing to enter into prolonged courtship, and if they are not able to mate soon after the initial encounter with a female, they leave her in search of another. Philanderer males may also desert the female after copulation.

We now can examine this situation to see if it is an evolutionarily stable strategy (see Chapter 7). Let the genetic payoff gained by each parent be +15 arbitrary points for each offspring successfully reared. The cost of rearing offspring is −20 points, and the cost of a prolonged courtship is −3 points. In a population in which all the females were coy and all the males faithful, a mated pair would share the cost of rearing each offspring. Each parent would gain +15 points per offspring reared. The average payoff for the first offspring is therefore $+15 - 10 - 3 = +2$. Although in real life the parents may not have to pay the cost of courtship for succeeding offspring, for the present let us assume that they do so. If a fast female now enters the population, she will not have to pay the courtship cost. In mating with a faithful male, her average payoff is $+15 - 10 = +5$, which is better than that of the coy female. Genes for fast female behavior, therefore, would start to spread.

In a population in which the proportion of fast females is high, the way is open for invasion by philanderer males. These are not successful in attempting to mate with coy females, but since they quickly abandon females that will not mate immediately, the philanderers lose little by such encounters. By mating with a fast female, however, the philanderer does well. He gains +15 points per offspring reared, but he pays neither the courtship nor the parental cost. His mate, the fast female, has to pay all −20 of the parental cost. As a result of mating with a philanderer, her average payoff is $+15 - 20 = -5$. This model is stable in that it does not evolve toward an extreme situation, but it is unstable in that it does not settle down at a particular mixture of male and female strategies; instead, it oscillates among alternative mixtures. Although on the basis of evolutionary theory, we might expect to find alternative female strategies represented within a single species, as suggested by the Dawkins (1976) model, these do not appear to be very common. Perhaps the models are too simple, or perhaps they lack some fundamental ingredient about which we do not yet know. There are, however, a few known instances of alternative female strategy.

Among ostriches, some females incubate eggs and some do not. The male ostrich defends a large territory and makes a rudimentary nest on the ground. He mates with a female, who starts to lay eggs in the nest. Within a few days, other females—who may or may not have mated with the resident male—begin to lay their eggs in the same nest. They do not, however, incubate or guard these eggs, but leave them in the care of the male or his mate. Eventually, a nest may contain 30 or 40 eggs, too many for one bird to incubate. Some eggs are pushed to the edge of the nest by the incubating female; they usually die from exposure to the hot sun, or they are taken by predators.

Brian Bertram (1979) discovered that the incubating ostrich female can recognize her own eggs and that she tends to push the eggs of other females out of her nest. She usually incubates all her own eggs and about 50 per cent of the eggs of other females. The incubating female allows other females to lay their eggs in her nest without attempting to discourage them. Apparently, the other females are not closely related to the incubating female, so her tolerant behavior cannot be explained in terms of kin selection. She probably benefits by having her own eggs surrounded by others. Predators that raid the nest, usually jackals and vultures, only take a few eggs because each is so large. The chances of the eggs of the resident female being taken are reduced by the presence of other eggs, particularly since hers are collected toward the middle of the nest. The same argument may apply to the mother ostrich surrounded by her brood of chicks: It may pay her to care for a few chicks of other hens because their presence reduces the chances of her own chicks being taken by a predator.

The alternative strategy for a female ostrich is to have her eggs incubated by another female, even though their chances of survival are low. If a large number of eggs were inserted into many nests, this strategy might pay, but it is evident that it is not the preferred strategy, for the parasite females often have no mate or have lost their clutch to predators.

For reasons discussed in Chapter 9, males usually compete for opportunities to mate with females. Such competition can take many forms, the most obvious being aggressive competition. If a male can prevent other males from gaining access to females, then he should have little difficulty in persuading the females to mate with him, since they have little alternative. In polygynous species such as red deer and elephant seals, dominant males may be able to collect a harem that they guard and defend against other males. Aggressive competition can be a risky business, involving injury or even death, and many other forms of male rivalry are found in the animal kingdom.

In cases in which males gain access to females by directly defending them against rivals, there is a tendency for the social organization of the species to be based upon the harem. This is greatly facilitated if the females have a tendency to be gregarious as a defense against predators, or to exploit feeding opportunities. In elephant seals, harem formation is facilitated by the fact that there is a scarcity of sites where the animals can rest out of water, so the animals tend to congregate in particular places. Antelopes and deer gain protection from predators by forming herds. Harem-owning males are often pushed to the limit in defending their females against rivals, and any tendency for the females to disperse would make life impossible.

In some species there is competition not for access to females but for fertilization. This is sometimes known as *sperm competition*. For example, male dungflies (*Scatophaga stercoraria*) compete for females and may manage to displace one another during copulation. Geoff Parker (1978) showed that when two males mate with the same female, the sperm of the second male fertilize most of the eggs. Parker irradiated males with cobalt^{60}, which prevents eggs fertilized by these sperm from developing. If an irradiated male mates with a particular female after a normal one has, only 20 per cent of the eggs hatch, whereas 80 per cent hatch when a normal male mates after an irradiated male. Somehow, the sperm of the second male displace most of those of the first.

Copulatory behavior in rodents is characterized by multiple intromissions and ejaculations, a pattern that seems to facilitate the transport of sperm within the female. A period of quiescence following ejaculation appears to be necessary for completion of the sperm transport and for successful implantation of the fertilized ova in the uterus (Adler, 1969). If a second male is able to copulate with the female during this period, the transport of the sperm of the first male may be disrupted, and the second male may succeed in fathering the offspring.

In species where sperm competition is possible, it is not surprising to find males taking precautions against it. Many mammals copulate in private, away from possible interference. The male dungfly sits on top of the female after copulation and guards her until the eggs are laid. Some insects cement up the genital opening of the female after they have copulated. Such mating plugs are thought to prevent subsequent matings in some water beetles, butterflies, and moths (Wilson, 1975). Effective mating plugs can be achieved by prolonged copulation. Male houseflies remain in the copulatory position for about an hour, although most of the sperm are transferred during the first 15 minutes. Some moths copulate for a full day. Following copulation in the fly *Johannseniella nitida*, the female eats the male, excepting his genitalia, which remain in place and serve as a mating plug (Wilson, 1975).

If a male is unable to gain access to females in competition with other males, he may be able to steal or sneak copulations. For example, male bullfrogs (*Rana catesbeiana*) compete for territories in ponds where the females come to lay their eggs. The females prefer certain parts of ponds, usually where the water is warm but vegetation not too dense. In such conditions the eggs develop quickly in a tight ball and are relatively safe from attack by leeches (Howard, 1978). The males compete vocally and physically for the best territories. The older, larger males usually win, and small young males may end up with no territory. The younger males adopt the strategy of sitting silently near a calling male and attempt to intercept females that he attracts. Similarly, young elephant seals may attempt to join the harem of a dominant male by behaving like a female. They then sneak copulation when the large bull is preoccupied with a rival (Le Boeuf, 1974). Similarly, young red deer stags may attempt to steal copulations from the harem of a dominant stag.

Alternative male strategies within a species may be practiced by males at different stages of development, as in the examples discussed here. In some species, individual males may use both strategies. In the tree frog (*Hyla cinerea*), for example, there is no difference in body size between the males that call for females and the silent satellite males that wait to intercept the females. About one in seven males is a satellite, but individuals may change strategy from night to night. The satellites have about the same mating success as the callers (Perril et al., 1978), in contrast to the case of bullfrogs, where satellite success is very much lower than that of the dominant callers. If we assume that males of the same size are free to choose between the two strategies, then we can expect the better strategy to depend upon the choice made by other males. Thus, if the majority of males plays the calling strategy, then the satellite strategy will probably pay off handsomely. If too many males play the satellite strategy, however, few females will be attracted to the region, and it will be better to play the calling strategy.

This is a situation that could be analyzed in terms of ESS theory. To do this we would have to know the cost of each strategy, as well as the benefits. The calling strategy might incur a cost due to the attraction of predators, while the cost of being a satellite might be connected with the lack of a territory. Howard (1979) found that calling bullfrogs tended to attract snapping turtles, but more information about the circumstances of tree frogs is needed before a full comparative analysis can be carried out. Similarly, territorial male field crickets (*Cryllus integer*) that call to females by rubbing their forewings together also attract unwelcome visitors (Cade, 1979). These included numerous satellite males that copulated with the incoming females and parasite flies (*Euphasiopteryx ochracea*), whose larvae develop in the body of the cricket, killing it in the process. Cade found that calling males were about five times as likely to be infected by these parasites as satellite males, but that they were also much more likely to obtain females.

Many species contain males that are never likely to become dominant or territorial. They may adopt the alternative strategy permanently. Thus, male ruffs mating at a lek adopt alternative strategies. The light-colored satellite males are tolerated by the territorial males, probably because they attract females to the lek. Each satellite male lurks near a particular territory and steals copulations while the owner is otherwise engaged. The two types of male are genetically distinct, and males do not change from one strategy to the other during their lifetime.

An interesting situation exists among bluegill sunfish (*Lepomis macrochirus*), studied by Mart Gross and Eric Charnov (1980) at Lake Opinicon, Ontario. These fish nest colonially, and each territorial male constructs a nest in which a female lays her eggs. He then fertilizes the eggs and subsequently guards the nest and takes care of the developing fry. These "parental" males have a light body color with a dark yellow-orange breast. Two other types of male can be identified. "Sneaker" males have a light coloration. They remain inconspicuous, close to the bottom of the lake. Just after the female has spawned, they may enter the nest quickly and fertilize some of the eggs. "Satellite" males have a dark body with darker vertical bars. In coloration and movement they mimic the females. They may enter a nest in a leisurely manner and may be aggressive to other fish of equal size. Gross and Charnov discovered that the three types of males become reproductively active at different ages. They established that sneaker males will develop into female mimics and become satellite males. Both types deceptively gain access to unfertilized eggs in the nest. They have no territorial behavior and provide no parental care. Territorial males do not become reproductively active until about seven years of age. They compete with each other for nest sites, and their parental investment is parasitized by the sneaker and satellite males.

Thus, male bullgill sunfish have alternative life histories. A male may mature early and adopt a parasitically dependent mating strategy, or he may delay maturation and become territorial and provide parental care for all the eggs fertilized in the nest, whether or not he is the father. Gross and Charnov (1980) argued that the equilibrium (or ESS) proportion of males adopting each strategy was such that the lifetime fitness of an individual is the same for either life history. On this basis they calculated that the equilibrium proportion of males at age 2 who adopt the sneaker-satellite life history should be equal to the proportion of eggs fertilized in each breeding season by all sneaker-satellite males in the population. They found that 21 per cent of males at age 2 develop into sneakers. The number of spawnings achieved by sneakers and satellites varied between 3 and 30 per cent, depending upon the depth of water. They thus obtained preliminary confirmation of their ESS model.

10.2 Sexual strategy in humans

In Chapter 8 we saw that Darwin (1871) thought that sexual selection was of paramount importance in human evolution but that he probably underestimated the importance of natural selection. While some of the secondary sexual characteristics of humans like differences in hair pattern, are attributable to sexual selection, others are probably due to natural selection, such as differences in strength and body size.

Sexual selection could be an effective evolutionary force in a polygamous human society, but in a monogamous society, few opportunities exist for choosing a new mate. The frequency of strict monogamy among humans is difficult to evaluate, and we must remember that it is reproduction outside marriage, not sexual activity itself, that is of evolutionary importance. Anthropological estimates of the proportion of monogamous societies range from 16 per cent to about 50 per cent. However, these estimates usually reflect the customary or official practice and may not indicate the true situation. It is probable that the extent of strict monogamy is rather low, and in comparison with other primates, this low extent is what one might expect. Only a few mammalian species are monogamous, including foxes, jackals, beavers, five species of New World monkey, and two ape species—the gibbon and the siamang (Passingham, 1982).

However, some features of human social relationships seem to indicate that a degree of monogamy is important. In most societies, the father stays with the mother when his children are young. He often has more than one child by a particular woman, and he usually helps to support her and to care for the children. It is not common for the father to desert his offspring as in most polygamous species. A similar situation exists in two other primate species. In the hamadryas baboon (*Papio hamadryas*) of Ethiopia, adult males have small harems that they guard from the advances of other males. They remain with their females for life and do not frequently change the members of the harem as do most other polygamous species. Young males acquire females by kidnapping infants and looking after them in a parental manner. These females are thought to remain with their captor and become members of his harem (Kummer, 1968). Mountain gorillas (*Gorilla gorilla*) live in groups in which fully adult males consort with a number of females. Particular females consort with the same male over many years (Harcourt, 1979).

The reproductive strategy characteristic of humans involves high investment in each of very few offspring. Infants are born one, occasionally two, at a time, at intervals of about two years. Reproductive maturity occurs late, and a woman will give birth to few offspring during her lifetime. These offspring must receive a considerable amount of parental care if the strategy is to be successful. It is difficult for the human mother to raise her children without help. This is a result of the helplessness characteristic of the human infant, compared with those of other primates (Passingham, 1982). Chimpanzee mothers, for example, nurse their infants for several years but retain sufficient freedom to manage unaided. The infant clings to its mother's fur, leaving her free to forage and to keep up with other members of the group. Chimpanzees are polygamous and the males have no parental role. In contrast, the human mother must hold her infant in her arms because it is unable to support itself. Even when able to walk, the child cannot keep up with other members of the group.

Studies of the Kalahari !kung bushmen living hunter-gatherer existence show that the women are severely encumbered by their children (Lee, 1972) (Fig. 9.1). The women provide about two-thirds of the calorie income by foraging for plant foods. Mongongo nuts, their most important plant food source, are in plentiful supply in the dry season but usually are situated some six miles from suitable campsites. The women make excursions to gather these nuts every few days, taking their children with them. The men take no part in food gathering but confine themselves to hunting. Lee shows that the weight carried by mothers on these excursions increases with the frequency of having babies, not only because there are more mouths to feed but also because the small children have to be carried. Nick Blurton Jones and Richard Sibly (1978) show that average birth spacing of four years is the optimum under the available conditions. Thus, the women maximize their reproductive success by spacing births widely and by foraging seldom. Bushman women have children more frequently when they do not have to carry them on foraging expeditions.

The helplessness of the human infant is due largely to its immature brain. The human brain is four times as large as would be expected for a primate of equivalent stature. The size of the brain at birth is very large compared to the size of the body (Fig. 10.5), but it is nevertheless only partly functional. It takes a baby twice as long as a gorilla or chimpanzee to reach the stage where it can support its body with its limbs. Although the newborn baby has a strong grasp reflex and can even support its own weight (McGraw, 1945), this ability soon disappears. The infant ape can hang onto its mother's hair, using hands and feet, but the human infant could not do this even if sufficiently strong, for the feet are the wrong shape for grasping, and the mother has too little hair.

Under the circumstances, we might expect a woman to exercise caution in choosing a mate, to try to ensure that he will be a good father. However, although female coyness is a feature of human courtship, the young woman does not always have much say in her choice of mate. In

Fig. 10.5 Growth of the brain in man and chimpanzee (After Passingham, 1975)

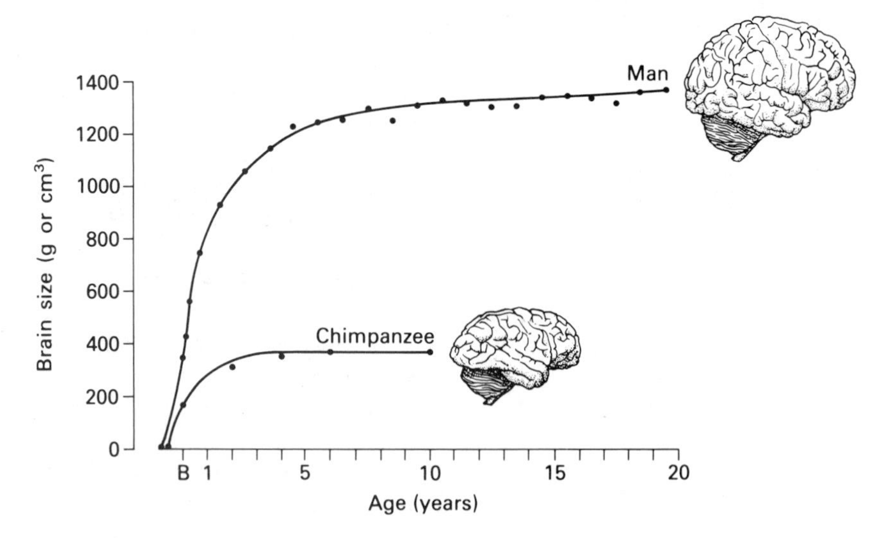

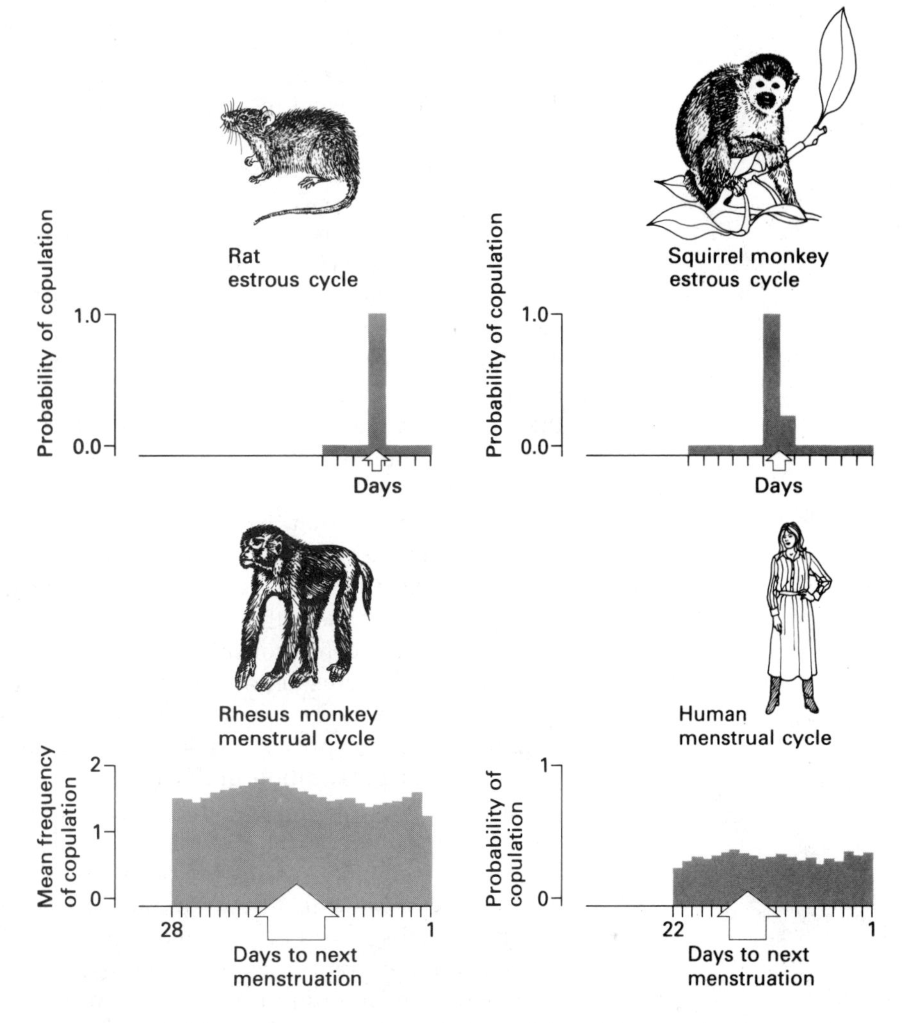

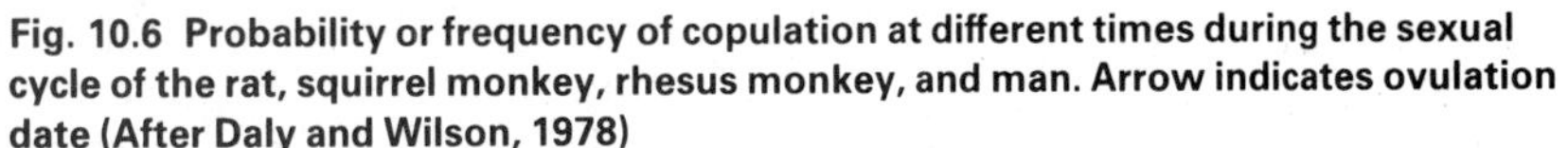

Fig. 10.6 Probability or frequency of copulation at different times during the sexual cycle of the rat, squirrel monkey, rhesus monkey, and man. Arrow indicates ovulation date (After Daly and Wilson, 1978)

many societies matters are arranged by a woman's parents or choice is limited by a shortage of men of appropriate social status. The method of mate selection varies widely from one human society to another but the freedom of choice of modern Western women is probably a relatively recent phenomenon.

The human female has evolved other mechanisms for encouraging male fidelity. In common with some other primates, but unlike other mammals, humans have a menstrual cycle, rather than an estrous cycle of sexual receptivity. The menstrual cycle, characterized by periodic bleeding due to the sloughing off of the uterine lining, occurs in all species of ape and some kinds of monkey. Animals with a menstrual cycle are sexually receptive most of the time, in contrast to those with an estrous cycle, which are receptive only at the time of ovulation (Fig. 10.6). Continual sexual receptivity helps to maintain the interest of the male partner. Moreover, the time of ovulation is concealed in women, unlike most mammals, in which it is advertised. This means that a man must copulate regularly with the same woman to ensure fertilization (Lovejoy, 1981). He must guard the woman against advances from other men if he is to be sure that he is the father of her offspring. The low and uncertain chance that a particular mating will lead to fertilization not only encourages continual male attentiveness but also reduces the benefits of opportunistic copulation with other women. The man has to run the risk of an aggressive encounter with another man for a payoff that is uncertain and unlikely (Halliday, 1980).

Although men have a vested interest in the welfare of their children, they might seem to have something to gain from promiscuity, especially if there is a chance that their children by other women will be cared for by other people. Many human societies attach considerable importance to the establishment of paternity. A family man, who has parental duties, has more to lose than a woman does if she has a child by another man. Her genetic relationship to the child is not in doubt, so her maternal care benefits her own genes. A man has no such certainty, and he usually guards his woman from other men. In some societies, virginity is highly prized, especially in brides. In others, adultery is punished and prostitution outlawed. In some societies, where female promiscuity is commonplace, the inheritance of property is through the sister of the man at the head of the household and not through the wife. The children of a man's sister are certain to be genetically related to him, while some of his wife's children may be unrelated (Halliday, 1980).

Male sexual strategy in humans involves rivalry in acquiring women and a certain amount of aggressiveness in protecting them from other men. This may account in part for the man's greater size and strength, though it is more likely that this has to do with division of labor within the family (Passingham, 1982). To raise his children successfully, the man usually is obliged to assist in their care, and it is in his interests to establish their paternity. The alternative male strategy is to exploit sexual opportunities without providing parental care for any ensuing children. In stable societies in which the women are protected, the opportunities may be few; but it is evident that such opportunities are often taken during warfare or other social disturbances.

It is not possible to give an entirely satisfactory account of human sexual strategy from a purely evolutionary viewpoint. Humans have tremendous capacity for innovation, many of whose products become culturally established. Consequently, we see large variation in sexual practice among different human societies.

10.3 Mating systems and social structure

A successful sexual strategy results in a particular kind of mating system, which has a profound effect upon the social organization of the species. Thus, a male may maximize his fitness by mating with many females and fathering a large number of offspring. Such *polygyny* is a good strategy provided the offspring survive to reproductive age. Their survival depends upon the ecological circumstances and upon the degree of parental investment by both male and female. A successful polygynous male has little time to care for his offspring. Either they must be self-sufficient or the female must be able to care for them without his aid. Polygyny will usually result in a few successful males and a large number of unsuccessful males. The harem may then become the basic feature of the social structure, as in many antelopes and deer. Males compete for possession of females and young males have to wait for a chance to steal copulations or to challenge a dominant male.

If we compare birds and mammals, we find that polygyny is much more common among mammals. Female mammals are specialized for parenthood, providing milk and elaborate maternal care. Male mammals are less well equipped, though they can help by providing food and guarding against predators. Male birds, in contrast, are just as well equipped to look after their progeny as female birds. Monogamy, in which each adult mates with only one member of the opposite sex, is often a necessary strategy when both parents are required to raise the young. More than 90 per cent of the bird species of the world are monogamous. In some the monogamy is perennial, in that a mated pair remains together for life, either on a seasonal or a continuous basis. Many migrant birds are monogamous during the breeding season but live separately for the rest of the year. Many gulls have a strong tendency to return to their exact nesting locality, where they join up with their partner each season. The pair remains together for life, unless a divorce takes place as a result of the failure of one partner to fulfill its duties (Hart, 1964). In swans and geese, a close association between partners may be maintained even outside the breeding season.

Polygyny, though not common in birds, occurs more frequently in birds, like pheasants, with precocial young, which require relatively little parental care. Altricial young have to be fed by the parents, and it is usually necessary for both parents to participate in rearing the brood successfully. *Polyandry*, in which one female mates with two or more males but in which males mate only with one female, has been described in the American jacana (*Jacana spinosa*) (Jenni and Collier, 1972). The female is more conspicuous, territorial, and dominant than the male. After courtship, the female lays eggs that are incubated solely by the male. The female then attempts to find other males to incubate further clutches of eggs.

In general, the formation of pair bonds is associated with parental

Fig. 10.7 Marmoset father carrying twins (*Photograph by Ron Garrison © Zoological Society of San Diego*)

Fig. 10.8 Pair of titi monkeys (*Photograph by Ron Garrison © Zoological Society of San Diego*)

care. In promiscuous species, in which both males and females mate with more than one member of the opposite sex, there are no pair bonds and parental care is minimal. In other species, the mating system is related partly to the ecological circumstances and partly to the needs of the young. This is well illustrated by comparing species of primate.

Although monogamy is not common among mammals, it does occur in 14 or more primate species (Clutton-Brock and Harvey, 1977). In most primates the males show little parental care, though they defend the family group in time of danger. In the monogamous species a much greater degree of male parental care is evident (Passingham, 1982). For example, marmosets (*Callithrix*) and tamarins (*Leontideus*) usually produce twins. The male carries these about (Fig. 10.7) and returns them to the mother only for nursing. Similarly, male titi monkeys (Fig. 10.8) spend more time with the infants than does the mother. In humans, the helpless baby is a burden on the mother because it cannot cling to her of its own accord. A toddler cannot keep up an adult walking pace and has to be transported by the mother. In primitive societies, this degree of dependence can restrict the frequency with which a woman has children because of the work required to provision the family, as discussed above.

10.4 Social organization of primates

The primates have a wide range of types of social organization, which reflect their diversity of life style and ecology. Many prosimian species, including lemurs, loris, and tarsiers, have a widely dispersed food supply. They are usually insectivorous, arboreal, and nocturnal. They have a solitary way of life and have social contact only during courtship and mating.

Monogamous species such as the white-handed gibbon (*Hylobates lar*) defend territories, like many monogamous birds. The young receive considerable care and remain with the parents until sexually mature.

The social organization of gelada baboons (*Theropithacus gelada*) is very similar to that of certain antelopes and appears to be based upon existence in a habitat subject to marked seasonal variation in food availability. Among impala (*Aepycerus melampus*), for example, there are bachelor herds made up of single males and harem herds made up of females under the control of a single male. Gelada baboons inhabit the mountain grassland of Ethiopia. During the dry season there are distinct types of herds, the harem herds, the all-male groups, and groups of juveniles. The all-male groups are the most widely dispersed, which has the effect of reducing competition for food between the groups. During the rainy season, when food is more plentiful, the groups aggregate in large herds within which the harems remain intact.

Primates with a more stable food supply tend to form multimale groups. For example, the mountain gorilla (*Gorilla gorilla beringei*) has a small home range and feeds on the leaves and stems of forest plants. These animals live in small groups, each led by a silver-backed male, but often including other males. They do not defend territories and may overlap with other groups.

The savannah-living baboons, such as the olive baboon (*Papio anubis*), usually live in troops made up of a number of males and females. The dominant males have privileged access to females, food, and water. They cooperate in maintaining their social position against rivals, and in protecting mothers and infants from interference by other members of the troop. The dominant males also cooperate in defense against predators. Rich habitats tend to attract large predators and male–male cooperation is a considerable advantage in such circumstances.

Comparative study (e.g., Brown, 1975) suggests that in harsh environments, baboon troops are dominated by a single male, whereas in rich

Fig. 10.9 Distribution of baboons in Africa. The male hamadryas baboon (*a*) has a large cape and many females per dominant male. The anubis baboon (*b*) is intermediate, and the male yellow baboon (*c*) has a small cape and few females per male (After Dorst and Dandelot, 1969)

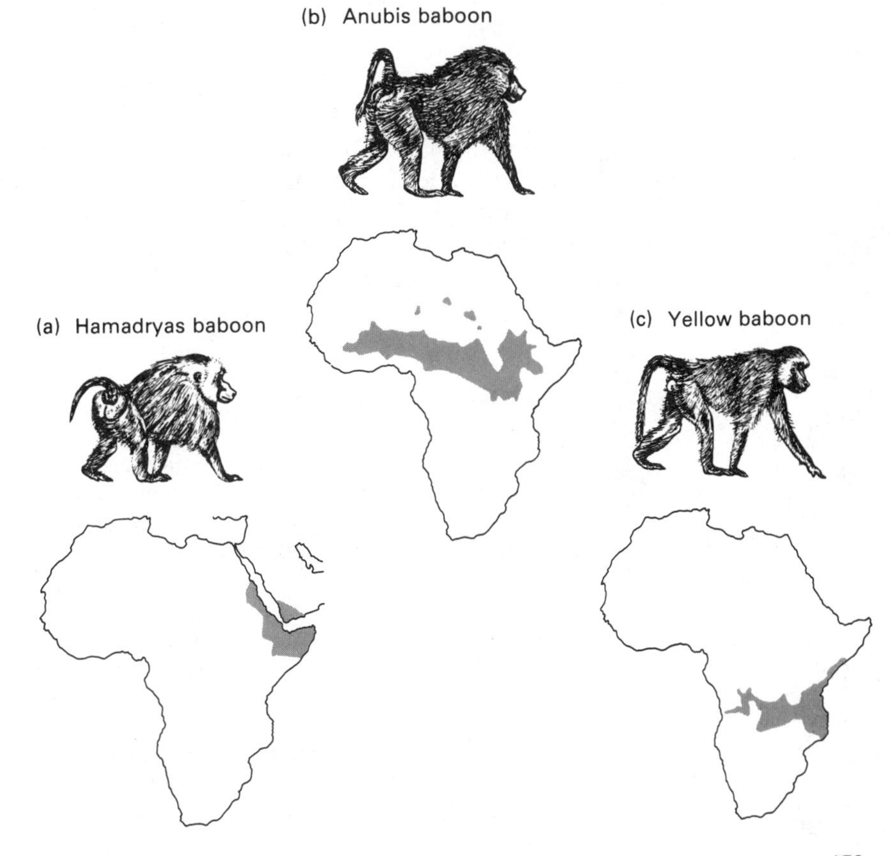

environments the troops have a number of cooperating dominant males. The hamadryas baboon (*Papio hamadryas*) of Ethiopia is typical of the former, while the yellow baboon (*P. cynocephalus*) represents the opposite extreme. The anubis (*P. anubis*) and chackma (*P. ursinus*) baboons are intermediate in these respects (see Fig. 10.9). The hamadryas baboon is subject to considerable sexual selection and the dominant males are distinguished by their large cape. Male rivalry is much less intense among the yellow baboons and there is much less sexual dimorphism.

Cooperation is a distinctive feature of human behavior. Early man probably hunted cooperatively and shared food (Isaac, 1978). Like other primates, humans have division of labor between the sexes. In chimpanzee societies, hunting is a male activity, while females collect termites much more frequently than males (McGrew et al., 1979). In the present-day human hunter-gatherer societies, the men do most of the hunting, while the women gather plant food (Lee and DeVore, 1968). Human babies are particularly helpless, so the women are often constricted in what they can do. Man is not the only species with this type of social organization. Female wild dogs (*Lycaon pictus*) remain with their pups in the den and are dependent upon the cooperative hunting of the males and upon the meat that is brought home to be shared.

Humans are unique among mammals in having division of labor within the sexes (Passingham, 1982). Different individuals play different roles, thus greatly increasing the effectiveness of cooperation. For example, Mbuti pygmies cooperate in hunting elephants. One man stops the elephant from the front, while another slices through one leg tendon, and then another, from behind (Turnbull, 1966). Such hunters communicate by gestures so as to remain unnoticed by the prey. Similarly, primitive man could have achieved complex cooperation, even without the use of language.

Points to remember

- The main evolutionary problem for males is to attract females in the face of competition from other males. The main evolutionary problem for females is to attract sexual partners that will endow their offspring with the greatest chance of survival and reproduction.

- Courtship involves a contest between male salesmanship and female sales resistance. Coyness helps the female to assess the suitability of the male as a potential mate and may also encourage the male to invest in the future offspring.

- Sexual strategies adopted by males may include aggressive competition, sperm competition and sneaked copulations. Different strategies may be employed at different stages of the life cycle.

- Sexual strategy in humans is closely related to the high degree of parental care. Female strategies include long periods of sexual receptivity which encourages continued attentiveness from the partner. Male strategies include rivalry in courtship and aggressiveness in protecting women. As a result of cultural innovation there is considerable variation in sexual practice among different human societies.

- Sexual strategy results in particular kinds of mating system which have profound effects upon the social organization of the species. A variety of examples can be found among the primates.

Further reading

Halliday, T. R. (1980) *Sexual Strategy*, Oxford University Press, Oxford.

PART 2 *Mechanisms of Behavior*

In the second part of this book we look at animals as machines. We focus our attention on the individual animal and ask how it adjusts to changes in the environment. The type of explanation we are seeking is a mechanistic one. How are we to explain, in terms of immediate causes, changes in the animal's behavior?

Ideally, we might wish to describe behavioral mechanisms in terms of the physiological hardware responsible—the events in the brain and other parts of the nervous system. However, our knowledge of physiology is not yet sufficiently developed to enable us to do this. Even if it were, physiological detail may not be the most appropriate vehicle for explaining complex behavior. Accordingly, we look at alternative ways of describing behavioral mechanisms: the physiological approach, the more abstract control-systems approach and the psychological approach. All three have their advocates, but they should not be regarded as rivals, because they complement each other, being suited to different aspects of behavior.

2.1 Animal Perception

In this group of three chapters we deal with the nervous and sensory systems of animals. Chapter 11 deals with the basic elements of the nervous, sensory and muscular systems. The nervous systems of invertebrates and vertebrates are then reviewed. The aim is to give a general picture of the physiological apparatus that is characteristic of species throughout the animal kingdom.

Chapter 12 discusses the major sensory systems, including chemoreception, hearing and vision. Three different ways of looking at perception are also discussed. Chapter 13 is concerned with ecological aspects of sensory systems, particularly the special sensory adaptations possessed by animals that operate in unfavorable environments.

Johannes Müller (1801–1858)

Johannes Peter Müller, the son of a poor shoemaker, introduced experimental physiology into Germany and is largely responsible for our picture of the body as a machine. Müller had many famous pupils, including du Bois Reymond, Helmholtz, Henle, Koelliker, Remak, Reichert, and Virchow. He was involved in many pioneering aspects of biology, especially sensory physiology and marine zoology. Müller published the monumental *Handbuch der Physiologie des Menschen für Vorlesungen*, which was translated into English by W. Baly and published in 1827 as *Elements of Physiology*.
Müller's most important contribution to the study of behavior is his doctrine of "specific nerve energies". Previously it was believed that each different environmental event or stimulus affected sensory nerves in a manner specific to that stimulus. Müller discovered that a given nerve always produces the same type of sensation regardless of its mode of stimulation. Thus, light falling upon the eye produces visual sensation, but so does mechanical stimulation such as a blow to the eye, and so does electrical stimulation of the optic nerve. If our ear could be attached to the optic nerve, auditory stimulation then would give rise to visual sensation. Thus, it is not the sense organ that is important, but the nerves that relay sensory messages to the parts of the brain that receive visual, auditory, tactile, and olfactory information and translate this into the relevant sensations. The sense organs are responsible for transforming the various forms of environmental stimulation, such as light, heat, and mechanical energy, into an electrical potential that can be registered by the receptor cells, that is, the nerve cells that are part of the sense organ and that connect to other cells of the nervous system.
The doctrine of specific nerve energies provides a key organizing principle for sensory physiology. It banishes all previous speculations as to the role of sense organs in perception, making it clear that these are primarily transducers of one form of energy to another. Although a certain amount of stimulus filtering may take place at the periphery, it is the brain that sorts and categorizes incoming information, purely on the basis of the intensity of stimulation of the peripheral sensory nerves. As it is now known, partly through the work of Müller's pupils, du Bois Reymond and Helmholtz, the nerve impulses that travel along a particular nerve axon have a measurable size and speed of propagation that are characteristic of that nerve cell, which means that the nerve cell can transmit messages about the intensity of stimulation only. Other information about the nature of the stimulus can be gained by the brain only by integrating the information from many nerve cells.

BBC Hulton Picture Library

11 Neural Control of Behavior

In this chapter we discuss the neural control of behavior, starting with an outline of general principles. We then review the types of nervous system that occur throughout the animal kingdom, and see how the gross organization of the nervous system is related to behavior.

11.1 Nerve cells

Nervous systems are made up of nerve cells, called *neurons*, which are specialized for transmitting information to one another. Each neuron has a cell body, containing the nucleus, and a number of branching protrusions, as shown in Figure 11.1. Usually, there are many short branches, called *dendrites*, and a single long protrusion, called the *axon*. The dendrites connect with other nearby neurons, while the axons convey messages over relatively long distances.

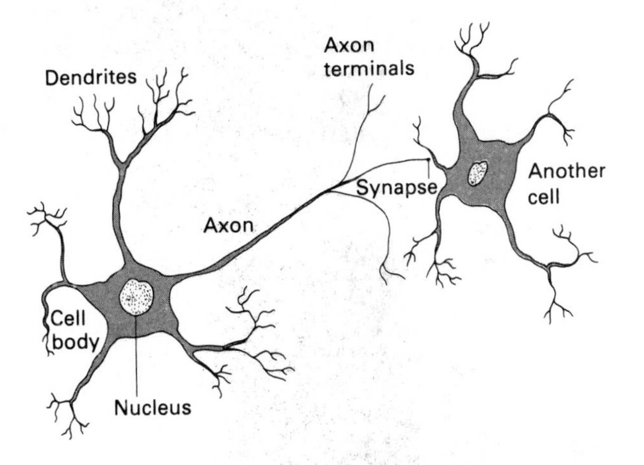

Fig. 11.1 One nerve cell forming a synaptic connection with another

The membrane of a neuron is usually polarized. That is, there is an electrical potential across it—called *resting potential* in the inactive neuron—which provides a stable state of readiness similar to that of an electrical battery that stores energy releasable upon demand. The resting

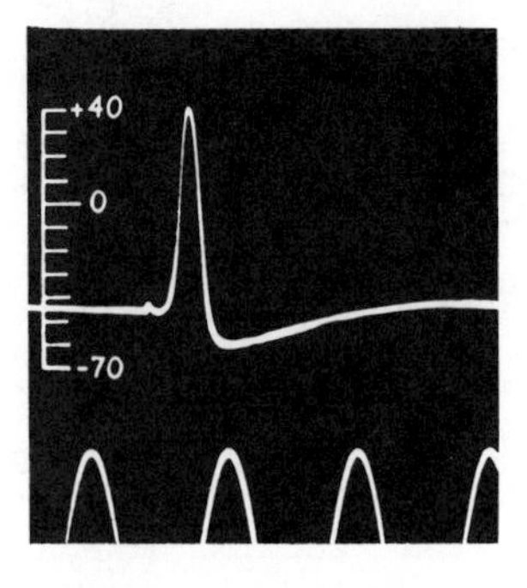

Fig. 11.2 Action potential spike recorded from the giant axon of a squid (From Hodgkin and Huxley, 1945)

potential results from unequal concentrations of K^+ ions on the inside and outside of the cell. When the cell is at rest the interior charge is negative relative to the outside. When the cell is depolarized its membrane potential is reduced towards zero, as shown in Figure 11.2. When the membrane potential is made more negative, the cell is said to be "hyperpolarized."

If the resting potential is depolarized beyond a certain threshold, then an *action potential* is propagated along the membrane. Action potential is a transient potential caused by systematic changes in the relative proportions of Na^+ and K^+ ions on each side of the membrane (see Fig. 11.3). The action potential passes down the axon as an electrical wave. It is always of the same amplitude (height), usually a function of the diameter of the axon. Larger axons propagate larger action potentials than smaller axons, and at a higher speed.

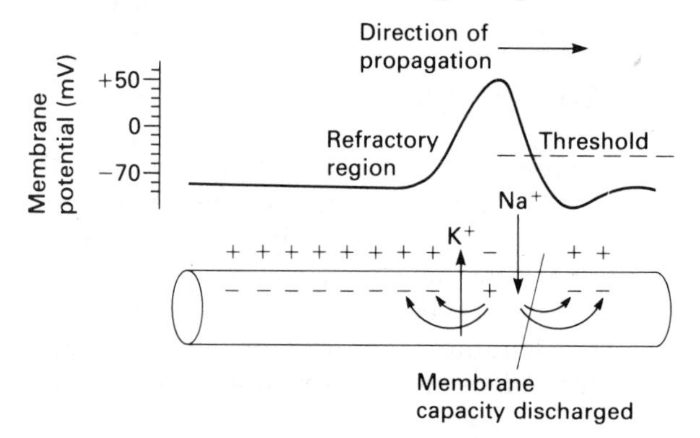

Fig. 11.3 Propagation of the action potential as a result of passage of K^+ and Na^+ ions through the axon membrane

After each action potential has passed along there is a refractory period during which the membrane recovers its normal ionic equilibrium and its normal resting potential. Since no other action potential can be generated during the refractory period; the refractory properties of the axon determine the maximum rate at which action potentials can be generated.

When an action potential occurs the neuron is said to "fire." The action potential often appears as a spike on an oscilloscope rigged up to measure membrane potentials by means of electrodes inserted into nervous tissue. The neuron fires in an all-or-none manner (there is either a full spike or there is none) at a maximum frequency determined by its refractory properties. The frequency of firing depends upon the intensity of stimulation of the neuron. Thus, the neuronal message is coded in terms of frequency (see Fig. 11.4).

The membranes of axons and dendrites do not make physical connections with other neurons, but come very close at junctions called *synapses*. Usually, very small quantities of chemical *neurotransmitters* are

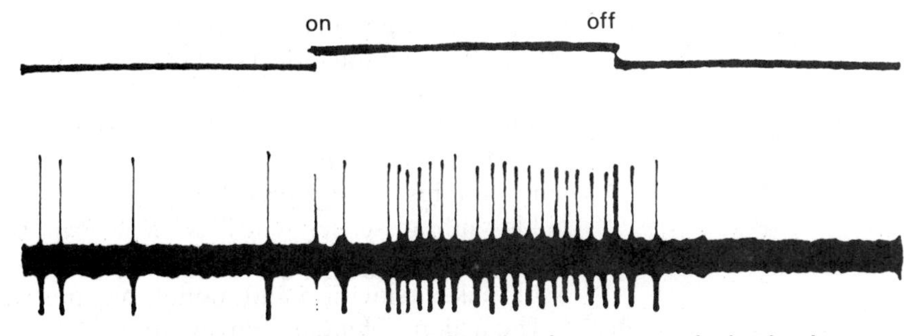

Fig. 11.4 Extracellular recording of nerve impulses from a neuron in the visual system of a cat. Note the increase in frequency when the stimulus is turned on (From Guthrie 1980)

released at the synapse, which influence the resting potential of the recipient membrane and hence the readiness of the receiving neuron to generate action potentials.

Neurons can be stimulated by other neurons, by injury, or by sensory receptors. The principles are the same in each case. The stimulation causes a change in the membrane potential and when this reaches a threshold an action potential is triggered. We now look at how this process is brought about in sensory receptors.

11.2 Sensory receptors

Sensory receptors are specialized nerve cells responsible for the transduction and transmission of neural information. Like ordinary nerve cells, sensory receptors have a cell body, dendrites, and one or more axons. Receptors are specialized according to the type of environmental energy to which they react. For example, *photoreceptors* contain pigments that are chemically modified by light and that give rise to an electrical potential when so stimulated. *Mechanoreceptors* undergo electrochemical changes as a consequence of deformation of the cell membrane. The transduction of energy usually takes place in the cell body, and a characteristic of all receptor cells is that the environmental energy is converted into a graded electrical potential, called *generator potential*, usually proportional to the intensity of stimulation of the receptor. When the generator potential reaches a certain threshold level, it triggers an action potential that travels along the axon of the receptor cell. This is the transmission part of the sensory process, and the information is usually coded such that the more intense the stimulus, the higher the frequency of the action potentials.

In the absence of stimulation, the generator potential falls gradually to its resting level. When it falls below the threshold level, no further action potentials are generated. When stimulation is resumed, there may be a short delay (the refractory period) while the generator potential rises from its resting level to the threshold level. When stimulation is intermittent, the generator potential will rise and fall rhythmically, giving rise to bursts of action potentials. If, however, the intermittent stimulation is of high frequency, the generator potential may not have time to fall during the gaps in stimulation, and the production of action potentials will be continuous. This explains why we are not able to distinguish between a constant stimulus and an intermittent stimulus of very high frequency. This flicker-fusion phenomenon holds with all senses, but it is most obvious in vision. The fact that a rapidly flickering light produces the same visual sensation as a constant light makes television and motion pictures possible.

The action potentials conveying sensory information are no different than any other nerve impulses. Their magnitude is determined by the size of the neuronal axon and their frequency by the intensity of stimulation. Each type of receptor sends impulses, either directly or indirectly, to a particular part of the brain. The sensations experienced depend not upon the type of receptor or the messages they send but upon the part of the brain that receives the message. The brain is also responsible for the localization of the sensation. In the case of pain, for example, the nerve fibers from the hand go to one part of the brain, those from the arm go to another, and so on. The "pain" experienced by the brain is referred to that part of the body from which the message came. An illustration of this phenomenon comes from reports of people who have had limbs amputated and who complain of pain sensations that seem to come from the missing or phantom limb. Irritation of the cut nerve endings sends impulses to those parts of the brain that were concerned with the amputated limb. The messages are interpreted by the brain as coming from the lost limb, and the sensations experienced depend upon which nerve is irritated. Sensations of heat, cold, or touch also may appear to come from the phantom limb.

11.3 Muscles and glands

The nervous system is responsible for controlling behavior and, to a certain extent, for controlling the animal's internal environment (see Chapter 15). This control is exercised by commands to muscles and glands.

Muscles are made up of complex protein molecules that are capable of contraction and relaxation. Nerve endings connect with muscles by means of synapses similar to those by which neurons connect with each other. The nerve impulses arriving at a neuromuscular junction set up electrical potentials that cause the muscle to contract. Muscular relaxation results from lack of stimulation. When the muscle contracts it becomes shorter, provided it is not prevented from doing so by being fixed at each end. A muscle can lengthen when it is relaxed, but only if it is stretched by the action of other muscles or by some extraneous force. Thus, muscles are usually arranged in opposing groups that act against each other. In some invertebrates, such as annelid worms, muscular contraction may be resisted by the hydrostatic pressure caused when the muscles squeeze part of the body cavity. This pressure causes the muscles to lengthen when they relax. In other invertebrates, such as arthropods, the muscles are housed inside a hard exoskeleton, which provides the necessary leverage for opposing sets of muscles (Fig. 11.5). In vertebrates the internal skeleton provides the leverage and the muscles are arranged so they pull against each other (Fig. 11.6). One set of muscles relaxes when the other contracts.

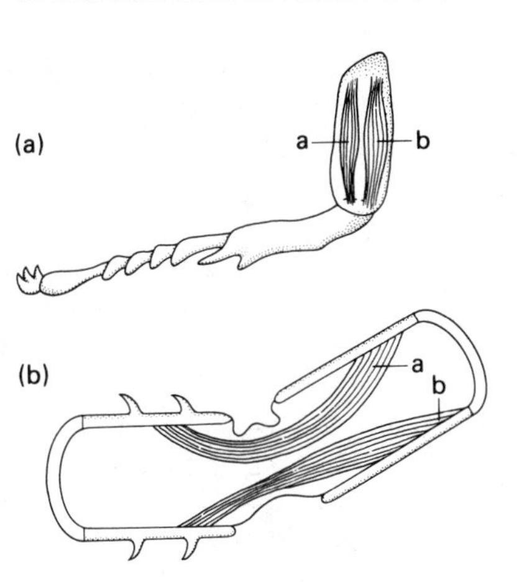

Fig. 11.5 Mechanical arrangement of muscle and skeleton in an insect's leg. The muscles are housed inside the skeleton, and in (*a*) muscle a is the flexor (bending the limb) and muscle b is the extensor (straightening the limb). In (*b*), where the muscles span the joint, the arrangement is the opposite. Muscle a is the extensor, and muscle b the flexor.

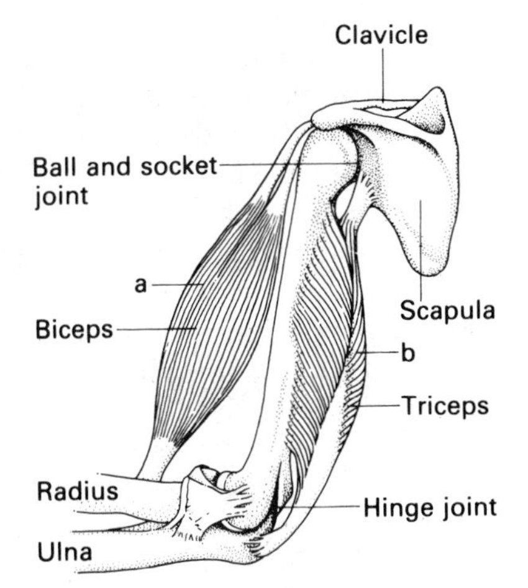

Fig. 11.6 Mechanical arrangement of muscle and skeleton in a human arm. The muscles are arranged outside the skeleton. Muscle a is the flexor and muscle b the extensor

Some glandular secretions are under nervous control. In vertebrates these include the salivary glands, which produce saliva, the adrenal medulla, which produces adrenalin, and the posterior pituitary gland, which produces several important hormones. The secretions of these glands can influence behavior indirectly through their influence upon the animal's internal state, as we see at the end of this chapter.

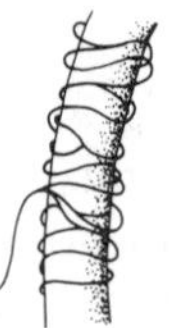

Nerve plexus around hair (touch)

End bulb of Krause (cold)

Ruffini ending (warmth)

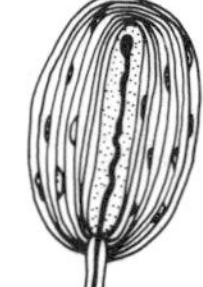

Pacinian corpuscle (deep pressure)

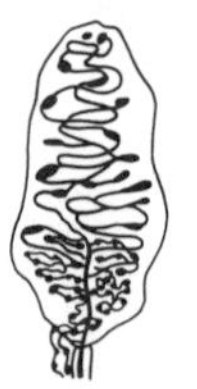

Meissner's corpuscle (touch)

Free nerve ending (pain)

Fig. 11.7 Some sensory receptors found in the skin and the senses with which they are associated (After Keeton, 1972)

11.4 The somesthetic system

It is important for an animal's brain to receive information about the state of the body. The positions of limbs, pressure on internal organs, the temperature of various parts of the body, and many other features are monitored by the *central nervous system* (*CNS*) via interoreceptors located at strategic points. This system, responsible for bodily sensations, is called the *somesthetic system*.

Numerous types of sensory receptors exist in the skin, skeletal muscles, and viscera of vertebrates. Some examples are shown in Figure 11.7. Invertebrates also have a wide range of receptors. Humans have five types of skin receptors, giving rise to sensations of touch, pressure, cold, heat, and pain. The pain receptors are numerous, being 27 times as common as cold receptors and 270 times as common as heat receptors. Some skin receptors show rapid sensory adaptation. In response to a step change in stimulation, the frequency of nerve impulses rises rapidly and then declines to its resting level. This means that the receptor is a good indicator of changes in intensity of stimulation but a poor indicator of the absolute level of intensity. This is an advantage in cases where the skin receptors give early warning of environmental changes that are likely to affect the body, like changes in temperature.

Receptors deep inside the body serve a wide variety of functions, including detection of changes in blood pressure, the tension of muscles, the amount of salt in the blood, etc. We are not directly aware of the information produced by the majority of interoreceptors. They do not give rise to sensations. Sometimes their effects combine to produce sensations of hunger, thirst, or nausea, but these are due to complex processes in the brain, which do not always refer the sensation to particular parts of the body. This is presumably because the action that has to be taken in response to hunger and thirst is much more indirect than action that is taken on response to peripheral touch or temperature change.

The orientation of animals in relation to gravity and external stimuli like light depends partly upon information about the spatial relationship of the various parts of the body. In mammals this information comes from the vestibular system and from receptors in the joints, muscles, and tendons. The joint receptors provide information about the angular position of each joint (Howard and Templeton, 1966). In the tendons of mammals are the Golgi tendon organ receptors, which are sensitive to tension. They send messages to the spinal cord and are involved in a simple reflex that acts to oppose increases in muscle tension.

Within the muscles are the muscle spindles, which are sensitive to changes in muscle length (Fig. 11.8). The muscle spindles consist of modified muscle fibers that have a spiral nerve ending, called the *primary* (or *annulospiral*) *ending*, wrapped around its middle. When the muscle increases in length, the muscle spindle is stretched and the primary endings send fast messages to the spinal cord. There may also be *secondary flower-spray endings*, which send slower messages. Many mammalian spindles have both primary and secondary endings, while others have only primary endings (Prosser, 1973). These spindles form part of a simple reflex that acts to oppose increases in muscle length.

The muscle spindles are contained within a *fusiform connective tissue*, and their muscle fibers are called *intrafusal fibers.* Ordinary muscle fibers are called *extrafusal fibers.* These are innervated by *alpha motor neurons,*

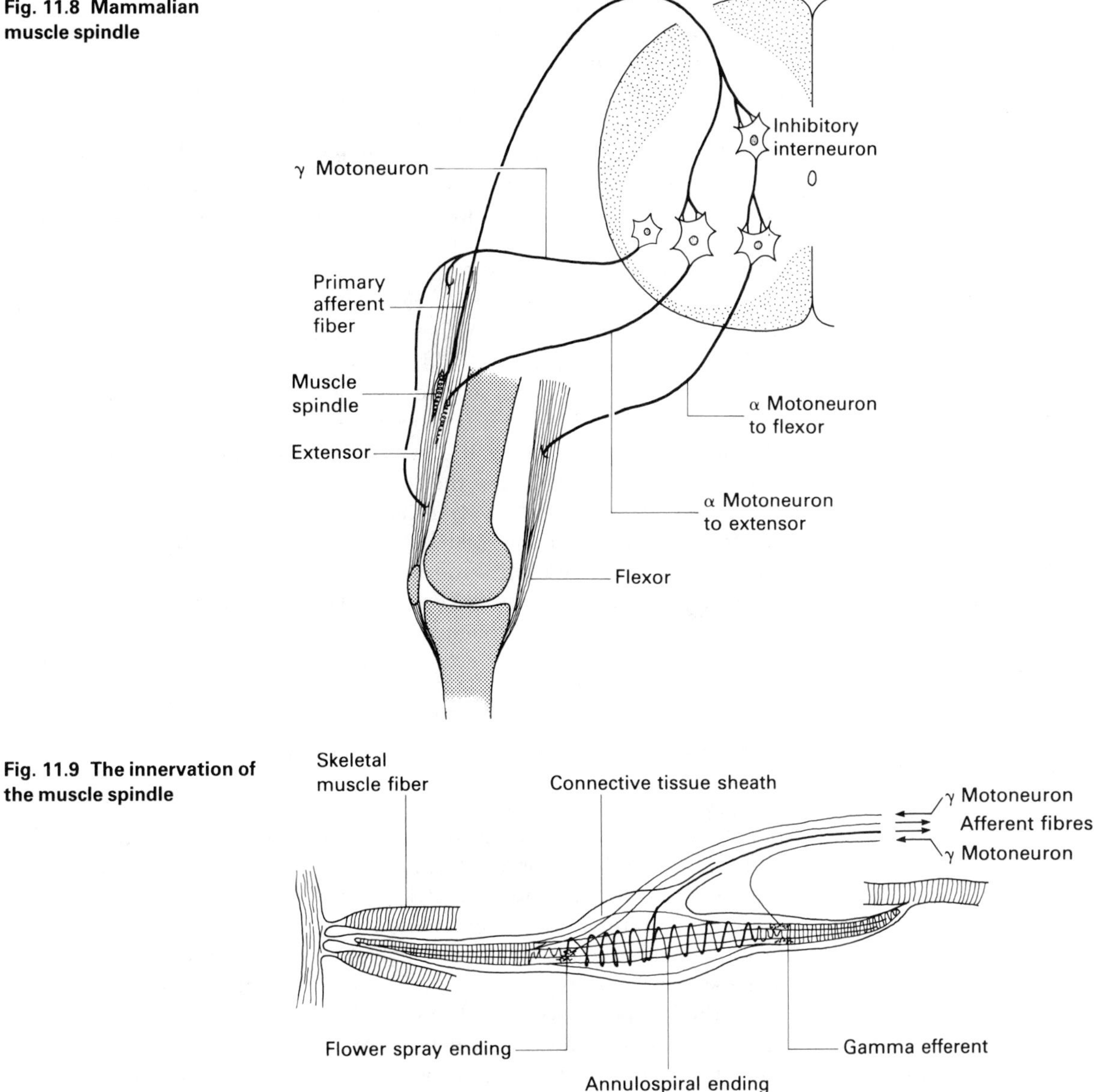

Fig. 11.8 Mammalian muscle spindle

Fig. 11.9 The innervation of the muscle spindle

which have their cell bodies in the spinal cord. In mammals the intrafusal fibers are innervated by smaller *gamma motor neurons*, which keep the spindle in a tonic state of activity so that less muscle stretch is required to activate the spindle. Since muscle spindles are arranged in parallel with the extrafusal muscle fibers, they tend to become slack when the muscle contracts. The gamma motor neurons can command the intrafusal fibers to take up the slack so that the spindle remains in a state of readiness (Fig. 11.9).

The muscle spindles of birds resemble those of mammals, with the intrafusal and extrafusal fibers in parallel. In the lizard *Tiliqua*, however, the muscle spindles appear to be in series with the extrafusal fibers (Prosser, 1973). In reptiles and amphibians there is no gamma motor neuron system, and both intrafusal and extrafusal fibers are supplied by alpha motor neurons. Fish have no muscle spindles, but they do have receptors in their fibers that are sensitive to their angular velocity.

Arthropods have numerous kinds of stretch receptors, of which there are two general types: (1) those that lie between exoskeletal elements and that respond to vibrations in the cuticle and (2) those that are attached to tendons and that can signal stretch and pressure changes (Prosser, 1973). Thus, crabs have receptors that signal both position and movement at a joint, and blowflies have stretch receptors in the gut that inhibit feeding when the gut is distended.

Most biologists would agree that one of the main evolutionary trends within the animal kingdom has been the increasing elaboration of the nervous system. In tracing the evolution of sensory processes, therefore, it seems sensible to use the complexity of the nervous system as a guide. However, we have very little direct evidence about the nervous system of animals in the past because soft nervous tissue seldom is preserved in fossilized form. Indirect evidence sometimes can be gained from skeletal remains, especially where well-preserved vertebrate skulls are available. Most of our conclusions about the evolution of the nervous system derive from present-day representatives of animal types known to have changed little over millions of years.

11.5 Nervous systems of invertebrate animals

A nervous system can be defined as an organized constellation of nerve cells and associated non-nervous cells (Bullock, 1977). This definition includes receptors but not effector organs such as gland and muscle. Throughout the animal kingdom, nerve cells have common attributes that readily distinguish them from other cells. These attributes include the graded electrical potentials that occur in receptor cells and at synapses and the impulse-like action potentials that carry information along the axons. Although not universal, these are the typical properties of neurons that are described above. Important exceptions to the general pattern occur in some primitive animals and in the protozoa.

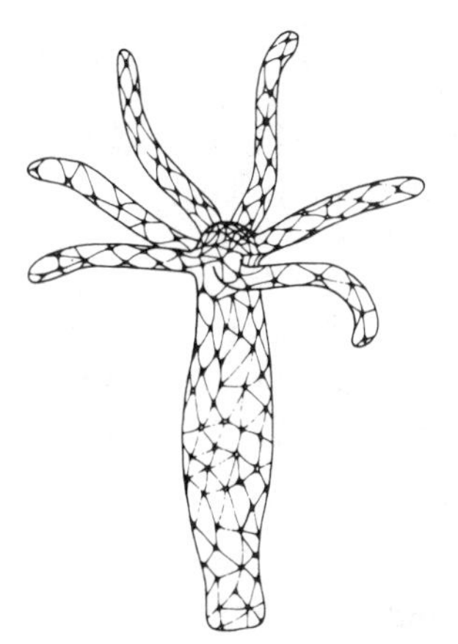

Fig. 11.10 The nerve net of Hydra (From *The Oxford Companion to Animal Behaviour*, 1981)

Protozoa, being single-cell organisms, cannot possess a true nervous system. Although they have undoubted sensory capabilities and exhibit rudimentary behavior, they do not appear to have specialized intracellular organelles that conduct excitation (Bullock, 1977). Instead, they seem to be organized along principles similar to those governing the physiology of neurons. Thus, the protozoan is like a receptor cell equipped with effector organelles.

The coelenterates include the simplest animals with a true nervous system. Nerve cells communicate with each other via recognizable nerve impulses and synapses. Coelenterates have no CNS, although nerve cells sometimes may be organized into simple ganglia. The nerve cells are often organized into *nerve nets* that permit diffuse conduction throughout the body, as in the case of Hydra (Fig. 11.10).

Hydra has receptor cells responsive to touch and to chemical stimuli. Photoreceptors also occur in some coelenterates. The receptors pass information across synapses to other nerve cells, but the nerve impulses are very slow compared to those of vertebrates. The nerve net coordinates the movements of the animal in a way that is not understood fully. The movements are slow and the behavioral repertoire is quite limited. Learning is not known to occur, except in the form of simple habituation. If the mouth of Hydra or of a sea anemone is touched repeatedly, it closes in a reflex manner at first, but the response gradually disappears with repeated stimulation. This habituation is known to be due to sensory adaptation and is not normally regarded as true learning.

Among the coelenterates, the medusae (jellyfish) show two important features of more advanced nervous systems. They have representatives of what were probably the first ganglia and the first sense organs. The ganglia are found in the marginal bodies, which innervate photoreceptors and statocysts. The marginal bodies contain four or more different types of neurons that interconnect with each other (Bullock et al., 1977). The *statocysts* are devices for detecting the direction of gravity. They contain a small pebblelike object, the *statolith*, whose position in a cavity can be detected by mechanoreceptors in the wall of the cavity, as illustrated in Figure 11.11. The receptors provide information about gravity because of the arrangement of the associated non-nervous tissue. This makes the statocyst a true sense organ. It is interesting that this fairly sophisticated sensory device is found in association with a relatively complex feature of the nervous system, the ganglia found within the marginal bodies. An association between specialized features of the nervous system and particular sense organs is a common phenomenon in the evolution of sensory processes.

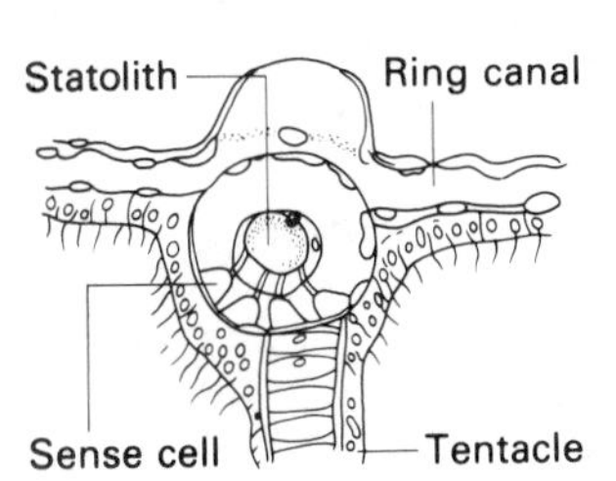

Fig. 11.11 A statocyst (gravity sensor) from a medusa (jellyfish). These are complexes of tissues and hence true sense organs

A further advance in nervous system organization is seen in the flatworms (Platyhelminthes). Unlike the coelenterates, but like most other invertebrates, the flatworms show bilateral symmetry and possess a head and tail. Sensory receptors tend to be concentrated in the head region instead of scattered over the body. The nervous system also

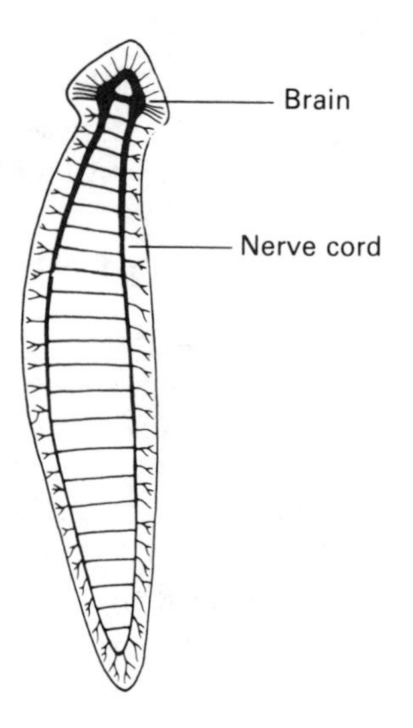

Fig. 11.12 The nervous system of a flatworm (From *The Oxford Companion to Animal Behaviour*, 1981)

shows concentration at the head, in the form of an anterior ganglion that constitutes a simple brain. From the anterior ganglion, two nerve cords run down the body, and these are linked by nerves in the form of a ladder, as illustrated in Figure 11.12. Nerve fibers pass from the nerve cords to all parts of the body in a netlike pattern. The anterior ganglion and nerve cords together form the CNS, while the network of nerve fibers constitutes the *peripheral nervous system.* This distinction is common to most invertebrates and all vertebrates and is seen in flatworms in its most primitive form. As a rule, the CNS contains most of the effector nerve cell bodies, while the peripheral nervous system contains the sensory receptors.

In flatworms such as *Planaria,* sensory cells on the head respond to touch, temperature, and the chemical composition of the water. They also have two eyes consisting of photoreceptors grouped together. The messages from the sensory cells are transmitted to the anterior ganglion. Conduction of nerve impulses along the nerve cords is faster than in nerve nets, and the behavior of flatworms is correspondingly more varied and definite than that of coelenterates. They are quick to detect the presence of food and to approach it. They avoid strong light and noxious chemicals and seem to be capable of rudimentary learning. If placed in a simple T maze, planarians can learn to turn to one side rather than the other to avoid being tapped with a rod. Learned responses to light stimuli also have been claimed, but these have proved to be controversial.

The more advanced invertebrates follow a general pattern based on the annelids, or segmented worms. These have a fairly complex nervous system typified by that of the earthworm *Lumbricus* (Fig. 11.13). The head region is characterized by a pair of supraesophageal ganglia, above the pharynx connected to another pair of ganglia below. These subesophageal ganglia are the first of a chain of ganglia along a central nerve cord, one ganglion for each segment of the worm. Many annelids have a system of giant fibers containing neuronal axons of large

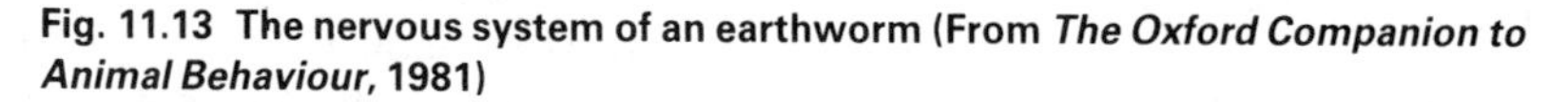
Fig. 11.13 The nervous system of an earthworm (From *The Oxford Companion to Animal Behaviour*, 1981)

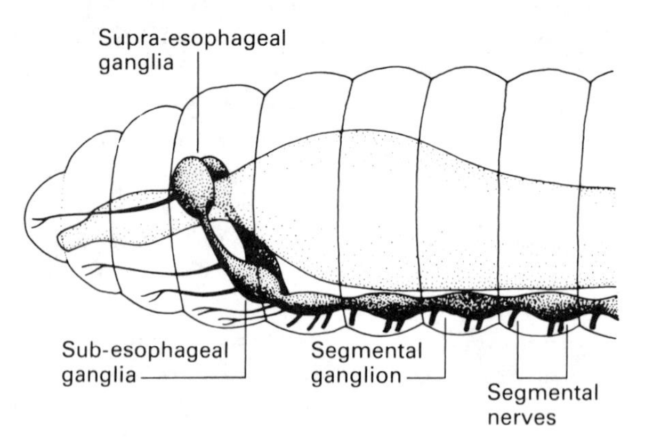

Table 11.1 Some types of arthropod skin receptors

Structure	*Morphological character*	*Function (known or presumed)*
Sensilla trichoidea	Sensory hairs and setae	Mechanoreceptors, proprioceptors, sound, contact chemoreception, humidity, olfactory in different places
Sensilla chaetica	Sensory spines and bristles	Mechanoreceptors, proprioceptors
Sensilla squamiformia	Sensory scales	Mechanoreceptors
Sensilla basiconica	Short, thick hairs; few to many neurons	Mechanoreceptors, contact chemoreceptors, olfactory, humidity, osmotic, temperature receptors
Sensilla coeloconica	Sunken cuticular cones	Olfactory or humidity receptors
Sensilla ampullacea	Sensory tubes	Olfactory receptors
Sensilla campaniforma	Cuticular domes	Directional strain gauges
Sensilla placodea	Cuticular plates and pore plates	Unknown; abundant on bee antennae

diameter and high conduction velocity. These are involved in rapid escape or withdrawal reactions to danger. Most annelid receptors are single elements that include chemoreceptors, mechanoreceptors, and photoreceptors. Proper sense organs appear in some species, examples being the taste buds in the skin of *Lumbricus* and the eyes of the free-swimming polychaete *Alciope*. Annelids have a fairly restricted repertoire of behavior patterns and are capable of rudimentary learning.

The gross plan of arthropods is similar to that of annelids, with a central cord and a pair of ganglia in each segment, plus cross connections called *commissures*. There is a dorsal anterior brain with circumesophageal connections to the central cord. In primitive arthropods, the segmental pattern can be seen clearly, but in more advanced forms there is considerable fusion of ganglia, as illustrated in Figure 11.14. This fusion of ganglia is characteristic of the evolution of the nervous systems of invertebrates and is associated with increasing complexity of the sensory systems and of behavior. Giant fiber systems occur in many arthropods, including shrimps, lobsters, scorpions, and some insects. They normally mediate a rapid tail flick or jumping movement, which forms part of the animal's escape behavior.

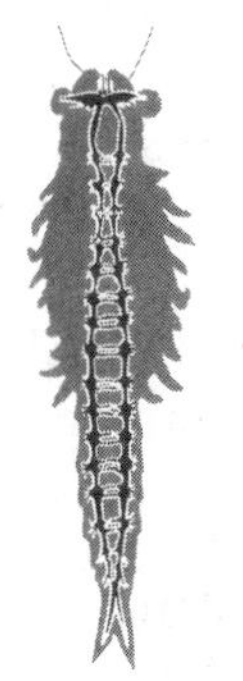

Fig. 11.14 Cephalization in Crustacea. Above: the brain and ventral cord of a fairy shrimp. Below: the brain and cord of a crab

Arthropods have evolved a greater variety of types of receptors than any other group, including vertebrates (Bullock, 1977). Table 11.1 gives a summary of the main types. The sensory neurons of these receptors have their cell bodies close to the sensory surface and not grouped into sensory ganglia. Some have few sensory neurons, and others have many, as illustrated in Figure 11.15.

Many types of mechanoreceptors, including statocysts, are common in crustaceans. *Chordotonal organs* are a distinct kind of internal mechanoreceptor that usually serve as proprioceptors responding to mechanical displacement (Fig. 11.16). They also may be involved in

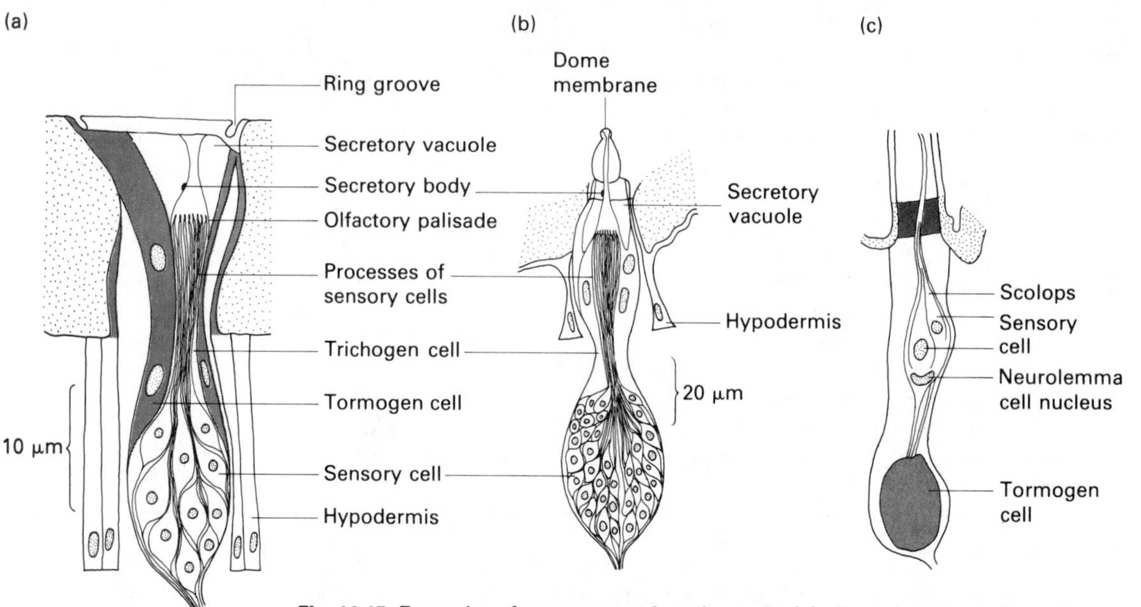

Fig. 11.15 Examples of sense organs in arthropods: (*a*) olfactory plate of a hornet (*Sensilla placodea*), (*b*) olfactory cone of a hornet (*Sensilla coeloconica*), (*c*) antenna hair of a moth (*Sensilla trichodea*) (After Bullock, 1977)

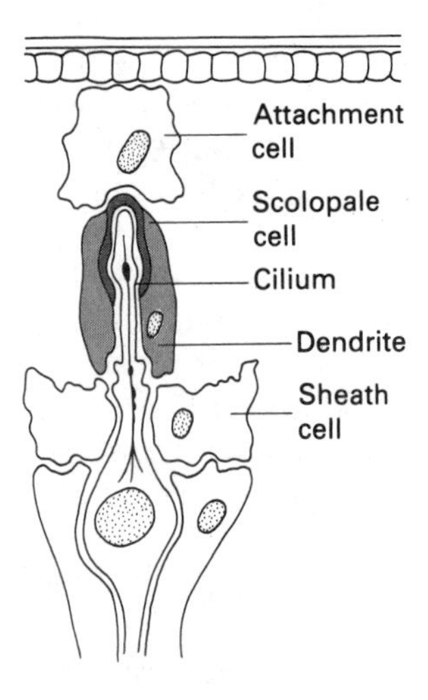

Fig. 11.16 Chordotonal mechanoreceptor of an insect (After Gray, 1950)

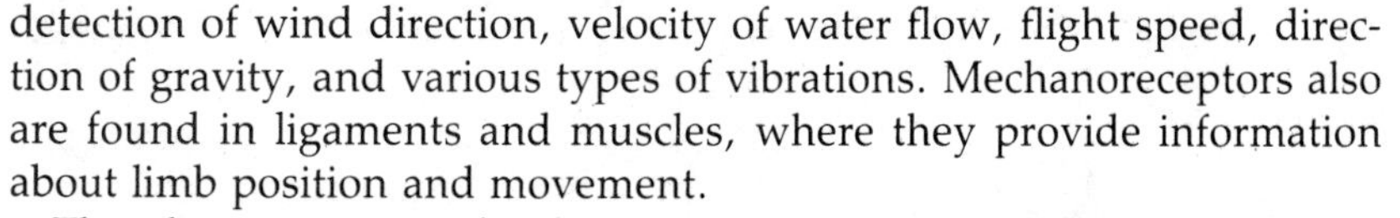

detection of wind direction, velocity of water flow, flight speed, direction of gravity, and various types of vibrations. Mechanoreceptors also are found in ligaments and muscles, where they provide information about limb position and movement.

The photoreceptors of arthropods are of particular interest, and one of these is illustrated in Figure 11.17. In general, the arthropods display a wide range of nervous complexity and demonstrate many evolutionary trends, including a progression from the segmented annelid pattern to the large brains that result from many fused ganglia. Arthropods have a remarkably wide repertoire of behavior patterns. Some species, such as the honeybee (discussed in detail in Chapter 23), have a complexity of behavior that rivals that of the vertebrates. Many arthropods are capable of simple learning, though this seems to be tied to specific situations.

These evolutionary trends are even more apparent among the mollusks, which have achieved the highest degree of sophistication of the CNS found in invertebrates. Among the gastropods, for example, there is an enormous range of complexity in nervous system organization. Thus, *Haliotis* (the abalones) have an annelidlike nervous system, while the land snail *Helix* has considerable fusion of ganglia, as illustrated in Figure 11.18. The most highly developed nervous systems are found in the cephalopods. These are the most active mollusks. They hunt mainly by vision and are capable of complex behavior, including recognition of complex objects, and rapid learning.

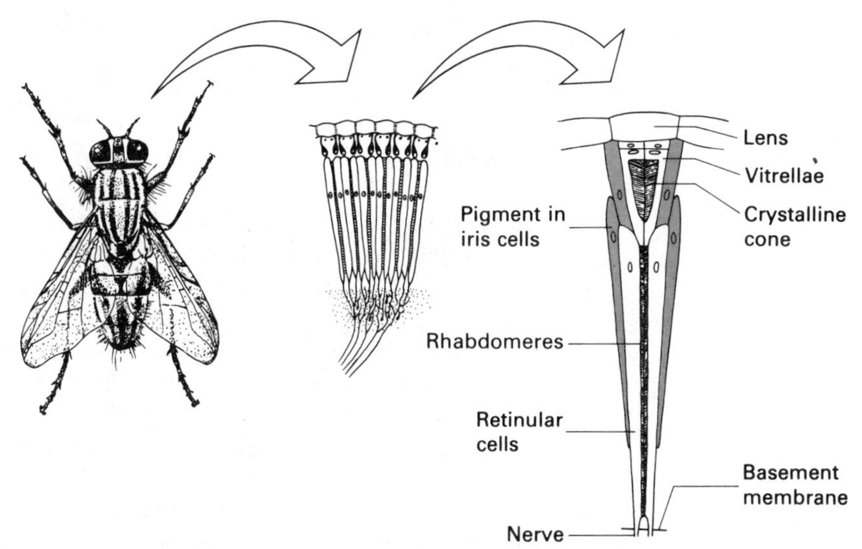

Fig. 11.17 The compound eye of a fly. Each eye is made up of numerous ommatidia. Each ommatidium is connected to a nerve cell which signals the intensity of light from the direction in which the ommatidium points. A single ommatidium does not produce an image, but as adjacent ommatidia point in slightly different directions, the CNS receives information about the distribution of light intensity over the visual field

The most studied representative of the cephalopods is probably the octopus. The octopus brain contains about 170 million nerve cells (compared with about 100,000 for the larger crustaceans) and is made up of about 30 different lobes, many of which have distinct functions (Fig. 11.18). The optic lobes make up more than half the nervous tissue of the brain. They are connected to a pair of large eyes that are the most advanced among the invertebrates and that rival those of vertebrates (see Chapter 12).

The mollusks vary considerably in the complexity of their behavior, but the cephalopods are the most advanced. This group has received

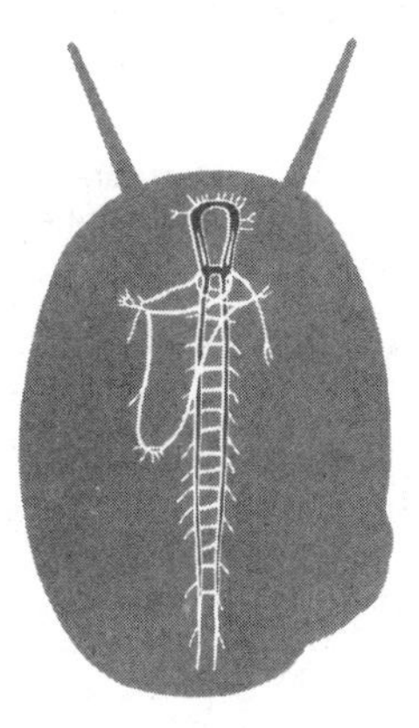

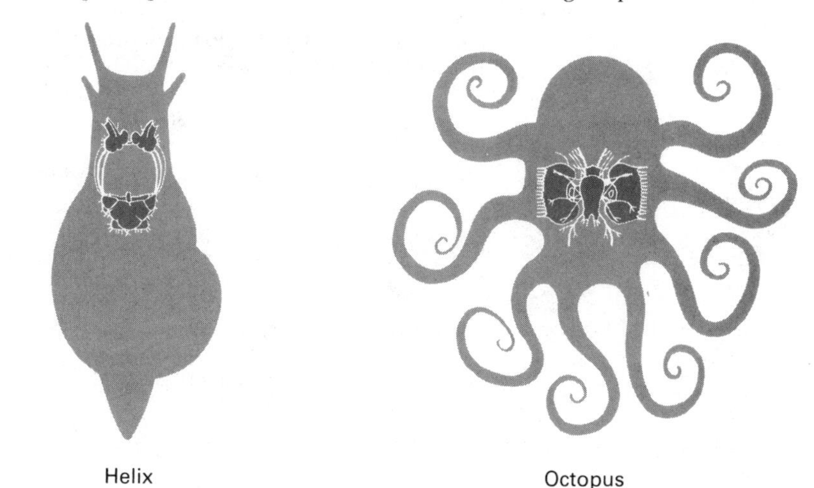

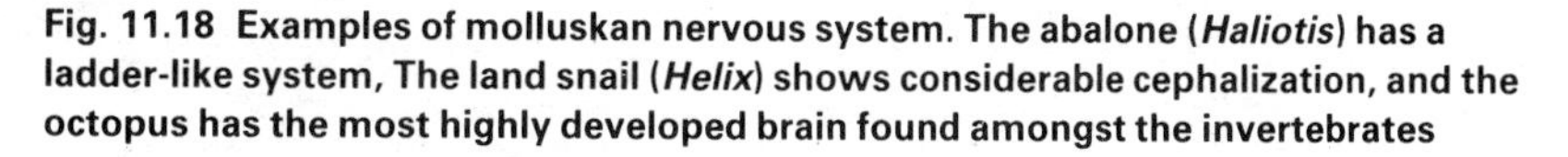

Fig. 11.18 Examples of molluskan nervous system. The abalone (*Haliotis*) has a ladder-like system, The land snail (*Helix*) shows considerable cephalization, and the octopus has the most highly developed brain found amongst the invertebrates

considerable study from behavioral scientists (e.g., Martin Wells, 1966) on account of their remarkable learning abilities.

11.6 The vertebrate nervous system

The organization of the vertebrate nervous system is distinct from that of invertebrates, though it is not always more complex. The vertebrate CNS develops embryologically from a dorsal neural tube to form a brain and single dorsal nerve cord, rather than the brain and dual ventral nerve cords characteristic of invertebrates. The evolution of the vertebrate brain is reflected in the embryonic development of the CNS in the individual, as illustrated in Figure 11.19.

The pattern of embryological development of the brain is remarkably constant throughout the vertebrates (Laming, 1981). As illustrated in Figure 11.19, the neural tube develops into three distinguishable parts: (1) the prosencephalon (forebrain), (2) the mesencephalon (midbrain) and (3) the rhombencephalon (hindbrain). These three hollow vesicles traditionally are associated with the olfactory, visual, and auditory senses, respectively. Each major division develops a secondary outgrowth: the telencephalon (cerebrum), optic tectum, and cerebellum, respectively, as illustrated in Figure 11.20. The CNS comprises the brain and spinal cord, which are enclosed within the bones of the skull and vertebral column. The peripheral nervous system consists primarily of the inflow (afferent nerves) and outflow (efferent nerves) from the CNS. The *somatic nervous system* carries sensory information to the CNS and commands from the CNS to the skeletal muscles responsible for bodily

Fig. 11.19 Embryonic development (from left to right) of the human brain showing the major divisions of the brain

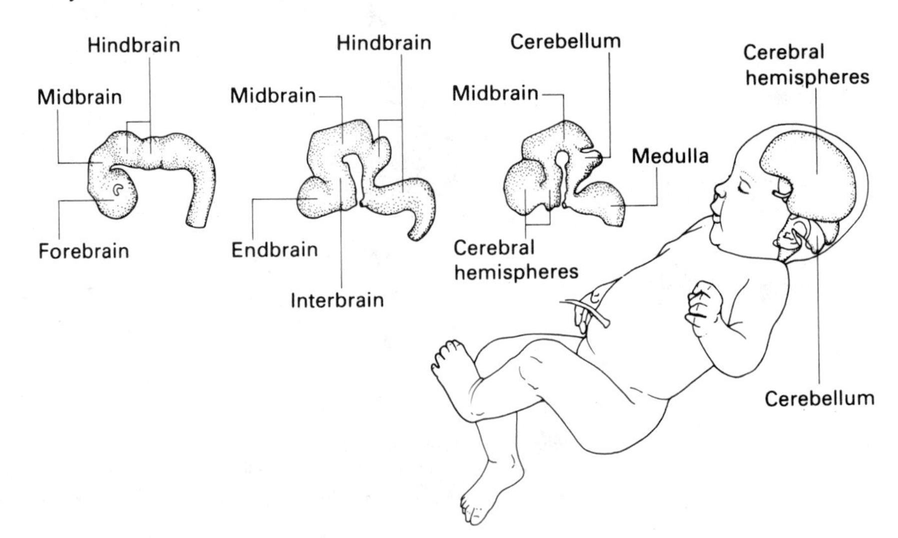

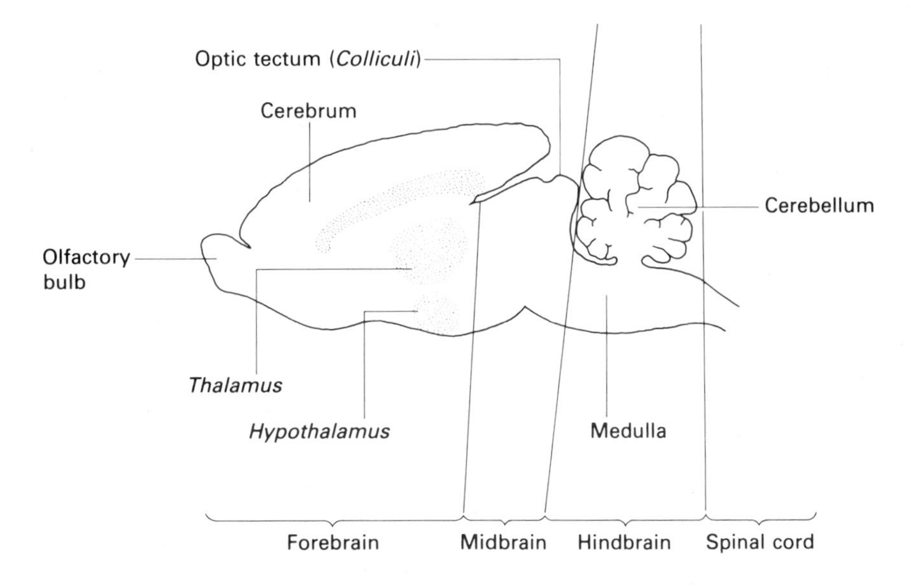

Fig. 11.20 One hemisphere of the rat brain (seen from the inside after section), showing the subdivisions of the forebrain, midbrain and hindbrain

movement. The *autonomic nervous system* supplies the internal organs with two types of innervation that have antagonistic effects. The *sympathetic nerve pathways* have an emergency function and become active under conditions of exertion or emotion. They have effects such as accelerating heart rate, dilating air passages to the lungs, reducing intestinal activity, and increasing the blood supply to the muscles and the brain. The *parasympathetic pathways* serve a recuperative function, restoring the blood supply to normal and counteracting the effects of sympathetic activity. The functioning of the autonomic nervous system is described in greater detail in Chapter 28.

It is fair to say that most of the interesting evolutionary trends within the vertebrates have to do with the CNS. The changes in the peripheral nervous system are more a reflection of the animal's particular anatomy and its adaptations to its particular niche than to any evolutionary "progress." The more ecological aspects of the senses are discussed in Chapter 13. Two important evolutionary trends that are worth discussing here are those of increasing brain size, and increasing tendency for behavior to be controlled by the higher centers of the brain.

Vertebrate brains vary considerably in size. However, direct comparison of brain size does not enable us to draw conclusions about intelligence or about the role of the CNS in controlling behavior. Larger animals tend to have larger brains because the larger the body, the greater the number of sensory fibers entering the brain and the more fibers leaving the brain to control the muscles. Therefore, it is the size of the brain in relation to the body that we must use as a measure. When we take account of differences in body size, we find that the brains of fish, amphibians, and reptiles are of similar sizes.

The brains of birds and mammals normally are much larger. Among the mammals, the brains of rodents and insectivores are relatively small, while those of ungulates and carnivores are much larger. The biggest brains are found among the primates and the sea mammals. The brains of the lower primates (prosimians) differ little from those of other mammals, but monkeys and apes have larger brains than any other land mammals. Even bigger brains, in relation to body size, are found among the dolphins and whales. But it is difficult to know what conclusions to draw in comparing animals specialized for life on land and at sea. The human brain is three times as large as would be expected for a non-human primate of the same body size. Among humans, however, considerable variation exists. The brain of a normal person can be as small as 1,000 cubic centimeters or nearly as large as 2,000 cubic centimeters (Coon, 1962). The norm for modern *Homo sapiens* is about 1,450 cubic centimeters. It is interesting that the fossil skulls of archaic man (*Homo sapiens neanderthalensis*), who lived some 45,000 to 75,000 years ago, indicate a slightly larger cranial capacity than that of modern man.

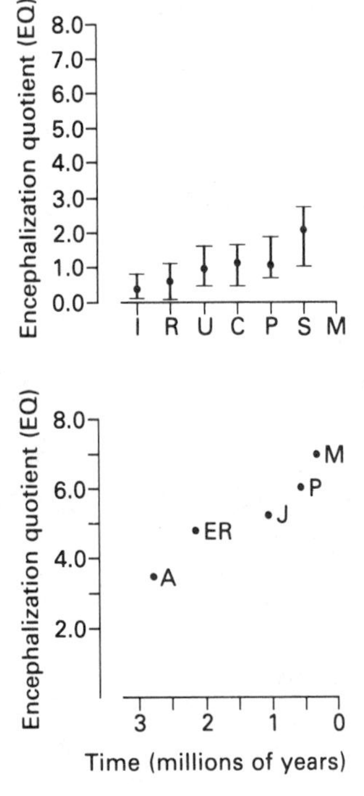

Fig. 11.21 Above: encephalization quotient for a variety of mammals. I, insectivores; R, rodents; U, ungulates; C, carnivores; P, prosimians; S, simians; M, man. Below: EQ in hominids plotted against time. A, *Australopithicus africanus*; ER, *Homo* skull from Kenya; J, *Homo erectus* from Java; P, *Homo erectus* from Peking; M, modern man (After Passingham, 1982)

The size of the brain gives a rough indication of the number of nerve cells in the brain. Larger brains have more nerve cells, and these cells tend to be larger and more loosely packed than those of smaller brains. In a large brain, each neuron tends to have a more complex set of dendrites and is open to influence from a greater number of other neurons. When we compare the brains of different vertebrates, after taking body size into account, can we see a definite evolutionary trend toward larger brains, implying greater and more sophisticated control of behavior?

The brains of animals have evolved as adaptations to the ecological niche, just as have other organ systems (Jerison, 1973). Animals that have specialized sensory systems and forms of behavior must have correspondingly specialized brain mechanisms. Contrary to popular belief, there is no progressive increase in brain size when animals in the sequence fish, reptile, bird, mammal are compared with each other (Jerison, 1973). Some fish brains are larger than those of reptiles of equivalent weight, and some bird brains are relatively larger than those of some mammals.

In charting the evolution of the brain and intelligence (see Chapter 27), we see no smooth progression from primitive to advanced animals (Hodos and Campbell, 1969; Hodos, 1982). Jerison (1973) calculated an encephalization quotient by relating the brain size of each species to the size expected for an average mammal of the same body weight (see Fig. 11.21). This measure shows some marked differences among different groups, but some anomalies occur. Thus, among the primates, some small monkeys are placed well above other primates. When brain size is related not to body size but to the size of the medulla (a more direct measure of the inputs and outputs of the brain), then the great apes appear to have the larger brains (Passingham, 1975). Encephalization quotients for fossil hominids (Fig. 11.21) show that brain size has increased over the past 3 million years (Passingham, 1982).

11.7 Vertebrate hormones

The nervous system has a parallel system of feedback and control of behavior in vertebrates. This is the *endocrine* (hormonal) *system*. Hormones are chemicals (usually peptides or steroids) that are produced by endocrine glands and released into the blood. Each hormone has specific functions and often affects specific target organs. The endocrine glands turn their secretions on and off at appropriate times. The control of hormone secretion is achieved either by direct nervous action or by the action of other hormones. There is usually a series of events that provides negative feedback regulating the level of the hormone in the blood. Hormones circulate in very small amounts, although most hormone molecules have a life of less than one hour in the blood. They

Table 11.2 Important hormones in mammalian behavior

Source	*Hormones*	*Principal effect*
Kidney	Angiotensin	Stimulates vasoconstriction and causes a rise in blood pressure. Stimulates thirst
Testes	Testosterone	Stimulates development and maintenance of male secondary sex characteristics and behavior
Ovaries	Estrogen	Stimulates development and maintenance of female secondary sex characteristics and behavior
	Progesterone	Stimulates female secondary sex characteristics and behavior and maintains pregnancy
Adrenal medulla	Adrenalin	Stimulates 'fight or flight' reactions
Anterior pituitary	Follicle stimulating hormone	Stimulates growth of ovarian follicles and seminiferous tubules of the testes
	Luteinizing hormone	Stimulates secretion of sex hormone by ovaries and testes
	Prolactin	Stimulates milk secretion by mammary glands
Posterior Pituitary	Oxytocin	Stimulates release of milk by mammary glands and stimulates contraction of uterine muscles

have to be secreted continuously if their presence is to be effective.

The most important hormones that influence behavior are indicated in Table 11.2. Some of these, such as follicle stimulating hormone (FSH) and luteinizing hormone (LH) have indirect effects, priming other endorine glands. Others have a more direct effect. There are three main ways in which hormones can affect behavior: (1) they can influence effectors, such as special structures involved in behavior; (2) they can influence peripheral sensory receptors and so modify the input to the brain; (3) they can affect the brain directly.

An example of the first type of effect can be seen in the mating behavior of the African clawed toad (*Xenopus laevis*). During the breeding season the male develops special nuptial pads on his forelimbs. These consist of a thick mass of small spines set into the skin that enable the male to grip the smooth and slippery pelvic area of the female during mating. The appearance of the pads depends upon the hormone testosterone, which is released from the testes at the appropriate time. At the end of the breeding season, testosterone production ceases and the pads disappear.

An example of the influence of hormones upon peripheral sensory receptors can be seen in the parental behavior of pigeons. Pigeons and doves feed their young on "pigeons milk," which is a proliferation of the lining of the crop. This substance is regurgitated in response to the begging behavior of the young. Its formation is under the control of the hormone prolactin. Daniel Lehrman (1955) showed that sensory stimulation of the enlarged crop is essential for inducing parental feeding. If the crop sensitivity is reduced by means of a local anesthetic, then the parental feeding is reduced. Prolactin is thus instrumental in inducing parental feeding behavior by its effect on the crop.

The direct action of hormones on the brain is the most important mode of behavioral influence upon behavior. It was first demonstrated (Harris et al., 1958) by implanting of minute quantities of artificial estrogen into certain regions of the brains of female cats. This could induce sexual behavior, even though the level of estrogens in the blood was well below the level normally required for such behavior.

Hormonal influences upon behavior may be slow and prolonged, or quick-acting and short-lived. As an example, let us consider the suckling of goats (Fig. 11.22). The luteinizing hormone (LH) stimulates the ovary to produce estrogens and progesterone, which are responsible for the development of the mammary glands at puberty, and for maintaining them in a state of readiness for milk secretion. During pregnancy there is a rise in the level of prolactin, and this stimulates milk secretion. When the kid suckles, the mechanical stimulation induces the hypothalamus of the brain to release oxytocin from the posterior pituitary gland. The rise in blood oxytocin triggers milk letdown. The oxytocin response is very rapid and the kid will normally receive milk within the first 40 seconds of suckling.

Fig. 11.22 Diagram of hormones involved in suckling

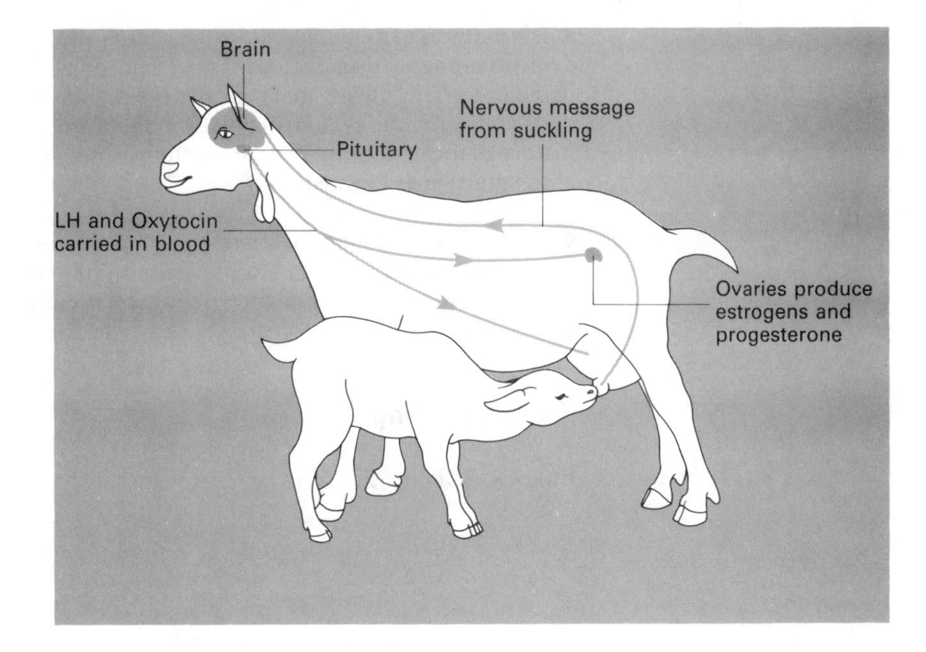

Points to remember

- Animal behavior is controlled by the nervous system, which is made up of specialized cells, called neurons. These operate on the same principles throughout the animal kingdom.

- The nervous system receives information from the environment through sensory receptors. These are nerve cells specialized as energy transducers. They may provide information about the animal's external environment and about its internal state.

- Behavior is effected through the action of muscles and glands. The muscles are responsible for animal movement, and the glands produce substances which may play a variety of behavioral roles, such as communication with other animals, protection against predators and adherence to the substrate.

- Muscular activity is controlled and coordinated by means of the somasthetic part of the nervous system. This provides feedback to the central nervous system about the position, tension and length of joints, tendons and muscles. The nervous commands to the muscles are modulated by this feedback of information.

- The nervous systems of invertebrates vary in plan from one phylum to another. Within a phylum there is usually a trend, from simple primitive forms to more complex advanced forms, which involves increasing cephalization and a concentration of neurons and sensory receptors in the anterior region of the animal. In some invertebrates this has resulted in a complex brain rivaling that of some vertebrates.

- The vertebrate nervous system has a basic plan common to all. It has evolved primarily by accretion of new mechanisms rather than modification of old ones. This means that it is possible to trace the evolution of the vertebrate nervous system in the developing embryo. The most distinctive feature of the vertebrate nervous system is the highly developed brain and associated sense organs.

- The vertebrate endocrine system produces hormones, some of which have important behavioral roles. Hormones may influence behavior by acting directly on the brain, by affecting peripheral organs or by influencing the production of other hormones.

Further reading

Bullock, T. H. (1977) *Introduction to Nervous Systems*, Freeman, New York.

12 Sensory Processes and Perception

The ability of animals to respond to changes in the external environment depends upon sensory processes designed to detect such changes. These range from simple discrimination of differences in temperature, or light intensity, to the recognition of complex patterns. In this chapter we look at the major sensory modalities of animals, and explore the nature of perception.

12.1 Chemoreception and thermoreception

Chemoreception is the capability of identifying chemical substances and detecting their concentration. It exists even among very primitive forms of life. In a technical sense, virtually every nerve cell is a chemoreceptor in that it reacts specifically to substances released by other nerve cells. The mechanisms of chemoreception involve the recognition of specific molecules by receptor sites on cell membranes. Whether this recognition occurs on a basis of chemical action, molecule shape, or both is not fully understood. Thus, for instance, we do not know what sugar and saccharine have in common that makes them both taste sweet to blowflies, rats, monkeys, and humans.

Chemoreceptors may be *exteroceptors* or *interoceptors*. The former detect the presence of chemicals in the external environment, while the latter detect substances circulating in the body fluids, such as carbon dioxide, nutrients, and hormones. Both taste and smell depend upon chemoreceptors. In the traditional sense, smell is concerned with the detection of low concentrations of airborne substances, while taste results from direct contact with relatively high concentrations of chemical substances. In both cases, however, the chemicals are presented to the receptor in solution, and the distinction is difficult to justify in some animals, like those that live in water. Nevertheless, in many animals there is a neurological distinction in that some nerves are concerned with olfaction, or the detection of low concentrations, while others convey *gustatory* messages from different receptors specialized for detecting high concentrations of chemicals. In the blowfly, for example, chemoreceptors on the antennae detect small quantities of airborne substances, and chemoreceptors on the tarsi (feet) are capable of detecting salt, sugar, and pure water. In vertebrates, the sense of taste is relayed via the facial (VII) and glossopharyngeal (IX) cranial nerves, while the sense of smell is transmitted by the olfactory nerve (I).

One of the most intensively investigated invertebrate olfactory systems is the perception of courtship pheromone by the silkworm moth (*Bombyx mori*). A *pheromone* is a chemical, or mixture of chemicals, that is released into the environment by an organism and that causes a specific behavioral or physiological reaction in a receiving organism of the same species. Pheromones are thus chemical messengers, and they are involved in what is probably the most primitive form of communication. The first chemically defined sex pheromone was that of the silkworm moth, and this chemical (a polyalcohol) has become known as "bombykol." It is secreted by an abdominal gland of the female, and by this means females can attract males over a distance of several kilometers. The chemical structure of bombykol is known, and it has been synthesized and presented to male moths or used to stimulate isolated antennae whose receptors were monitored with electrodes (Schneider,

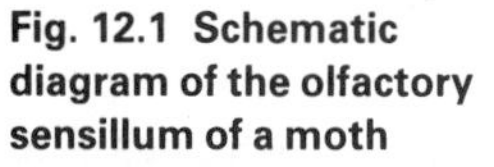

Fig. 12.1 Schematic diagram of the olfactory sensillum of a moth

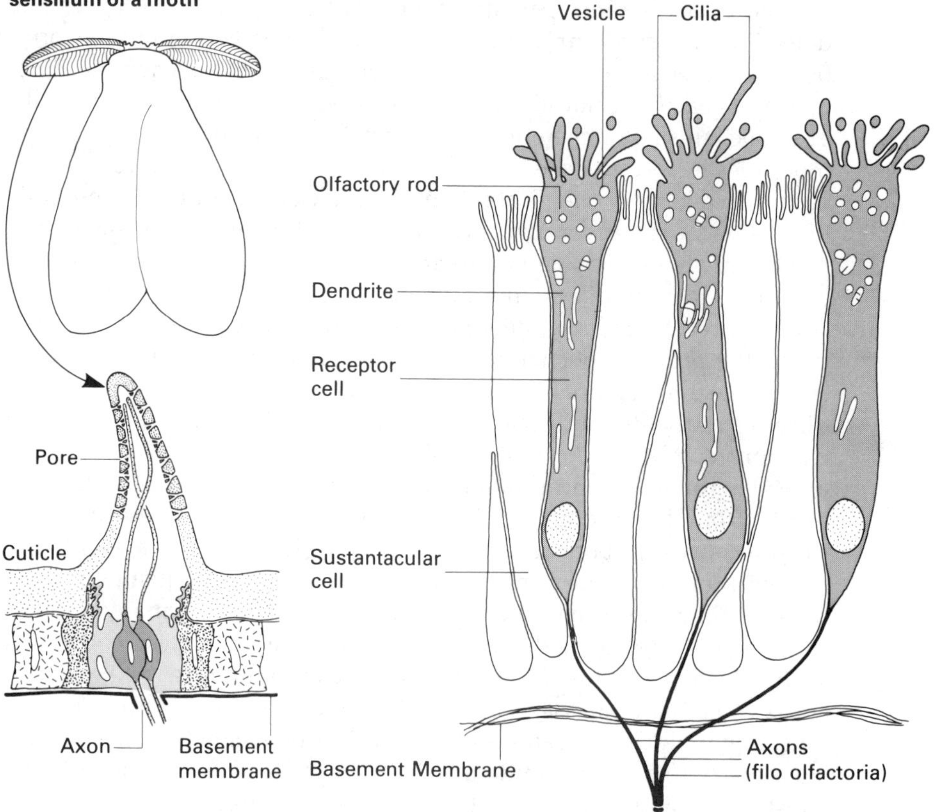

Fig. 12.2 Schematic representation of the olfactory mucosa of a rabbit. Three receptor cells are surrounded by supporting cells

1969; Payne, 1974). A full physiological response is shown only to bombykol, while a lesser response occurs to certain very close chemical relatives of bombykol. This demonstrates very great specificity. Even more remarkable is the finding that a single molecule of bombykol is sufficient to trigger a full response (Kaissling and Priesner, 1970). The behavioral response of the male silkworm moth is to fly upwind upon detecting bombykol molecules; when he finds the female he copulates with her. This type of orientation to chemical stimulation is common in insects.

Insects have various types of olfactory sensilla, as illustrated in Figure 12.1. These usually have numerous minute pores in the surface, which terminate in fluid-filled tubules. Dendrites of the receptor cells extend into the sensilla, and the receptor axons travel directly to the brain. Airborne pheromone molecules enter the pores of the sensillum and pass through the pores into the fluid-filled interior, where they come into contact with the receptor membrane.

Fig. 12.3 Distribution and structure of human taste buds

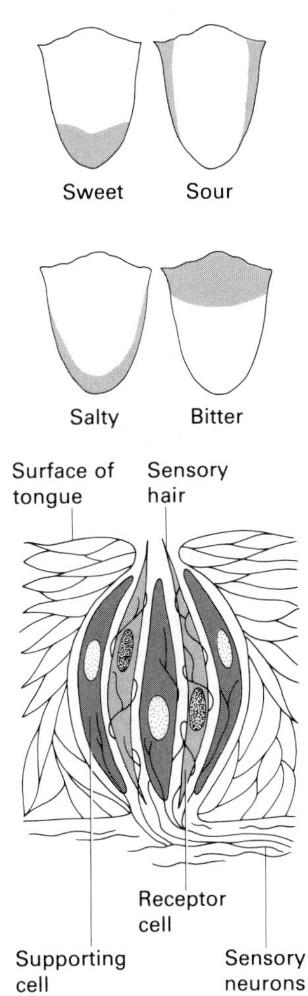

In vertebrates the olfactory receptors are primary sensory neurons with dendrites that extend as cilia into a mucous layer, as illustrated in Figure 12.2. The axons of these neurons go to the olfactory bulb, where they synapse with secondary neurons whose axons form the olfactory tract that enters the forebrain. The taste receptors consist of sense cells that usually are arranged in clusters called *papillae*. The sense cells are in close contact with sensory nerve fibers that remain fixed, while the taste cells are replaced every few days.

The basic tastes in mammals are acid, bitter, salt, and sweet. In humans, different parts of the tongue are sensitive to different tastes. Acid or sour tastes affect the sides of the tongue toward the back, and bitter tastes affect the extreme back of the tongue. Salty tastes affect the sides toward the front, and sweet tastes affect the tip of the tongue (see Fig. 12.3). The flavor of food depends upon both taste and smell. This can be demonstrated easily by asking a person to close his or her eyes and tell the difference between small pieces of apple and onion placed on the tongue. Most people find the discrimination easy if they are allowed to breathe through the nose but impossible if they are asked to hold the nose while taking the test.

There have been many attempts to identify basic odors equivalent to the four basic tastes. The most widely accepted classification is that of John Amoore (1963), illustrated in Table 12.1. While an undergraduate at Oxford University, Amoore noted great dissimilarities in the chemical structure of substances that smelled alike. He examined over 600 organic compounds that had well-described smells, and in 1952 he published his stereochemical theory of olfaction, in which all odors are described in terms of the seven different primary odors given in Table 12.1. Amoore's theory classifies chemicals in terms of their molecular shape and size. He postulates seven basic types of receptors, each with characteristic receptor sites into which certain shaped molecules can fit. Thus,

Table 12.1 Primary odors with chemical and more familiar examples

Primary odor	*Chemical example*	*Familiar substance*
Camphoraceous	Camphor	Moth repellent
Musky	Pentadecanolactone	Angelica root oil
Floral	Phenylethylmethyl ethyl carbinol	Roses
Pepperminty	Menthone	Mint candy
Ethereal	Ethylene dichloride	Dry-cleaning fluid
Pungent	Formic acid	Vinegar
Putrid	Butyl mercaptan	Bad egg

After Amoore, 1963

camphoraceous-smelling molecules are roughly spherical, while musky-smelling molecules are disk shaped. Amoore's theory has received some support as a result of subsequent research (Amoore, 1964), but it remains controversial.

Somewhat akin to chemoreception is *thermoreception*, which probably occurs in most animals, but has been studied in relatively few. Nerve endings sensitive to temperature are known to occur in a variety of insects. For example, the cockroach *Periplaneta* has thermoreceptors on the legs that perceive ground temperature and some on the antennae that perceive air temperature. Fish have thermoreceptors in the skin, lateral line, and brain, making them very sensitive to temperature changes. Catfish have been shown to respond to changes in temperature of less than 0.1 °C. Many reptiles have a well-developed temperature sense, with thermoreceptors in the brain as well as in the skin. Pit vipers have special pits on the face that are sensitive to infrared radiation and are shaped so as to give the animal a directional temperature sense.

Birds are thought to have few thermoreceptors in the skin, except on the tongue and bill of some species. In pigeons (Columbidae), there are thermoreceptors in the brain that influence behavior and plumage adjustment and others in the spinal cord that control shivering and panting. In mammals, distinct heat and cold receptors are distributed in the skin, with the heat receptors usually deeper than the cold receptors. There are also receptors deep in the body—in veins, for example—that can initiate shivering even though the temperature at skin and brain receptors is kept constant. Thermoreceptors in the spinal cord influence shivering, panting, and blood flow, and these functions are repeated by thermoreceptors in the hypothalamus. In general, the most sophisticated forms of thermoregulation are found in mammals, and the brain receives information from many parts of the body. The integration of this information leads to appropriate activation of various mechanisms of warming and cooling.

12.2 Mechanoreceptors and hearing

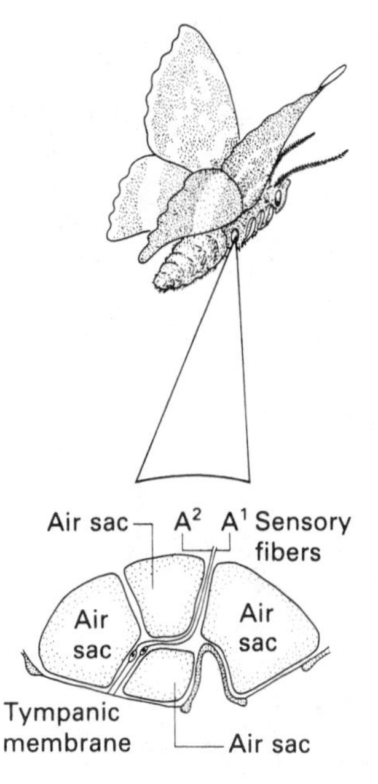

Fig. 12.4 The ear of a noctuid moth. Vibrations of the tympanic membrane are detected by the A_1 and A_2 sensory neurones

Sound results from minute changes in pressure that originate from a vibrating source within a medium such as air or water. The receptors that detect sound are basically mechanoreceptors that show rapid adaptation (recovery) and are thus sensitive to vibration.

Vibration-sensitive hairs and receptors in the limb joints have been described for many arthropods. Thus, blowflies have receptors in some joints of the antennae, called "Johnson's organs." These organs can follow movements up to 500 hertz. In mosquitoes, the same kind of organ signals direction of sound. The backswimmer *Notonecta* locates prey by detecting waves on the surface of the water. The receptors are in the legs and are maximally sensitive to vibrations between 100 and 150 hertz. Web-spinning spiders can discriminate live from dead prey by vibrations of the web. Such vertebrates have mechanoreceptors capable of picking up vibrations of the substrate. These have been found in the skin of snakes and the leg joints of cats and ducks (Prosser, 1973).

The auditory systems of animals, though diverse, have certain features in common. For example, there is a peripheral device for converting sound pressure to vibratory motion. Sensory receptors convert this motion into nerve impulses that can be decoded by the CNS. One of the simplest types of peripheral auditory systems is found in noctuid moths. There are two ears, each composed simply of a tympanic membrane on the side of the thorax, and two receptor cells embedded in a strand of connective tissue, as illustrated in Figure 12.4. This remarkably simple ear enables the moths to hear the ultrasonic cries of hunting bats. By means of an elegant series of experiments, Kenneth Roeder (1963; 1970) showed how this is achieved.

One receptor, called the A_1 cell, is sensitive to low-intensity sounds and responds to cries from bats about 30 meters away, too far for the bat to detect the moth. The frequency of impulses from the A_1 cell is proportional to the loudness of the sound, so the moth can tell whether or not the bat is approaching. By comparing the time of arrival and intensity of the stimulus at the two ears, the moth can determine the direction of approach. The difference occurs because the moth's body shields the sound from one ear more than the other. The relative altitude of the bat also can be determined: when the bat is higher than the moth, then the sound reaching the moth's ears will be interrupted intermittently by the beating of the moth's wings, whereas this will not happen when the bat is below the moth.

The A_1 cells give the moth early warning of an approaching bat and may enable it to fly away from the bat before it is detected. By heading directly away from the bat, the moth presents itself as the smallest possible target, since its wings are edge-on rather than broadside-on to the bat, and can do this simply by turning, equalizing the sound reaching the bat's two ears. If, however, the bat detects the moth, the

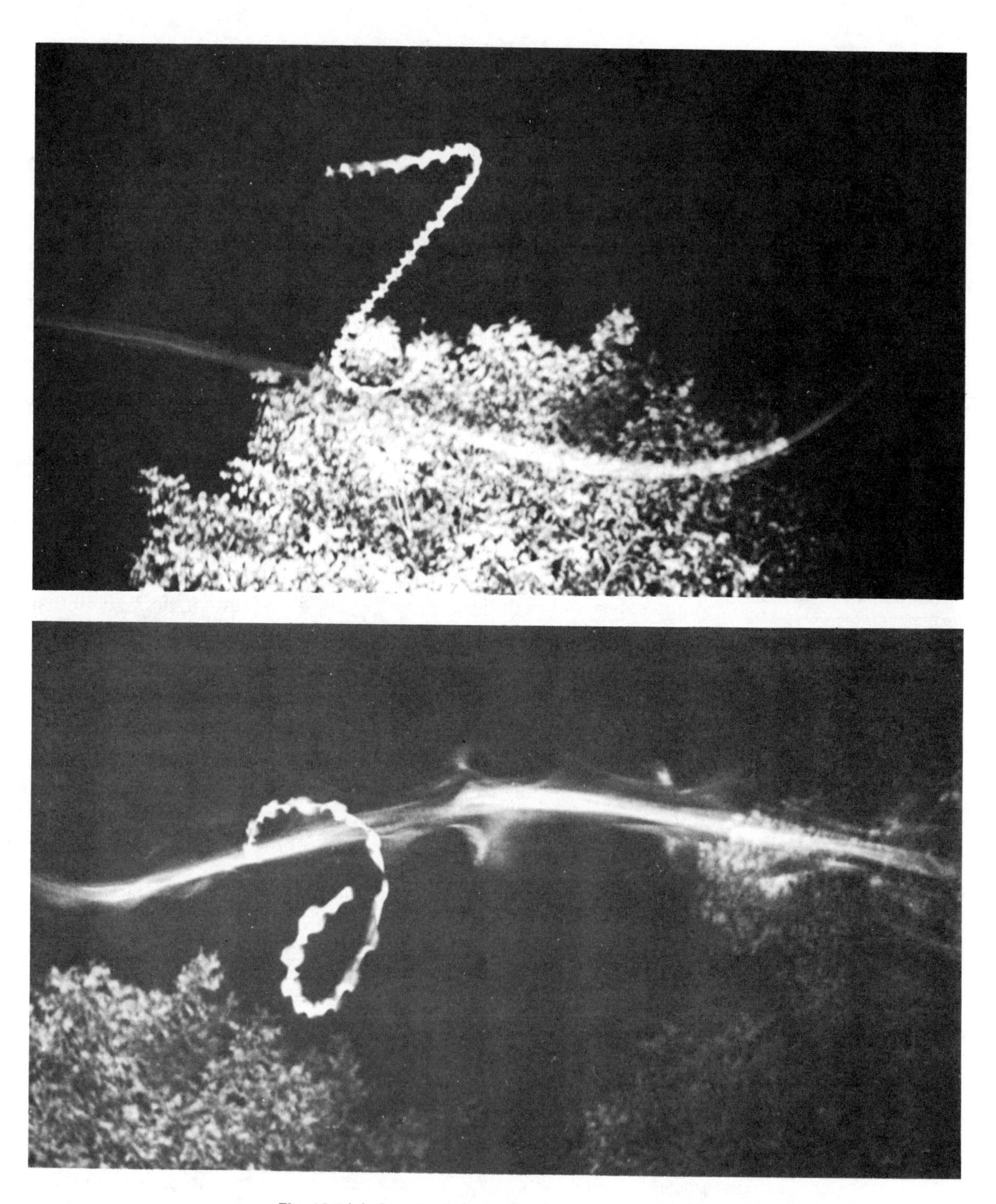

Fig. 12.5 (*a*) **Streak photograph of a moth escaping from a Red bat** (*b*) **streak photograph of a Red Bat catching a moth** (*Photographs: Frederick Webster*)

moth cannot escape simply by outflying the bat because the bat is a much faster flier. Instead, the moth employs evasive action when the bat comes within two or three meters (Fig. 12.5).

The A_2 cell only produces nerve impulses when the sound is loud. It starts responding when the bat is nearby, and its impulses probably cause disruption of the flight control mechanisms of the CNS. The moth consequently flies erratically and drops toward the ground. By means of such evasive action, moths have been observed to escape just as the bats come within striking distance. Thus, Roeder's (1963; 1970) work on hearing in noctuid moths is a beautiful illustration not only of the workings of a simple ear but also of the fact that an animal's sensory apparatus often is finely tuned to the circumstances of its ecology.

There are many properties of sound to which an animal might respond. As sound travels through a medium, particles move back and forth, creating oscillating pressure waves. The extent or amplitude of the waves determines the intensity or loudness of the perceived sound. The speed of transmission of sound depends upon the density of the medium through which it travels and is independent of the intensity of the sound. In air, sound travels at about 340 meters per second. It travels faster in hot air than in cold air. In water, sound travels about four times as fast as in air.

If we picture sound as a waveform, as in Figure 12.6, the distance between successive peaks, called the *period*, is inversely related to the *frequency*, or number of peaks in a unit of time. Sound frequency is measured in hertz or cycles per second. The simplest sound is a pure tone, which has a single frequency perceived subjectively as *pitch*. Natural sounds are seldom pure tones and are made up of a number of frequencies mixed together. When a complex sound is analyzed into its component frequencies, the result is called a *sound spectrum*. A hearing organ may be responsive to a wide range of frequencies, an example being the tympanal organ of a locust, which responds to frequencies between 1,000 and 100,000 hertz. When a sound receptor responds to a narrow band of frequencies, it is said to be "sharply tuned."

The antennal (Johnson's organ) receptor response of the male mosquito *Aedes aegypti* responds to sounds between 150 and 550 hertz, which corresponds to the wing tone of the female. The wing tone of the male is of a higher frequency and is not detected by the Johnson's organ (Haskell, 1961).

In general, vertebrates' ears are more versatile than those of invertebrates. Thus, few, if any, invertebrates can discriminate between two frequencies, unless the animals happen to have two different kinds of receptors tuned to different frequencies (Haskell, 1961). For the vertebrate ear, however, such a discrimination presents no problem. This is partly due to the structure of the ear and partly to the analyzing role of the CNS.

The human ear (Fig. 12.7), like most mammalian ears, is divided into

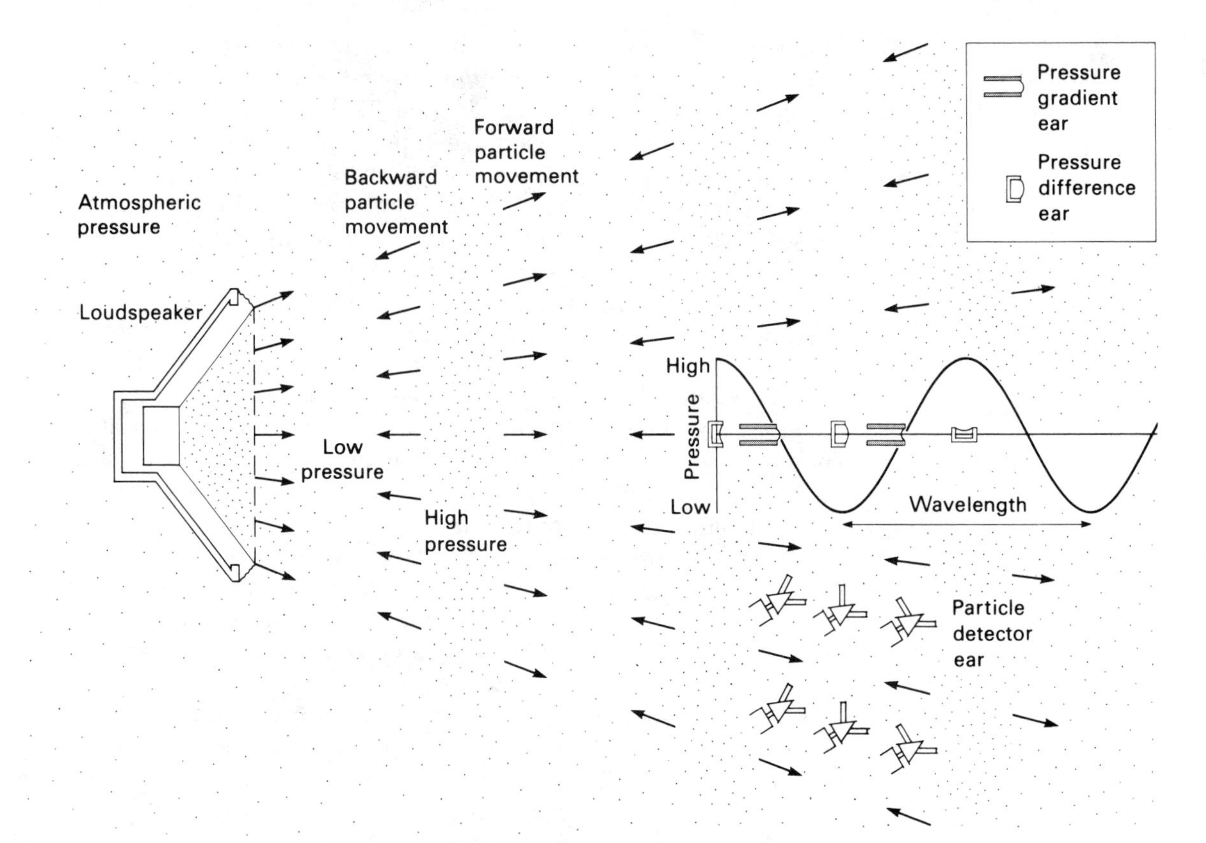

Fig. 12.6 Different types of ear respond to different aspects of sound. Particle-detector ears, found in bees, mosquitoes and some fish, are stimulated by air molecules moving from high to low pressure areas. Pressure difference ears, found in mammals, and some birds, fish and insects, have a sealed chamber that provides a reference pressure, and a membrane that is deformed by changes in the ambient pressure. Pressure gradient ears, found in reptiles, amphibians, and some birds, fish and insects, measure the difference in pressure between the two ends of a tube by means of a membrane in the tube. They are maximally responsive when aligned along the axis of sound transmission (After Gould, 1982)

three parts: the outer ear, the middle ear, and the inner ear. The *outer ear* consists of the *pinna* (ear flap) and the *auditory canal*, which terminates in the *tympanic membrane* (eardrum). The *middle ear* consists of a chamber on the inside of the tympanic membrane that is connected to the *pharynx* (mouth) by the *Eustachian tube*, an arrangement that makes possible the equalization of air pressure between the outer ear and middle ear. Passengers in airplanes may experience pain in the ears when they change altitude quickly, as in takeoff or landing, a pain due to unequal pressure on either side of the tympanic membrane. Passage of air through the Eustachean tube equalizes the pressure, which is facilitated by chewing, swallowing, or yawning. The middle ear contains three small bones—the *malleus* (hammer), the *incus* (anvil) and the *stapes* (stirrup)—that connect the tympanic membrane to another membrane, the *oval window*, in the wall of the middle-ear chamber.

The *inner ear* is a labyrinth of interconnected fluid-filled chambers and canals. It has two distinct parts: the *vestibular apparatus*, which is concerned with the sense of equilibrium, and the *cochlea*, which is a coiled tube that is the organ of hearing. Inside the cochlea are three canals: the *vestibular canal*, which begins at the oval window, the tympanic canal, which begins at the round window and connects with the *vestibular canal* at its other end, and the *cochlear canal*, which lies between the other two canals. The boundary between the cochlear canal and the tympanic canal is formed by the *basilar membrane*, which carries the *organ of Corti*. The organ of Corti carries rows of receptor cells with sensory hairs at their apexes. The hairs project into a gelatinous *tectorial membrane*. Dendrites of sensory neurons terminate on the surfaces of the hair cells, and when vibrations of the basilar membrane cause the sensory hairs to move, deformations of the hairs give rise to generator potentials in the hair cells, and these stimulate the sensory nerves.

Vibrations in the air are caught by the pinna and pass down the

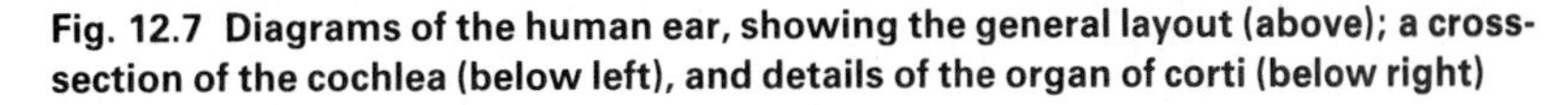

Fig. 12.7 Diagrams of the human ear, showing the general layout (above); a cross-section of the cochlea (below left), and details of the organ of corti (below right)

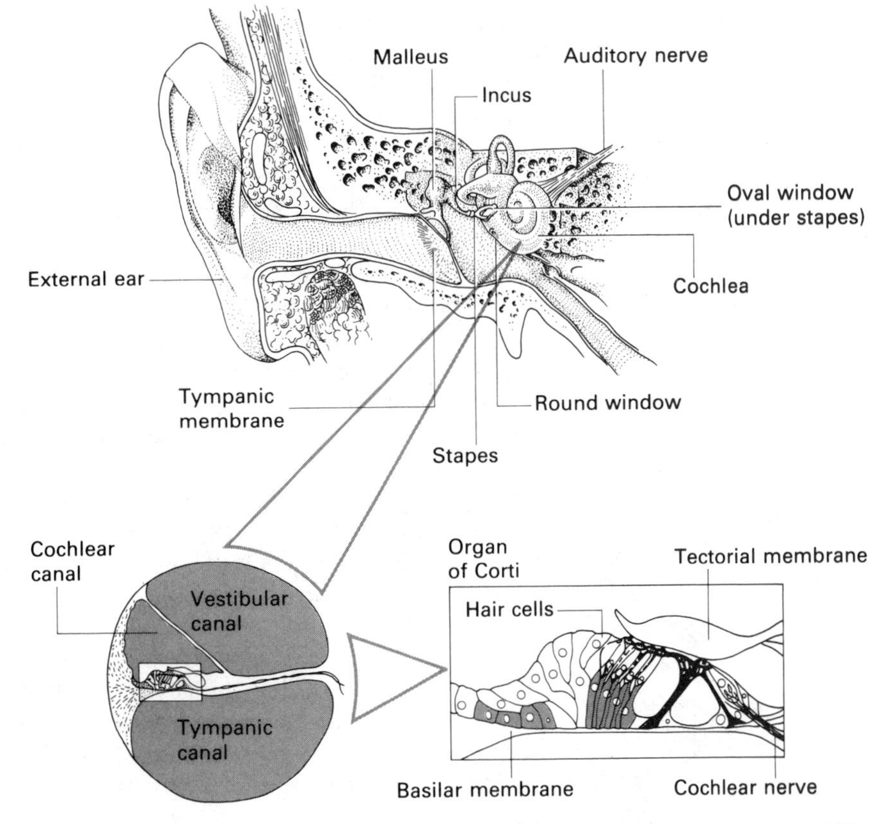

auditory canal, causing the tympanic membrane to vibrate at the same frequency. These vibrations are transmitted across the cavity of the middle ear by the three ear bones. The bones are arranged as a lever system that diminishes the amplitude of the vibrations but increases their force. Moreover, the vibrations of the large tympanic membrane are transmitted to the much smaller oval window, with the result that the sound pressures at the tympanic membrane are magnified by about 22 times at the oval window, thus enhancing the detection of faint sounds.

The movement of the membranes of the oval window causes a corresponding movement of the fluid in the cochlea. When the open window bows inward, fluid is pushed from the vestibular canal to the tympanic canal, causing the round window to bow outward and to release the pressure in the cochlea. The movement is then reversed during a complete cycle of vibration. The movements of the fluid in the cochlea occur at the same frequency as those of the outside air. As the cochlear fluid vibrates, a traveling wave is set up in the basilar membrane, and as it moves up and down, it deforms the hair cells as they push against the tectorial membrane. As the hair cells are bent they stimulate the sensory neurons.

The point in the cochlear tube with the maximum amplitude of membrane displacement varies with the frequency of the sound stimulus. As early as 1867, Helmholtz had postulated correctly, on anatomical grounds, that the high-frequency waves are focused near the base of the cochlea and that low-frequency waves have their maximal effect near the apex. Our current understanding of the operation of the cochlea is due to the work of communications engineer Georg von Békésy (1952; 1960), awarded the Nobel Prize for his work. He watched events inside the cochlea by removing the fluid and replacing it with a suspension of coal and powdered aluminum. By reflecting flashes of intense light from this suspension, he was able to observe events like the travelling wave of the basilar membrane. He found that the membrane is stiffer at the base, favoring high-frequency vibrations, and less stiff near the apex, favoring low-frequencies. Thus, particular frequencies perturb particular parts of the basilar membrane, and each part stimulates different receptors in the organ of Corti.

The nerve fibers of these receptors synapse in the *spiral nucleus,* and the secondary neurons make up cranial nerve VIII. Each signals a particular frequency of sound to the cochlea nucleus of the brain.

The structure of the ear is not the same in all vertebrates. For example, fish and cetaceans (dolphins and whales) have no outer ear, and fish also lack a middle ear with eardrum and ossicles. Since the tissues of fish have about the same density as water, vibrations arriving at the head can pass directly to the inner ear. Some fish, however, have another mechanism that serves a function analogous to that of the inner ear—the gas-filled swimbladder, which may have a bony connection to the inner ear that provides considerable improvement in hearing ability. In amphibians and reptiles, the tympanic membrane is the outermost part of the ear, but in birds there is an external canal (*auditory meatus*) that leads from the body surface to the eardrum. In birds, a rod-shaped bone, the *columella,* extends to the inner surface of the tympanic membrane and joins the stapes. In amphibians and reptiles, these bones form part of the jaw, though they have some role in hearing in some species.

The lateral-line organs of fish and aquatic amphibians are sensitive to vibrations, including low-frequency sound, and they are made up of modified hair sensilla that respond to the flow of water in the lateral line canal or on the body surface.

12.3 Vision

Vision is based upon detection of electromagnetic radiation. The electromagnetic spectrum covers a wide span, of which the visible spectrum is a very small proportion, as illustrated in Figure 12.8.

The energy content of electromagnetic radiation is inversely proportional to the wavelength. The long wavelengths contain too little energy to activate the photochemical reactions that form the basis of photoreception. The short wavelengths contain so much energy that they damage living tissue. Most of the shortwave radiation from the sun is absorbed in the ozone layer of the atmosphere, and it is doubtful that life could have evolved on earth if this had not been the case. All photobiological responses are confined to a narrow band of the spectrum between these two extremes.

Fig. 12.8 The electromagnetic spectrum (above) wavelengths in meters, with the visible portion enlarged (below)

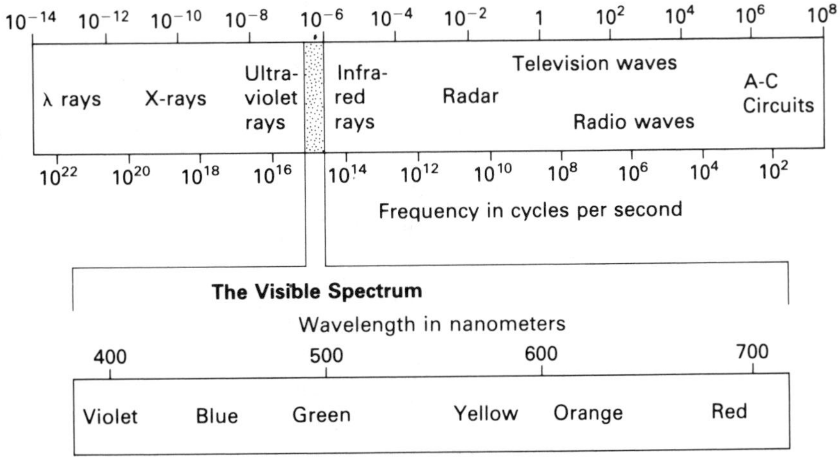

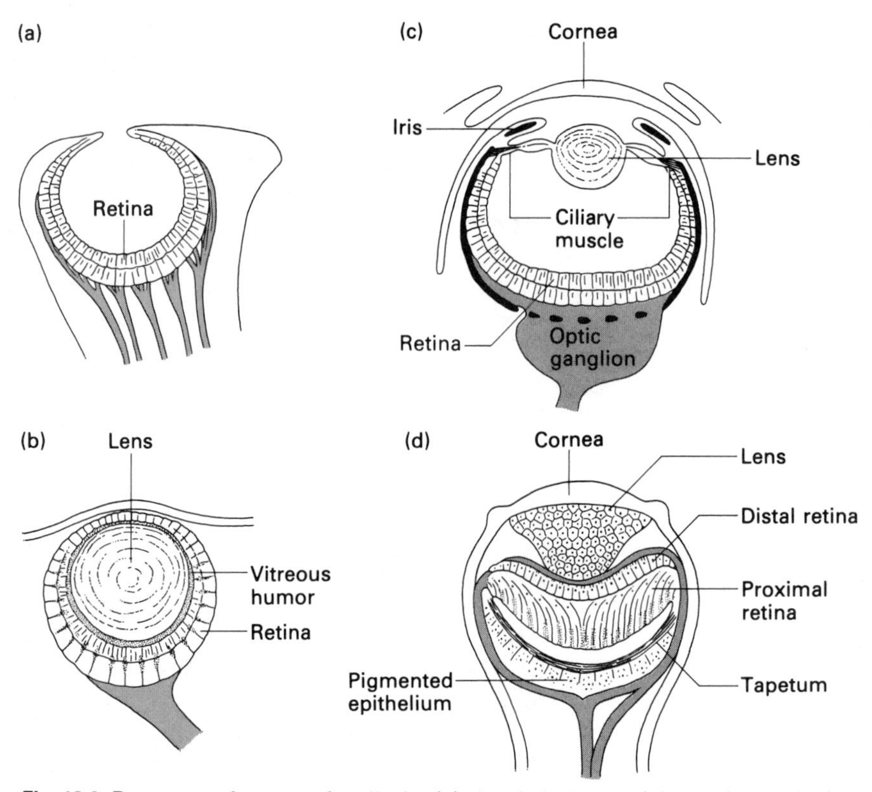

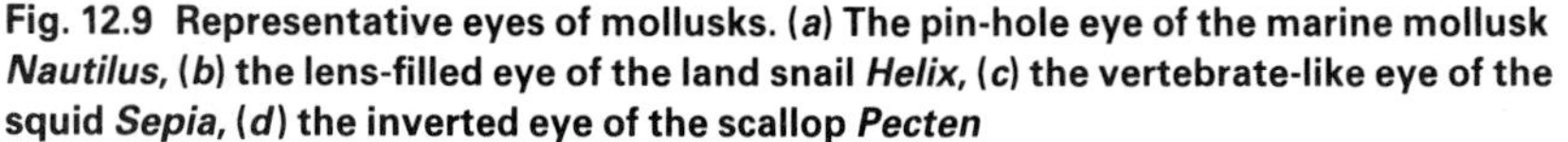

Fig. 12.9 Representative eyes of mollusks. (*a*) The pin-hole eye of the marine mollusk *Nautilus*, (*b*) the lens-filled eye of the land snail *Helix*, (*c*) the vertebrate-like eye of the squid *Sepia*, (*d*) the inverted eye of the scallop *Pecten*

Photoreceptor cells contain a pigment that is bleached by the action of light. The bleaching involves changes in the shape of the pigment molecules, and unlike the bleaching with which we are familiar in everyday life, it is reversible. The bleaching process leads to electrical changes in the receptor membrane that are not fully understood (Prosser, 1973).

Although photoreceptor cells may be scattered over the body surface, as in the earthworm (*Lumbricus*), they usually are gathered together in clusters. The most primitive type of eye is made up of a cluster of receptors arranged at the bottom of a depression or pit in the skin. Such an eye can give a rough indication of the direction of incident light. As a result of shadows cast by the sides of the pit, light coming from the side will illuminate one part of the pit while the rest remains relatively dark. These differences in illumination can be detected by an array of photoreceptors at the base of the pit, forming a rudimentary retina. The pinhole eye of the mollusk *Nautilus* (Fig. 12.9) is a development of the pit eye, the rim having grown inward and the photoreceptor layer forming a retina. The pinhole eye works exactly like a pinhole camera; that is, light from each point reaches only a very small area of the retina, so an inverted image is formed.

Fig. 12.10 Section through the human eye

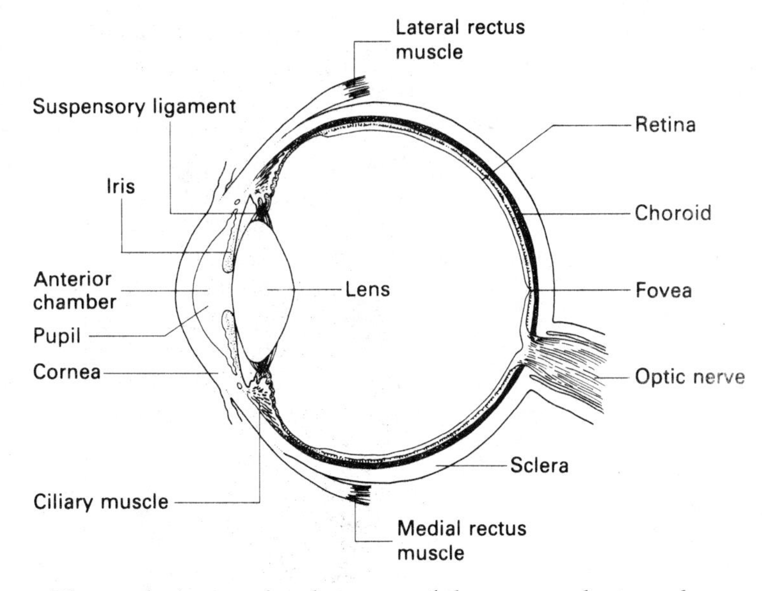

The evolutionary development of the eye can be traced among living mollusks, as illustrated in Figure 12.9. From the pinhole eye of *Nautilus*, the eye evolved a protective layer, presumably to exclude dirt. Inside the eye a primitive lens formed, as in the snail *Helix*. This type of eye is also found in spiders. There are some variations, such as the eye of the scallop *Pecten*, which has an inverted retina and a mirrorlike *tapetum* (see Chapter 13). The eye of the cuttlefish *Sepia* (Fig. 12.9) is very similar to that of vertebrates. The shape of the lens can be changed by ciliary muscles, and there is an iris diaphragm capable of regulating the amount of light falling on the retina.

The eyes of vertebrates, of which the human eye is a good example, conform to a single basic plan, although some ecological adaptation occurs, as we see in Chapter 13. Figure 12.10 shows a horizontal section through the human eye. It is surrounded by a tough coat, the *sclera*, that is transparent at the front of the eye, where it is called the *cornea*. Just inside the sclera is a black lining, called the *choroid*, that reduces both transmission of light through the sides of the eye and reflections within the eye. Inside the choroid is the light-sensitive *retina*, which we discuss in detail later. At the front of the eye there is no choroid or retina. A large *lens* divides the eye into *anterior* and *posterior chambers*, which are filled with *aqueous* and *vitreous fluids*, respectively. In front of the lens is the *iris*, a muscular diaphragm with an aperture called the *pupil*. The iris controls the size of the pupil and the ultimate amount of light entering the eye. The lens is circumscribed by the *ciliary muscle*, which changes the shape of the lens. When the muscle contracts, the lens becomes more convex, bringing close objects into focus on the retina. When the ciliary muscle relaxes, the lens becomes more flattened, bringing more distant objects into focus.

The vertebrate retina, unlike that of cephalopods like *Sepia*, is inverted. The photoreceptors lie next to the choroid, and light reaches them after passing through layers of neurons, mostly *ganglion cells* and *bipolar cells*. The ganglion cells are adjacent to the vitreous fluid, and their axons pass over the inner surface of the retina to the *optic disk*, or blind spot, where they form the *optic nerve* and leave the eye. The bipolar cells are neurons that link the ganglion cells to the photoreceptors, as indicated in Figure 12.11.

The photoreceptors are of two types, *rods* and *cones*. The rods, more elongated than the cones, are very sensitive to low levels of illumination and have only one type of photopigment, *rhodopsin*. Rod vision is therefore colorless. It also provides poor definition (visual acuity) because many rods are connected to a single ganglion cell. Consequently,

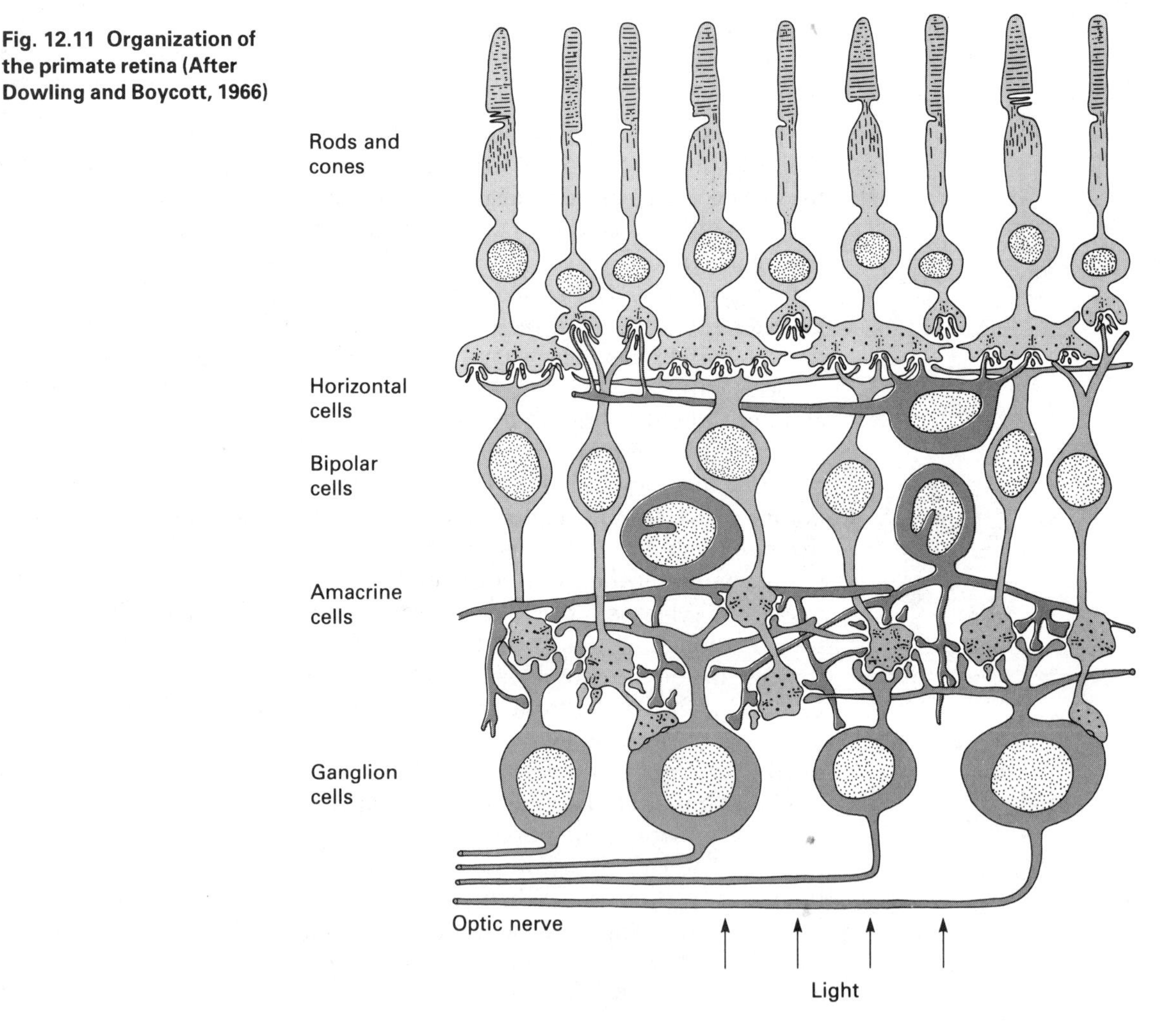

Fig. 12.11 Organization of the primate retina (After Dowling and Boycott, 1966)

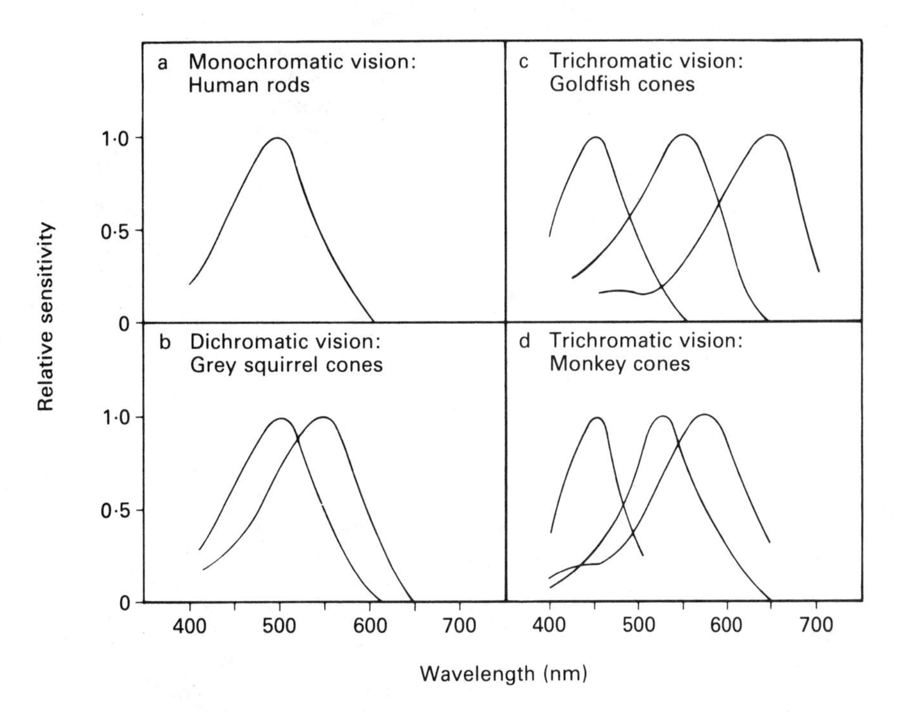

Fig. 12.12 Typical receptor mechanisms for different types of color vision (From *The Oxford Companion to Animal Behaviour*, 1981).

a single fiber in the optic nerve receives information from many rods, an arrangement that increases sensitivity at the expense of acuity. Rods are predominant in nocturnal species, where sensitivity is at a premium.

The cones are maximally sensitive to high levels of illumination and provide sharp vision. Only a few cones report to each ganglion cell. Cones may be of more than one type, the photopigments absorbing in different parts of the spectrum. Cones thus provide the basis for color vision. Cones are most sensitive to the wavelengths most strongly absorbed by their photopigments. Vision is *monochromatic* when there is only one active photopigment, as with twilight vision in humans, when only the rods are operating (Fig. 12.12).

Dichromatic vision can occur when two photopigments are active, as in the grey squirrel (*Sciurus carolinensis*) (Fig. 12.12). Each wavelength stimulates the two types of cones in a specific ratio, in accordance with their relative sensitivities in that part of the spectrum. Provided that the brain can recognize these ratios, it should be possible for the animal to discriminate wavelength from intensity. Note, however, that particular ratios occur in more than one part of the spectrum and that some wavelengths therefore will be confused. This also occurs in certain forms of color blindness in humans. The wavelength that stimulates both types of cones equally (where the absorption curves cross) is indistinguishable from white light and is known as the "neutral point" of the spectrum. Its existence has been demonstrated behaviorally in the grey squirrel (Muntz, 1981).

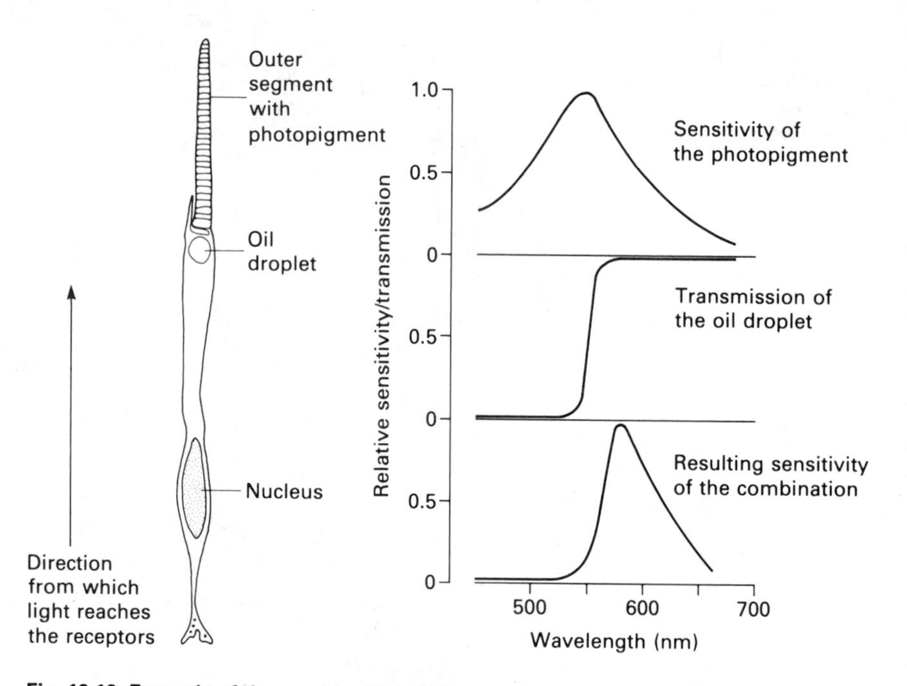

Fig. 12.13 Example of the combination of pigment and oil droplet in the cone of a bird (From *The Oxford Companion to Animal Behaviour*, 1981)

Confusion is reduced in visual systems with three receptors, or *trichromatic vision* (see Fig. 12.12), which is found in many species, including humans. Some confusions still occur, and it is possible, for example, to give the impression of any particular color by various mixtures of three monochromatic lights, suitably adjusted in intensity and saturation. The visual perception of color photography and color television would be impossible if this were not the case.

More than three receptor types are found in many birds and reptiles. In addition to various photopigments, the cones of these animals often contain colored oil droplets that act as filters and combine with the photopigment to determine the spectral sensitivity of the receptor as illustrated in Figure 12.13. The oil droplets usually are not distributed evenly over the retina but are concentrated in certain areas.

In 1825, the Czech physiologist Jan Purkinje noted that red flowers seem brighter than blue flowers in the daytime but that as twilight falls, they appear to fade before the blue flowers do. This change in the spectral sensitivity of the eye, called the *Purkinje shift*, was shown by Schultz in 1866 to be due to a change from cone to rod vision during *dark adaptation*. The change in sensitivity during dark adaptation can be measured, in humans, by determining the threshold of just-visible light at intervals after entering a dark room. During dark adaptation, this threshold falls progressively, as illustrated in Figure 12.14. The kink in the curve is due to the change from cone to rod vision. The contribution

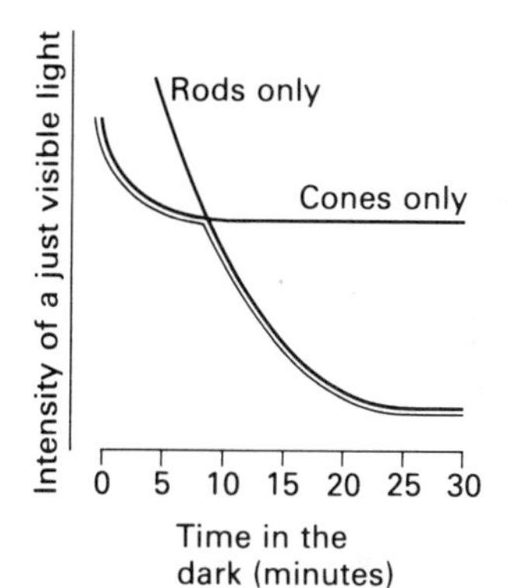

Fig. 12.14 Graph showing how the intensity of a just-visible brief light decreases as we adapt to darkness (double line). The steep single line shows what happens when there are only rods in the retina. The shallow single line shows what happens when only cones are illuminated (i.e. when the light is restricted to the fovea) (From *The Oxford Companion to Animal Behaviour*, 1981)

of the cones can be determined by shining a very small light on the *fovea*, which has no rods. The contribution of the rods can be seen in "rod monochromats," individuals who have no cones in the retina, a rare condition. As we can see from Figure 12.14, the rods are much more sensitive to light than the cones. However, they contain only the photopigment rhodopsin, the sensitivity maximum of which is toward the blue end of the spectrum. This is why blue objects appear to be brighter than objects of other colors during twilight vision.

The range of light intensity over which the vertebrate eye can operate is enormous; there is a billion-fold change in sensitivity. This is achieved by various mechanisms, which vary from species to species. In many fish, amphibians, reptiles, and birds, choroid pigment migrates between the outer segments of the receptors during high illumination and retracts as light levels fall. The outer segments of the cones also may move in these animals. In some fish and amphibians, the outer segments of the rods also move in the opposite direction. The level of light reaching the retina can be controlled by contraction of the pupil. This reflex is well developed in eels and flatfish, nocturnal reptiles, birds, and mammals (Prosser, 1973).

To produce a sharp image on the retina, the light passing into the eye must be refracted so as to focus at the retina. This is accomplished at the cornea and at the lens. In the human eye the cornea provides about twice the refraction of the lens.

The problem is that the cornea is a fixed distance from the retina, and thus, some accommodation is required if objects at various distances are to be brought into focus. The necessary adjustment is provided by the lens. In fish the lens is nearly spherical, with a high refractive index and short focal length. This is necessary because the refractive index of water is about the same as that of the cornea, so that no refraction occurs at the surface of the eye. The lens has a fixed shape, and accommodation is achieved by altering the distance between the lens and the retina. In land vertebrates, accommodation is controlled by the ciliary muscles that alter the shape of the lens. When focusing on nearby objects, the lens is rounded, and when focusing on distant objects it is flattened, as illustrated in Figure 12.15. Animals that live both in and out of water cannot see well in both media. The eyes of the frog, crocodile and hippopotamus are placed high in the head so that the animal can see over the surface of the water while keeping the body submerged. In the so-called four-eyed fish (*Anableps anableps*), each eye is divided into two by a band of skin. The upper portion of the eye protrudes out of the water while the lower portion remains submerged. The single lens in each eye is oval and is shaped in such a way that the lower eye can focus underwater while the upper eye can focus on objects above the surface of the water.

The eyes have a field of view that is largely dependent upon their position in the head. In vertebrates, the field of each eye is about 170

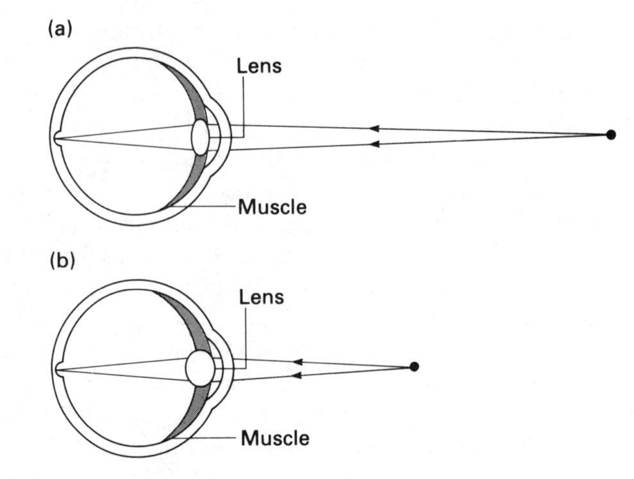

Fig. 12.15 Focusing the eye. The lens is flattened for viewing distant objects, and is more rounded for viewing nearby objects

degrees (Duke-Elder, 1958). Species vary in the degree of overlap of the fields of view of the two eyes. In general, predatory animals have a large overlap at the front of the head and a blind area behind, whereas prey animals have little overlap and a smaller blind area, as illustrated in Figure 12.16.

Where the two visual fields overlap, *binocular vision* is possible. The advantage of binocular vision is that it permits more accurate depth perception and judgment of distance than monocular vision. This is an important attribute for animals that rely on such information in capturing their prey. The advantage of a wide field of view is that movements can be detected readily, even if they occur behind the animal. This is

Fig. 12.16 Visual fields in (*a*) the squirrel, (*b*) the cat, and (*c*) the night monkey (After Kaas *et al*, 1972)

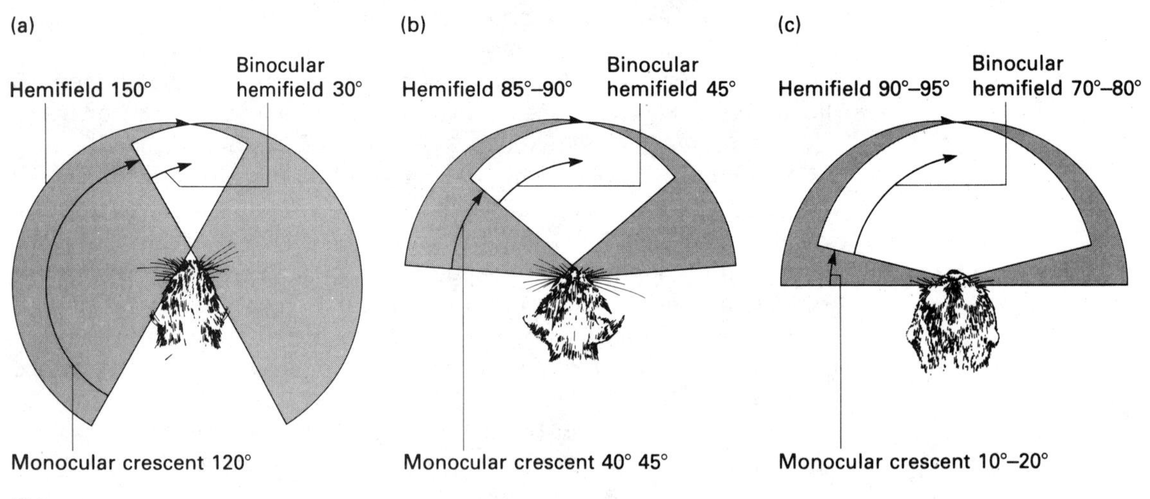

obviously important for animals that need to be vigilant for approaching predators.

During visual examination, when acuity is important, the image is brought into focus on the fovea. In order for both eyes to focus, there must be some convergence of the two lines of sight. The nearer the object under scrutiny, the greater the necessary convergence. The direction of the two lines of sight is adjusted, via the *extraocular muscles*, until the two retinal images coincide and a single image is registered by the brain. If, at the same time, the degree of convergence of the two eyes is noted by the brain, then information about the distance of the object is available. It is not possible, however, for the retinal images of nearby objects to coincide exactly. Depending upon the distance between the eyes, there will be a difference between the positions of the two images. This retinal disparity also provides important information about the distance of objects. The judgment of distance and depth is a complex matter, involving many cues in addition to those provided by convergence and retinal disparity.

Although the basic layout of the eye is the same for all vertebrates, there are a number of variations and specializations, some of which are discussed in Chapter 13. Particularly relevant to this discussion, however, is the variation in foveae. Humans have a single, centrally placed fovea that is circular in shape, contains only cones, and is the part of the retina that provides greatest acuity of vision. In the cheetah and many birds, the fovea is elongated in the horizontal plane. This band type of fovea seems to be associated with animals that live in open country or that fly over the sea (Meyer, 1977; Hughes, 1977). Arboreal mammals, such as the cat and squirrel, have a disc-shaped fovea, as do nocturnal mammals, such as the hedgehog and mouse. For such animals, the vertical direction is likely to be as important as the horizontal (Hughes, 1977).

Most birds have some kind of fovea, and about half of those examined have more than one (Meyer, 1977). Many birds have one fovea that serves binocular vision and another that serves the lateral field of view. The adaptive significance of the various arrangements of foveae in birds is not understood fully. Their main functions probably are connected with the complex visual tasks associated with flight, especially those involved in taking prey on the wing and in landing (Lythgoe, 1979).

12.4 Sensory judgments

Psychologists traditionally have made a distinction between sensation and perception. *Sensations* are the basic data of the senses, the raw material from which knowledge is gained. Red and blue are examples of color sensations. *Perception* is a process both of interpretation of sensory information in the light of experience and of unconscious inference. For example, in looking at a landscape we use color as a guide in judging the distance of hills and other features of the terrain. This is well known to landscape painters, who typically use shades of blue to give an impression of distance.

For more than a hundred years, psychologists have intensively studied perception in humans, mainly in the laboratory and usually by asking subjects to respond verbally to various perceptual situations. The study of sensation and perception in animals is hindered by the fact that they cannot speak to us and that we have difficulty imagining how experimental situations are interpreted by animals. Nevertheless, many of the methods developed by human psychologists have application in the study of animal behavior.

Gustav Fechner (1801–1887) believed that sensations cannot be measured directly because they are experiences that are entirely private. The magnitude of a physical stimulus can be measured, and a person can be asked to rate the magnitude of the received sensation. The problem is that if two people give different ratings to the same range of stimuli, we cannot tell whether they received different sensations or whether they merely rated them differently. Fechner saw that, although sensations cannot be compared to physical stimuli, sensations can be compared to each other. A person can compare two sensations and judge whether they are the same or different. One way of doing this is to progressively alter the intensity of a stimulus presented to an experimental subject until the subject reports detecting a change. The difference between the original intensity and the repeatedly changed intensity is called the *just-noticeable difference*, a physical quantity that can be measured over a wide range of physical stimulus intensity. For example, a subject might be asked to compare two weights repeatedly, one a standard weight—say, 100 grams—and the other taken from a series of comparison weights. The subject would report on whether the latter was heavier than, equal to, or lighter than the former. From repeated tests of this type, the experimenter would gain an estimate of both the *uncertainty interval* within which the subject was not sure whether the test weight was different than the standard and the *difference threshold*, which is the extent to which the stimulus intensity has to be altered to produce an altered sensation.

In 1834, the German physiologist E. H. Weber proposed that the difference threshold is a constant ratio of the standard stimulus. In other words, if a change in weight from 100 to 110 grams (10 per cent) were the minimum detectable difference, then the difference between 10 and 11 grams, also a 10 per cent change, should be equally as detectable. If I is the magnitude of the standard stimulus and ΔI the increment in stimulus intensity required to produce just-noticeable difference, then:

$$\frac{\Delta I}{I} = \text{constant}$$

This is known as *Weber's Law*, and the fraction $\Delta I/I$ is called the *Weber fraction*.

Fechner believed that Weber's Law was the key to the measurement of subjective experience. He and his co-workers carried out numerous experiments to test the law. In general, the law holds true for the normal range of stimulus intensity but tends to break down at the two extremes of the intensity range. Fechner was interested particularly in the relationship between the physical intensity of the stimulus and the subjective magnitude of the resulting sensation. He proposed the following formulation, which has become known as *Fechner's Law*:

$$S = k \log I$$

where S is the subjective magnitude, I is the stimulus intensity, and k is a constant. An interesting result of this approach is that different sensory modalities have different typical Weber fractions, as illustrated in Table 12.2. The smaller the fraction, the keener the sense. Thus, the discriminating power of vision is the greatest, and that of taste is the least of the human senses tested.

Table 12.2 Representative (middle-range) values for the Weber fraction for the different senses

Sensory modality	*Weber fraction* ($\Delta I/I$)
Vision (brightness, white light)	1/60
Kinesthesis (lifted weights)	1/50
Pain (thermally aroused on skin)	1/30
Audition (tone of middle pitch and moderate loudness)	1/10
Pressure (cutaneous pressure)	1/7
Smell (odor of India rubber)	1/4
Taste (table salt)	1/3

This entire line of thought and research is known as *psychophysics*. Various alternatives to Weber's and Fechner's Laws have been proposed, and some researchers claim that a *power law* offers a more accurate interpretation of the data (for a review, see Kling et al., 1971). However, there have been more fundamental objections to the classic psychophysical approach.

The classic psychophysical experiment requires the subject to respond in a yes-no manner. Its interpretation assumes that there is a fixed criterion for saying "yes I do" detect a change or "no I do not"; the subject is assumed to have no response bias. The early investigators were aware that the subject's attitude could alter the judgments made in an experiment. They attempted to eliminate the problem by using only carefully trained subjects or by introducing an occasional catch trial and making allowance for false response. A fundamental improvement was obtained by the introduction of methods derived from the theory of signal detectability, orginally developed to cope with problems in radio and telephone communications (Swets et al., 1961).

The theory of signal detectability, usually called *signal-detection theory*, assumes no fixed yes-no criterion. It maintains that whether or not a subject says yes depends upon the effect of the stimulus in relation to the prevailing variation (noise) in the same stimulus dimension, the expectation of the observer, and the potential consequences of the observer's decision. For example, suppose a subject is required to say whether or not a test sound is louder than a standard sound. The four possible outcomes are illustrated in Table 12.3. Suppose we decide to reward the subject with a payment of 25¢ for each correctly detected loudness difference (hit), but we impose a fine of 25¢ for failure to detect a difference (miss). We also pay 10¢ for each correctly identified non-difference (correct negative) and impose a fine of 5¢ for claiming a difference when there was none (false alarm). The consequences of the subject's responses now can be represented as a payoff matrix, as shown in Table 12.4. Suppose the subject works for 100 trials, on half of which there is a genuine loudness difference and on half of which there is no difference. By consistently saying yes the subject would earn $12.50 by being correct 50 times and would be fined $2.50 for being incorrect 50 times. By consistently saying no the subject would earn $5 by being correct 50 times and would be fined $12.50 for the 50 incorrect responses. Thus, by consistently responding positively the subject would gain $10, and by consistently responding negatively the subject would lose $7.50. It is not surprising that, with this particular payoff matrix, the subject would be biased toward saying yes.

Table 12.3 The four possible outcomes of a stimulus detection experiment

		Response	
		Yes	*No*
Stimulus	*On*	Hit	Miss
	Off	False alarm	Correct rejection

Table 12.4 Payoff matrix. The dollars earned and lost as a result of responding yes and no when the stimulus is on or off in 100 trials.

		Response	
		Yes	*No*
Stimulus	*On*	$12.50	−$12.50
	Off	−$2.50	$5.00

Another form of bias is generated by the subject's expectations. For example, Linker et al. (1964) found that in a taste experiment, the tendency to respond positively depended partly upon how often the stimulus was presented. When the stimulus was presented on 90 per cent of occasions, they obtained a high proportion of hits and a high proportion of false alarms. When the stimulus was presented on only 10 per cent of occasions, they obtained low proportions of hits and false alarms.

Signal-detection theory assumes there is no such thing as a zero stimulus. The subject is assumed to know that sensory events are contaminated by background noise. The task is not, therefore, simply to say whether or not the stimulus is present but to discriminate between noise plus signal and noise alone. Figure 12.17 illustrates the manner in which this is done. The decision criterion is set by the subject's response bias, which in turn is determined by the payoff matrix. Thus, one way of

Fig. 12.17 The basis of signal detection theory. (*a*) shows the relative frequency of sensations of various magnitudes when noise alone is present. (*b*) shows a similar normal frequency distribution for signal plus noise. (*c*) shows that when (*a*) and (*b*) are combined, the boundary between "no" and "yes" responses depends upon the position of the decision criterion

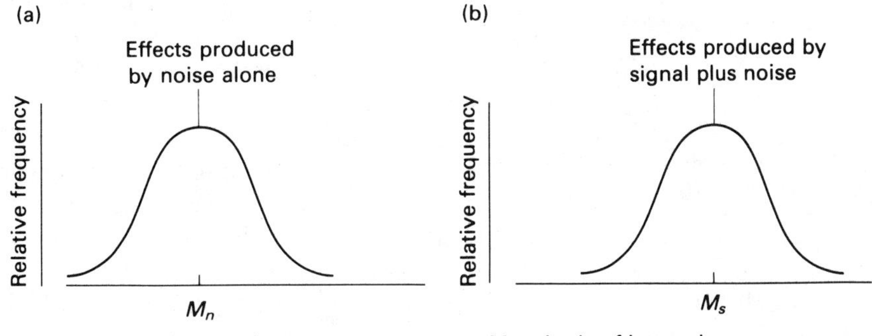

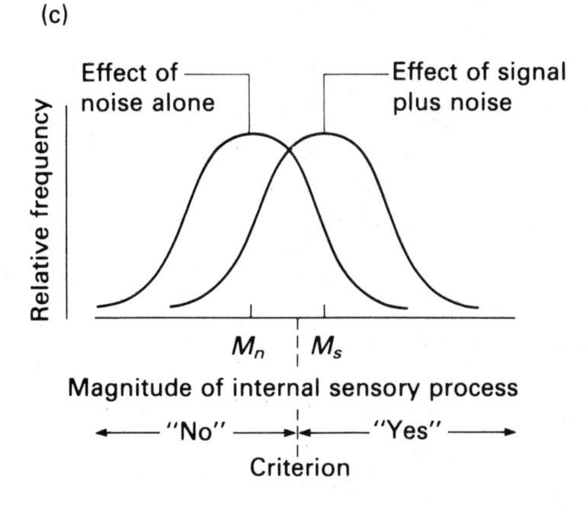

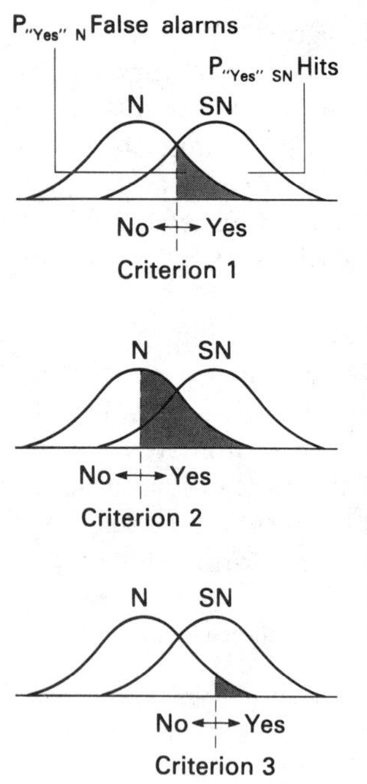

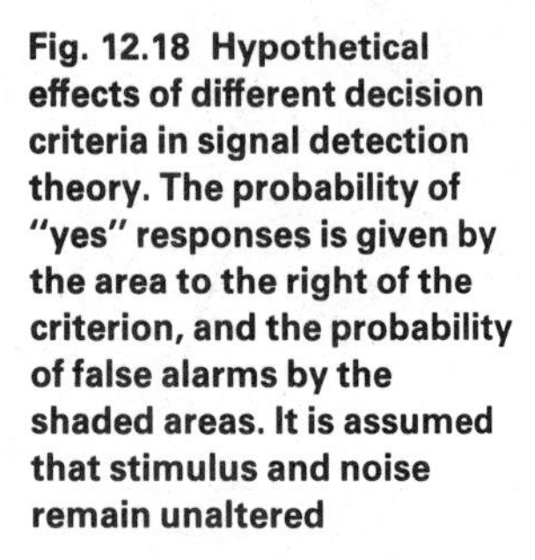

Fig. 12.18 Hypothetical effects of different decision criteria in signal detection theory. The probability of "yes" responses is given by the area to the right of the criterion, and the probability of false alarms by the shaded areas. It is assumed that stimulus and noise remain unaltered

testing the theory is to hold the stimulus situation constant and to manipulate the payoff. Another kind of test is to maintain a constant response bias and to alter the stimulus properties. The effect of this kind of experiment on response bias is fairly reliable. When the subject is more conservative, there is a reduction in both the false alarms and hits, and vice versa (Fig. 12.18). When the stimulus magnitude is increased, the number of false alarms diminishes and the number of hits increases.

Psychophysical methods have been applied to a number of aspects of animal behavior. The first study of this type seems to have been that of Donald Blough (1955), who investigated dark adaptation in pigeons. Blough trained pigeons to peck at two response keys just below an illuminated window. The only light in the apparatus came from this source. The birds were trained to peck key A when the stimulus was visible to them and key B when it was not visible. Pecks at key A automatically resulted in dimming of the stimulus, while pecks at key B increased the brightness of the stimulus. Once the bird has been trained, it is then tested in the following way.

Initially, the stimulus window is brightly lit and the bird pecks only at key A. Each peck makes the stimulus less bright, but the bird continues to peck key A until the stimulus becomes so dim it is below the pigeon's absolute threshold. At this point the bird begins to peck key B, which causes the stimulus to become brighter. When the stimulus again becomes visible to the bird, it returns to key A. Over the course of an hour, the bird alternately pecks keys A and B and the stimulus brightness fluctuates in the region of the bird's absolute threshold. The effect of this procedure is to trace out the pigeon's *dark adaptation curve.* At first the bird is light adapted, but as the stimulus becomes dimmer, it starts to be dark adapted. Thus, the pigeon's threshold progressively changes in the manner illustrated in Figure 12.19. Notice that this dark adaptation curve exhibits the characteristic kink that occurs at the change from cone to rod vision (see Fig. 12.14).

12.5 Stimulus filtering

Anatomical and physiological investigations of sense organs and of associated parts of the nervous system can yield valuable information about an animal's sensory capabilities. However, they do not on their own provide conclusive evidence of what an animal does or does not perceive, so behavioral confirmation is usually desirable. Moreover, the demonstration that certain sensory information is available to the CNS does not tell us how it is used.

Much more information is potentially available to an animal than it possibly could register and respond to. In some way the animal has to be selective, to respond to those events that are important and to ignore others. The term for this phenomenon is *stimulus filtering,* the idea being

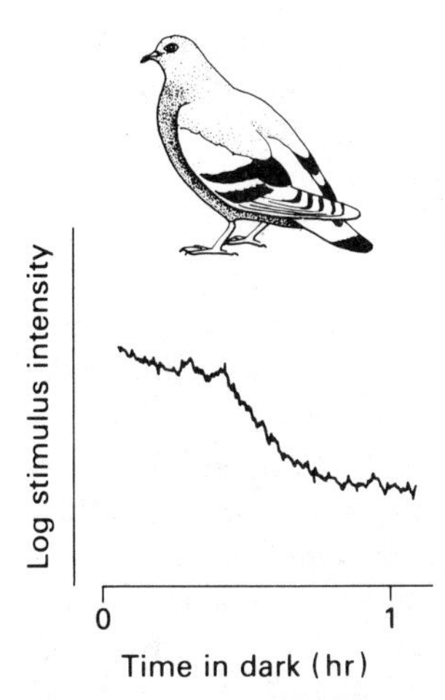

Fig. 12.19 Dark adaptation curve of a pigeon obtained by a psychophysical method (After Blough, 1955)

that, at various stages in the causal chain between stimulus and response, certain stimuli are filtered out and do not influence the animal's behavior.

In a sense, some filtering is inherent in the limited capabilities of the sense organs. For example, the human ear does not respond to sounds with a frequency higher than 20 kilohertz. The human eye filters out the infrared and ultraviolet parts of the spectrum, although these are detectable by some other animals (see Chapter 13). An interesting example comes from work on the tree frog *Eleutherodactylus coqui.* This frog is named after the distinctive *co-qui* call of the male, which serves to attract females and to repel males. The ear membrane is tuned differently in males and females. The males hear only the *co* note, and the females hear only the *qui* part of the call. Similarly, insect receptors often are highly specialized, responding only to a narrow range of stimulation, as in the case of olfaction in the silkworm moth (*Bombyx mori*).

Once a stimulus has been detected, it may be automatically categorized in a way that filters out other aspects of the stimulus. For example, Jerry Lettvin et al. (1959) showed that the photoreceptors of the frog's retina are connected together to form a receptor field, as illustrated in Figure 12.20. Some of these, termed "bug detectors," respond preferentially to small, dark, moving objects. When feeding, the frogs respond more to such objects than to other stimuli.

Selective responsiveness is common among animals. An animal responding in a particular situation utilizes only part of the potential information available. For example, David Lack (1943) observed that male European robins (*Erithacus rubecula*) often attack other red-breasted robins that trespass on their territory. They also will attack a stuffed robin placed in the territory, but only if it has a red breast. It appears that the red breast is a powerful stimulus for labeling another robin as an intruder. Indeed, Lack showed that a territory owner would vigorously attack a bunch of red feathers, just as if it were an intruding robin (see

Fig. 12.20 The bug detectors of a frog are unresponsive to movement of a field of dots (*a*), but are responsive to movement of a single dot (*b*), especially if the movement is irregular (*c*)

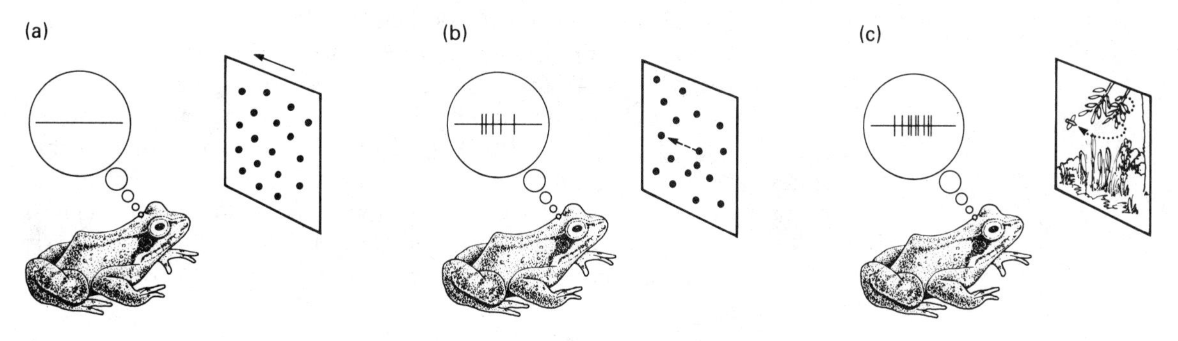

Fig. 20.1). No one doubts that a robin is capable of distinguishing between a bird and a bunch of feathers, but in the territorial defense situation robins appear blind to all attributes except the red breast. This type of stimulus is called a *sign stimulus*.

Sign stimuli may change with changes in the animal's internal state. Thus, herring gulls (*Larus argentatus*) will steal and eat another gull's eggs. The raiding gull recognizes the eggs by their shape. For the incubating gull, however, the size and coloration of the egg are most important in retrieving eggs that have rolled from the nest, and shape is relatively unimportant. Thus, the birds will retrieve readily spherical or cylindrical objects of the same size and coloration as a normal egg. Once the bird has settled down, however, shape becomes important again, and if the eggs do not have rounded edges, the birds will not settle down to incubate (Baerends and Drent, 1970). Thus, the eggs provide three different sets of sign stimuli appropriate to three different kinds of behavior.

Another interesting example of sign stimuli comes from the work of Gordon Burghardt and his co-workers on the feeding behavior of garter snakes (*Thamnophis*) and water snakes (*Natrix*), which live in rivers and ponds in the United States and feed on small fish and worms. They detect their prey by taste and smell and have chemoreceptors located in paired pits in the roof of the mouth, called "Jacobson's organ." The snake flicks out its tongue, picks up chemicals from the prey, and then places the tip of the tongue in the taste pits. Burghardt tested various species of garter snake with cotton swabs containing solutions from extracts of prey such as fish, frogs, salamanders, and worms. He found that the different species had strong preferences for the type of prey on which they would normally feed in nature. These preferences were present in newborn snakes that had never fed, and they could not be changed by manipulation of the mother's diet or by force-feeding the young snakes on artificial food (Burghardt, 1970). The preferences do change, however, when the snakes switch to another prey. Thus, snakes with a natural preference for minnows would prefer goldfish once they had become used to feeding on them (Burghardt, 1975). Apparently the chemical characteristics of certain prey act as sign stimuli toward which the snake has a genetically determined bias and that is modifiable as a result of experience.

While some sign stimuli may be the result of peripheral filtering, as with the frog's bug detectors, it is evident that most must be due to events within the CNS. The early ethologists postulated the existence of an *innate releasing mechanism* (*IRM*), which they thought was responsible for the recognition of sign stimuli. For reasons discussed in Chapter 20, the IRM concept is no longer popular among ethologists. Nevertheless, ethologists agree that some kind of central filtering mechanism is responsible for the fact that many species show preferential responsiveness to sign stimuli.

Ethological research also has drawn attention to a type of stimulus filtering that is transitory compared with the relatively permanent recognition of sign stimuli. This is the *searching image* concept first proposed by Jakob von Uexkull (1934). We are all aware of the perceptual phenomenon of suddenly seeing something that we previously overlooked. When we look at a photograph of camouflaged insects, we may not see any at first, but then we suddenly may see one. Thereafter, the insects seem easy to pick out. We have developed a searching image for the insect.

A number of studies demonstrate the existence of such searching images in animals. Harvey Croze (1970), for example, trained carrion crows (*Corvus corone*) to search for bait hidden under mussel shells of various colors that were widely distributed over the ground. The crows had to turn over each shell to see if there was any food underneath. They tended to concentrate their search on one color at a time and overlooking shells of other colors, even though they had experience of gaining food under all types of shells. A single sample, handed by the experimenter, was sometimes enough to trigger the search for a particular color of shell. The crows behaved as if they had a searching image for a particular color of shell that lasted for a certain amount of time but that could be readily switched to another color.

The most thorough study of searching images is that of Marian Dawkins (1971a; b) on domestic chicks in a laboratory situation. The chicks were fed for three weeks on either orange- or green-dyed grains of rice on a white background. Then they were tested for their ability to detect grains on a background of a different color (conspicuous grains) and on a background of the same color (cryptic grains), as illustrated in Figure 12.21. Dawkins found that although the chicks at first were unable to detect the cryptic grains, they showed a marked improvement with experience. She also found that the experience of eating conspicuous grains decreased the chicks' ability to see cryptic food. It appears, therefore, that the ability to detect cryptic food cannot be due simply to improvement with experience.

In further experiments, Dawkins tested chicks that were already

Fig. 12.21 Rice presented to chickens on various backgrounds. (*a*) green rice on a green background, (*b*) green rice on an orange background (*c*) orange rice on an orange background (*d*) orange rice on a green background (From Marian Dawkins, 1971a)

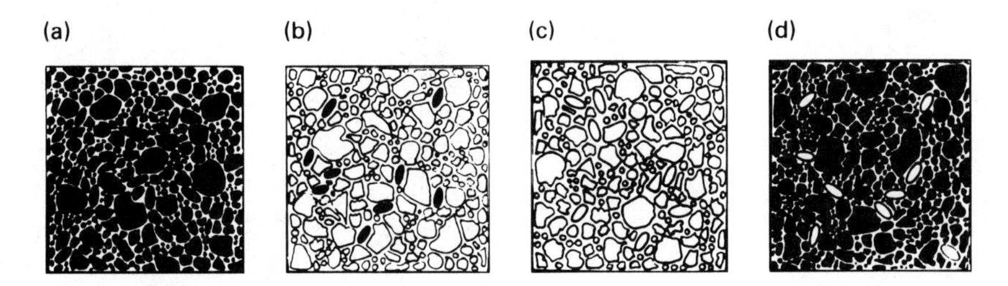

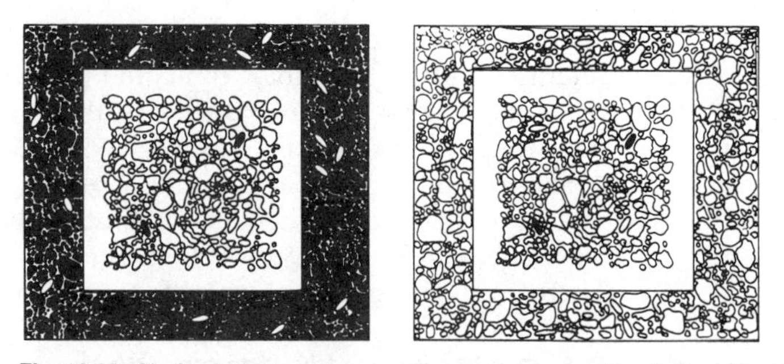

Fig. 12.22 Choice between two cryptic orange grains (top left and bottom right of test square) and two conspicuous green grains, amid conspicuous orange sample grains (left) and amid cryptic orange sample grains (right) (From Marian Dawkins, 1971b)

feeding on particular sample grains (Fig. 12.22). She found that when the sample grains were conspicuous, the chicks were more responsive to test grains that were distinguishable by color. When the sample grains were cryptic, the chicks responded more to test grains distinguishable by non-color cues such as texture and shape. These results can be interpreted in terms of selective attention. Thus, when the sample grains were conspicuous by virtue of their color, the chicks were attending to color and consequently found it easier to detect test grains on the basis of color. When the sample grains were cryptic, they were exactly the same color as the background, and the chicks attended to non-color cues. As a result, they found it easier to detect test grains on the basis of cues other than color. Dawkins's hypothesis is supported by numerous demonstrations of selective attention in other animals.

An animal may be able to discriminate between two stimuli, but it may not normally do so until it has learned to tell the difference. Such learning is called *discrimination learning*. Appropriate stimuli for investigating discrimination learning in rats would be a black rectangle versus a white rectangle. It is usual to present these on a grey background, and the experiment would incorporate all the normal controls discussed below. Instead of limiting the animal to a single cue, which is essential when investigating sensory capabilities, we might study how the availability of two ways of solving the problem influences the learning process. Thus, the rectangles could be of different sizes or of different orientations.

Let us consider the case of brightness and orientation differences, so that if the black rectangle is presented vertically, the white one is horizontal. Now we have to decide which stimulus is to be rewarded and which is not. Here we run into the problem of confounding the rat's preferences and learning ability. Thus, it may be that rats find it easier to learn about brightness than about orientation but that they also prefer black to white. If we reward the black stimulus, the rat already would be biased toward the correct choice and would learn the problem more easily than if we rewarded the white stimulus. Thus, the rat's initial preferences easily could confound the investigation of its learning abilities. One way of overcoming this difficulty is to divide 20 rats into two groups of 10, one rewarded for choosing black and the other rewarded for white. However, the same problem arises with respect to the orientation cue, so each group should be divided again into halves, one of which is rewarded for choosing the vertical rectangle and the other for choosing the horizontal rectangle. Table 12.5 shows that, with respect to the configuration of the rewarded stimulus, the experiment has a balanced design.

Table 12.5 Design of discrimination experiment

Group	*Number of rats*	*Orientation of stimulus*	*Color of stimulus*
1	5	H	B
2	5	V	B
3	5	H	W
4	5	V	W

B = black; W = white; H = horizontal; V = vertical.

To solve this type of learning problem, rats have to learn which stimulus to approach to obtain reward and which to avoid. However, reward could be associated with many characteristics of the stimuli. Thus, the rats could learn to associate reward with the shape, size, position, orientation, color, or brightness of the stimuli. In the present case, the rats never could solve the problem on the basis of shape, size, position, or color because the rewarded and unrewarded stimuli do not differ in these respects, whereas they do differ in brightness and orientation. But how are the rats to discover which stimulus characteristics to associate with reward?

Stuart Sutherland and Nick Mackintosh (1971) have shown that many animals have to learn two things in solving problems of this type: (1) which aspects of the stimuli to attend to, and (2) which of the two manifestations of that aspect is rewarded. For example, a rat that was rewarded on a trial in which it had attended to the brightness of the stimuli would be more likely to attend to brightness in the future; and if the rat happened to choose the black stimuli, it would be more likely to choose black on those future trials during which it attended to brightness. Rats that learned to attend to brightness or orientation or both could learn successfully to solve the problem outlined here, whereas rats that attended to other aspects of the stimuli, such as size or position, would be able to obtain only 50 per cent reward. Therefore, the rats attending to brightness or orientation or both always will be more rewarded.

To determine which aspects of the stimuli the rats have in fact learned to attend to, unrewarded *transfer tests* are given. During transfer tests, the rats are presented with stimuli differing in orientation or brightness, but not both. Thus, on half the trials the stimuli are black or white rectangles differing only in orientation, and on the other half the stimuli are horizontal or vertical rectangles differing only in brightness. Rats that have learned to attend to orientation only are able to solve the first problem but not the second, while those that had learned to attend to brightness only are able to solve the second problem but not the first. Rats that have learned to attend to both brightness and orientation are able to solve both types of transfer tests.

Overall, the evidence strongly suggests that selective attention occurs in many animal species. In many cases the alternative cues are embodied in the same objects, and this means that the selective attention must be controlled centrally and not simply due to the orientation of the head or some other peripheral phenomenon. Selective attention is a type of stimulus filtering in which aspects of the stimulus situation, although registered by the animal's sense organs, are not registered in a way that can influence the animal's learning or behavior.

There is little difference in principle between the searching image concept investigated by Dawkins (1971a and b) and the selective attention concept proposed by Sutherland and Mackintosh (1971). Indeed, if the theories of selective attention are correct, we would expect to find circumstantial evidence for it in the behavior of animals in their natural environment. The ecologist Luke Tinbergen (1960) noticed that great tits (*Parus major*) hunting for prey in pine forests in the Netherlands did not take a new type of food as soon as it became available. Even if the new prey appeared gradually, it often was ignored for several days, and then there was an abrupt change in the predation level. Tinbergen postulated that the new prey was overlooked until the birds had formed an appropriate searching image. John Krebs (1973) reviews other similar examples.

12.6 Complex stimuli

In animals, as in humans, we can describe sensory abilities in terms of the physical attributes of the stimuli, but this does not tell us about the animal's subjective assessment of the stimulus. It is important for ethologists to know what importance animals attach to the various aspects of a stimulus situation. The most successful attempts to do this have been those in which the animal is persuaded essentially to reveal its internal standards of comparison among stimuli. As an example, we first consider the work of Gerard Baerends and Jap Kruijt (1973) on egg recognition in herring gulls, birds which lay three eggs in a shallow nest on the ground and which will retrieve dummy eggs that resemble real eggs in certain respects (Fig. 12.23).

Baerends and Kruijt set out to discover the features by which herring

Fig. 12.23 A herring gull retrieving an egg into its nest (*Photograph: Gerard Baerends*)

gulls recognize an egg in this type of situation. They removed two eggs from the nest and placed two dummy eggs on the rim of the nest. The dummies were of various sizes and could differ in color, pattern, or shape. The size of a dummy egg was measured in terms of the area of its maximal projection (that is, the maximal shadow obtainable when the egg is turned around in a parallel beam of light). All the dummies used in these tests were measured and their sizes expressed in terms of a standard series ranging from sizes 4 to 16. A real herring gull egg would be about size 8.

When two identical dummy eggs are placed on the rim of the nest, the gull usually has a marked position preference, retrieving the left before the right or vice versa. To overcome this positional bias, a series of tests were run in which position was titrated against size (Baerends and Kruijt, 1973) (Fig. 12.24). If the bird tends to respond to the egg on its right in preference to the one on its left when both eggs are the same size, the experimenter places a larger egg on the left so that the bird is equally motivated to respond to either. By this means, the position preference is given a value r, which is the ratio of the sizes of the

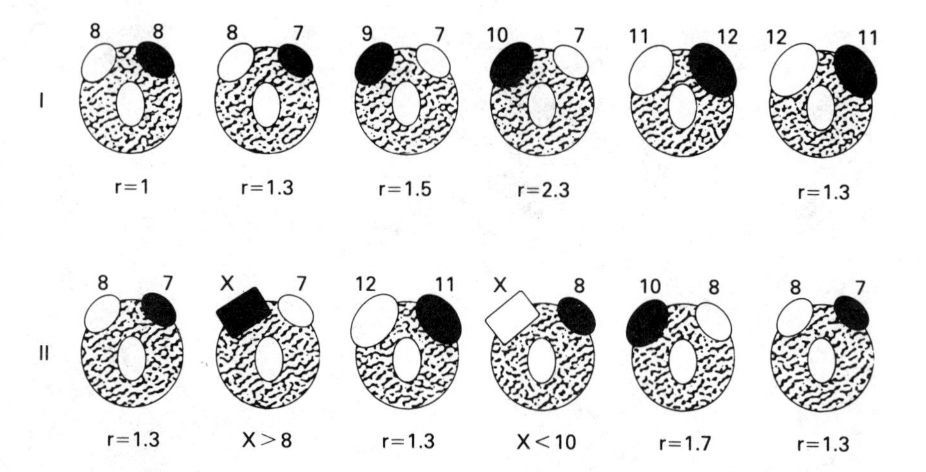

Fig. 12.24 The titration method of determining the value of an egg dummy. The circle represents the nest with one egg in the nest bowl and two dummies on the rim. The numbers 7, 8, 9, 10, 11, 12 refer to the size of the dummies. *r* is the ratio of the sizes of the dummies on the nest rim. *X* is the model to be measured. The dummy chosen in each trial is indicated in black.

I. Determination of the value of the gull's position preference. The first (left) shows that the right-hand egg is preferred. The next (to the right) shows that this preference remains when a smaller egg is substituted on the right-hand side. The sequence of tests shows that the value of the position preference lies between $r = 1.3$ and $r = 1.5$.

II. Determination of the value of model *X*. The control tests show that the position preference remains unchanged. The experimental tests show that the value of *X* lies between sizes 8 and 10 (After Baerends and Kruijt, 1973)

dummies on the nest rim. Once *r* has been established for a particular bird, it is possible to introduce a test dummy that differs from a real egg in shape or coloration. Figure 12.24 shows that the titration procedure can yield accurate *r* values for any test dummy.

The results of large numbers of titration tests are given in Figure 12.25. Four types of dummy eggs—green-speckled, plain green, plain brown, and cylindrical brown-speckled—are compared with the standard-size series of dummy eggs with the same brown-speckled coloration as real eggs. Apparently, the gulls prefer a green background color to the normal brown background. They also prefer eggs with speckles to eggs without speckles. More important, however, is the finding that the differences among eggs of different sizes remain the same when other features like coloration are changed. This means that each feature adds a specific contribution that is independent of the contribution of other features; that is, features such as the size, shape, and coloration of the eggs are additive in their effects upon the animal's behavior. Such a state of affairs had been proposed long ago (Seitz, 1940) and called the *Law of heterogeneous summation*, which states that the independent and heterogeneous features of a stimulus situation are additive in their effects upon behavior.

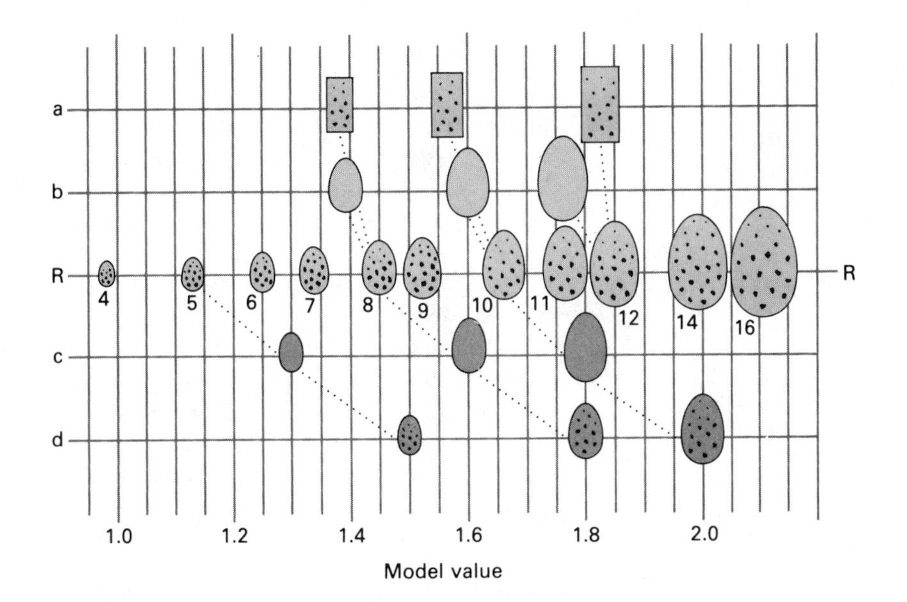

Fig. 12.25 Results of the titration tests. The average values for various dummies are shown in relation to the standard size series R. The dummies used were (*a*) brown, speckled, block-shaped; (*b*) brown, unspeckled, egg-shaped; (*c*) green, unspeckled, egg-shaped; (*d*) green, speckled, egg-shaped. The numbers 4 to 16 indicate egg size. Equal distances between points on this scale imply equal *r* values (After Baerends and Kruijt, 1973)

Another study that comes to the same general conclusion is that of Walter Heiligenberg and his co-workers on the stimuli eliciting aggression in cichlid fish. Heiligenberg and his co-worker (Leong, 1969) studied the quantitative effect of sign stimuli on the attack readiness of the cichlid fish *Haplochromis burtoni*. Heiligenberg's method was to place an adult male fish in an aquarium together with numerous juvenile fish. The male is free to attack the juvenile fish, but they easily escape and fighting does not occur. A dummy of a rival adult male is presented behind glass at the initiation of each trial. Before the presentation, the attack rate of the resident male toward the juvenile fish is recorded for 15 minutes. During the presentation, usually 30 seconds long, the resident male usually quietly watches the dummy. After the presentation, the attacks on the juvenile fish are recorded for 30 minutes. The rate at which the adult male attacks juvenile fish usually increases markedly just after presentation of a dummy and then wanes to its prepresentation level. This increment in attack rate is used as an index of the effectiveness of the dummy in increasing the aggressiveness of the resident fish.

Because the attack rate fluctuates considerably, presumably reflecting the male's internal state, statistical methods have to be used to compare the observed attack rate with that expected from the immediately preceding observation period. Using this approach, Leong (1969) found that different aspects of the color pattern of territorial males painted on

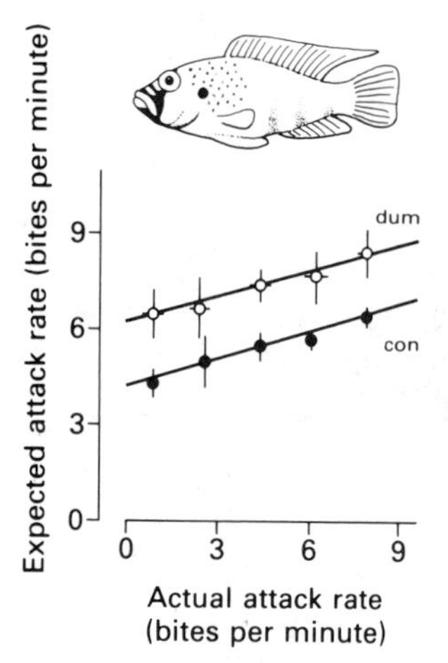

Fig. 12.26 Expected attack rates of male cichlid fish as a function of previous attack rate. These results show that, compared with the control (con) situation, presentation of a dummy rival fish (dum) has an additive effect upon the attack rate (After Leong, 1969)

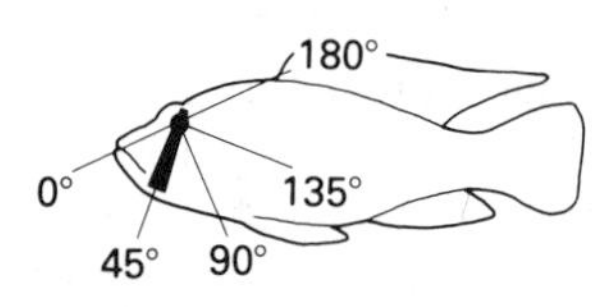

Fig. 12.27 A dummy male cichlid fish fitted with a black "eye bar" made of metal foil, which can be rotated around the center of the eye (After Heiligenberg et al, 1972)

dummies were additive in their effects upon the attack rate, as illustrated in Figure 12.26. Similarly, Heiligenberg et al. (1972) showed that different angular orientations of the black eye bar of a male *Haplochromis burtoni* were additive in their effects upon attack rate and that the orientation of the whole body also made a contribution (see Fig. 12.27).

These experiments show that the color markings of a male can increase or reduce the aggressiveness of a rival male in an additive manner, in agreement with the law of heterogeneous summation. However, Heiligenberg (1976) claims that the body posture has a multiplicative effect upon attack rate. Moreover, Eberhardt Curio (1975), working on the antipredator behavior of the pied flycatcher (*Ficedula hypoleuca*), also obtained some evidence of multiplicative relationships among stimulus aspects of models of two natural predators, the red-backed shrike (*Lanius collurio*) and the pygmy owl (*Glaucidium passerinum*). Thus, it appears that the law of heterogeneous summation does not hold universally.

Where stimulus summation operates within a relatively restricted range of features, such as those pertaining to egg recognition, the additive components can be combined into a single index called the *cue strength* (McFarland and Houston, 1981). Thus, all those additive cues involved in egg recognition as shown in Figure 12.28 can be combined into a single cue strength for egg-recognition features and represented along a single dimension, as in Figure 12.29. Other independent features of the egg retrieval situation, like the distance of the egg from the nest, must be represented along a different dimension. Suppose we now measure the gull's tendency to retrieve eggs as a joint function of the stimulus properties of the egg and the distance from the nest. We would find that certain combinations of these two factors would give rise to the same retrieval tendency. The line joining all points giving rise to a certain motivational tendency is called *motivational isocline* (see Chapter 15.6).

When independent factors combine multiplicatively, as in the example illustrated in Figure 12.29, then the motivational isocline is hyperbolic in shape. The evidence for multiplicative relationships comes mainly from situations where disparate aspects of the stimulus situation are involved in determining a single response (McFarland and Houston, 1981). In the case of the cichlid fish *Haplochromis burtoni*, for example, we can see that the color markings on a rival fish are additive in their effect in eliciting aggression in another fish, as Heiligenberg found. They combine to indicate the potential threat from the rival. The body posture of the other fish is a separate matter, since it signifies the aggressive intention of the rival. The head-down posture often is adopted just prior to attack. Thus, a possible interpretation of Heiligenberg's (1976) study is that a multiplicative relationship exists between the stimuli indicating the aggressive potential of a rival and those indicating his aggressive intention.

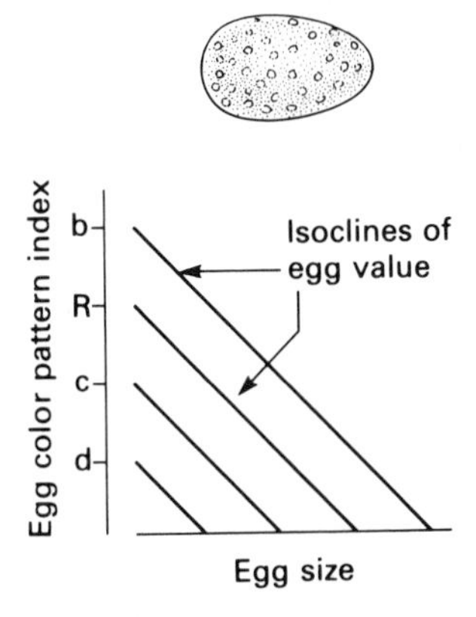

Fig. 12.28 Cue space for egg value. The parallel isoclines indicate that egg color pattern and egg size are additive in their effects, and could be represented by a single index of egg value

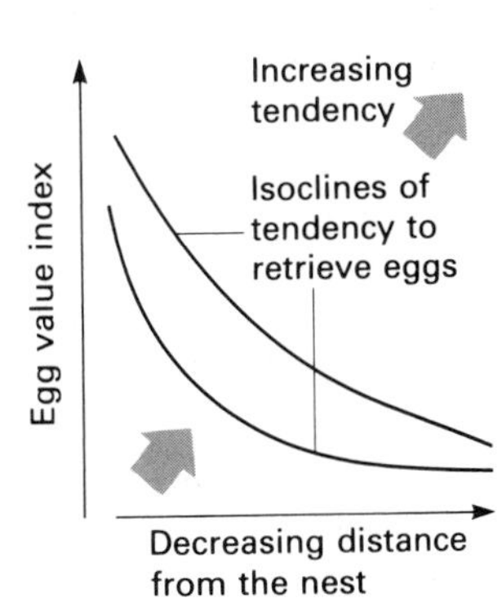

Fig. 12.29 Cue space for egg retrieval. The hyperbolic shape of the isoclines indicates that the egg value index combines multiplicatively with distance of the egg from the nest

In the interpretation of responses to complex stimuli the main difficulty is to interpret the animal's responses in terms of the animal's perception of the stimuli rather than in the experimenter's terms. The problem is that the experimenter is obliged to use an arbitrary scale of measurement without knowing how this corresponds to the subjective evaluation made by the animal. Some ways of overcoming this difficulty have been suggested by Houston and McFarland (1976) and McFarland and Houston (1981).

Points to remember

● Sensory processes can be divided into different modalities, each having characteristic sense organs. However, the sensation perceived depends upon the part of the nervous system that is activated and not upon the type of sense organ stimulated (Müller's Law of specific nerve energies).

● Chemoreception and thermoreception both provide information about the situation inside the body as well as about changes in the external environment.

● Hearing is a form of mechanoreception designed to pick up vibrations in air and water. It may be based upon a simple mechanism involving few neurons, as in the special-purpose ears of moths, or it may involve a complex mechanism, as in the sophisticated hearing of humans.

● Vision ranges from simple discrimination of light and dark to the formation and resolution of images. It is widespread in the animal kingdom, but species vary enormously in the type of eye and visual capability.

● Sensory judgments made by humans are not simply a matter of minimal discriminable differences between stimuli but depend upon the ability to separate signal from noise.

● Stimulus filtering, the process of classifying the information that impinges upon the animal, takes place at many levels. It may occur peripherally, as part of the design of sense organs, or it may occur centrally, as a form of selective attention.

● Complex stimuli may be evaluated by animals in different ways according to their internal state. In some cases specific sign stimuli elicit particular responses; in others it is the summation of heterogeneous factors that is important (law of heterogeneous summation).

Further reading

Ewert, J. P. (1980) *Neuroethology*, Springer-Verlag, Heidelberg.

13 Ecology of the Senses

When the animal kingdom is viewed as a whole, we can see evolutionary trends in sensory mechanisms. However, evolution results from natural selection, and natural selection is a form of adaptation to current circumstances. It should not surprise us, therefore, to find that many of the differences among animals are due not so much to ancestry but to their ecological circumstances.

Different sense modalities are best suited to different habitats, and particular modalities may be modified to suit the animal's way of life. This is true not only of the relationship between the animal and its physical environment but also in respect to its social environment. The effectiveness of communication depends upon the relationship between the modality used and the medium through which the information is transmitted.

In the case of the visual modality, the directionality of light permits greater accuracy of stimulus localization and spatial pattern recognition than is possible with any other modality. It is no accident that the predatory animals, such as the octopus, hawk, and cat, have highly developed visual apparatus compared with other mollusks, birds, or mammals. The visual modality also permits relatively durable means of communication in that animals may evolve particular surface patterns or may construct artifacts that convey information on a long-term basis. The main disadvantage of vision is that it works well only under certain conditions. Visual signals cannot bypass obstacles in the way that auditory and olfactory signals can. Vision is restricted in dark or murky conditions, although many animals have special adaptations that enable them to overcome this disadvantage to some extent.

Sound has two particular properties that make it an important vehicle for communication. First, it is possible to make rapid temporal changes in sound signals. Their transient nature permits rapid exchange of information, which may be important in a highly mobile species. Sound signals also can be turned off when necessary. For example, the house cricket (*Acheta domestica*) is well camouflaged visually. The male advertises its presence by prolonged bouts of chirping. When it senses danger it simply stops chirping and remains immobile, thus rendering itself difficult to locate by a predator. The second important property of sound is that it can be raised in intensity above the background level of the environment. This is not possible with visual signals, except for those few species that emit light. The loud vocalizations of birds and monkeys of the tropical forest are examples of utility of sound in penetrating obstacles and background noise.

Olfaction has some of the advantages of both vision and hearing. Scents can be deposited as a durable message and are used widely in this way be terrestrial insects and animals. Like auditory signals, chemical messages can be transmitted around obstacles and can be raised in intensity above the background level. The sensitivity of olfactory receptors, which we saw in our discussion of the silk moth pheromone bombykol (Chapter 12), makes chemical signals particularly suitable where communication over very long distances is required. The relative durability of chemical signals sometimes can be a disadvantage, and the inadvertent deposition of chemical clues is exploited by many predators. Each modality has its advantages and disadvantages, and many animals make use of all three modalities, deploying them according to circumstances. There are also many other, more specialized senses, which we discuss in the remainder of this chapter.

13.1 Visual adaptations to unfavorable environments

As we saw in Chapter 12 an animal adapted for seeing in bright light is likely to have good acuity, color vision, and movement perception. Such typical diurnal eyes are relatively insensitive to low levels of illumination. Animals adapted for seeing in dim light have greater sensitivity, but achieved at the expense of detailed pattern and color vision.

The need to see in dim light occurs in nocturnal animals and in animals that inhabit deep water and caves. These conditions are not strictly comparable since the spectrum of available light is shifted toward the longer wavelengths at night and toward the shorter wavelengths in deep water. There is no evidence that nocturnal animals have an increased sensitivity to long wavelengths (Lythgoe, 1979). However, some do have adaptations that enhance sensitivity to light. Eyes with a large pupil and lens pick up more light than small eyes. Such eyes are found in the opossum, house mouse, and lynx (Fig. 13.1). In other nocturnal animals, such as owls and bushbabies, the limitation on lateral expansion of the skull has led to tubular extension of the eye. Tubular eyes also are found in some deep-sea fish (Fig. 13.1). The eye of the tawny owl is somewhat more sensitive than a human eye (Martin, 1977), but not enough for efficient night hunting. As we shall see owls also rely on other senses. Many deep-sea animals are adapted specifically to the prevailing light conditions. The visual pigments of deep-sea fish have absorption maximums that coincide with the wavelength of maximum water transparency (Denton and Warren, 1957; Munz, 1958). The deep-diving beaked whale (*Berardius bairdi*) has visual pigments

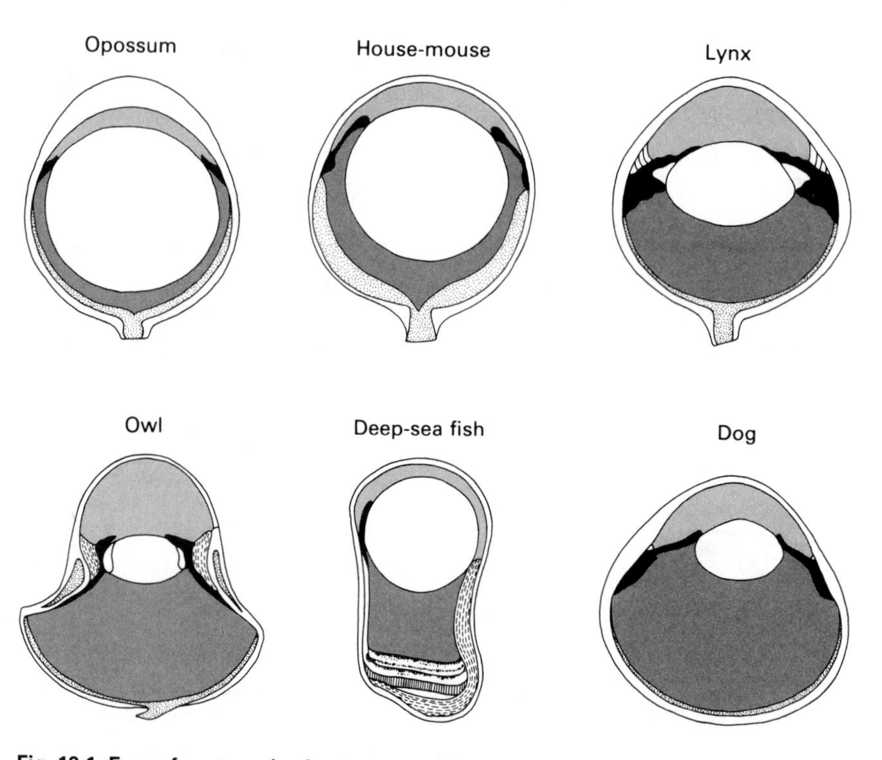

Fig. 13.1 Eyes of nocturnal animals compared with those of a dog which has both day and night vision (After Tansley, 1965)

with absorption maximums of shorter wavelength than the shallow-diving grey whale (*Eschrichtius gibbosus*). Similarly, deep-sea crustacea possess visual pigments of shorter-wavelength maximal absorption than shallow-water crustacea (Goldsmith, 1972).

In addition to the properties of the visual pigments, other specialized features enhance vision in dim light. Typically, nocturnal animals have a greater proportion of rods than cones, and some—such as the dogfish (*Scyliorhinus*) and bushbaby (*Galago*)—have few or no cones. Many rods may be connected to the same bipolar cell, thus increasing the sensitivity to light at the expense of acuity (see Chapter 12). Many nocturnal vertebrates have light-reflecting bodies, called *tapeta*, near the photoreceptors that cause the eyes to glow when caught in a beam of light. Light that has passed through a photoreceptor without being absorbed is reflected back, thus increasing the probability of absorption. In some fish, the tapeta can be covered by migrating pigment granules. In the spur dogfish (*Squalus acanthias*), for example, the dark-adapted tapetum reflects 88 per cent of the incident light, whereas after light adaptation—when the tapeta's reflections have been screened out—only 2.5 per cent of the incident light is reflected (Nicol, 1965). Animals that live in very dim light may adapt in various ways, as we have seen. Some animals live in turbid media, in which visual contrast is reduced as a result of light scattering by the suspended particles. For such animals, little can be done to rectify the situation (Lythgoe, 1979).

There comes a point at which visual conditions are so difficult that vision has to be abandoned as a primary sense. Most animals that live in caves, the very deep sea, or turbid water have rudimentary eyes that are the result of regression. For example, the cave salamanders (*Typhlotriton* and *Proteus*) have eyes at the larval stage but no eyes as adults. If these salamanders are reared in light, the adults develop normal eyes (Lythgoe, 1979). The blind cave fish (*Astyanax mexicanus*) has eyes in the juvenile form that degenerate in the adult. Sadoglu (1975) made genetic crosses between these fish and a surface-living species of *Astyanax* that has normal vision. He found that genes are responsible for the degenerative condition of the eye in the cave-living species. The cusk eels (Ophidiidae) usually live at greater depths and have regressed eyes. Some species (e.g., *Lucifuga subterranea* and *Stygicola dentatus*) have evolved secondarily to life in caves.

Among the mammals, moles and bats provide the most obvious examples of visual degeneration. Moles have very small eyes, covered with skin in some species. The fruit-eating bats (Megachiroptera) have well-developed vision, but the nocturnal bats (Microchiroptera), particularly those species that catch insects on the wing, have very poor vision. Obviously, they have to rely on other senses to forage for food.

13.2 Senses that substitute for vision

Animals that live in dim light and that have poor vision have to rely on other senses. Thus, the bottom-living cat fish and cusk eels have various sensory barbels with which they probe the substratum. These are supplied with numerous touch receptors and chemoreceptors. These senses, however, cannot substitute for vision in providing information about the size and position of objects in the environment. Fish with neuromast and lateral line organs that are sensitive to vibration can detect the presence of moving objects and may obtain some information about stationary objects from water movements reflected from them (Schwartz, 1974; Pitcher et al., 1976). The electromagnetic senses and developments in the sense of hearing, however, have provided animals with the best substitutes for vision.

Many lower animals are able to orient themselves in artificial electric fields, but little is known about the sensory basis of this behavior. A number of species of fish make use of electrical sensitivity in their normal orientation and communication, and scientists know a considerable amount about their electrosensory systems. Sensitivity to magnetic fields also has been demonstrated in a number of animals. Certain bacteria, for example, orient toward magnetic north and will respond to a magnet in the laboratory (Blakemore, 1975). Examination under an electron microscope reveals that these bacteria have chainlike structures containing crystals of magnetite, which also have been found in the abdomen of honeybees and in the pigeon retina. The bacteria in the northern hemisphere follow the declination of the earth's magnetic field, and this steers them down into the anaerobic mud, their normal habitat. Those in the southern hemisphere have the polarity reversed. Magnetically oriented behavior has been studied also in bees and pigeons, and some researchers have claimed that humans are sensitive to magnetic fields (Baker, 1981).

Fishes use electricity in three principal ways: (1) The so-called "strongly electric" fish, such as the electric ray (*Torpedo*) and electric eel (*Electrophorus electricus*), produce electric shocks capable of stunning prey but do not possess an electric sense. (2) Electrosensitive fish, such as the dogfish (*Scyliorhinus*) and sharks, do not produce electricity. Dogfish are capable of detecting prey, even when buried in sand, by the local distortion of the earth's electric field. The sense organs responsible are the ampullae Lorenzini, which are distributed widely over the body surface, especially near the head. (3) The so-called "weakly electric" fish (Gymnotidae and Mormyridae) generate their own electric fields and are sensitive to electrical changes in the environment. These fish are usually nocturnal and live in turbid water, where vision is not practicable. They have two types of electrosensitive receptors: *ampulla* receptors, which respond to slowly changing electric fields, and *tuberous* receptors, which only respond to rapidly changing fields. Some species possess only one type of receptor; others have both. These species generate weak electric fields by means of electric organs, which are modified muscles or neuronal axons. The electrical discharges typically are pulsed at up to 300 pulses per second. Some fish can vary the pulse rate as a means of communication with other fish or as a part of a jamming avoidance response that reduces interference from the fields generated by other

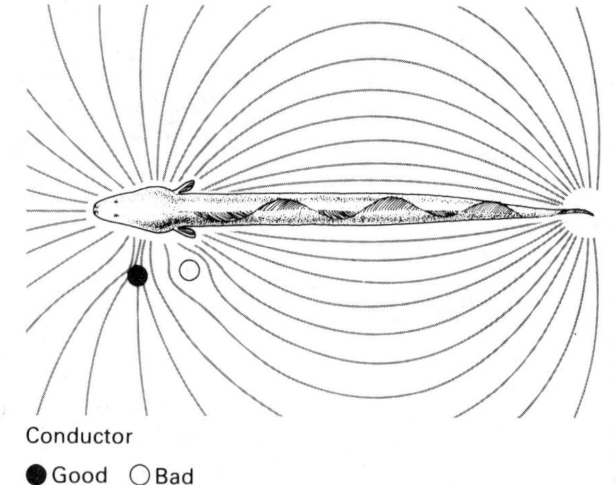

Fig. 13.2 The electric field of *Gymnarchus*. The field on the animal's right is undistorted. The field on the left is distorted by a good conductor (black) and by a poor conductor (white). The animal can detect the presence of these two objects by sensing the effects they have on the electric field

members of the species. In other words, when one fish is subject to electrical interference from another fish it can change its pulse rate to reduce the interference. The electroreceptors also are used to locate objects in the surrounding water by the distortions they cause in the electric field. Some fish, like *Gymnarchus*, can discriminate between good and poor electrical conductors such as a metal rod and a plastic rod, as illustrated in Figure 13.2. A more detailed account of the mechanisms of electroreception in weakly electric fish can be found in Ewert (1980).

Hearing as a substitute for vision occurs in a number of species, some of which have developed highly sophisticated and specialized extensions of normal hearing. All these specializations depend upon an ability to locate sound accurately. Comparison of the sounds reaching the two ears is the most important method of sound localization in vertebrates. A person with one ear can turn the head and scan for the direction of maximum sound intensity because the head blocks some of the sound and causes a sound shadow. With two ears, simultaneous comparison is possible, and this is much more rapid and accurate. If the ears are sufficiently far apart, there will be differences in arrival time and phase of the sounds coming from a particular direction. Thus, small animals rely on intensity comparisons alone, whereas humans use both monaural and binaural methods and for a long time were thought to be the species with the greatest ability to locate the source of a sound. However, as a result of the pioneering work of Roger Payne (1962), we now know that the hearing abilities of the barn owl (*Tyto alba*) far exceed those of humans.

The barn owl is a nocturnal hunter. It can locate and capture a freely moving mouse in total darkness. It even can determine the direction of the animal's movement and thereby aligns its talons with the long axis of the mouse's body. The barn owl is particularly sensitive to the differences in time of arrival of sounds at the two ears. This helps it to determine the azimuth (direction in the horizontal plane) of the sound. Sound intensity differences give further clues to horizontal location. In this respect, the barn owl achieves about the same accuracy as humans, but it is some three times as accurate in determining the elevation (direction in the vertical plane) of the sound source. This accuracy is achieved largely as a result of the owl's facial structure and the positional asymmetry of the ears (Fig. 13.3). The right ear is directed slightly upward and the left ear slightly downward. The right ear is more sensitive to high-frequency (3 to 9 kilohertz) sounds from above the midhorizontal plane of the owl's head, and the left ear to high-frequency sounds from below. As the sound source moves downward, the high-frequency components of the sound become louder in the left ear and softer in the right. When the sound source moves upward, the opposite occurs. This gives precise information about the elevation of the sound source because the owl uses low-frequency components of the sound to locate sounds in the horizontal plane and high-frequency

Fig. 13.3 Facial structure of owl with some of the surface feathers removed to show the assymetrical arrangement of the ears (After Knudsen, 1981)

components in the vertical plane. It does not confuse the two types of information, even though both depend upon comparison of the sounds reaching the two ears (Knudsen, 1981).

The most sophisticated substitute for vision is echo-location, in which the animal utters high-frequency sounds and detects the presence of objects by the echoes produced. The principle is the same as that used in military sonar. Simple forms of echo-location occur in shrews (*Blarina sorex*), oilbirds (*Steatornis caripensis*), and the Himalayan cave swiftlet (*Collocalia brevirostris*) that roost and nest in dark caves. More advanced forms of echo-location are found in dolphins (*Tursiops*) and other marine mammals, but it reaches its pinnacle in the bats (Chiroptera).

The pioneering work of Donald Griffin (1958) has led to a large amount of research into the mechanism of echo-location in bats. We now have a considerable knowledge of the physiology of the bat's vocal and auditory apparatus and of the brain mechanisms involved in echo-location. The student can find more detailed accounts of these in Ewert (1980) and Guthrie (1980).

Echo-locating bats emit bursts of ultrasonic sound pulses of short duration (5 to 15 milliseconds) and high frequency (20 to 120 kilohertz), which are beyond the auditory range of humans. The brief pulses permit accurate timing of the echoes so that the bat can determine the distance of the object producing the echo. Natural sounds produced by wind and other animals are usually of low frequency, so by emitting high-frequency cries, the bats are unlikely to be subjected to interference. Experiments in the laboratory have shown that artificial sounds above the 20-kilohertz frequency range cause flying bats to become disoriented. Another advantage of high frequencies is that they can be beamed precisely, thus allowing resolution of small objects. Bats produce their ultrasonic cries from a specialized larynx and emit them from the lips, as in the naked-backed bat (*Pteronotus*), or from specially shaped nostrils, as in the horseshoe bat (*Rhinolophus*) and leaf-nosed bat (Phyllostomidae).

Bats also have many special adaptations that enable them to detect time and to localize the echoes that result from their ultrasonic vocalizations. Most bats that catch insects on the wing have large external ears shaped to enhance directional sensitivity. As each loud outgoing sound is produced, the sensitivity is reduced by special muscles in the inner ear. If the outgoing pulses are very short, as in the vespertilionid bats, there is no overlap between the end of the outgoing pulse and the beginning of its echo. Since echoes arrive more quickly from nearby objects, the pulses are shortened progressively as an object is approached, so that zero overlap is preserved.

Fig. 13.4 Echolocation of a horseshoe bat approaching a stationary object. As the bat nears the object its velocity declines (top). The bat adjusts the frequency of its cries (middle) in such a way that the frequency of the reflected sound (bottom) remains constant

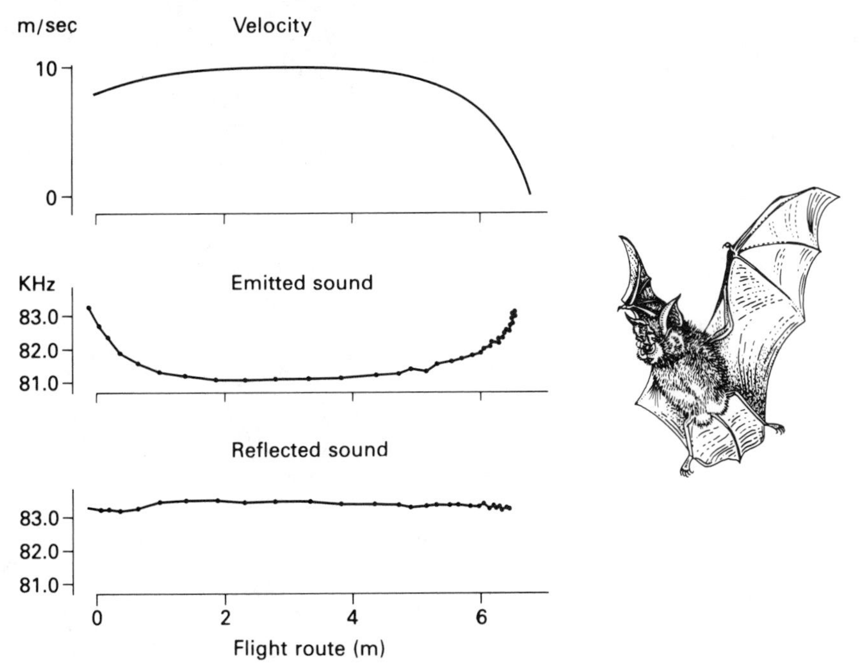

In other bats there is overlap between outgoing pulses and returning echoes, so they require other means of enhancing echo detection. For example, the greater horseshoe bat (*Rhinolophus ferrumequinum*) adjusts the frequency of its cries in such a way that the frequency of the returning echoes is maintained within a narrow range, as illustrated in Figure 13.4. For a bat flying toward an object, the perceived frequency of the echo is always higher than that of the outgoing sound. This is due to the *Doppler effect* that results from the relative motion of the bat and the object: The faster the bat flies, the higher the apparent frequency of the echoes. To compensate for this effect, the bats alter their outgoing frequency in order to maintain the perceived echo frequency as constant as possible. By this means the bat can obtain an estimate of its flying speed and the direction and relative speed of flying prey.

As a substitute for vision, the echo-location abilities of bats are impressive. Laboratory studies show that a bat with a 40-centimeter wing span can fly in total darkness through a 14- by 14-centimeter grid of nylon threads only 80 μm thick (see Fig. 13.5). Bats also have been

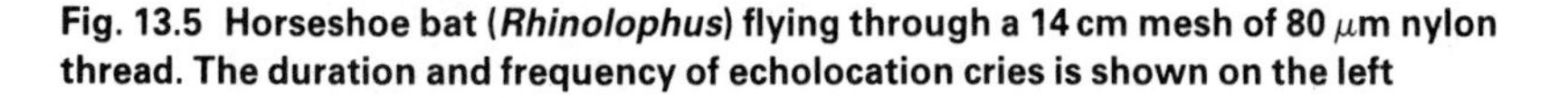

Fig. 13.5 Horseshoe bat (*Rhinolophus*) flying through a 14 cm mesh of 80 μm nylon thread. The duration and frequency of echolocation cries is shown on the left

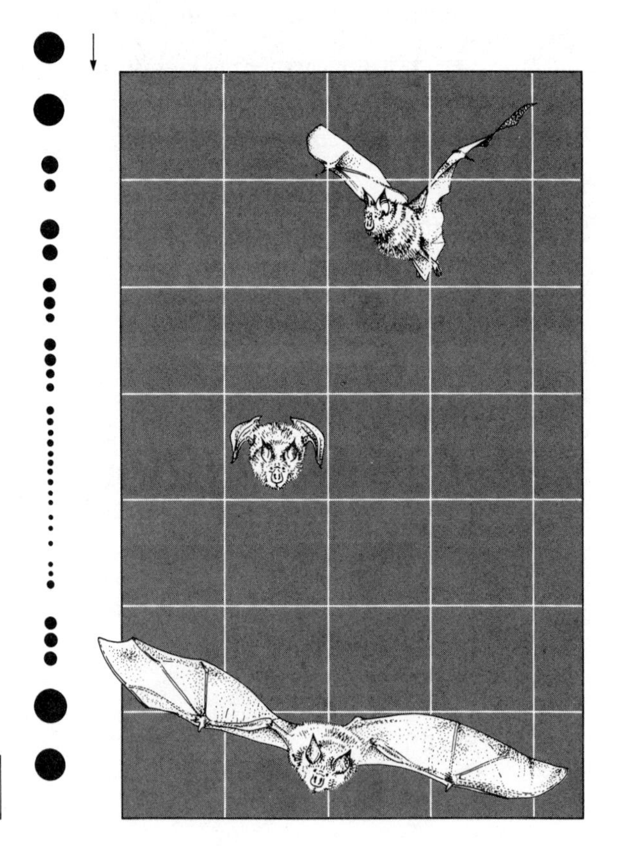

trained to catch small food particles thrown into the air in complete darkness and to discriminate edible from inedible objects on the basis of small differences in shape (Simmons, 1971). The little brown bat (*Myotis lucifugus*) can capture very small flying insects such as fruit flies and mosquitoes at a rate of two per second.

13.3 Visual recognition of prey and predators

Most predators encounter a large number of different prey species that they have to discriminate from non-prey. The three most commonly used cues are size, movement, and shape.

Predators that have a choice between prey items differing only in body size usually take the largest. This is the more energy-efficient strategy. However, as size increases there usually comes a point beyond which the stimulus is no longer regarded as prey. For example, when the common toad (*Bufo bufo*) is presented with "prey" stimuli of different sizes, it responds positively to items within a specific size range, but actively avoids larger stimuli, as shown in Figure 13.6.

How does the toad judge the size of the prey item? The simplest way is by the size of the retinal image, measured by degrees of visual angle. For an object of constant size, the size of the visual angle changes with its distance from the eye: Nearby objects look larger than far away objects. To select prey of a single size it is necessary for the toad to judge the absolute size of a visual object by taking into account both its retinal size and an estimate of its distance. Midwife toads (*Alytes obstetricans*) gradually attain size constancy during development (Ewert and Burghagen, 1979). Just after metamorphosis they prefer a certain angular-size prey dummy, almost regardless of the distance. Six months later they prefer objects of a specific absolute size, regardless of distance (Fig.

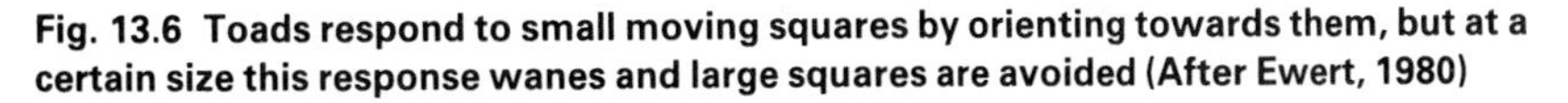

Fig. 13.6 Toads respond to small moving squares by orienting towards them, but at a certain size this response wanes and large squares are avoided (After Ewert, 1980)

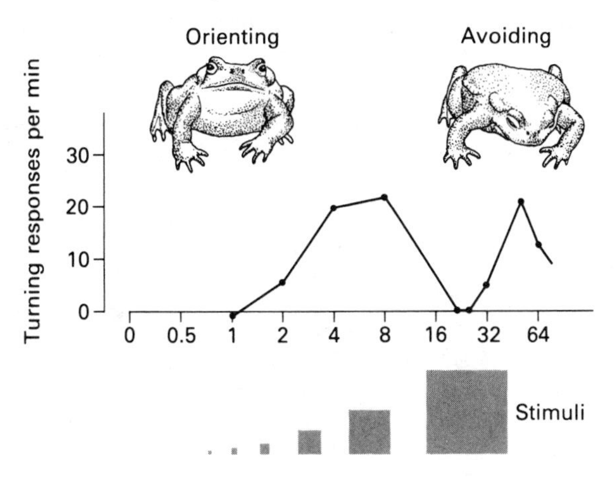

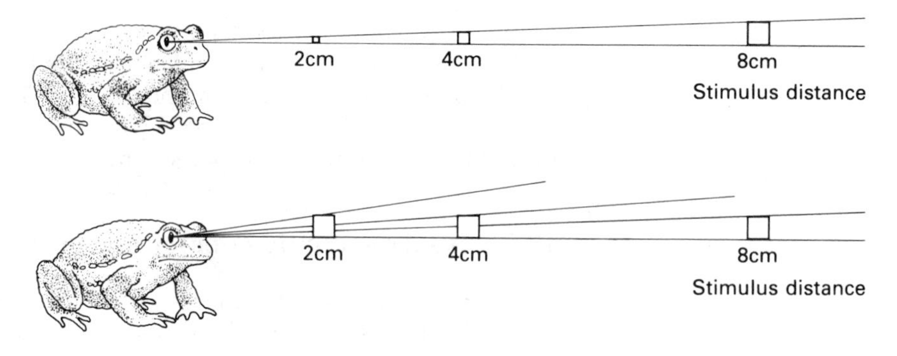

Fig. 13.7 Young toads (above) judge size by the visual angle, but older toads (below) can make size judgements independently of the visual angle (After Ewert, 1980)

13.7). In the meantime they have somehow learned to incorporate distance judgements into their size assessments. In some cases there is a bias towards larger prey, which is not a result of active preference but of better detectability. Thus, rainbow trout (*Salmo gairdneri*) catch large crustaceans more often than small ones, because the large prey are seen at a greater distance (Ware, 1972).

Movement of the prey-stimulus is essential for prey recognition in some species, such as frogs and toads. The common cuttlefish (*Sepia officinalis*) will normally only attack prawns that are moving. But, if a prawn is taken from a cuttlefish that has just captured and paralyzed it, then it will immediately be attacked again, even if it is presented motionless (Messenger, 1968). Some predators prefer prey that move in an irregular pattern. Thus, dragonfly larvae prefer prey moving in a zigzag pattern (Etienne, 1969), and Bluegill sunfish (*Lepomis gibbosus*) attack a wriggling dummy prey-fish rather than one that is smoothly moving (Gandolfi et al., 1968). Sometimes, it is shape in relation to movement that is important. Thus, toads that are presented with a dark moving stripe on a white background most readily attack when the stripe is moving along its own axis, in a worm-like fashion (Fig. 13.8).

Prey recognition by shape is such a complex matter that it is difficult to

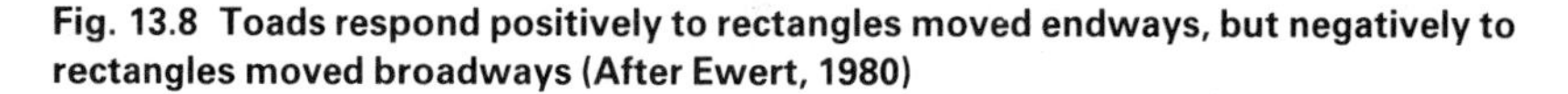

Fig. 13.8 Toads respond positively to rectangles moved endways, but negatively to rectangles moved broadways (After Ewert, 1980)

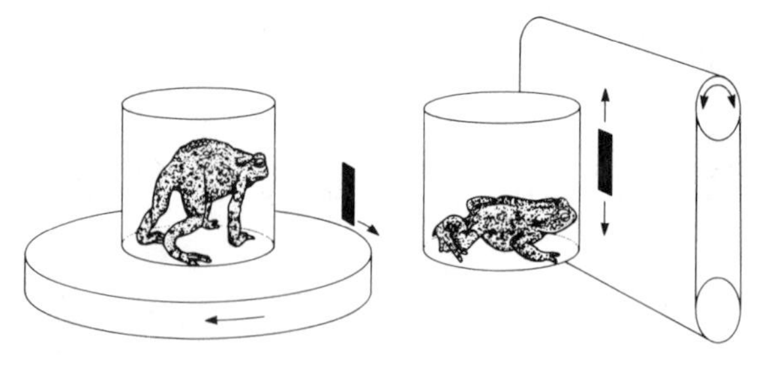

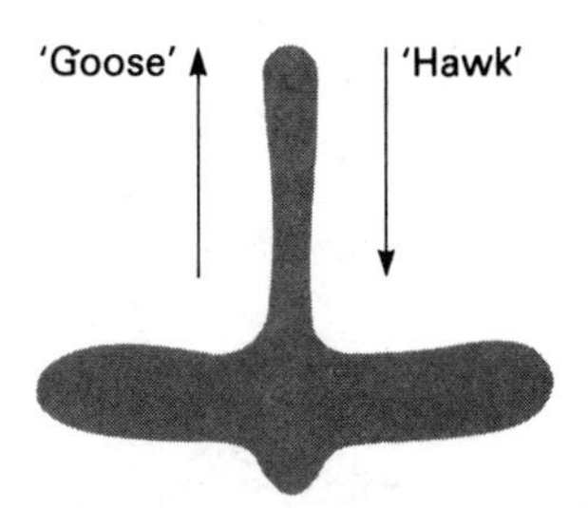

Fig. 13.9 This silhouette resembles a hawk when moved in one direction and a goose when moved in the other

give any simple rules. In a series of behavioral studies, Mike Robinson (1970) offered insect prey to captive rufous-naped tamarins (*Saguinus geoffroyi*), which are highly insectivorous New World monkeys. Some experiments involved presenting a mantid that was normal, headless, two-headed, etc. The results suggested that identification of the head was important in prey recognition. Experiments using stick insects as prey showed that the monkeys tended to overlook stick insects without legs but quickly seized insects with visible legs or small sticks with insect legs attached. Thus, the heads and appendages of insects are important cues for prey recognition by tamarins. To counter this tendency, the insect prey employ a variety of protective devices to conceal their appendages or disrupt the body outline. In experiments with abstract two-dimensional patterns, Robinson found that bilateral symmetry is one feature of prey that is probably commonly recognized by predators.

Similar principles apply to recognition of predators. For example, a silhouette of a hawk when passed above ducklings or goslings induces fear responses when moved in one direction but not the opposite direction (Fig. 13.9). The short neck and long tail is characteristic of a hawk, whereas the long neck and short tail resembles a flying goose (Tinbergen, 1951). Toads avoid snakelike shapes with a raised head (Fig. 13.10). A leach walking jerkily along is regarded as prey when its frontal sucker is on the ground, but if it lifts its sucker in the air it is taken for an enemy (Ewert, 1980).

As we have seen, behavioral studies (reviewed by Ewert, 1980) indicate that toads show prey-catching behavior toward small, elongated objects presented horizontally but not to similar objects presented vertically (see Fig. 13.8). The toad normally feeds on insects, larvae, worms, etc. Its prey-catching behavior involves orientation of the head and body, visual fixation of the prey, snapping at the prey by extension of the neck and tongue, swallowing the prey, and wiping the snout with the forelegs (Fig. 13.11). A small moving object is necessary to elicit the prey-catching behavior. Large moving objects elicit defensive behavior. Although various purely behavioral methods can be used to investigate sensory process physiological investigation can also yield valuable information about the functioning of sense organs and the type of

Fig. 13.10 Responses of a toad to simple models of (*a*) a snake, (*b*) abstract pattern, (*c*) leech with raised head, (*d*) leech with lowered head. The first three are avoided, while the last is investigated (After Ewert, 1980)

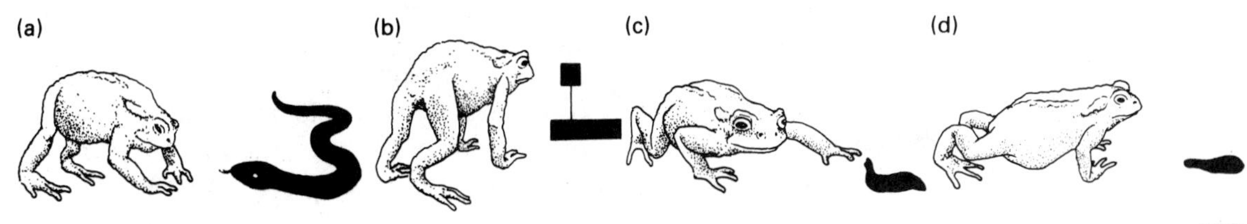

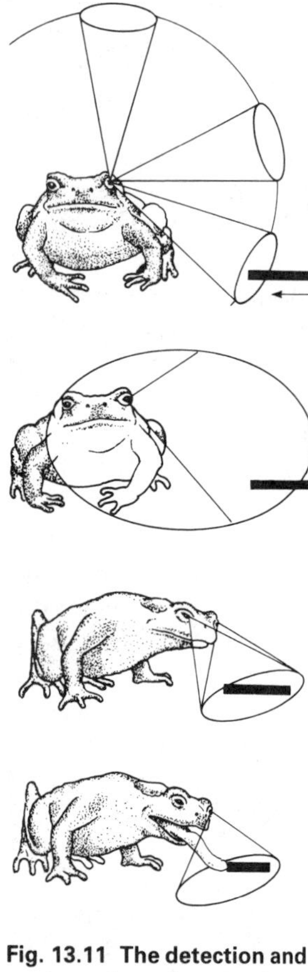

Fig. 13.11 The detection and capture of prey by a toad. From top to bottom: the prey enters the toad's lateral visual field; the toad turns towards the prey, bringing it more towards the centre of the visual field, and then into the binocular field; the toad grasps the prey with its tongue (After Ewert, 1980)

information they send to the brain. To discover how the brain makes use of such information, however, a combination of the behavioral and physiological approaches is required. This approach has been used by Jorg-Peter Ewert and his co-workers in their extensive investigations of prey and enemy recognition in toads.

Physiological studies show that some of the prey recognition is carried out at the retinal level. Jerry Lettvin and his co-workers (1959) made electrical recordings from the frog optic nerve while objects were moved across the visual field. They found four main types of response that appear to correspond to the four types of ganglion cells in the retina. These cells were found to be detectors for (1) a stationary boundary, (2) a dark, convex moving object, (3) change of contrast or movement, and (4) dimming. The toad appears to have three types of ganglion cells (corresponding to the frog types 2, 3, and 4) that send their axons along the optic nerve to the optic tectum of the brain (Ewert and Hock, 1972). The information provided to the brain thus includes the angular size and velocity of the object, its degree of contrast with the background, and the general level of illumination. This information, however, is not sufficient for the toad to discriminate prey from non-prey.

The fibers of the optic nerve can be traced by nerve degeneration studies to the various parts of the brain, including the optic tectum and the thalamus pretectum (Fig. 13.12). The retina of one eye projects topographically to the surface layers of the opposite optic tectum. This projection can be seen as a map in which each area of the visual field corresponds to a particular area of the optic tectum. Electrical stimulation of the optic tectum in the freely moving toad produces an orientation response toward the corresponding part of the visual field, causing the toad to behave as if the appropriate part of the optic tectum had been stimulated by sight of a prey object. When the retinal projection to the thalamic pretectal area is stimulated electrically, the toad responds with avoidance behavior.

Surgical destruction of part of the brain (the thalamic pretectal region) results in animals that react with prey-catching behavior to any moving object. Destruction of the optic tectum abolishes all reaction to moving stimuli, including the avoidance behavior. These results led Ewert (1980) to postulate that the projection from the retina to the thalamus pretectum is instrumental in eliciting avoidance behavior but that some excitatory input from the optic tectum also is required. The projection from the retina to the optic tectum provides a basis for prey-catching responses to all moving stimuli, but responses to large or enemylike stimuli are inhibited by the thalamus pretectum. Thus, prey catching occurs only to appropriately small stimuli.

Ewert's hypothesis is confirmed by physiological studies in which electrical recordings are made from neurons in the optic tectum and thalamus pretectum in response to stimulation of the retina and of other

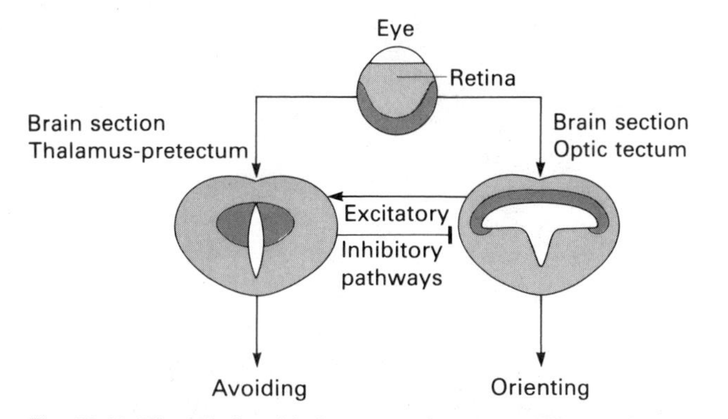

Fig. 13.12 The relationship between the parts of the toad's brain involved in the control of avoiding and orienting towards visually detected objects (After Ewert, 1980)

relevant parts of the brain (Ewert, 1980). Not only is this research of great interest in illustrating the respective roles of the retina and brain in stimulus filtering, but also it is a beautiful example of what can be achieved by a judicious combination of behavioral and physiological methods.

Points to remember

- Animals living in dim light often have special visual adaptations, such as tubular shaped eyes, a tapetum and visual pigments designed to maximize sensitivity under the prevailing conditions.

- Animals that live, or have to operate, under conditions too dark for vision may have degenerate eyes and may have other sensory systems that substitute for vision. These include the echolocation system of bats, the auditory location system of owls and the electrosensitivity of some fishes.

- The visual recognition of prey and predators often involves detection of key stimuli (sign stimuli) which enables the animal to make a rapid decision and a rapid response.

Further reading

Gould, J. L. (1980) 'The case for magnetic-field sensitivity in birds and bees (such as it is),' *American Scientist*, **68**, 256–67.
Lythgoe, J. N. (1979) *The Ecology of Vision*, Clarendon, Oxford.

2.2 The Animal and the Environment

Here we consider the mechanisms used by animals to regulate their relationships with the environment. Chapter 14 begins with the coordination and orientation of the body and progresses from simple orientation to navigation. Chapter 15 considers the animal in relation to its internal environment. The concept of homeostasis is discussed in relation to thermoregulation, feeding and drinking. In Chapter 16 we discuss the ways in which animals are able to adapt their physiology and behavior to changing environments. The roles of biological clocks in reproductive physiology, hibernation, migration and daily routines are discussed in some detail.

Claude Bernard (1813–1878)

BBC Hulton Picture Library

Claude Bernard studied medicine from 1834 to 1843 and then began research at the College de France, as a pupil of Magendie. His main discoveries were made between 1840 and 1859, after which he did little experimental work, instead devoting himself to developing his theories and to writing.
Bernard's major work was the *Introduction à l'Etude de la Médicine Expérimentale*, published in 1865. Bernard had previously published various *Leçons*, but these were usually notes taken by pupils during his lectures and published under his supervision. Bernard published various other contributions culminating in his *Leçons sur le Phénomènes de la Vie Communs aux Animaux et aux Végétaux*, published shortly after his death in 1878.
Prior to Bernard's publications, the body had been regarded as a collection of organs, each with its separate functions. Bernard showed that the various physiological activities are interrelated and that the body should be thought of as a single complex and sophisticated machine.
From 1848 onwards, while working at the College de France, Bernard tested animals on various diets. He discovered that their blood always contained sugar, whether or not the animals had recently eaten. Even dogs fed solely on meat had sugar in the blood, and Bernard showed that this was secreted into the blood from the liver. He demonstrated that the liver stores some of the sugars made available by the digestion of food, and releases sugar into the blood when none is available from the diet. Bernard also did important research on digestion, and he documented the regulation of blood supply to the different parts of the body.
Claude Bernard is regarded as the founder of modern experimental physiology. One of his often quoted dictums is: "Why think when you can experiment. Exhaust experiment and then think." His experimental work was always directed at a particular hypothesis. Commenting on the "splendid work of Laplace" and noting that statistics were not much used by biologists, he remarked, "Of course, I expect that in another hundred years or so, everybody will not only be using but abusing statistics and will rely on this method to salvage work undertaken without the benefit of any working hypothesis."
Nonetheless, Claude Bernard is remembered mainly for his theoretical concepts. His most famous dictum—"The stability of the internal environment is the condition for free and independent Life"—was first mentioned in 1859 in one of his *Leçons*, and elaborated in his posthumous work of 1878. Bernard realized that the blood provided an internal milieu for all bodily tissues, and that the state of the blood was influenced partly by internal processes and partly through changes in the external environment. Animals able to maintain the stability of the internal environment had far greater ecological freedom than those at the mercy of environmental fluctuations.
The idea of regulation of the internal environment was implicit in Bernard's thinking, though it was not explicitly discussed by scientists until the next century. This basic idea has had profound effects upon modern thinking on physiology and animal behavior.

14 Coordination and Orientation

In this chapter we look at the animal's spatial relationships with its external environment. In responding to environmental changes, animals have to coordinate and orient their movements in relation to external stimuli. We look first at the subject of coordination and then at orientation. We conclude with a discussion of navigation, the most complex form of spatial orientation.

14.1 Coordination

As we saw in Chapter 11, the effective coordination of movement and locomotion depends upon the information received by the CNS about the position, tension, and so on of the muscles, information that is provided by internal sense organs, including muscle spindles, tendon organs, and joint receptors. In addition to providing information about the relative positions of the limbs or other organs of movement, these sense organs also enable animals to make reflex responses to changes induced either by outside agents or by the movement of the animal itself.

Reflex behavior is the simplest form of reaction to stimulation. Reflexes are normally automatic, involuntary, and stereotyped. A sudden change of tension in a muscle may result in automatic postural adjustment. A sudden change in the level of illumination may result in a reflex withdrawal response. Reflexes may be relatively localized, involving a single limb or other parts of the body, as illustrated in Figure 14.1. Sometimes, however, the whole animal may be involved in the reflex response. For example, the startle response of humans requires reflex coordination of numerous muscles. Similarly, the reflex withdrawal responses of invertebrates such as polychaete worms and various mollusks involve the whole body. As we saw in Chapter 11, such reflexes are triggered by giant nerve fibers that carry very fast messages to all the muscles involved so that they contract suddenly and simultaneously.

Although normally automatic, most vertebrate reflexes can be subject to interference at the synapses that occur within the CNS (see Fig. 14.1). In the case of postural reflexes, there is often interference from other,

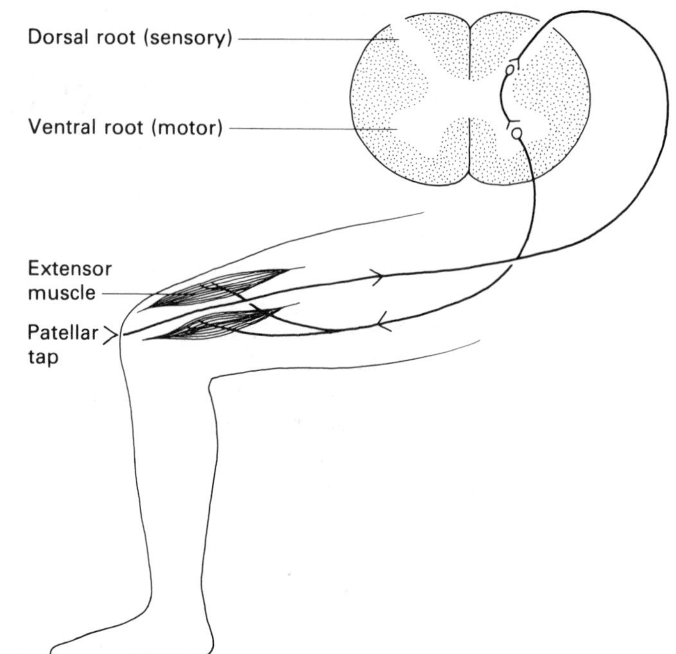

Fig. 14.1 Diagram of a simple reflex—the knee jerk reflex. A tap just below the knee cap stimulates receptors in the tendon which send messages to the spinal cord. These are relayed to the motor nerves serving the extensor muscles which reflexively contract and cause a kicking action

incompatible reflex mechanisms. When one reflex utilizes the same muscles as another, the reflexes are incompatible in that both cannot occur simultaneously. Such pairs of reflexes are also neurologically incompatible in that stimulation of one reflex inhibits performance of the other. The inhibition is usually *reciprocal*—providing the most elementary form of coordination—so that one activity completely suppresses, or is completely suppressed by, the other. This type of reciprocal inhibition is typical of walking and other types of locomotion.

In general, muscular coordination is achieved by two main processes: central control and peripheral control. In the case of *central control*, precise instructions are issued by the brain and are obeyed by all the muscles involved. The coordination of swallowing movements in mammals seems to be of this type (Doty, 1968). Central control is important in the coordination of many skilled movements, which require rapid muscular activity. For example, the cuttlefish *Sepia* catches small crustaceans by means of two extensible tentacles (Fig. 14.2). The control of attack falls into two phases. The first is a visually guided system in which the prey is brought into focus binocularly and movements of the prey are followed by movements of the cuttlefish in such a way that visual error is reduced to zero. When this dead-reckoning phase is complete, the tentacles are ejected suddenly and the prey is seized. This

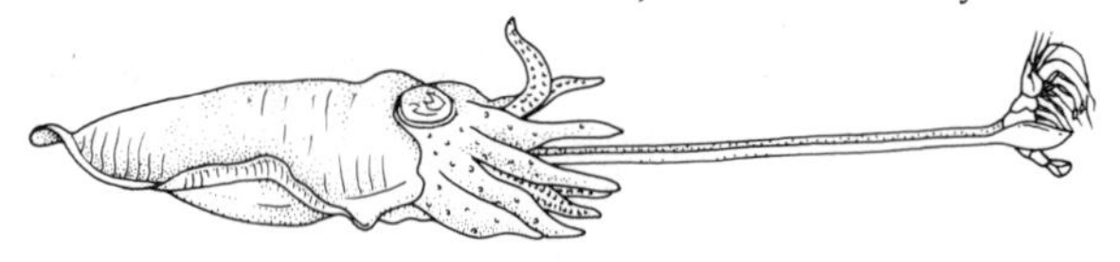

Fig. 14.2 A cuttlefish catching a shrimp

final phase is so rapid (about 30 milliseconds) that there is no time for the tentacles to be guided visually onto the prey. This can be demonstrated by turning off the lights during tentacle ejection, which does not affect prey capture. However, the prey is missed if it moves during ejection, showing that the cuttlefish is unable to correct its strike using visual feedback (Messenger, 1968).

Locomotor rhythms often are generated centrally. Electrical recording from motor neurons and interneurons in insects has demonstrated central control rhythms for walking (e.g., Pearson and Iles, 1970; Burrows and Horridge, 1974), swimming (e.g., Kennedy, 1976), and flight (e.g., Wilson, 1968) in arthropods. The precise patterns of behavior, however, often are influenced by reflexes and feedback from the periphery. Thus, removal of proprioceptive input reduces stroke frequency in flying locusts, and wind stimulation increases motor neuron discharge (Wilson, 1968). It appears that the fast walking movements of the cockroach are influenced less by peripheral stimuli than are the slow walking movements of their relatives the stick insects (*Carausius*) (Wendler, 1966).

Peripheral control of coordination is achieved through sense organs in the muscles and other parts of the body that send information to the brain and that thereby influence the instructions issued from the brain to

Fig. 14.3 Central versus peripheral control of movement. Movement of the eyeball (above) is under central control. The eyeball is not normally subject to load, so detailed motor commands from the brain can be followed quickly and precisely. If the eyeball is subjected to a load (e.g. from finger pressure), its displacement is not corrected. Peripheral control is important in the movement of limbs (below), because these are often subjected to loads. Displacement of limb position is monitored by sensory receptors and fed back to the central control center, and the displacement is corrected

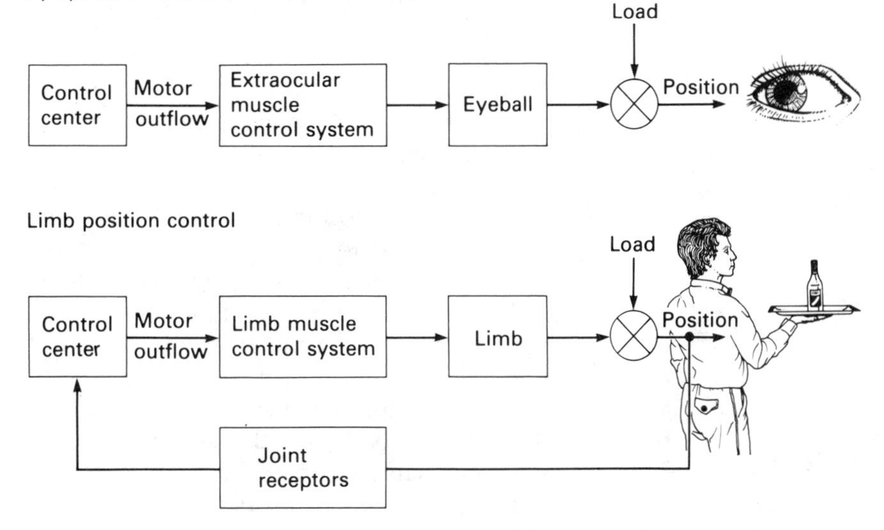

the muscles, as shown in Figure 14.3. Peripheral control usually acts in cooperation with central control. For example, in the coordination of swimming movements in fish, the brain provides rhythmic signals that pass down the spinal cord in waves, coordinating the rhythmic movements of the fins and tail. In the dogfish (*Scyliorhinus*) the rhythm disappears if all nerves leading from the muscles to the brain are cut. If some nerves are left intact, however, the rhythm persists. Thus, it appears that some peripheral feedback is necessary to trigger the centrally produced rhythm (Gray, 1950). In cartilaginous fish (Chondrichthyes), including the dogfish, and in all other sharks and rays, the fins show little independent rhythmic movement, but in bony fish (Teleostei) the fins can beat at different frequencies under some circumstances. Holst (1939; 1973) showed that the rhythms of different fins can influence each other, a feature he called "relative coordination." Sometimes the rhythm of one fin attracts and dominates that of another so that they fall into step. In other cases, the amplitudes of fin movements summate so that the movements become smaller when the fins are out of step with each other and larger when they are in step.

In fishes and amphibians, locomotion appears to be mostly under the control of endogenous spinal rhythms. There is no evidence of specialized motor control exerted by the forebrain. Removal of the forebrain in fishes produces no change in posture or locomotion (Bernstein, 1970). In frogs and toads, removal causes a general decrease in spontaneous movement, but electrical stimulation of the forebrain has no specific motor effects. The midbrain plays some role in motor control in fishes and amphibians. As we saw in Chapter 13, electrical stimulation of the tectum in toads causes turning of the head as well as food snapping and swallowing. In higher vertebrates the higher brain is capable of exerting greater control of movement, but the automatic aspects of locomotion are still primarily controlled by the brain stem and spinal cord.

In mammals, the *corticospinal* tract is the most important pathway involved in the voluntary control of movement. It begins in the motor cortex and proceeds through the midbrain and brainstem to the spinal cord. This system, sometimes called the *pyramidal system*, is present in all mammals except the very primitive monotremes (e.g., platypus and spiny anteater). In the marsupial possum (*Trichosaurus*) the pyramidal axons run only to the midthorax, where they innervate the forelimbs. The hindlimbs are innervated by an extrapyramidal system.

The *extrapyramidal system* includes all non-reflex motor pathways not included in the corticospinal or pyramidal systems. It is thought to be more primitive than the pyramidal system. In animals with little or no cerebral cortex, the basal ganglia of the extrapyramidal system are the most important centers of motor control. They are particularly well developed in birds, which have virtually no cerebral cortex but a larger striatum than mammals. Thus, it appears that as the cerebral cortex evolved, it brought into being a second source of motor coordination

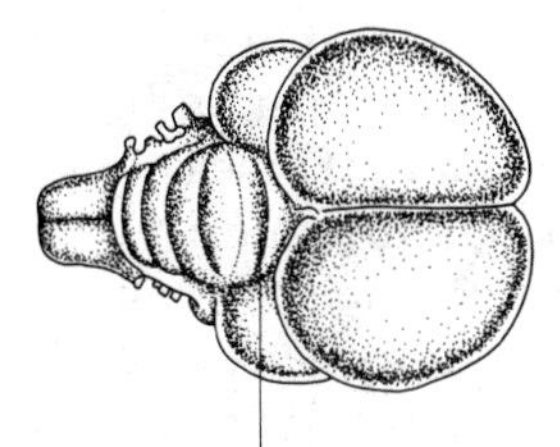

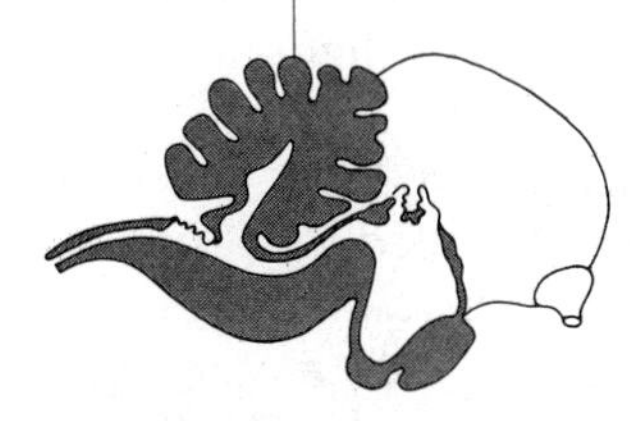

Fig. 14.4 Brain of a pigeon, showing the large cerebellum

that acts through the pyramidal system.

In monkeys, neurons in various parts of the corticospinal pathways change their firing patterns during voluntary movements of the eyes, hands, and legs (Evarts, 1968). Artificial stimulation of the motor cortex causes responses of single muscles or of single motor units within a muscle. More gross stimulation elicits discrete movements of whole limbs. It is possible to map the surface of the motor cortex according to the parts of the body moved in response to electrical stimulation. A similar map can be obtained for the sensory cortex in that the various parts of the body are represented differentially according to their sensory importance for individuals of the species concerned. Voluntary skilled movements are initiated via the cortical pyramidal neurons, while reflex maintenance of motor responses and of posture is controlled by nearby extrapyramidal neurons. There are estimated to be 1 million pyramidal neurons in humans (Prosser, 1973).

Another part of the vertebrate brain that is important in coordination is the cerebellum (Fig. 14.4). The cerebellum is not directly involved in postural reflexes or motor control. It serves as a monitor and coordinator of neural events involved in orientation and balance and in various other refined aspects of motor control. The basic neuronal organization of the cerebellum is similar in all classes of vertebrates, and the cerebellum has undergone less evolutionary change than any other part of the brain. The cerebellum receives information from the senses of vision, hearing, touch, equilibrium, and the condition of muscles and joints. It also has connections with the motor areas of the cerebral cortex. In addition to connections through the thalamus to the motor cortex, there are two-way connections to the sensory areas of the cortex, which are important in the maintenance of posture. Upright posture cannot be maintained without visual, vestibular, and proprioceptive information. The cerebellum combines visual and vestibular information about equilibrium with the state of contraction of the muscle spindles concerned. It sends appropriate instructions to the muscles, particularly through the gamma efferent fibers (see Chapter 11). The cerebellum thus exerts a modulating and refined control over the muscular contractions involved in the maintenance of posture and in the complex coordinated movements required for locomotion.

14.2 Spatial orientation

The orientation of the whole animal in space may be based on very simple principles but may also involve very complex mechanisms. The simple principles can be seen most easily in certain invertebrate species. Gottfried Fraenkel and Donald Gunn (1940) proposed a system of classification based on the work of earlier writers, which system has provided the basis of more recent discussions and reviews (e.g., Adler, 1971; Kennedy, 1945; Hinde, 1970).

The simplest form of spatial orientation is *kinesis*, in which the animal's response is proportional to the intensity of stimulation but is independent of the spatial properties of the stimulus. For example, common woodlice (*Porcellio scaber*) tend to aggregate in damp places beneath rocks and fallen logs. They move about actively at low humidity levels but are less active at high humidity levels. They consequently spend more time in damp conditions, and their high activity in dry conditions increases the chances that they will discover a damp place. Similar behavior is shown by the ammocoete larva of the lamprey, which varies its swimming activity in accordance with the light intensity (Jones, 1955).

The type of kinesis in which a relationship exists between the speed of locomotion and the intensity of stimulation is called *orthokinesis*. Another type of kinesis, shown by the flatworm *Dendrocoelum lacteum*, is *klinokinesis*. Here the rate of change of direction increases as the light intensity increases (Ullyott, 1936; Fraenkel and Gunn, 1940; Hinde, 1970).

In a number of types of orientation, usually grouped together as taxes, the animal heads directly towards or away from the source of stimula-

Fig. 14.5 Klinotaxis in a maggot. Notice the side-to-side movement of the head. When the light is moved from below to the side, the maggot turns away (After Mast, 1911)

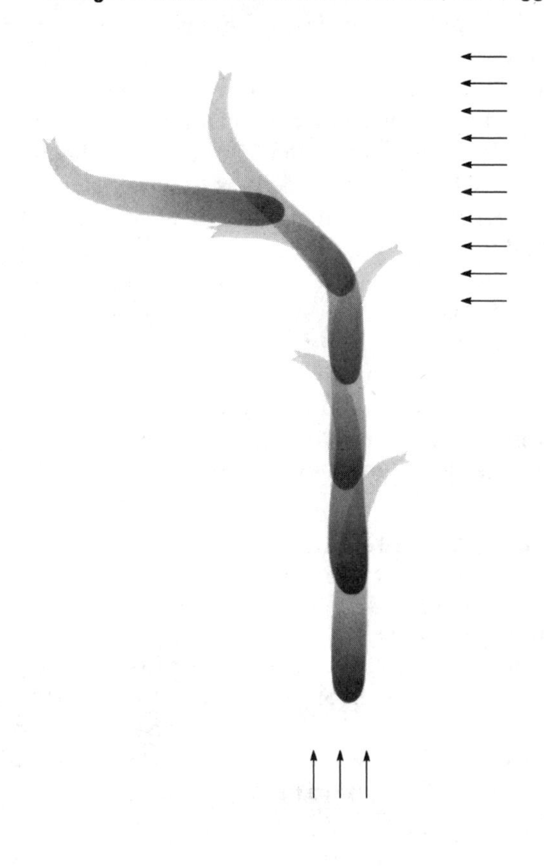

tion. For example, when the larva of the housefly (*Musca domestica*) has finished feeding, it seeks out a dark place to pupate. At this stage it will crawl directly away from a light source, and it is said to show "negative phototaxis." The maggot has primitive eyes on its head that are capable of registering changes in light intensity but that are not able to provide information about the direction of the light source. As the maggot crawls it moves its head from side to side (Fig. 14.5). When the light on the left is brighter than that on the right, the maggot is less likely to turn its head towards the left. It thus tends to crawl more toward the right, away from the light source. In response to an increase in the level of illumination, the maggot increases its rate of head turning. If an overhead light is turned off every time the maggot turns its head to the right and is turned on every time it turns to the left, then the maggot turns away from the illuminated side, describing a circle toward the right. Thus, although the animal has no directional receptors, it can perform a directional response. Similar behavior is shown by the single-eyed organism *Euglena* (Fraenkel and Gunn, 1940).

Orientation by successive comparison of stimulus intensity requires turning movements. Usually it is called *klinotaxis*. Many animals show klinotaxis in response to gradients of chemical stimulation. Simultaneous comparison of the intensity of stimulation received at two or more receptors enables the animal to strike a balance between them. It can then achieve *tropotaxis*, which enables it to steer a course directly toward or away from the source of stimulation. For example, the pill woodlouse (*Armadillidium vulgare*), which lives under stones or fallen trees, shows a positive phototaxis after periods of dessication or starvation. With its two compound eyes on the head, the animal is able to move directly toward a light source. When one eye is blacked out, however, it moves in a circle. This shows that the two eyes normally provide a balance of stimulation. When presented with two light sources, the woodlouse often starts off by taking a medium course but then heads toward one light. This happens because equal stimulation of the two eyes can be achieved either by steering between two sources or by moving directly toward one. Deviations from one source tend to be self-correcting, and lateral sources of light tend to be ignored because the eyes are shielded at the back and sides. Deviations from two sources set wide apart, therefore, may result in loss of contact with one.

Eyes that are capable of providing information about the direction of light by virtue of their structure are capable of *telotaxis*, a form of directional orientation that does not depend upon simultaneous comparison of the stimulation from two receptors. When there are two sources of stimulation, the animal moves toward one and never in a median direction, showing that the influence of one of the stimuli is inhibited. An example is depicted in Figure 14.6.

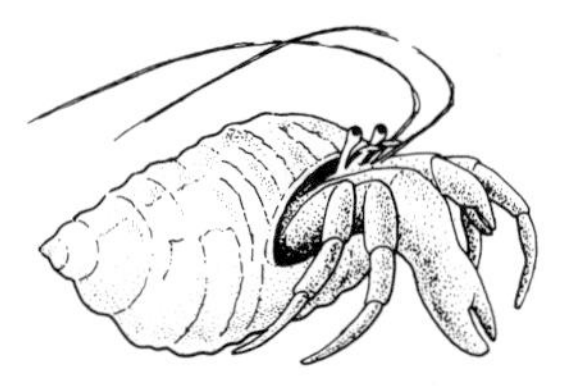

Fig. 14.6 Telotaxis: tracks of hermit crabs presented with two lights. Each part of the track is directed toward one light only (After Fraenkel and Gunn, 1940)

Menotaxis is a form of telotaxis (Hinde, 1970) that involves orientation at an angle to the direction of stimulation. An example is the light-

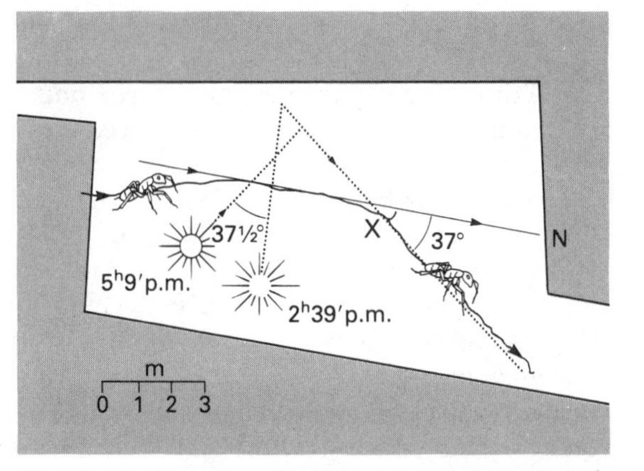

Fig. 14.7 Light compass: an ant (*Lasius niger*) returning to its nest N, with the sun shining from the right at an angle of about 90° to the animal's path. At the place X the ant was imprisoned for 2.5 hours. When released it deviated from its former path by an angle which was the same as that through which the sun had traveled during this period, so that the sun's rays again made an angle of 90° which the animal's path (after Brun, 1914)

compass response shown by homing ants. These animals are guided, in part, by the direction of the sun. If the apparent direction of the sun is changed slowly by means of a mirror, then the ants change course accordingly (Fig. 14.7). It was thought (Brun, 1914) that if ants (*Lasius niger*) were confined in a dark box for a few hours in the middle of their homeward journey, they would maintain the same angle to the sun when released. However, it later transpired (Jander, 1957) that the ants compensated for the movement of the sun and continued in the same direction upon being released. Such time-compensated compass reactions have been shown to occur also in the beetle *Geotrupes sylvaticus*, the pond skater *Velia currens*, and the honeybee *Apis mellifera* (see Saunders, 1976, for a review).

The type of orientation achievable in a given situation depends jointly upon the nature of the external cues and the sensory equipment of the animal. An animal with only a stimulus-strength sensor is limited to successive measurements of stimulus strength in different localities. If the external cues are inherently directional, then a single receptor shielded on one side can provide directional information. Thus, a shielded photoreceptor is useful in this respect, but a shielded chemoreceptor is of no advantage because chemical stimuli are not inherently directional. With two receptors, simultaneous comparison can be used to detect gradients (Fig. 14.8). With many receptors arranged in the form of a *raster* (a row or mosaic arrangement), more sophisticated types of orientation can be achieved (Fig. 14.8). Examples of rasters are the lens eyes of vertebrates and the compound eyes of arthropods (see Chapter 12).

Spatial orientation is often achieved by a combination of methods. For

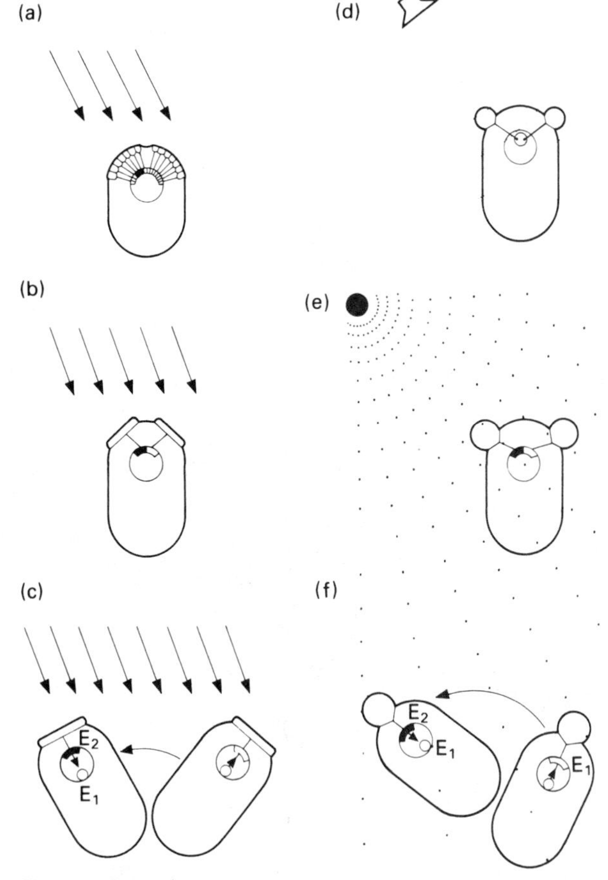

Fig. 14.8 Schematic representation of some of the basic principles of sensory orientation. (*a*) The direction of stimulation (e.g. light) is registered by a raster of sensory receptors. (*b*) Direction registered by simultaneous comparison between two receptors. (*c*) Only one receptor is available and the animal makes successive comparisons by moving its body. (*d*) Time of arrival of stimulation (e.g. sound waves) is compared by two receptors. (*e*) A gradient of stimulation (e.g. chemical) is registered by comparison between two receptors. (*f*) A gradient is registered by a single receptor as the animal moves to sample different localities (From *The Oxford Companion to Animal Behaviour,* 1981)

example, some moths are attracted to females as a result of the airborne pheromone released by the female. The scent is windborne, so the flying moth must orient with respect to the wind. Flying animals normally use visual cues to monitor their progress with respect to the ground. The flight path of the animal is affected by the wind direction, giving a resultant track (Fig. 14.9). Experiments with moths show that the track angle changes with the scent concentration. When the scent is absent, the animal flies backward and forward without progressing upwind (i.e., with a track angle of 90 degrees). When scent is detected in the wind, the track angle increases and the animal zigzags upwind. The

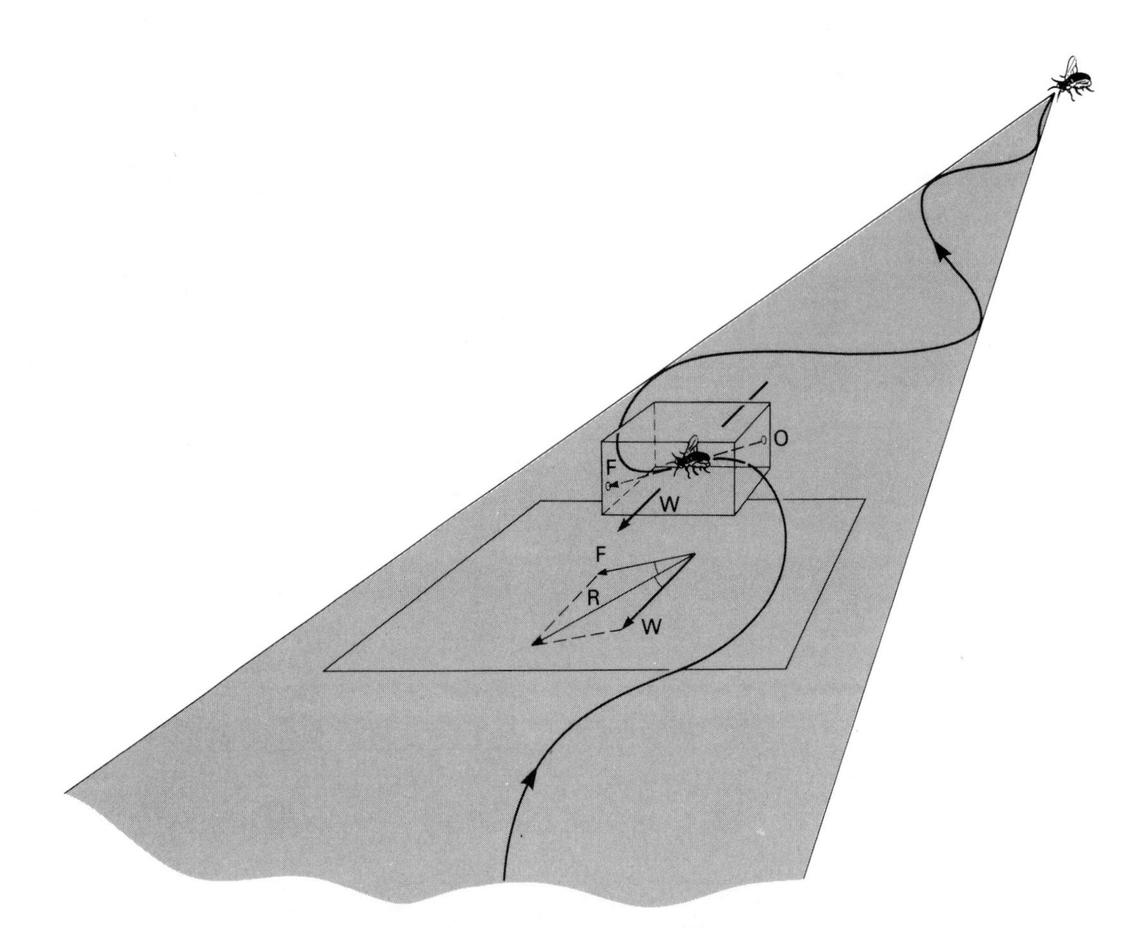

Fig. 14.9 Male moth flying upwind in response to pheromone released by a female. The flight path F is affected by the direction of the wind W, giving a resultant track R. The track angle is the angle between the wind direction W and the flight direction R with respect to the ground

changes of direction are related to the borders of the scent trail, as illustrated in Figure 14.9. When the scent concentration drops below a certain level, as at the edge of the scent plume, the animal turns in a direction opposite to that of its previous turn. This turning behavior is not related to wind direction, but depends upon internal reference, or *idiothetic* information. Thus, the flying moth uses a combination of visual, amenotaxic (wind), and idiothetic orientation mechanisms in searching for a mate.

14.3 The reafference principle

A sophisticated orientation system must be able to distinguish between stimulation from the outside world and stimulation caused by the animal. In the case of human vision, for example, movement of objects

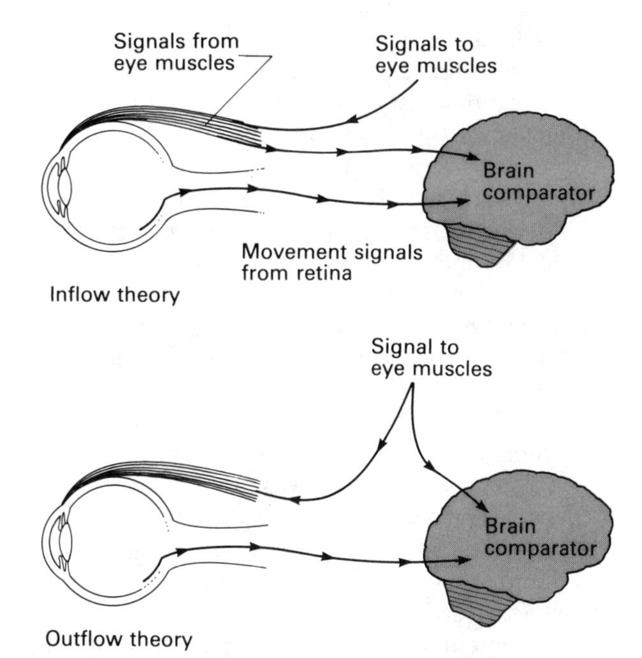

Fig. 14.10 Inflow and outflow theories of eyeball movement control

in the outside world causes movement of the image on the retina, which we perceive. Voluntary movement of the eyes, however, also produces movement of the image on the retina, yet we do not perceive such movements. Somehow the brain distinguishes between the movement of the retinal image that is independent of the animal and movement that is due to movement of the eyeball.

Two theories have been proposed to account for this phenomenon: the outflow theory and the inflow theory. The *outflow theory*, originally due to Hermann von Helmholtz (1867), maintains that instructions to the eye muscles to move the eyeball are accompanied by parallel signals to a comparator in the brain. Here they are compared with the incoming visual signals, as shown in Figure 14.10. *Inflow theory*, due to Charles Sherrington (1918), maintains that receptors in the extraocular muscles send messages to the brain comparator whenever the eyes are moved (Fig. 14.10). In both theories, the brain comparator assesses the two incoming signals and determines whether the visual signals correspond to the movement that would be expected on the basis of the other signal. If the two signals do not correspond, then some of the movement must have been due to outside causes.

The extraocular muscles contain muscle spindles, and their existence would seem to support the inflow theory of Sherrington (1918). However, it appears that these muscles are not involved in providing a sense of eye position (Merton, 1964; Howard and Templeton, 1966). Helmholtz (1867) used evidence from mechanical manipulation of the eyeball and apparent motions induced by attempts to move the eye when the extraocular muscles are paralyzed to argue that there is no position sense in the eye muscles.

As a matter of common observation, when the eyeball is displaced in its socket by finger pressure, the visual axis is shifted (as can be seen from the double image); it remains shifted as long as the finger pressure is maintained. The eyeball does not push back against the finger in an attempt to retain its previous position, as might be expected if extraocular muscle spindles were involved in the control of eye position (McFarland, 1971). Moreover, when the eyeball is displaced with the finger, movement is perceived, which would not be expected on the basis of the inflow theory. The muscle spindles should be stimulated however the eyeball is moved, and the brain comparator then should cancel the movement of the image on the retina. The evidence therefore seems to favor the outflow theory.

The outflow theory was extended and generalized by Erich von Holst and Horst Mittelstaedt (1950) (see also Holst, 1954). According to their *reafference principle*, the brain distinguishes between *exafferent* stimulation (stimulation that results solely from factors outside the animal) and *reafferent* stimulation (that which occurs as a result of the animal's bodily movements). Motor commands not only cause patterns of muscular movement but also produce a neural copy (the "efference" copy) that corresponds to the sensory input that could be expected on the basis of the animal's behavior. The brain then makes a comparison between the efference copy and the incoming sensory information (Fig. 14.11). All reafferent information should be cancelled by the efference copy so that the output of the comparator will be zero. Exafferent information will not be cancelled, however, and will be passed on by the comparator to another part of the brain.

Von Holst and Mittelstaedt (1950) showed that the fly *Eristalis*, when placed inside a cylinder painted with vertical stripes, shows a typical *optomotor reflex*, that is, turning in the direction of the stripes when the cylinder is rotated. Such reflexes do not occur when the fly moves of its own accord, although the visual stimulation is similar. When the head of the fly is rotated experimentally through 180 degrees (see Fig. 14.12), the

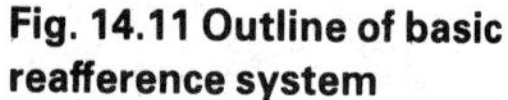
Fig. 14.11 Outline of basic reafference system

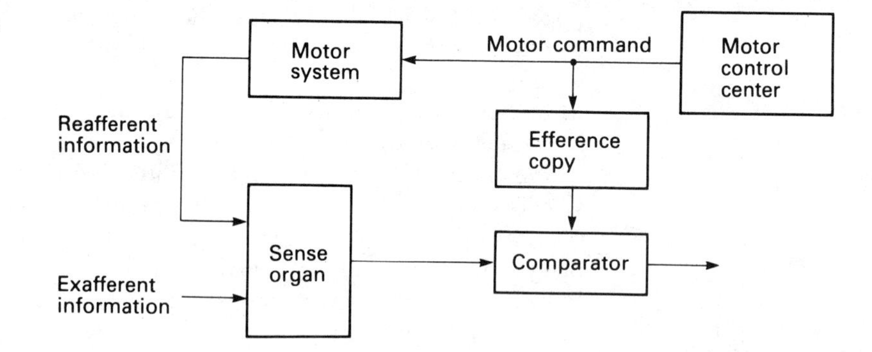

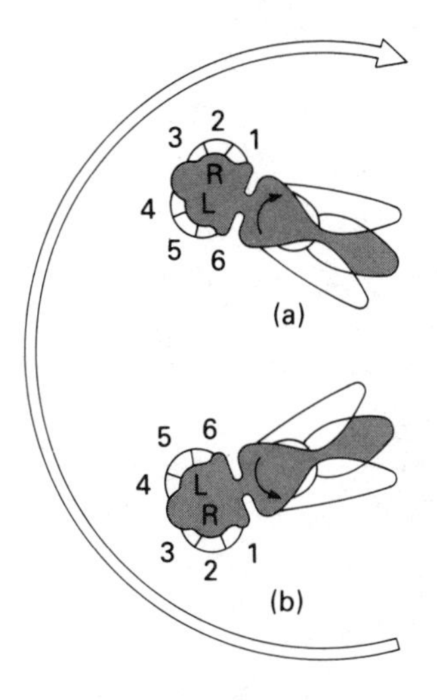

Fig. 14.12 The fly *Eristalis* in a rotating cylinder. L, left eye; R, right eye; (*a*) head in normal position, (*b*) head rotated through 180° (After von Holst, 1954)

optomotor reflex is reversed as expected. However, when the fly attempts to move of its own accord, it goes into a spin and its movements appear to be self-exciting. These results can be explained in terms of reafference theory. Normally, the output of the comparator determines bodily movement, and when the fly moves of its own accord, the output is zero and no movement occurs. The optomotor apparatus provides exafferent stimulation that is not cancelled by an efference copy, and the fly responds in a reflex manner. When the head of the fly is reversed, exafferent stimulation has the same effect as before but in the opposite direction. In the case of reafferent stimulation, however, the perceived movement is reversed in sign, and instead of being subtracted from the efference copy, it is added to it, with the result that the comparator now has a magnified output, whereas normally it would be reduced to zero. The more the animal responds, the greater the reafferent stimulation and the greater the amplification of the animal's response. Consequently, the fly tends to spin faster and faster.

The reafference principle is important not only with respect to vision but also in the control of limb position, posture, etc. For example, we can tell the difference between the arm movements involved in shaking the branch of a tree and those—which may be identical—produced while holding passively onto a branch being moved by the wind.

14.4 Navigation

The most complex form of spatial orientation is navigation. Navigation requires not only a compass, or directional sense, but also some kind of map. To illustrate this, we can consider an experiment designed by Robin Baker (1981) to test the possible use of magnetic information by humans. Baker transported a group of students in a small bus from a particular starting point (home) to a particular secret destination. Upon arrival, the students were requested to point in the direction of home. To be able to do this accurately, they would have to have some kind of compass, and Baker was looking for evidence of an internal magnetic compass. However, such a feat also would require a knowledge of the relative position of the two locations. Even if the students had possessed an accurate magnetic compass, it would not be possible for them to point toward home without some kind of map.

Three types of orientation are important in navigation: (1) *pilotage*, or steering a course using familiar landmarks; (2) *compass orientation*, the ability to head in a particular compass direction without reference to landmarks, and (3) *true navigation*, the ability to orient toward a goal such as a home or breeding area without the use of landmarks and regardless of its direction. On their long-distance migrations, birds probably use all three types of orientation. Thus, Perdeck (1958; 1967) captured juvenile and adult starlings as they passed through Holland on

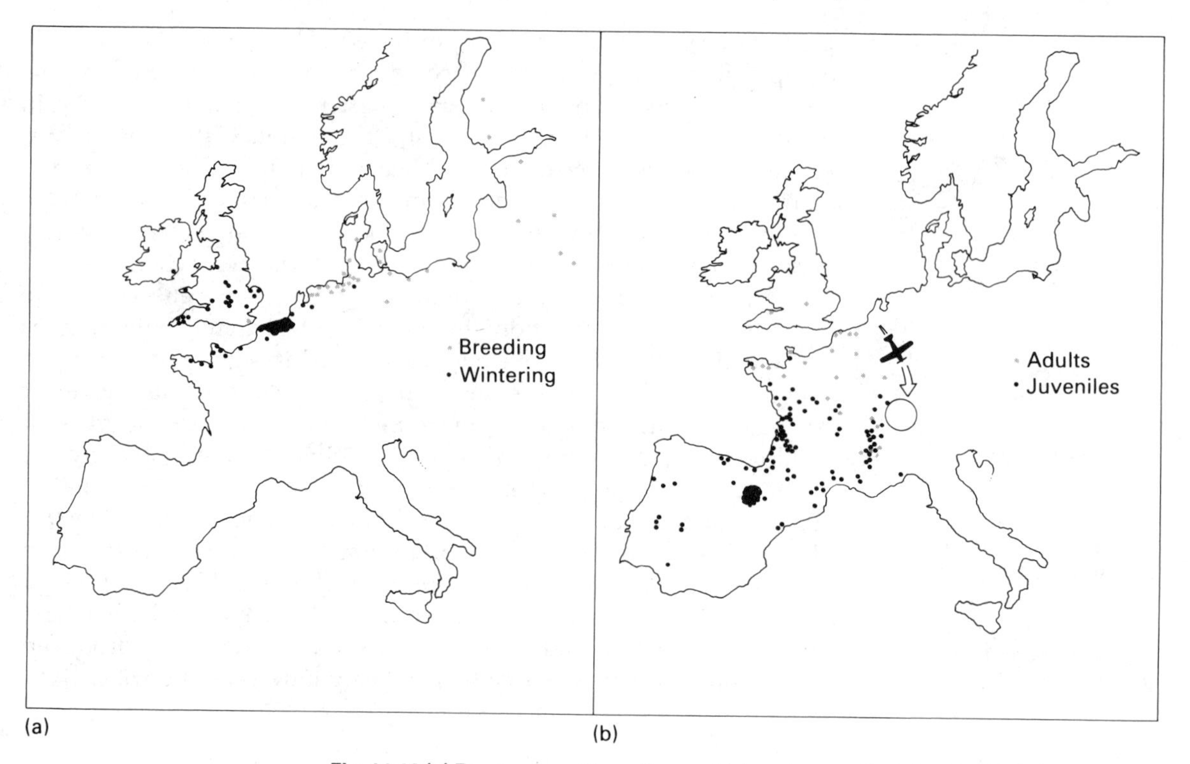

Fig. 14.13 (*a*) Recoveries of breeding and wintering starlings banded in Holland. (*b*) Recoveries of adult and juvenile starlings flown from Holland to Switzerland, and released there during the fall migration (After Perdeck, 1958)

their first autumnal migration and then banded and transported them by plane to Switzerland, 750 kilometers southeast of their normal migration route. Normally, the starlings migrate from the breeding grounds in the Baltic to the wintering grounds in Belgium, southern Britain and northern France (see Fig. 14.13). After being displaced to Switzerland, the juvenile starlings were recaptured in Spain and southern France (Fig. 14.13), indicating that they maintain the normal southwesterly direction of migration after displacement. Adult starlings, however, were recaptured in their normal wintering grounds. Thus, the adult starlings had corrected for the displacement, whereas the juvenile starlings had maintained the compass orientation characteristic of their breeding population. The juveniles of many migrating species reach their winter habitat for the first time on the basis of innate information about the direction and distance of the target area. They are unable to correct for displacement because they lack the map component necessary for true navigation (Schmidt-Koenig, 1979).

Orientation by landmarks may play an important part in navigation, particularly as the destination is approached. A classic example of this comes from Arthur Hasler's (1960) studies of migrating salmon. Pacific

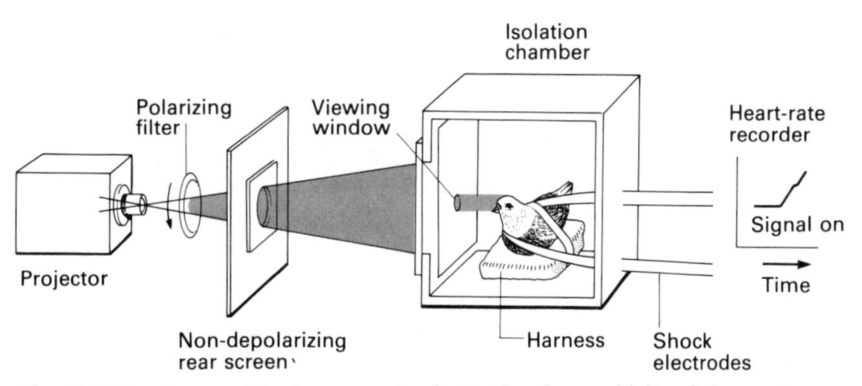

Fig. 14.14 Cardiac conditioning apparatus for testing the sensitivity of pigeons to various stimuli (polarized light in this example) (After Schmidt-Koenig, 1979)

salmon (*Oncorhynchus*) hatch in streams of the western United States and Canada. After the smolt stage of development, they swim downstream to the Pacific Ocean. After two or three years at sea, they become sexually mature and migrate back to the exact stream where they were spawned. The journey to the coast is probably accomplished by means of a sun compass. Once at the coast, however, they have to select the correct river and the correct tributary stream within the river system. After many years of research, Hasler and his co-workers discovered that during the smolting period the fish become imprinted upon the olfactory characteristics of their native stream. During their return journey they are able to discriminate between the water coming from their native stream and that from other tributaries. They essentially are recognizing landmarks, but because the scent of their birthplace flows all along the migratory route, it is not necessary for the young salmon to remember landmarks all along the outward journey. In fact, if salmon are transferred to another stream after the period of imprinting, they return to their native tributary and not to the stream down which they passed on their outward migration.

Animals are known to possess compasses of various types, based upon features of the geophysical environment such as the magnetic field of the earth. To be sure that a particular physical feature is used as a compass, it is necessary to show that the animal can detect the phenomenon and that it can use it for orientation under natural conditions. Because of the uncontrolled variability of the outside world, testing for sensory capability is best done in the laboratory. A favorite method used by those interested in navigation is the cardiac-conditioning method illustrated in Figure 14.14. By this and other methods researchers have shown that pigeons are sensitive to the following stimuli:

Ambient pressure. The pigeon is sensitive to atmospheric-pressure changes in the region of 1 to 10 millimeters of water, equivalent to a change of less than 10 meters in altitude (Kreithen and Keeton, 1974; Delius and Emmerton, 1978), a sensory ability that may provide the pigeon with an accurate physiological altimeter.

Infrasound. Sound with a frequency of less than 10 hertz is called *infrasound.* Humans cannot hear it, pigeons can hear sounds as low as 0.06 hertz. Infrasound travels a very long distance, and natural sources of infrasound, like surf, could be used by pigeons as a navigational aid (Yodlowski et al., 1977; Kreithen, 1978).

Odor. Birds have long been thought to have a poor sense of smell, but experiments using cardiac-conditioning (e.g., Henton et al., 1966; Shumake et al., 1969) confirm the results from physiological methods in showing that the olfactory sense in pigeons is sufficiently good to be used in navigation. Similar results have been obtained from other bird species (Schmidt-Koenig, 1979).

Magnetic compass. Scientists long had thought that the energy involved in geomagnetic effects would be too low to be detected by animals. This now is known to be incorrect, and responses to magnetic fields have been shown to occur in many species.

Although there were early indications of magnetic sensitivity in birds (e.g., Merkel and Wiltschko, 1965), the possibility was long subject to doubt because of successive failures to demonstrate magnetic sensitivity in cardiac-conditioning experiments. However, positive results have been obtained in pigeons that were free moving rather than strapped down (Bookman, 1978).

Miniature magnets have been found in bacteria, honeybees and pigeons. Although these bees and pigeons are known to be sensitive to magnetic fields, it is not known how magnetic information is made available to the nervous system. There is evidence that the internal clock of bees is influenced by magnetic phenomena (Gould, 1980). In birds, it appears that it is the declination of the earth's magnetic field that is important in providing a directional sense, and there is speculation that this is involved in the map sense of homing pigeons (see following discussion).

Sun compass. That pigeons can use the sun as a compass was discovered by Gustav Kramer (1951). He trained starlings in a circular cage to look for food in a certain compass direction. All visual landmarks were excluded, and only the sun and sky were visible. The birds were able to maintain a particular compass direction throughout the day, showing that they could compensate for the movement of the sun.

The starling's internal clock (see Chapter 16.3) can be reset experimentally (Hoffman 1954) by confining the bird in a lightproof room and exposing it to artificial photoperiods. Klaus Schmidt-Koenig (1958; 1960; 1961), in experiments with homing pigeons, tested the effects of shifting the clock 6 hours forward, or 6 hours backward, or by 12 hours. The

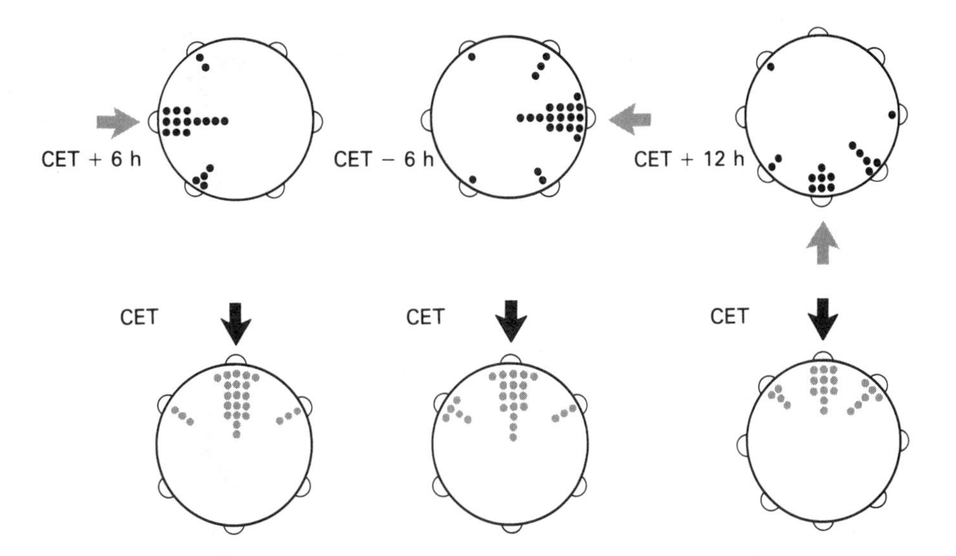

Fig. 14.15 Results of clock-shift experiments with homing pigeons. The animals were trained in a circular cage to look for food in a particular compass direction. They were tested by counting the pecks at food dishes around the periphery (each dot indicates one peck). CET is Central European Time. The arrows indicate the direction expected under the conditions of the experiment. The top row shows experiments, the bottom row shows simultaneous controls (After Schmidt-Koenig, 1960)

results illustrated in Figure 14.15, show that the bird's orientation in the testing apparatus is exactly that expected if it were using the sun as a compass.

The pigeons' sun compass has been shown to be sufficiently accurate for navigational purposes, provided the birds make corrective measurements at intervals during their journey. However, the compass direction of the sun, in relation to local time, is primarily of use in determining longitude. It is the sun's altitude (or azimuth) that alters with changes in the latitude of the observer. There is some evidence, from operant conditioning experiments, that pigeons can make fairly accurate measurements of changes in azimuth. There is some evidence that pigeons estimate the sun's altitude by measuring shadows rather than the sun directly (McDonald, 1972, 1973). Shadows may magnify the sun's movement by a factor of six, but it is not known whether pigeons are able to use this information out of doors. Andy Whiten (1972) trained pigeons to relate sun altitude to the northerly or southerly direction of home. His experiments show that pigeons can associate sun altitude with north-south direction. They do not demonstrate that such abilities are used in navigation.

Some workers, notably Geoffrey Matthews (1955; 1968) and Colin Pennycuick (1960), have attempted to account for the pigeons' navigation in terms of the sun's movement. In theory, there is sufficient information to determine longitude (from the time-compensated com-

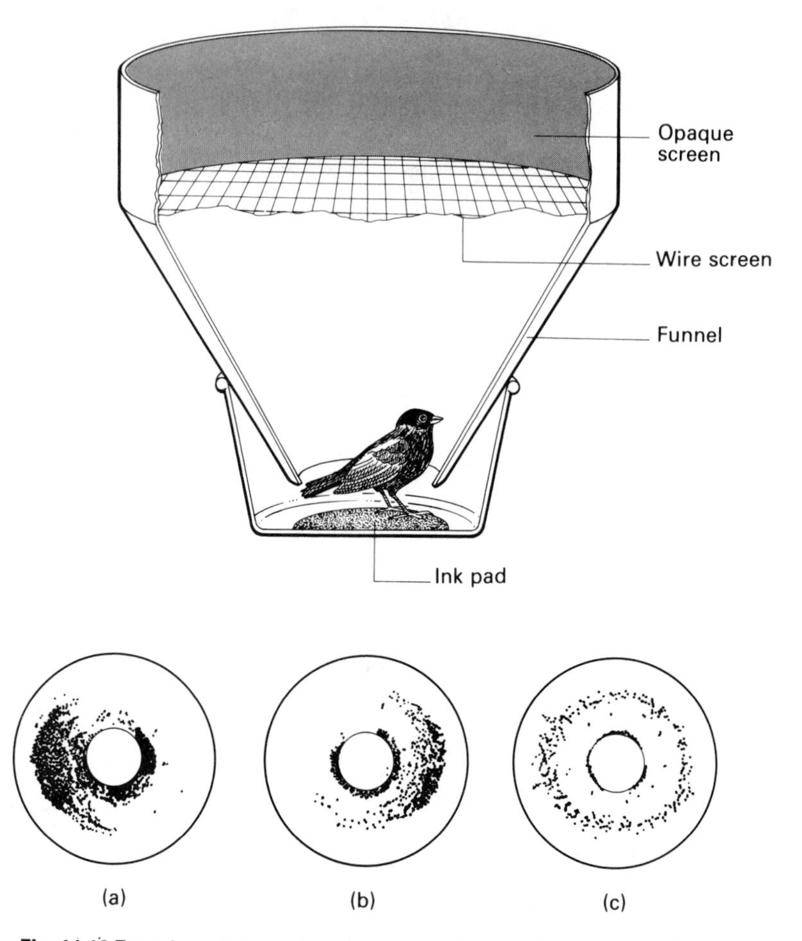

Fig. 14.16 Experimental cage for measuring migratory restlessness. When the bird attempts to leave the cage, it produces inky footprints on the blotting paper lining the funnel. Some examples of records are shown below (After Schmidt-Koenig, 1979)

pass direction of the sun) and latitude (from the sun's azimuth in relation to time) and so to construct the equivalent of a map. However, such theories require very considerable visual acuity and accuracy of measurement of sun motion in relation to time. Most scientists doubt that pigeons are capable of such accuracy (Schmidt-Koenig, 1979). Moreover, the ability of a flying pigeon to determine the altitude of the sun has never been demonstrated, and this is an essential ingredient of sun navigation theories.

Star compass. If songbirds are caged during the period when they normally would be migrating, they show a typical directional migratory restlessness; at night this directionality is related to the stars (Sauer and Sauer, 1955). The birds were oriented under a natural starry sky and under a planetarium sky. The techniques for recording migratory restlessness were later improved (Emlen and Emlen, 1966), using the apparatus illustrated in Figure 14.16. Steve Emlen (1967) demonstrated that the directionality of indigo buntings (*Passerina cyanea*) was the same under a natural sky and a stationary planetarium sky. The birds would follow a shift in the planetarium sky, and Emlen (1972) discovered that their orientation was linked to the rotation of the sky rather than specific constellations.

To a terrestrial observer, the stars appear as if on the inside of a spherical surface, called the *celestial sphere*. At any given time, the stars form a definite pattern on the celestial sphere, a pattern that moves so as to give the impression that the celestial sphere is spinning. In reality, the earth is spinning about its polar axis. The point in the northern sky about which the celestial sphere seems to rotate is called the *north celestial pole*. The star Polaris lies very close to this pole, and for practical purposes this star can be used to determine the observer's northerly direction.

Emlen (1972) showed that buntings hand raised without any sight of the sky were unable to orient during migration. Birds exposed routinely, during the period between fledging and the autumn migration, to a planetarium sky rotating around Polaris showed the normal southerly orientation. However, northerly orientation was shown by birds exposed to an experimental planetarium sky made to rotate around Betelgeuse, in the constellation Orion, which appears in the southern sky for an observer in northern latitudes. Apparently the newly fledged birds learned which part of the sky rotated least and oriented away from that part during their migratory restlessness.

Cardiac-conditioning methods also have been used to investigate the perception of the night sky by birds. There is evidence that mallards (*Anas platyrhynchos*) learn to recognize particular star patterns (Wallraff, 1969). Although star patterns provide potential maplike information, the evidence so far indicates that the stars are used purely as a compass (Schmidt-Koenig, 1979).

Polarized light. In normal, unpolarized light, the wavelike vibration occurs equally in all planes. In polarized light, the vibrations are greater in one plane. When unpolarized sunlight is scattered by atmospheric molecules, polarization occurs, greatest for the light scattered at an angle of 90 degrees to the rays of the sun, which means there is a pattern of polarization in the sky that changes according to the position of the sun.

Karl von Frisch (1967) discovered that the orientation of honeybees, as indicated by their dance (see Chapter 23), depended upon the position of the sun, even when the sun was obscured by clouds. He showed it was necessary for the bees to see only a small portion of sky and that the essential information was confined to the ultraviolet portion of the spectrum. When ultraviolet light from the sun was passed through a polarizing filter, von Frisch found that the orientation of the bee dance changed in accordance with the angle of polarization.

Cardiac-conditioning experiments (see Fig. 14.14) clearly show that pigeons are able to perceive rotations in the plane of polarization of light (Kreithen and Keeton, 1974; Delius et al., 1976), but it is not known how they interpret this information. Although the pattern of polarization of the sky can provide an indication of the sun's position even when the sky is overcast by clouds, it appears that pigeons do not use their sun compass under such conditions (Schmidt-Koenig, 1979).

Points to remember

- Coordination of movement involves both central control and peripheral control. Peripheral control is largely preprogrammed, but central control may involve feedback (closed-loop control) or it may be predictive (open-loop control).

- The spatial orientation of simple animals is based upon kineses and taxes which vary in type in accordance with the animal's sensory capabilities. Navigation, a more complex form of orientation, requires the equivalent of both map and compass.

- A wide variety of sensory modalities are known to be involved in navigation, but the way the information is coordinated is not well understood.

Further reading

Fraenkel, G. S. and Gunn, D. L. (1940) *The Orientation of Animals*, Clarendon, Oxford (Dover Books, 1961).

Schmidt-Koenig, K. (1979) *Avian Orientation and Navigation*, Academic Press, London.

15 Homeostasis and Behavior

The notion of physiological stability was implicit in Claude Bernard's (1859) concept of the internal environment. He observed that the level of sugar in the blood remained constant, even when the animal was deprived of food or had just consumed meat or food containing sugar. He postulated that there was some process of regulation and control to maintain the constancy of the internal environment. He also realized that an animal able to regulate its internal environment in the face of fluctuations in the external environment has greater freedom to exploit a variety of potential habitats.

Animals can be roughly divided into *conformers*, which allow their internal environment to be influenced by external factors, and *regulators*, which maintain their internal environment in a state largely independent of external conditions. The processes by which regulators control their internal state come under the general heading of *homeostasis*.

15.1 Homeostasis

The term *homeostasis* was used first by the American physiologist Walter Cannon (1932), who wrote the following: "The coordinated physiological processes which maintain most of the steady states in the organism are so complex and so peculiar to living beings—involving, as they may, the brain and nerves, the heart, lungs, kidneys and spleen, all working cooperatively—that I have suggested a special designation for these states, homeostasis." Cannon envisaged a situation in which sensory processes, monitoring the internal state of the body, initiated appropriate action whenever the internal state deviated from a preset, or optimal, state. For example, when human body temperature rises above 37°C, cooling mechanisms such as flushing and sweating are brought into action. When the temperature falls below the optimal level, warming mechanisms like shivering come into play. By employing a number of such finely tuned mechanisms, humans are able to achieve a precise thermoregulation and thermal homeostasis.

Although regulatory mechanisms of the types envisaged by Cannon now are known to be widespread in the animal kingdom and to involve a great variety of physiological processes, they are not the only types of processes involved in the control of the internal environment. For example, until recently it was thought that increased drinking by animals in response to high environmental temperature was a result of dehydration arising from cooling responses such as sweating and panting, a view entirely consistent with the theory of homeostasis outlined previously. Evaporation of water is necessary to maintain thermal homeostasis in a hot environment, and this upsets the fluid balance of the body, the restoration of which requires increased drinking. However, we now know that in species such as the rat and pigeon, drinking occurs as a direct response to the temperature change in *anticipation* of any change in fluid balance arising from thermoregulation and not in response to it, as the regulatory theory would require (Budgell, 1970a; 1970b). In other words, the animals drink in order to have water available for thermoregulation.

The role played by behavior in the control of the internal environment varies considerably with species and with circumstances. Drinking behavior, for instance, is essential for the maintenance of homeostasis in many species, the physiological mechanisms of water conservation being unable to prevent lethal dehydration after prolonged deprivation. However, some species—such as the Mongolian gerbil (*Meriones unguiculatus*) and the parakeet or budgerigar (*Melopsittacus undulatus*) (Cade and Dybas, 1962)—are able to survive indefinitely without water. Because of the great efficiency of their water conservation mechanisms, they can live on the water contained in the seeds they eat. Others, like aquatic species, need no special behavior to obtain the water necessary for the maintenance of homeostasis.

In many aspects of homeostasis, the role of behavior is normally negligible. However, experiments involving surgical interference with the normal physiological mechanisms responsible for homeostasis show that animals often have the capacity for appropriate behavior, even though it is not used in their normal lives. For example, the work of Curt Richter (1942–1943) shows that if the thermal homeostasis of rats is interfered with by the removal of the thyroid gland, they respond by building warmer nests and by engaging in other forms of behavioral thermoregulation, given the opportunity by the experimenter. Similarly, removal of the adrenal gland, which is involved in salt balance, causes rats to shift their preferences toward saltier food and water.

In other cases, there may be no special organs involved. Thus, rats fed a vitamin-deficient diet are able to select food containing the required vitamins even though they cannot taste the presence of the vitamins in the food. As we see in Chapter 18, they are able to learn which food makes them feel better (Revusky and Garcia, 1970; Rozin and Kalat, 1971). Clearly, the mechanisms by which internal stability is maintained are many and various, and it is incorrect to think of homeostasis as necessarily implying a simple regulatory feedback process. Moreover, it is wrong to think that homeostasis simply implies constancy of the internal environment. Many intermediates exist between conformers and regulators, and one species may be able to regulate one body function but not others. This is clearly illustrated by reference to thermoregulation in animals.

15.2 Thermoregulation

Most animals have an optimum body temperature at which they function most efficiently. Below this temperature, their metabolism progressively slows down, muscular activity diminishes, and the animal may become torpid. Above the optimum temperature, metabolic rate rapidly increases, and this may be expensive to maintain. Moreover, there is an upper limit to the temperature at which bodily processes remain viable. For most species this limit seems to be in the region of 47°C.

Most animals are able to influence their own body temperature to some extent either by employing specialized physiological mechanisms or by appropriate behavior. In both cases it is necessary for the animal to be able to detect the environmental temperature, its own body temperature, or both, by means of various sensory processes usually grouped together under the heading *thermoreception* (see Chapter 12).

The metabolic reactions of the body produce heat continuously, and the more active the animal, the greater the rate of heat production. In a cold environment, however, the heat production from normal metabolism and activity may not be sufficient to counteract the cold, and the animal may produce heat by increasing metabolic rate and taking action to prevent the loss of body heat. Many invertebrates are cold blooded in that their body temperature tends to conform with that of the environment. Because the rate of metabolic reactions is determined by the temperature at which they occur, such animals are forced to reduce their activity when the body temperature falls. However, some invertebrates, such as the common woodlouse (*Porcellio scaber*) and millipedes (Myriapoda) are stimulated into extra activity by falling temperatures and thus are able to maintain a body temperature higher than that of the environment. In warm-blooded animals, the rate of heat production can be raised by increasing muscular activity, as in shivering, and by direct effects of thyroid hormones on metabolic rate. Food intake also can serve to increase heat production because heat is released during digestion. Many animals increase food intake in response to cold.

Animals that gain heat primarily from external sources like sunlight are usually called *exothermic*, while those that gain heat primarily from internal processes are usually called *endothermic*. Exothermic animals sometimes enhance the warming effects of sunlight by color change, since dark-colored objects absorb more radiant heat than light-colored objects. Some species of lizard are able to change their color in accordance with their thermal requirements. For example, the desert iguana (*Dipsosaurus dorsalis*) has dark coloration in the early morning, when its body temperature is low, but becomes progressively paler as its temperature rises, reaching the halfway stage of color change at about 40°C. Similarly, some turtles expose their black feet to sunlight to increase the rate of heat gain.

The metabolic reactions of the body produce heat continuously, and animals can become overheated easily, especially when they are active. Overheating also can occur in especially hot environments or when heat dissipation is impaired. Because the lethal body temperature of many animals is not much above their normal body temperature, cooling mechanisms have to be especially rapid and effective.

There are four main ways of losing heat from the body: conduction, convection, radiation, and evaporation. *Conduction* is the transfer of heat through solids and liquids, between regions differing in temperature. It can occur within the body tissues or between the body and an external object like the ground. Heat conduction can be reduced by insulation provided by layers of fat within the body and by layers of air trapped in hair or feathers at the body surface. *Convection* is the transport of heat in a fluid medium. Heat loss occurs as a result of the circulation of warm blood from the interior of the body to the cooler surface tissues. The control of blood flow to the periphery thus provides an important means of temperature regulation. *Radiation* is a form of heat transfer that does not depend upon material mediation but that can occur in a vacuum. Heat loss by radiation is roughly proportional to the temperature difference between the animal and its environment. Animals both gain and lose heat by radiation. An animal's color makes little difference to its heat loss by radiation, but it does affect heat gain, as we saw earlier. Thus, black animals gain more heat by absorption of radiation than white ones, which have greater reflectivity. *Evaporation* of water from moist areas of the animal's body surface involves heat loss and is an important means of cooling in many species.

Animals are able to control the amount of heat lost in these four ways to a variable extent, depending upon the species. To some extent they can control the rate at which the body loses heat by conduction by altering their insulation. Long-term regulation can be effected by increasing deposition of fat and hair growth in winter and by decreasing these in summer. Short-term changes can be achieved by raising or lowering hair and feathers (see Fig. 15.1), thereby controlling the amount of trapped air, and by altering body posture in relation to the prevailing weather. Surface insulation is influenced by wind and wetness. Wet hair increases heat loss by conduction. When a mammal lies or sits on the ground, the hair is compressed and holds less air. Heat then is conducted into the ground, particularly when the ground is wet. Cows frequently are observed to sit down before rain comes, in order to rest without losing as much heat as they would once the ground is wet.

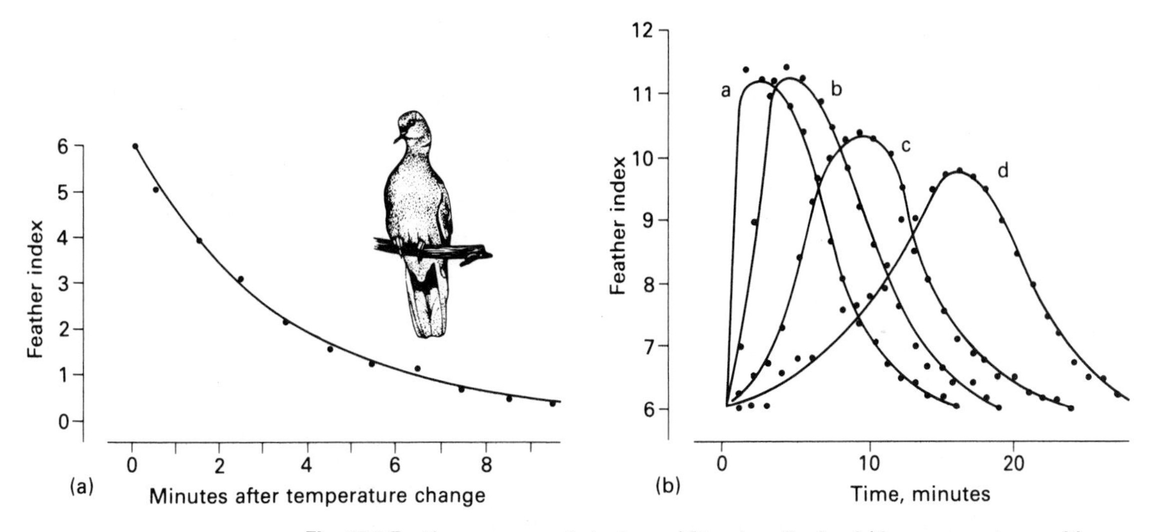

Fig. 15.1 Feather movements in doves (*Streptopelia risoria*) in response to a sudden increase in environmental temperature (*a*), and in response to drinking 10 cc cold water (*b*). The feather index is a measure of feather position averaged over the whole body (see Fig. 22.10). The curves a, b, c, d are the responses to drinking at different rates (After McFarland and Budgell, 1970)

Some animals can enhance radiant heat losses by behavioral means. For example, certain fiddler crabs (*Uca*), ground squirrels (*Citellus*), and other burrowing animals make sorties between their cool burrows and the warm external environment. In this way they can cool off by radiation if overheated. The effective surface area of the body can be reduced by curling up or by huddling with other members of the same species, thus lowering the amount of heat lost by radiation.

Evaporative water loss through the skin is uncontrolled in amphibians, reptiles, and birds; in mammals it is regulated by the sweat glands, which are present in all higher mammals except rodents and lagomorphs (rabbits). In humans, the sweat glands on the palms are controlled emotionally while those on the rest of the body normally respond to thermal control. Sweating is controlled by thermoreceptors in the brain and not by those in the skin. Thus, humans usually sweat during exercise but not necessarily when sitting by a hot fire. Rats and some other animals enhance evaporative cooling by moistening the body surface with saliva or by wetting themselves with water, as elephants do.

Respiratory evaporation is controlled to some extent in most animals. Crocodiles, snakes, and some lizards gape widely when hot. The desert iguana (*Dipsosaurus*) pants like a dog. Because animals lose so much water in respiratory evaporation, they tend to use it only in emergencies. Birds and mammals pant only when their body temperature approaches the lethal level. Flying birds generate a lot of heat and rely primarily

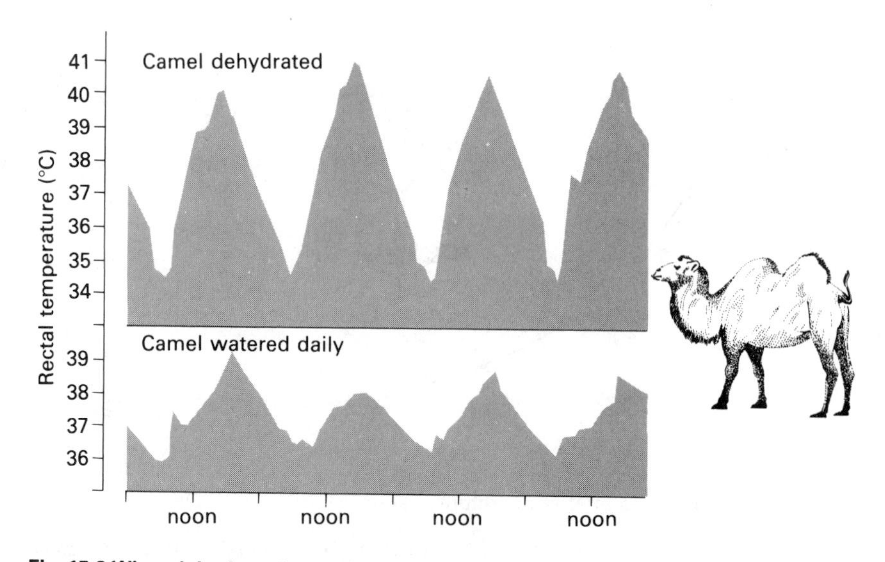

Fig. 15.2 When dehydrated, camels allow their body temperature to rise during the day, so that they do not have to lose water in thermoregulating. During the cold desert night the camel loses the excess heat that it has stored during the day (After Schmidt-Nielsen, 1964)

upon respiratory evaporation to dissipate it. The camel does not pant at all but relies on radiant cooling during the night. Camels do not store water to a greater extent than other species and cannot afford to expend water in keeping cool (see Fig. 15.2).

Exotherms are able to regulate their body temperature only to a limited extent. Amphibians can keep cool by evaporation of water from the body surface. The leopard frog *Rana pipiens* can maintain a body temperature of 36.8°C at an environmental temperature of 50°C. Reptiles have a more limited ability to cool themselves and tend to avoid very hot conditions. The Namib desert lizard *Aporosaura anchietae* burrows into the sand when the midday temperature climbs above 40°C (see Fig. 15.3).

True thermal homeostasis is found in birds and mammals, which are endotherms, able to maintain a constant body temperature despite fluctuations in environmental temperature. Their high metabolic rate provides an internal source of heat, and their insulated body surface prevents uncontrolled dissipation of this heat. Birds and mammals maintain a body temperature that is usually higher than that of their surroundings. The brain receives information about the temperature of the body and is able to exercise control over the mechanisms of warming and cooling. When the brain temperature gets too high, the cooling mechanisms are activated, and if it gets too cool, then heat losses are reduced and warming mechanisms are called upon. This feedback principle is the same as that for a thermostatically controlled electric heater (Fig. 15.4).

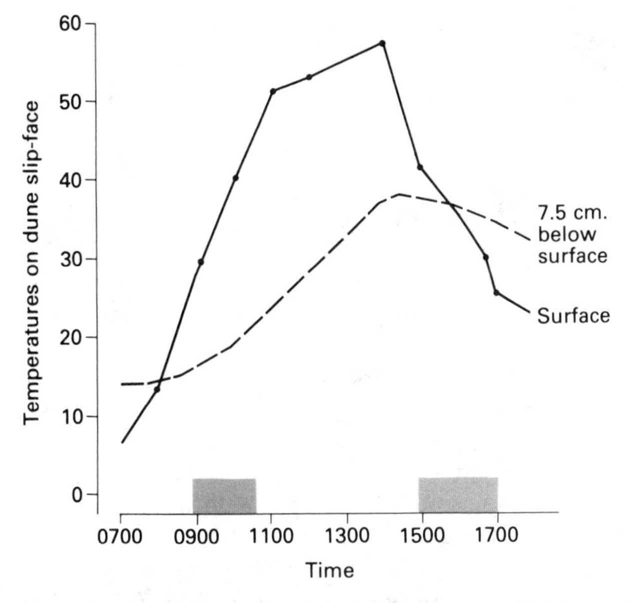

Fig. 15.3 The lizard *Aporosaura* lives in sand dunes, which have marked fluctuations in surface temperature. The lizard remains under the surface when the surface temperature is too cold, or too hot. This usually means that it has one period of activity on the surface in the morning, and another in the afternoon. Active periods are shown as shaded blocks (After Louw and Holm, 1972) (*Photograph: Gideon Louw. See also back cover*)

Fine control of body temperature occurs in humans, who have an early warning system consisting of numerous thermoreceptors in the skin. On the basis of this type of information, people are able to take anticipatory action and so forestall any undue fluctuations in body temperature. Controlled temperature changes do occur in birds and

Fig. 15.4 The feedback principle involved in a simple thermostatic electric heater. When the temperature reaches a certain preset value, the switch opens, the heater goes off, and the temperature falls. When it has fallen below the preset value, the heater is turned on again

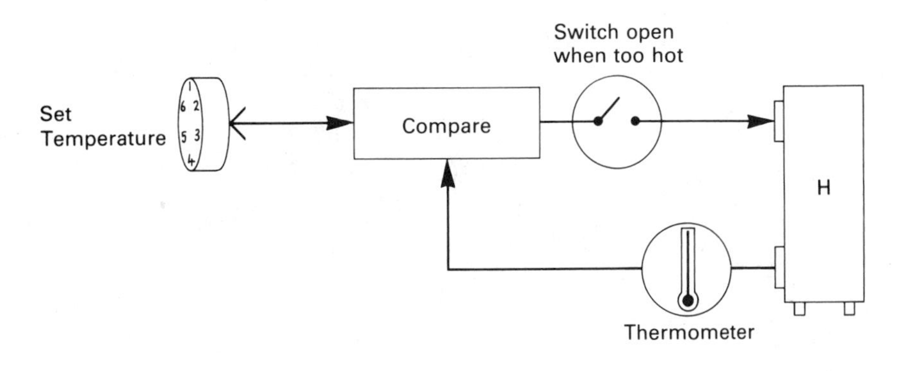

mammals, often on a diurnal basis. The average human body temperature is 36.7°C in the early morning and 37.5°C in the late afternoon. Many endotherms tolerate some internal temperature fluctuation, probably for energy conservation.

15.3 Water balance

All animals require water to maintain their metabolic processes. Animals continuously lose water by a variety of means, including excretion and evaporation from the body surface. In vertebrates, the water of the body is located in distinct compartments, as illustrated in Figure 15.5. The losses occur from the vascular compartment. For example, the lungs are a source of water loss by evaporation. The lungs have a rich supply of blood vessels, and gas exchange takes place across thin membranes moistened by the plasma in which the gases dissolve. Exhaled air nearly always contains more water than inhaled air, and this water comes from the vascular compartment. In many animals, respiratory evaporation is an important means of cooling. As mentioned earlier, birds rely heavily on respiratory evaporation as a means of cooling during flight. Thermoregulation, which involves sweating or spreading saliva over the body surface, also necessitates a certain amount of water loss.

The kidneys filter the plasma of the blood, remove waste products, and excrete them in urine. Thus, some water loss due to excretion is inevitable, but we shall see that some savings are also possible here. Water is also lost in defecation, but in times of water shortage, many

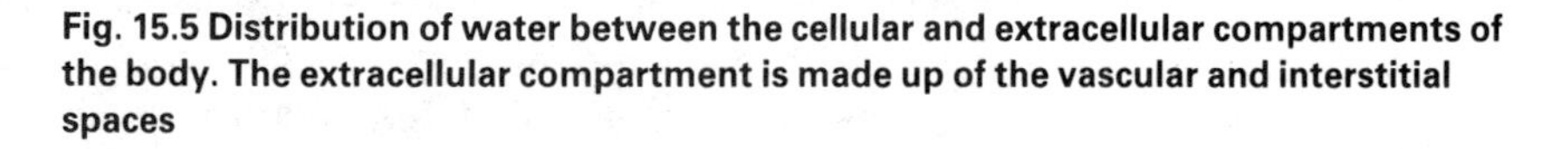

Fig. 15.5 Distribution of water between the cellular and extracellular compartments of the body. The extracellular compartment is made up of the vascular and interstitial spaces

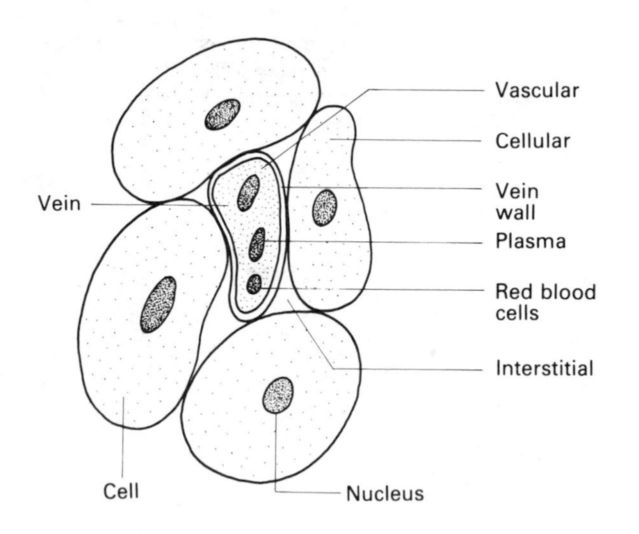

vertebrates are able to reabsorb water from the small intestine and, thus, excrete very dry feces. Loss of water from the vascular compartment results in some redistribution among all the compartments of the body, but the lost water must be replaced eventually.

Most animals obtain water by drinking, but amphibians absorb water through the skin, as do some insects. Some animals, especially those that have infrequent access to water, are able to drink enormous quantities. Whereas a very thirsty man can drink 1 liter (~1 US quart) of water in a minute, and possibly 3 liters (~3 US quarts) in 10 minutes, a 352-kilogram (778-pound) male camel has been observed to drink 104 liters (~27 US gallons) in one go. After water enters the mouth, it passes through the esophagus into the stomach and then enters the intestine. From here it may enter the blood by osmosis. If the concentration of salts in the intestinal fluid is less than that of the blood, then water will pass into the blood. However, if the salt concentration is higher in the intestine, as it may be after ingestion of salty food or water, then water may pass the other way, and there may be temporary dehydration of the vascular compartment.

Loss of water from the vascular compartment results in an increase in the concentration of salts in the extracellular compartment, which induces some redistribution of water between the cellular and extracellular compartments, resulting in the body cells undergoing some dehydration and shrinkage. These changes are detected by specialized cells in the brain, called *osmoreceptors*. Opinions differ as to whether the osmoreceptors are sensitive to changes in the extracellular fluid or whether they measure their own shrinkage during dehydration (see Toates, 1980, for a review of this problem). The osmoreceptors are known to occur in the hypothalamic region of the brain, and their stimulation has two main effects: an increasing tendency to seek water to drink, and the activation of various water conservation mechanisms.

There are osmoreceptors in the hypothalamus that control the release of antidiuretic hormone from the pituitary gland, lying just below the hypothalamus (Fig. 15.6). The presence of antidiuretic hormone in the bloodstream leads to a decrease in the amount and an increase in the concentration of urine excreted by the kidneys. Damage to the pituitary gland or associated areas of the hypothalamus results in diabetes insipidus, the symptoms of which include excessive urination and consequent thirst. Antidiuretic hormone is thus an important factor in water conservation.

Other conservation mechanisms include increased reabsorption of water in the small intestine, so that less is lost in the feces, and a reduction of food intake. Because the waste products of digestion and food metabolism have to be excreted, some water loss is inevitable, but this can be reduced by eating less. Laboratory studies show that water losses in doves deprived of food are only about a quarter of their normal level (McFarland and Wright, 1969). There is also some evidence that

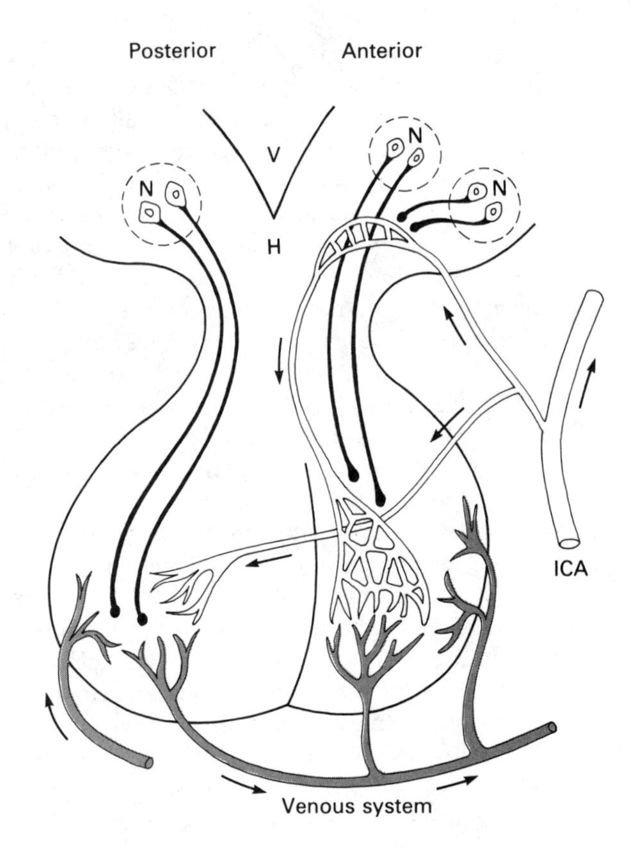

Fig. 15.6 Diagram of the human pituitary gland. The hypothalamus H receives its blood supply from the internal carotid arteries ICA, which also supply the anterior and posterior pituitary. Neurosecretory neurones N shown originating in the hypothalamus are responsible for the release of hormones into the venous system. V is the third ventricle of the brain (After Wilson, 1979)

doves can control their respiratory water losses (Wright and McFarland, 1969). Water loss resulting from thermoregulation sometimes can be reduced through behavior such as seeking a cool place and reducing heat production due to exercise and food consumption. When a camel is short of water, it allows its body temperature to rise and stores heat in the fatty tissues of its hump during the daytime. During the cold desert night this heat is dissipated by radiation, without any loss of water. Contrary to popular belief, camels do not store water in their humps, although metabolism of the fatty tissue would, of course, release some water (Schmidt-Nielsen, 1964).

Animals may become dehydrated as a result not only of intracellular dehydration but also of reduction in the volume of the extracellular fluid. Hemorrhage and other forms of blood loss do not alter the animal's osmotic balance, but the lost fluid has to be replaced. Animals

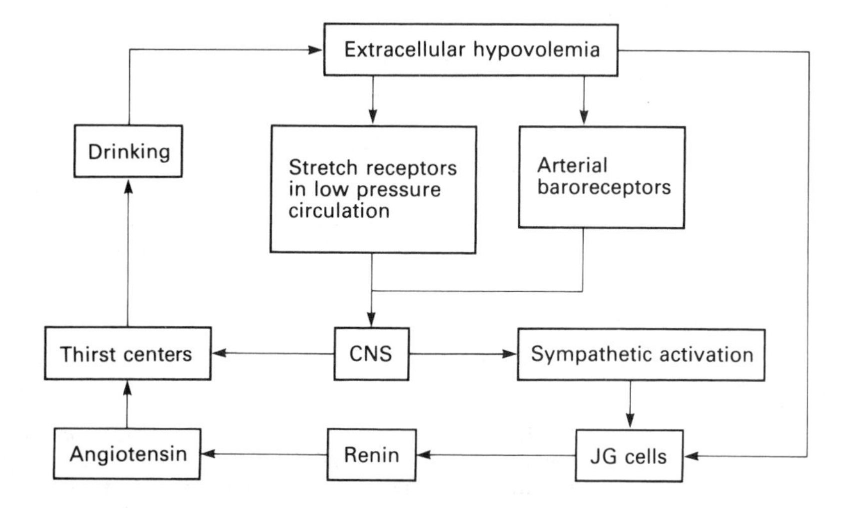

Fig. 15.7 Extracellular thirst. Hypovolemia (loss of extracellular volume) monitored by blood-pressure receptors, causes secretion of renin by the JG (juxta-glomerular) cells. This is converted into angiotensin, which promotes drinking (After Fitzsimons, 1971)

have various mechanisms for detecting such a loss. The hormone *renin* is produced by the kidneys in response to a reduction in renal blood flow. Renin is released into the blood, where it stimulates production of another hormone, *angiotensin.* Angiotensin has two main effects: It acts on the vascular system to maintain normal blood pressure and circulation, and it is a powerful thirst stimulus (animals injected with very small amounts of angiotensin will stop whatever they are doing and seek water [Fitzsimons, 1976]). The main components of the complicated mechanisms involved in extracellular thirst are illustrated in Figure 15.7.

Maintenance of water balance is bound up intimately with thermoregulation and with eating. Many animals greatly reduce their food intake when dehydrated, an important means of water conservation because food intake normally necessitates considerable water loss in the excretion of waste products, as we saw earlier. The main water conservation mechanisms are indicated in Figure 15.8. It is important to realize that any method of water conservation is disadvantageous to the animal in some way. It may interfere with normal thermoregulation or it may reduce the intake of energy. The different mechanisms are given different emphasis in different species, in accordance with the animal's normal ecological circumstances. Thus, the camel sacrifices a stable body temperature to conserve water, whereas a pigeon sacrifices food intake to maintain a relatively stable body temperature (see Fig. 15.2).

At first sight, the mechanisms that maintain water balance would seem to be relatively straightforward examples of homeostasis: The animal detects displacements from the normal amounts (volume) and

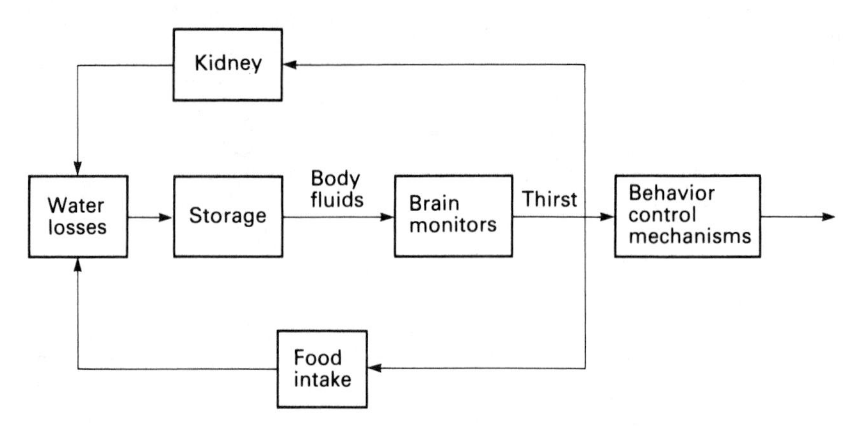

Fig. 15.8 Water conservation mechanisms operate when drinking is not possible. As thirst increases, antidiuretic hormones from the pituitary gland promote water retention in the kidneys, so that less water is lost in urine. Another important means of water conservation is reduction in food intake, because food digestion and excretion normally involve water loss

concentration (osmosity) of extracellular water and takes action to remedy the situation, either by drinking or by reducing the rate of water loss through various conservation mechanisms. However, the situation is not quite so simple because of interactions with other systems and because drinking often occurs when the animal is not dehydrated.

15.4 Energy and nutrients

Animals require food that can be digested to provide energy and certain specific nutrients and vitamins for growth and repair of tissues and to combat parasites and disease organisms.

The cells of the body obtain their energy primarily in the form of glucose dissolved in the extracellular fluid. In the process of metabolism, energy is made available to the cell, with water, carbon dioxide, and heat as by-products. The glucose in the extracellular fluid may come directly from digestion or may be released by the liver from its store of glycogen.

The cells of the body also can obtain energy from metabolism of fatty acids. The cells of the nervous system are the exception, for they can utilize only glucose. It is important for the glucose level of the blood to be maintained within fairly narrow limits in order to ensure an adequate supply of energy for the cells of the nervous system. The availability of glucose to cells is controlled by the hormone *insulin*. The cells of the nervous system can absorb glucose in the absence of insulin, but other cells require insulin to transport glucose across the cell wall. During times of glucose shortage, as in fasting, the level of insulin in the blood falls to such an extent that the blood glucose is effectively available only

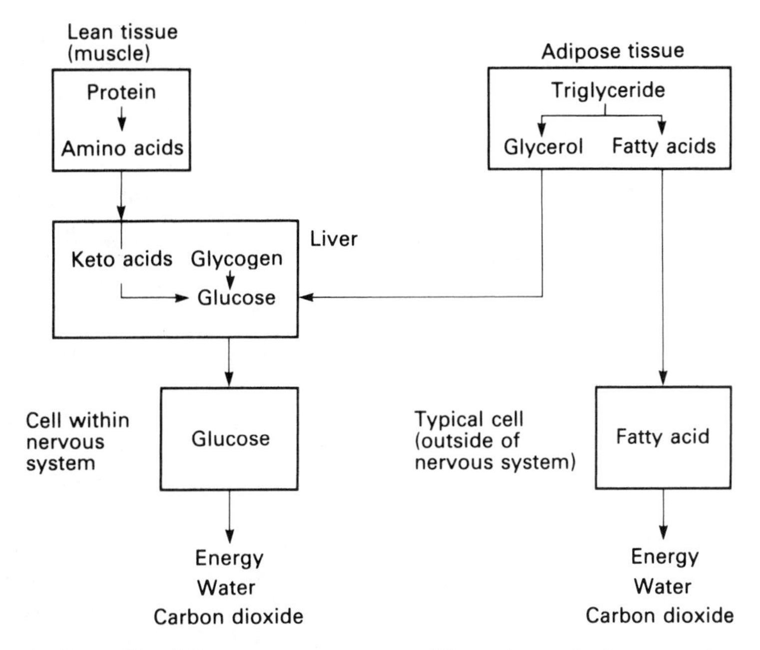

Fig. 15.9 The main sources of energy available during fasting (After Toates, 1980)

to the cells of the nervous system. The other cells have to depend upon metabolism of fatty acids to obtain energy. During fasting, glucose is derived from the body reserves, which include glycogen stored in liver and muscle, fat stored in various parts of the body, and in the last resort, the protein of muscle and other tissues. Fat is broken down to glycerol and fatty acids, and the glycerol is converted to glucose in the liver. Protein is broken down into amino acids, and these are metabolized by the liver to produce some glucose. The sources of energy available during fasting are summarized in Figure 15.9.

The other major source of energy is food. Species differ greatly in the amount and types of food they require. Small animals that have a high metabolic rate, like songbirds (Passeriformes), quickly suffer an energy shortage when they are unable to obtain food. To maintain its body weight and normal activity, the great tit (*Parus major*) must feed every few minutes. At night, when feeding is impossible, some birds become torpid, lower their body temperature, and conserve energy. Other animals are able to draw upon energy reserves and can live for long periods without food. For example, incubating jungle fowl (*Gallus*) do not eat for many days and eat little even if food is placed next to the nest (Sherry et al., 1980). Animals with low metabolic rates at low temperatures, such as fishes, reptiles, and hibernating mammals, may not need to eat for weeks.

Food that enters the mouth may be stored immediately, as in the crop of pigeons and the cheek pouches of hamsters, or it may enter the stomach and intestine and undergo the process of digestion. Enzymes in the digestive tract break foodstuffs down into their basic components.

Thus, fat molecules are split into glycerol and fatty acids by the action of lipase, and the enzymes trypsin and chymotrypsin split specific amino acid bonds in proteins. The processes of digestion and the enzymes involved vary considerably from one animal species to another. Some animals—like the flatworm *Planaria*—do not have a complete digestive tract, whereas others—like vertebrate herbivores—have very complex digestive systems enabling them to cope with plant materials like cellulose that other animals are unable to digest. The products of digestion are absorbed into the bloodstream, partly by diffusion and partly by active transport across the wall of the intestine.

The process of digestion and the pattern of eating often are intimately related. Many animals vary their food intake according to the nutritive value of the products of digestion. A variety of mechanisms are involved in this type of regulation, the simplest of which involves direct detection of the substance, as is thought to be the case with sodium. A large and fairly constant quantity of sodium in the body fluids is vital. Sodium is involved in many physiological processes of fundamental importance, including the propagation of nerve impulses (see Chapter 11). Sodium is available in nature in the form of sodium chloride (common salt), but its distribution is patchy. It is not surprising, therefore, that animals should develop a specific appetite for sodium. Animals can detect sodium in the diet in two main ways. First, salt is a primary aspect of taste in most vertebrates (see Chapter 12.1). Second, sodium has profound effects on the body fluid compartments, as discussed earlier, and sodium depletion results in secretion of the hormone aldosterone from the pituitary gland, which causes sodium to be reabsorbed from the urine passing through the kidney.

Sodium appetite appears to be innate, but many animals are adept at learning and remembering the location of sources of sodium. For example, rats can be given a choice of pure water or salty water as a reward in a maze. Rats deprived of water developed a preference for the place where water was available in the maze. When these rats later were depleted of sodium but not of water, they immediately switched to the salty water, showing that they remembered the location of sodium, even though it was sampled at a time when they were thirsty and averse to salty water (Krieckhaus, 1970).

Animals are not able to directly detect many essential vitamins and minerals either by taste or by their levels in the blood. Nevertheless, deficient animals develop strong preferences for foods containing the missing substances. For many years, such specific hungers posed something of a problem for scientists attempting to explain how the animals knew which food contained the beneficial ingredient. There were also reports in the medical literature of children eating coal and other unusual substances. These habits were linked to dietary deficiencies for particular essential minerals like cobalt, but how the child learned what to eat remains unexplained.

2 h fasting

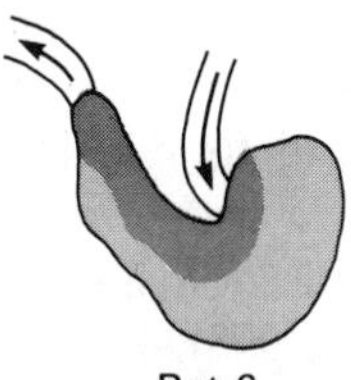

Rat 6

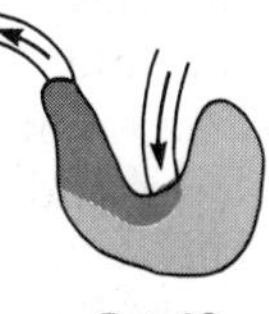

Rat 18

24 h fasting

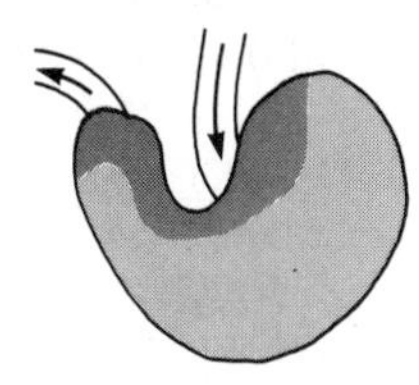

Rat 9

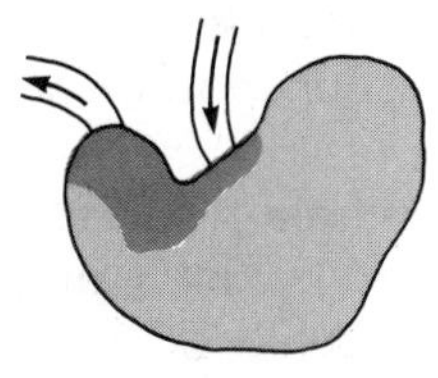

Rat 4

Fig. 15.10 Rat stomachs showing distribution of blue food (dark) eaten after normal food (pale) (After Wiepkema et al, 1972)

Thiamine-deficient rats show an immediate marked preference for a novel food, even when the new food is thiamine deficient and when the old food has a thiamine supplement (Rodgers and Rozin, 1966). The preference is short lived. If consumption of a novel food is followed by recovery from dietary deficiency, however, then the rat rapidly learns to prefer the novel food. Such rapid learning on the basis of the physiological consequences of ingestion enables the rat to exploit new sources of food, and, thus, to find out which contains the required ingredients. Detailed study of the feeding behavior of rats shows that they normally tend to avoid novel foods or to sample them a little bit at a time. The intervals between meals are sufficiently long to enable the animal to assess the consequences of sampling new foods. Scientists also have found that when a rat eats, some of the food is passed rapidly through the stomach into the intestine, bypassing food from a previous meal that is still in the stomach (Fig. 15.10). In this way a rat can test the qualities of the food it has eaten most recently (Wiepkema et al., 1972).

Paul Rozin and James Kalat (1971) and Rozin (1976a) point out that a vitamin-deficient diet is like a slow-acting poison. Rozin (1967) noted that rats on a thiamine-deficient diet were reluctant to eat their familiar food but ate avidly from a new food, similarly deficient. The aversion to the familiar food persisted even after the rats had recovered from the deficiency. Rats that become sick after eating a poisoned food also show an aversion to such food and show a more than normal interest in novel foods. Thus, there is a close affinity between the rat's responses to a deficient diet and its behavior toward a toxic diet. This is discussed in Chapter 18.

An animal's dietary requirements are many and various. In the rat they include water, nine essential amino acids, some fatty acids, at least 10 vitamins and 13 minerals (Rozin, 1976). Despite a large amount of scientific work, the question of what physiological factors initiate eating remains somewhat of a mystery. Walter Cannon (1932) thought that stomach contractions and other peripheral factors were responsible. However, when the nerves from the stomach are cut, or when the stomach is removed for medical reasons, human eating behavior is largely unaffected. Several theories have proposed that receptors within the brain are sensitive to the presence of nutrients in the blood. Substances such as glucose, amino acids, or fats might be used as indices of dietary requirements. The level of glucose in the blood increases during digestion, but it also changes in other circumstances, such as during autonomic arousal and in anticipation of meals. In 1967, Jean Mayer proposed that the brain responds to the *utilization* of glucose rather than to its *availability*, and others have proposed similar mechanisms (Toates, 1980). The suggestion is supported by the results of recent research, but the evidence remains inconclusive.

Some scientists doubt that direct monitoring of the nutrient state of the body is of prime importance in hunger. Although the fact that animals are able to respond to dietary excesses or deficits is suggestive of direct monitoring and regulation, there are alternative possibilities. Animals suffering from a particular deficiency tend to sample a wider range of food and to learn quickly to select appropriate foods. The evidence suggests that such learning is based upon general sensations of sickness and health rather than upon detection of specific deficiencies (Rozin and Kalat, 1971; Rozin, 1976a). Curt Richter (1943; 1955) showed that rats left to their own devices with a large variety of purified nutrients from which to choose select a balanced diet from this cafeteria.

How does the rat know which food to select at a particular time? The answer is that for some constituents of the diet, such as water, sodium, and possibly glucose, there is different monitoring by taste and by sensory processes in the brain that detect the presence of each of these substances in the blood. For other dietary elements, particularly the vitamins and minerals, there is no direct monitoring. The animal learns what to eat in order to avoid sickness. How this learning occurs we discuss in Chapter 18. In the case of amino acids and fats, the situation remains unclear. Direct monitoring and homeostatic control of these substances have been proposed (see Booth, 1978, for details), but the evidence remains weak.

15.5 Motivational systems

Homeostatic requirements place behavioral demands upon animals, as we have seen. At any given moment an animal must assess its total internal state, incorporate its knowledge of likely future demands and of the new demands to be created by any given course of action, and then choose what to do next.

The traditional view of motivation is built upon a simple feedback principle. A change in the animal's internal state is sensed by the brain and leads to a buildup of drive to perform the appropriate behavior. The drive gives rise to appetitive and consummatory behavior. *Appetitive* behavior involves a search for suitable external stimuli; when these are encountered, *consummatory* activity, like eating or drinking, takes place. The consequences of the consummatory behavior reduce the drive, either directly or by diminishing the internal or external stimuli that led to the drive. The consummatory behavior then ceases. For example, dehydration of the body tissue is sensed by the brain and results in a buildup of thirst drive; the drive induces the animal to search for water (appetitive behavior); and when it finds water it drinks (consummatory behavior). The consequences of drinking may reduce thirst directly via short-term satiation mechanisms such as the sensation of water in the mouth or loading effects of water in the gut (Rolls and Rolls, 1982). The short-term satiation mechanisms are complemented, or bypassed in some species (Rolls and Rolls, 1982), by absorption of water into the bloodstream, thus alleviating the dehydration that gave rise to the thirst drive. The animal stops drinking either as a result of the short-term satiation mechanisms or because its state of water balance no longer gives rise to thirst.

The term *drive* was introduced by Robert Woodworth (1918) as an alternative to William McDougall's (1908) concept of *instinct* (see Chapter 20). Woodworth distinguished between the energizing (*drive*) aspects of motivation and the *directing* aspects of motivation. *Primary* drives resulted from tissue needs, and other (*secondary*) drives were derived from learned habits. Similar notions of drive were developed by the early ethologists. Thus, Konrad Lorenz (1950) outlines the ethological view in terms of three successive processes: (1) accumulation of action-specific energy giving rise to appetitive behavior; (2) appetitive behavior striving for and attaining the stimulus situation activating the innate releasing mechanism; and (3) setting off of the releasing mechanism and discharge of endogenous activity in a consummatory action. Lorenz postulated "that some sort of energy, specific to one definite activity, is stored up while this activity remains quiescent, and is consumed in its discharge."

The basic idea of drive as an urge to perform particular activities was common to ethologists and to various schools of animal psychology in the United States. During the 50 years since its introduction, the drive concept has been subjected to numerous quasi-philosophical discussions and analyses. Controversy hinged around questions such as whether drives were inherently purposive (Thorpe, 1956; Peters, 1958), whether drives were general or specific (see Hinde, 1970; Bolles, 1967), and whether drive could be said to energize behavior (Bolles, 1967; McFarland, 1971). In recent years, a tendency to drop the concept of drive has arisen for reasons we discuss below.

The classical view of hunger and thirst as homeostatic drives implies that feeding and drinking are initiated as a result of monitored changes in the animal's physiological state. Feeding and drinking are said to be under negative feedback control because food and water intake serve to diminish the deviations in physiological state that initiated the behavior. However, in addition to occurring in response to physiological changes, feeding and drinking often occur in anticipation of such changes. Many animals have distinctive meal patterns repeated day after day under constant conditions. Just as humans experience hunger at habitual mealtimes, so the feeding tendency of animals can be governed by time of day. In conditions in which the environment changes little from day to day, animals quickly establish a daily routine of activities and eat at particular times even when food is continuously available. Physiological processes may become attuned to the routine. In humans, for example, the liver may cease to mobilize glycogen just before a meal is due. This leads to a fall in blood sugar level in anticipation of the increase that will occur when the meal is digested. Experiments have shown that such

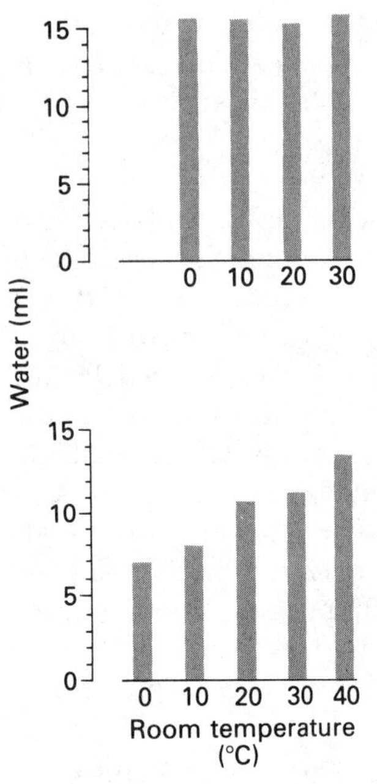

Fig. 15.11 Effects of room temperature on drinking in doves. Above: the effect of temperature during water deprivation upon subsequent drinking. Below: the effects of temperature during drinking (Data from Budgell, 1970a)

physiological adjustments can undergo conditioning in relation to the time of day.

We have seen that the long-term consequences of food intake tend to lead to increased thirst. Instead of drinking as a result of such thirst, many animals drink in advance and thus anticipate the dehydrating effects of food ingestion (Fitzsimons and Le Magnen, 1969). Similarly, we have seen, thermoregulation often involves water loss, but instead of drinking in response to thermally induced dehydration, some animals drink in advance and thus have water available for thermoregulation (McFarland, 1970a). For example, Phil Budgell (1970a) found that doves (*Streptopelia risoria*) deprived of water for two days at different temperatures drank the same amount when offered water in a test chamber at a temperature of 20°C. However, doves deprived for two days at the same temperature (20°C) drank different amounts when tested at different temperatures (see Fig. 15.11). Similar results have been obtained with rats (Budgell, 1970b), showing that these animals drink in direct response to environmental temperature changes before any thermally induced dehydration occurs. The term *feedforward* is used for situations in which the feedback consequences of behavior are anticipated and appropriate action is taken to forestall deviations in physiological state (McFarland, 1971; Toates, 1980).

There may not always be a one-to-one relationship between a particular activity and its physiological consequences. In some cases, such as behavioral thermoregulation, a particular activity may have consequences that are measurable along a single dimension, such as a change in body temperature. More commonly, however, behavior has consequences that affect a number of different aspects of the animal's state. For example, food intake will alter an animal's physiological state in a number of respects, depending upon the constitution of the ingested food. The consequences of such behavior are said to be ambivalent.

In 1972 Richard Sibly and I showed that in order to be able to cope with the ambivalent consequences of behavior, animals must be capable of adaptive control by which the animal can change the properties of its regulatory mechanisms to suit environmental circumstances. A number of instances of such adaptive control are known. Thus, I found (1971) that doves learned to recalibrate their short-term estimate of the hydrating value of their drinking water after experience of changes from pure water to salty water or vice versa. We have already seen how rats can learn to avoid or select particular foods as a result of physiological consequences occurring hours after ingestion (see also Chapter 18). Curt Richter (1943) showed that when certain physiological regulators were surgically removed, the animals themselves made an effort to maintain a constant environment of homeostasis. Thus, rats given a suitable environment could learn to compensate behaviorally following removal of the adrenal glands, thyroid glands, parathyroid, and pancreas. In this way, animals deprived of physiological regulation resorted to behavioral measures.

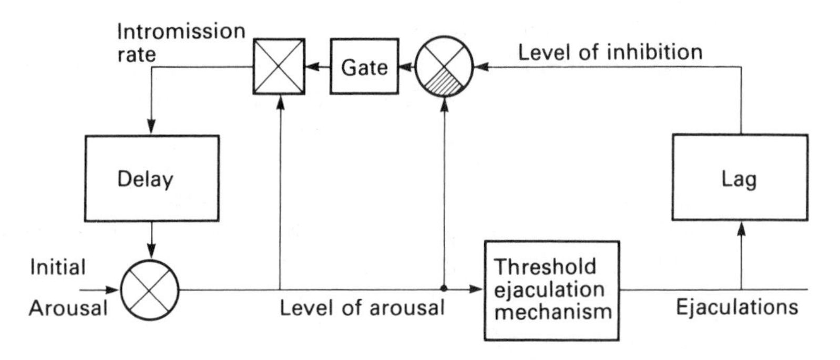

Fig. 15.12 This control model for sexual behavior in male rats is little different in principle from those proposed for the control of feeding or drinking (After McFarland and Nunez, 1978)

In order to account for homeostasis, in both its physiological and behavioral aspects, it is necessary to go beyond simple negative feedback concepts. The maintenance of physiological state within narrow limits may be achieved by a combination of negative feedback, feed forward, and adaptive control.

Traditionally, a distinction has been made between homeostatic aspects of behavior, such as feeding, drinking, and temperature regulation, and non-homeostatic aspects of behavior, such as aggression and sexual behavior, a distinction that has been challenged in recent years (e.g., Hogan, 1980; Davis, 1980). The control of sexual behavior is not fundamentally different than that of feeding or drinking (McFarland and Nunez, 1978; Toates, 1980). It involves control principles that are essentially similar and that can be represented in models employing the same terminology and concepts, as we can see from Figure 15.12.

The interactions among the physiological processes that give rise to behavior are complex. In the case of an apparently simple activity—drinking for instance—several different factors can contribute to its initiation, maintenance, and termination (Rolls and Rolls, 1982). In the case of feeding, the situation is much more complex (Booth, 1978). To handle this type of complexity, behavioral scientists have adopted the techniques of control systems theory used by engineers to describe and analyze complicated machines. By making quantitative models of the various components of a system, like the drinking control system, it is possible to carry out a computer simulation of the system as a whole. Such a simulation can be used to make quantitative predictions testable in experiments, and the results of the experiments then used to update the hypotheses, in this way developing increasingly refined models. Control systems theory has been applied to a variety of behavior, including feeding, drinking, thermoregulation, and sexual behavior (McFarland, 1971; 1974; Booth, 1978; Toates, 1975; 1980).

Compared with the precision and rigor of control systems theory, the concept of drive is vague and confusing. Moreover, the new approach has uncovered conceptual problems that effectively make the concept of drive no longer viable. Two of these problems were pinpointed by

Robert Hinde (1959; 1960), but the theoretical concepts necessary to handle the issues were not then current in the field of animal behavior. One misconception (Hinde, 1959) is that drives can be considered as unitary variables. When we think of the hunger drive, we imagine a quantity measurable along a single scale. Thus, we think of an animal as having low or high hunger. However, as we have seen, there are many different aspects of hunger. An animal may have a specific hunger for, say, salt or thiamine. It may have a protein deficiency or a shortage of readily available energy. These different aspects of hunger can influence the animal's feeding behavior, so it is misleading to think of hunger as a unitary variable. An alternative formulation (McFarland and Sibly, 1972) represents hunger and other so-called drives in terms of vectors. This approach paved the way for the state-space representation of motivational systems, which we discuss below.

A second misconception (Hinde, 1960) is of drive as an energizer of behavior. Although this view has been very influential in psychology and ethology, it involves concepts that cause difficulties once the subject develops beyond the stage of vague analogy. Many early psychologists, anxious to develop a mechanistic psychology, based their views on fundamental misconceptions of force, power, and energy, as used in physics. They equated drives with energy, as did the early ethologists (see above). They thought that some kind of motivational energy accumulated during deprivation of food, water, sexual behavior, etc. and that this accumulated energy, or drive, governed the intensity of the subsequent behavior. The fundamental problem with this view is that energy, as used in physical sciences, is a concept of capacity and cannot itself be a causal agent (see McFarland, 1971). By pursuing an incorrect analogy, theories of drive became unworkable and contradictory (Bolles, 1967).

The third major problem with the concept of drive arose from attempts at classification. Some psychologists sought to identify a drive for every aspect of behavior, an approach that is implicit in early ethological theories. Thus, one might postulate a sex drive responsible for energizing sexual behavior, or one might divide this into a courtship drive, a copulation drive, and an ejaculation drive. The problem is: How many drives should there be? Some psychologists advocated a single general drive, but this suggestion has not proved to be satisfactory either (Hinde, 1970; Bolles, 1967). Once the concept of drive is abandoned and attention is focused upon the changes in motivational state that underlie behavior, then the problem of enumerating the factors involved can be tackled (McFarland, 1974).

15.6 Motivational state

At any particular time, the animal is in a specific physiological state,

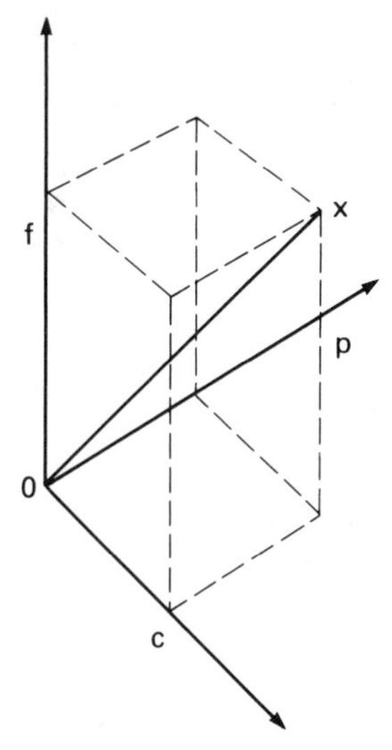

Fig. 15.13 Hunger represented as a multidimensional vector quantity. f = fat, p = protein, c = carbohydrate, O = origin, x = hunger state (After McFarland and Sibly, 1972)

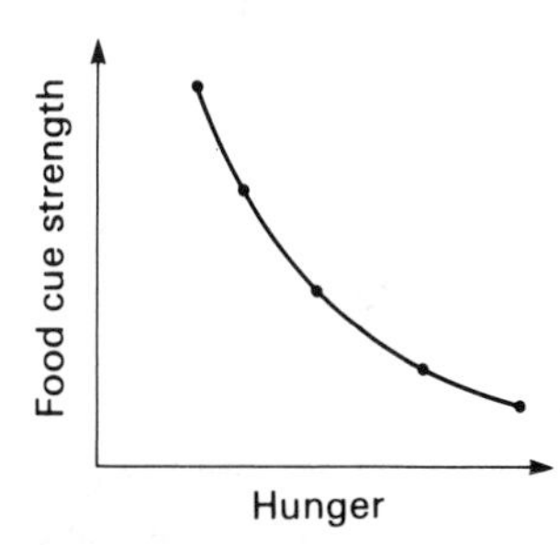

Fig. 15.14 A motivational isocline relating hunger and strength of food cues. The line joins those states (points) that lead to the same strength of feeding tendency

which is monitored by the brain. The behavior we see is determined by the brain in accordance with this monitored state and in combination with the animal's perception of environmental stimuli. The combined physiological and perceptual state, as represented in the brain, is called the "motivational state" of the animal. It includes factors relevant to incipient activities, as well as to the animal's current behavior. It is thus fundamentally different from the old concept of drive it has come to replace.

An animal's physiological state can be represented as a point in a physiological space. The axes of the space are the important physiological variables, such as temperature, and the space is bounded by the tolerance limits for these variables. In a similar way, the animal's motivational state can be represented by a point in a motivational space. The axes of the space are the important motivational stimuli, such as the degree of thirst or the strength of some external stimulus. The state-space concept is important not only in describing motivational systems but also in providing a link between mechanistic and design aspects of decision making.

One advantage of the state-space representation is that the components of non-unitary aspects of motivation like hunger can be portrayed readily, as is shown in Figure 15.13. Another advantage is that the combined effects of internal and external stimuli can be represented in a simple way. In Chapter 12, we saw how quantitative measures of the motivational effectiveness of external stimuli can be combined into a *cue-strength* index, and that the relationships among different stimuli can be represented by the motivational isocline. For example, the tendency to seek food that arises from a high hunger and a low strength of cues indicating the availability of food could be the same as the tendency due to the combination of a low hunger and a high cue strength for food availability. The points representing these two different motivational states lie on the same motivational isocline, as illustrated in Figure 15.14. The shape of a motivational isocline gives an indication of the way in which the different factors combine to produce a given tendency. For example, the internal and external stimuli controlling the courtship of the male guppy (*Lebistes reticulatus*) seem to combine multiplicatively (see Fig. 15.15), although as we saw in Chapter 12, we have to be careful about our scales of measurement before coming to any firm conclusion.

A major difference between the state-space approach to motivation and the traditional drive concept is that the modern approach makes no assumptions about the ways in which different motivational factors combine or about the relationship between motivation and behavior. The early psychologists and ethologists tended to assume that motivational factors combined in a particular way. A common assumption was that internal and external factors combine multiplicatively so that the tendency to perform an activity would be zero if internal drive were high and if there were no relevant external cues and would be zero also if

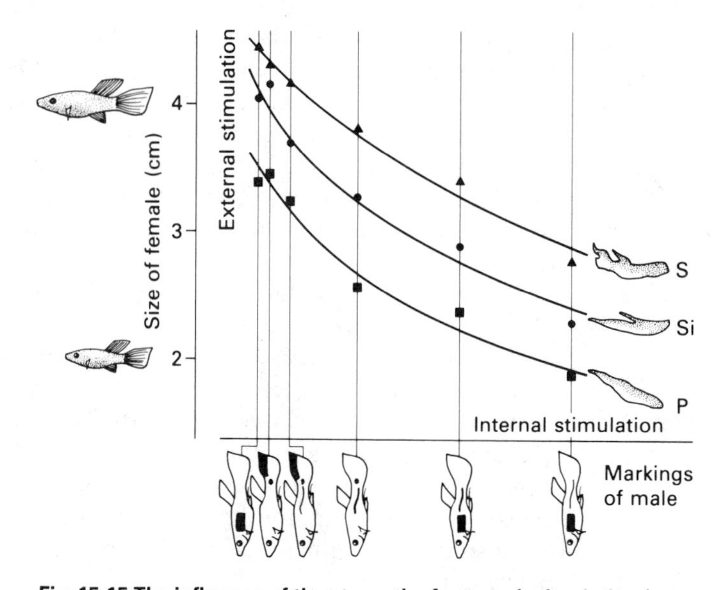

Fig. 15.15 The influence of the strength of external stimulation (measured by the size of the female) and the internal state (measured by the color pattern of the male) in determining the courtship behavior of male guppies. Each curve (isocline) represents the combination of external stimulus and internal state that produces the sigmoid courtship postures of increasing intensity P, Si, S (After Baerends et al, 1955)

drive were negligible but if the external cues were strong. While this may be true in some cases, the modern approach is to regard this question as an entirely empirical matter, not subject to any particular doctrine.

In the traditional view, behavior is driven from within, and there is a direct causal relationship between the strength of drive and the properties of the ensuing behavior. In the modern view, the relationship between motivational state and behavior is not direct. Although particular combinations of factors can give rise to a given tendency, there is not necessarily a direct relationship between the tendency and the observed behavior. For example, an animal may have a tendency to feed that is stronger than any other tendency, yet in the interests of some long-term strategy, it may postpone feeding; or an animal may feed even though its feeding tendency is not the strongest tendency. These possibilities have been suggested by observations of animal behavior. To evaluate them is a complicated task, and many questions currently are unresolved. In this chapter we consider only some aspects of the problem. Others are discussed in Chapter 25.

In general, the focus of research on this topic has shifted away from attempts to directly establish the causal chain linking motivation and behavior, partly because of conceptual difficulties with the calibration and measurement of causal factors (McFarland, 1976; Houston and

McFarland, 1976; McFarland and Houston, 1981) and partly because of a realization that animals are probably more mentally complicated than hitherto thought. The focus of most current investigation is on the consequences of behavior rather than its immediate causes. What determines the consequences of an animal's behavior? How do the consequences influence the behavior of the animal? To what extent is an animal able to take account of the probable consequences of future behavior? To what extent is it able to assess the costs and benefits of different courses of action?

The consequences of an animal's behavior are the result of interaction between behavior and environment. The consequences of foraging behavior, for instance, depend partly upon the foraging strategy used and partly upon the availability and accessibility of food. Thus, a bird foraging for insects may search in appropriate or inappropriate places, while the insects may be abundant or scarce at the particular time. Similarly, the consequences of courtship display, in terms of the reaction of the partner, depend partly upon the type and intensity of the display and partly upon the motivation and behavior of the other animal.

The consequences of an animal's behavior influence the motivational state of the animal in a number of different ways. Two important consequences of foraging are the expenditure of energy and the intake of food. Energy and other physiological commodities such as water are expended in all behavior to a degree that depends upon the level of activity, the weather, etc. Such expenditure must be taken into account because it influences the animal's state. Food intake may alter the animal's state in a number of ways. The presence of food in the mouth may increase appetite temporarily through a positive feedback mechanism (Wiepkema, 1971); it may have a satiating effect (negative feedback); or it may lead to the rejection of the food as a result of its unpalatability. The presence of food in the gut may have a short-term satiating effect, and it may influence other physiological factors such as temperature and water balance. The food in the gut is digested, and nutrients are absorbed into the bloodstream. The changes in the blood that are consequent to feeding are complex and may influence many aspects of the animal's physiological state. The point to remember is that all aspects of behavior have consequences of equivalent complexity, which may include effects upon other animals and upon features of the external environment, such as a nest. While it is relatively easy to pinpoint the consequences of feeding because of the vast amount of research that this aspect of behavior has attracted, we should not forget that the effects of behavior upon other animals, and upon the environment, influence the motivational state of the animal in a manner equivalent to the consequences of feeding (McFarland and Houston, 1981).

The consequences of behavior can be represented in a motivational state space. The motivational state of the animal is portrayed as a point

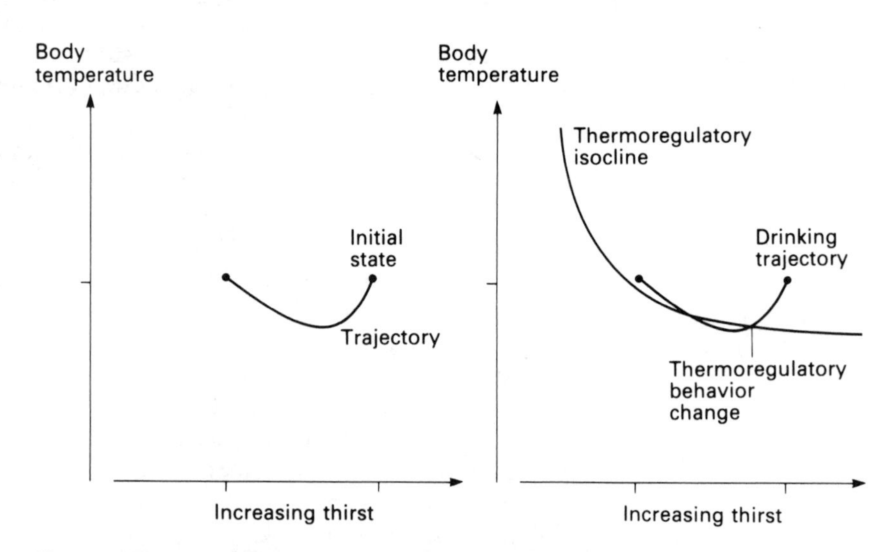

Fig. 15.16 The consequences of behavior described as a trajectory in motivational space. Left: the consequences of drinking cold water (a short-term reduction in body temperature and a long-term reduction in thirst). Right: where the trajectory cuts a motivational isocline, changes in behavior can be expected

in the space. As a consequence of the animal's behavior, the state changes, and so the point describes a trajectory in the space, as illustrated in Fgure 15.16. On a two-dimensional page it is possible to portray the trajectory in only a single plane of the space, but we can imagine that the changing state describes an equivalent trajectory in multidimensional space. The advantage of this type of representation is that it enables us to portray very complex changes in state relatively simply. In the case of an animal drinking cold water, for example, we can see that decisions to change behavior depend partly upon the trajectory and partly upon the position of the relevant motivational isoclines, as shown in Figure 15.16.

Points to remember

● Homeostasis, the maintenance of an internal steady state, is important for animals that live in a variable environment. It is not, as used to be thought, achieved primarily through negative feedback mechanisms, but by a combination of feedback, feedforward and adaptive control.

● Most animals have some form of thermoregulation. In ectotherms the primary source of heat is outside the body, while in endotherms it is generated by metabolic processes.

● All animals have some obligatory water loss for which they must compensate. Many animals drink more water than is necessary for homeostasis and are able to cut down on their water losses when water is not available. This may, however, involve curtailment of other aspects of behavior.

● All animals require both energy and nutrients, and many are able to learn how to obtain essential substances, even though they can not directly detect them in their food.

● The idea of specific drives responsible for different aspects of behavior has been superseded by the notion of motivational state. This is influenced both by external factors and by the consequences of the animal's own behavior.

Further reading

Bolles, R. C. (1967) *Theory of Motivation*, Harper and Row, New York.
Toates, F. M. (1980) *Animal Behaviour–A Systems Approach*, Wiley, Chichester.

16 Physiology and Behavior in Changing Environments

In the last chapter we looked at mechanisms whereby animals maintain their internal environment at steady state through physiological and behavioral adjustment. In this chapter we consider the steps animals must take to deal with the physiological demands of maintaining survival and reproduction in variable environments.

16.1 Tolerance

Life very probably originated in the sea (Whitfield, 1976; Croghan, 1976). Compared with other biomes, those of the ocean environment are relatively stable. Fluctuations in physical characteristics such as temperature and oxygen tension are small so that the internal environment of many marine invertebrates is subject to little disturbance. Such animals are usually conformers in that their bodily condition tends to conform with that of the environment, and they cannot live in variable conditions. For example, the salinity of the body fluids of many marine invertebrates is identical to that of seawater (Baldwin, 1948; Barrington, 1968). Other animals, known as *regulators*, maintain their bodily functions in a condition relatively independent of environmental fluctuations, an ability that was a prerequisite for the invasion of fresh water and of dry land.

Each species has a characteristic ability to tolerate extreme values of environmental factors such as temperature and humidity. For example, many marine invertebrate animals are affected by variation in the salinity of the water because their body fluids normally have much the same salt concentration as seawater and their body tissues are adapted to function efficiently in such a medium. If they become immersed in a less saline medium, water enters their tissues via osmosis. In a more saline medium, the opposite process occurs. Animals that cannot control the passage of water into their bodies are limited in their range of living conditions by the degree of tolerable salinity.

An animal's habitat is restricted by the animal's tolerance range. In estuaries, for example, different species of the amphipod *Gammarus* are restricted to different parts of the estuary due to differing salinity tolerance. As illustrated in Figure 16.1, *Gammarus locusta* has a high

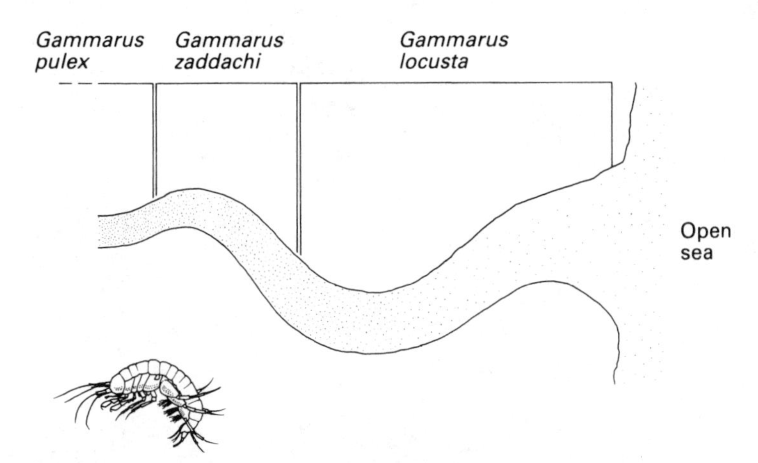

Fig. 16.1 The distribution along a river of three closely related species of the amphipod crustacean *Gammarus*, relative to the concentration of salt water. The degree of freshness of the water is indicated by the degree of shading (From *The Oxford Companion to Animal Behaviour*, 1981)

tolerance of salt water and is found at the mouths of estuaries. *Gammarus zaddachi* has moderate tolerance of salt water and usually is found in the stretch of river between eight and 12 miles from the sea. *Gammarus pulex* is a true fresh-water species and does not occur at all in parts of the river showing any influence of the tide or salt water. In this example each species can tolerate only a restricted range of salinity, yet each has been selected towards a salinity spectrum uninhabited by any other member of the genus.

When measures of survival, or fitness, are plotted against important environmental variables, bell-shaped curves like those illustrated in Figure 16.2 usually result. Only a few animals can survive the extreme high or low values and must cluster around some central range. Such

Fig. 16.2 Tolerance of three species of marine wood-boring isopods to different constant chlorinities (After Reish and Hetherington, 1969)

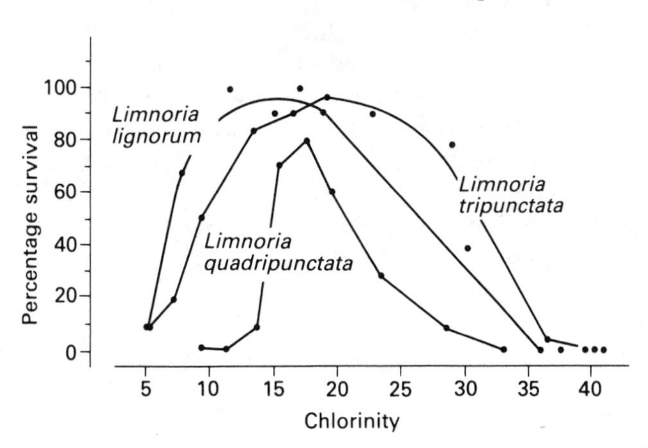

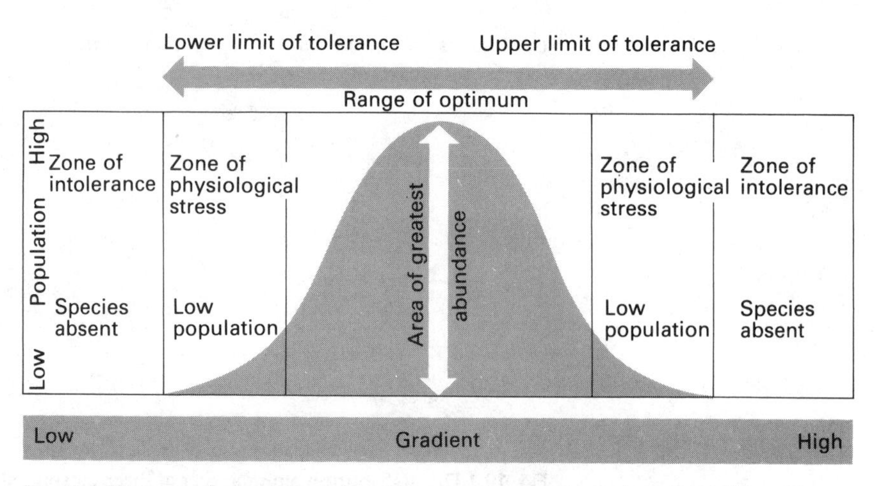

Fig. 16.3 Diagram of population abundance along an ecological gradient (From *The Oxford Companion to Animal Behaviour*, 1981)

curves show not only the limits and range of tolerance of the species but also the optimum values of the environmental variables (Fig. 16.3). Similar considerations apply to the tolerance of animals along all sorts of environmental gradients.

In the case of conformers, tolerance often will be directly reflected in the animal's physiological state. Thus, if an environmental temperature of 40°C is lethal, it is because an internal temperature near 40° causes biochemical disruption. However, the limits of tolerance for a given factor are influenced by the values of other environmental variables. Environmental factors in combination may kill an animal at intensity levels that if taken separately would not be lethal. For example, for the American lobster *Homarus americanus*, a temperature of 32°C is lethal when the salinity is 30 parts per thousand and when the oxygen content is 6.5 mg per liter. If oxygen content decreases to 2.9 mg per liter, the thermal limit is lowered to 29°C. In one study, lobsters were subjected to 27 combinations of temperature, salinity, and oxygen content (McLeese, 1956). From his results, McLeese was able to construct the 3-dimensional diagram illustrated in Figure 16.4 which shows how the interaction of factors affects the tolerance limits of the species.

Adaptation at the tolerance limits, usually called *resistance adaptations*, are greatly influenced by exposure time, the rate of change in environmental factors, and the history of the individual. The physiological mechanisms involved in resistance adaptation require a certain amount of time to adjust to a changing situation. If the environmental change is very sudden, death may occur, whereas if the same change takes place more gradually, it might result in survival. The processes involved in resistance adaptation vary from very rapid regulation to slow acclimatization. Thus, very gradual changes in the environment may enable the individual to acclimatize to new conditions.

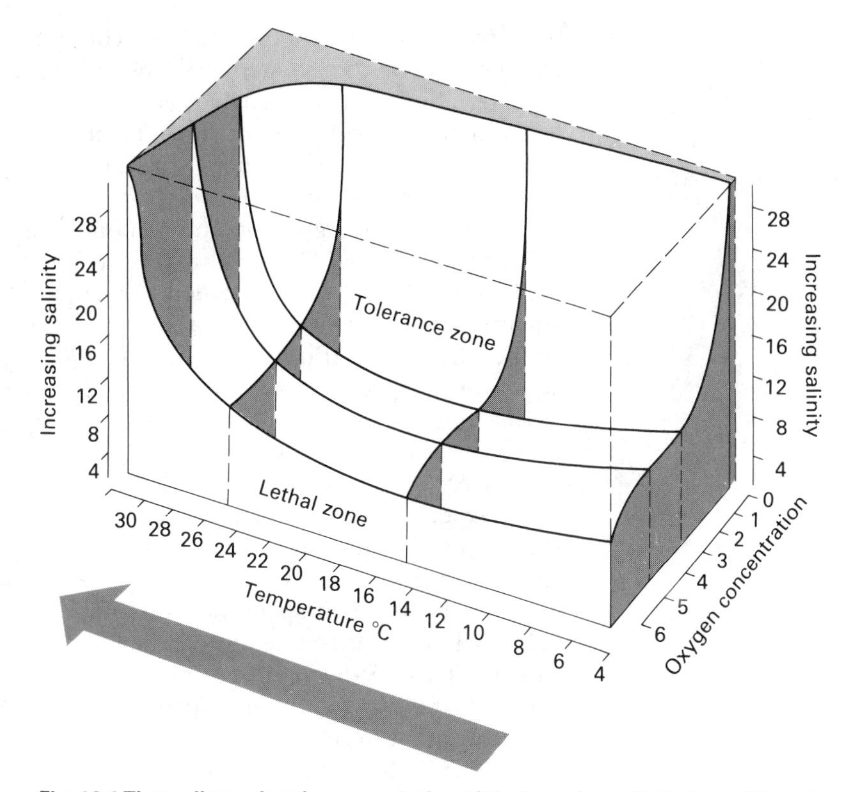

Fig. 16.4 Three-dimensional representation of the boundary of lethal conditions for the American lobster *Homarus americanus* for various combinations of temperature, salinity and oxygen (After McLeese, 1956)

Some species have the ability to change their range of tolerance through the processes of acclimatization. For example, the small tree lizard *Urosaurus ornatus* normally has a critical thermal maximum of 43.1°C. By maintaining these animals in the laboratory for a period of seven to nine days at a temperature of 35°C, compared with the more normal temperature of 22–26°C, it was found that the average lethal temperature could be raised to 44.5°C.

16.2 Acclimatization

Acclimatization is a form of physiological adaptation by which an animal is able to alter its tolerance of environmental factors. Usually, the term *acclimation* is used in laboratory studies in which the adaptation occurs to a single factor like temperature. The term *acclimatization* is reserved for the complex of adaptive processes that occur under natural conditions.

Acclimatization often occurs in response to seasonal changes in climate. Thus, seasonal shifts in the upper lethal temperature of fresh-water fish often are directly correlated with changes in the temperature of the habitat (Fry and Hochachka, 1970). Because behavioral adjustments can make acclimatization unnecessary and vice versa, there is a wide variety of combinations actually used by animals.

The temperature preferences of fish often are related to their state of acclimatization. Many fish species seem to have highly developed behavioral thermoregulation and respond to a temperature gradient by selecting a particular temperature (Fry and Hochachka, 1970). Individual goldfish (*Carassius*) can be trained to maintain the temperature of their aquarium water by actuating a valve to introduce cold water as the temperature rises (Rozin and Mayer, 1961). Such fish maintain the aquarium temperature at about 34°C with considerable precision.

To a large degree, the phenomenon of temperature selection explains the distribution of fish in nature, and fish usually prefer the temperature to which they are acclimatized (Fry and Hochachka, 1970). Such an arrangement makes biological sense. If slow changes in physiological state due to acclimatization were not followed by corresponding changes in behavioral preferences, then acclimatory processes could be opposed by behavioral mechanisms. For example, acclimatization to cold might be counteracted by a behavioral tendency to select a warmer climate. If a short-lived opportunity to select a warm environment were exploited to the full, then much of the work done in establishing acclimatization to cold would be undone. The obvious alternative is for animals to prefer conditions to which they are acclimatized. The picture is complicated, however, by the phenomenon of *anticipatory adjustment*,

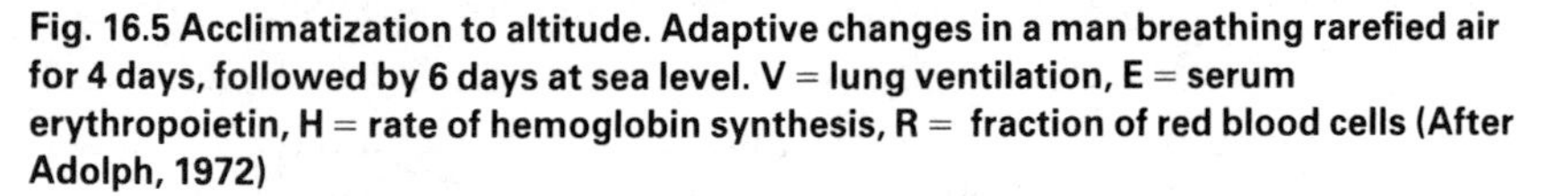

Fig. 16.5 Acclimatization to altitude. Adaptive changes in a man breathing rarefied air for 4 days, followed by 6 days at sea level. V = lung ventilation, E = serum erythropoietin, H = rate of hemoglobin synthesis, R = fraction of red blood cells (After Adolph, 1972)

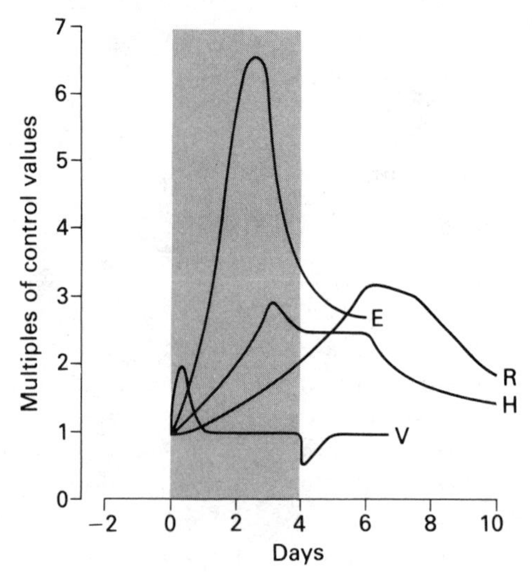

by which the animal responds to some aspects of the environment or to its own biological clock and is thereby able to predispose itself to climatic changes.

Acclimatization normally is regarded as a relatively slow process, compared to the rapid physiological adjustments many animals are able to make in response to sudden changes in environmental conditions. In reality, however, usually there will be a spectrum of adaptive processes, ranging from rapid physiological reflexes to slow acclimatization. For example, when a person is transported suddenly from low to high altitude, the oxygen level of the blood drops because of the rarefied air. This drop initially is counteracted by an increased rate of breathing. However, this type of physiological response involves high energy expenditure. Less costly, though slower, forms of physiological adaptation (see Fig. 16.5) then supplant the rapid physiological response, and these in turn are replaced by more long-term forms of acclimatization, like the manufacture of more red blood cells. Even this, however, is not without cost, since energy is required to manufacture the extra cells, and their presence in the blood increases its viscosity and the work the heart has to do in pumping the blood around the body. When the person is returned to low altitude, therefore, all the adaptive processes occur in reverse, as illustrated (Fig. 16.5). This type of reversibility is characteristic of physiological (as opposed to genetic) adaptation.

As we can see from Figure 16.5, acclimatization and regulation together form a spectrum of adaptive processes, ranging from homeostatic regulation involving quick-acting processes that restore physiological imbalance within a day, to acclimatization, which may take days, or weeks, to attain physiological steady state. However, the processes of acclimatization are complementary to those of regulation. In acclimatization to altitude, for example, the rapid breathing required at first is gradually slowed as true physiological acclimatization takes place (Fig. 16.5). This means that the effects of regulation and acclimatization upon physiological state sum like vectors, together achieving a desired state (such as a given rate of oxygen transport), with the contributions of each changing as acclimatization proceeds. In cases where there is also a contribution from behavior, the result will be the vector sum of three processes.

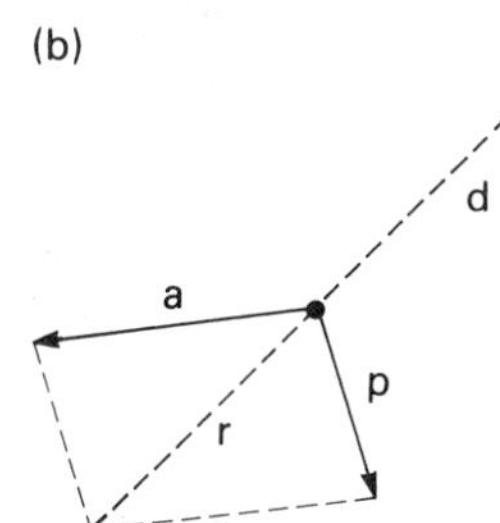

Fig. 16.6 Vectorially additive processes in physiological regulation. The acclimatization vector a sums with the regulation vector p to give the resultant r, which opposed the drift in physiological state d. The resultant may be made up of a large contribution from p and a small contribution from a, as in (*a*), or a small contribution from p and a large contribution from a, as in (*b*). (After Sibly and McFarland, 1974)

When vectorially additive processes occur at very different rates as illustrated in Figure 16.6, the state of the slow processes can be regarded as the "goal" of the faster processes (Sibly and McFarland, 1974). The state of acclimatization now can be regarded as the optimal point of physiological regulation, and the same argument applies in discussing behavior. For example, if a man is transported from a cold to a hot climate, he has various possible alternative means of adjustment. He can carry on with his normal behavior, exposing himself to the sun and relying on sweating and other physiological responses to maintain a normal body temperature. After a few weeks he could acclimatize to the heat and so would sweat to a lesser extent. Alternatively, he could alter his behavioral routine and seek the shade. He then would rely less on physiological adjustments like sweating but would take longer to acclimatize to the new conditions. Thus, as is often the case, the physiological and behavioral solutions to the immediate problem are alternatives. Their effects are complementary and therefore vectorially additive (Sibly and McFarland, 1974; McFarland and Houston, 1981).

As a consequence of this arrangement, we can expect that the effect of acclimatization upon behavior is to alter the goal of behavior. There is some evidence that such alteration does occur. For example, the golden hamster (*Mesocricetus auratus*) prepares for hibernation when the environmental temperature drops below about 15°C. This preparation involves a number of physiological changes, amounting to a form of acclimatization to the cold, and makes hibernation physiologically possible when other conditions are favorable, including the availability of nest material and sufficient food for the animal to set aside to store. Temperature preference tests, conducted in the laboratory, show that the hamsters develop a marked preference for cold environmental temperatures during the prehibernation period of acclimatization. They prefer an 8°C environment to one at either 19° or 24°. Following arousal from a period of hibernation, the situation is reversed, and the hamsters actively prefer the warmer environments (Gumma et al., 1967). As we saw in the case of temperature preferences in fishes, it makes biological sense for behavioral preferences to be linked to the state of acclimatization.

Acclimatory changes can occur as a direct response to environmental changes, but they are also influenced by circannual rhythms. For example, the marine polyp *Campanularia flexuosa* has an annual cycle of growth, development, and longevity, a rhythm that persists under constant laboratory conditions. Seasonal changes in various metabolic factors are known to occur in mammals, including man (Reinberg, 1974), and are thought to be due to endogenous circannual rhythms (Senturia and Johansson, 1974). Such rhythms would enable the animal to adapt physiologically in anticipation of seasonal environmental changes.

16.3 Biological clocks

Animals commonly confront environmental changes that are cyclical in nature, such as days, tides, and seasons. Many maintain some sort of internal rhythm, or clock, to predict the onset of the periodic changes and to prepare for them.

There are three basic ways of synchronizing physiology and behavior to cyclic changes in the environment. (1) There may be a direct response to various changes in external (*exogenous*) geophysical stimuli. (2) There may be an internal (*endogenous*) rhythm that programs the animal's behavior in synchrony with the exogenous temporal period, particularly a 24-hour or a 365-day period. (3) The synchronization mechanisms may be a combination of (1) and (2).

An animal could use many features of the external environment to gain information about the passage of time. The apparent motions of the sun, moon, and stars, as seen by a terrestrial observer, provide indications of the time of day, the season of the year, etc, and many animals are known to make use of this type of information. Thus, honeybees native to Brazil use the sun as a compass in foraging. They can be trained to search for food in a particular compass direction. When they are moved from one locality to another, they continue to forage in the same compass direction, irrespective of the time of day. Thus, bees native to Brazil are capable of compensating for the anticlockwise motion of the sun. However, northern hemisphere bees transported to Brazil are initially unable to make the appropriate compensation. In the northern hemisphere the sun appears to move in a clockwise direction, and the bees have to learn to adjust to the changed conditions in Brazil (Lindauer, 1960; Saunders, 1976).

Similar abilities are well documented in fishes (Hasler and Schwassmann, 1960) and in birds (Schmidt-Koenig, 1979). There is also evidence that some animals respond to the motions of the moon (Papi, 1960) and stars (Schmidt-Koenig, 1979). In addition, it is possible that animals can obtain time cues from factors such as changes in temperature, barometric pressure, and magnetic phenomena.

The term *circadian* (*circa* means "about," *dies* means "day") is used to describe endogenous rhythms that usually fall short of a 24-hour periodicity. The term *circannual* is used to refer to the endogenous rhythm that is usually less than 365 days. Many animals maintain rhythmic activity when isolated in the laboratory, which suggests that they possess an endogenous clock. However, it is always possible that the animals are responding to some exogenous factor undetected by the experimenter, and it is necessary to establish suitable criteria for judging whether a clock is truly endogenous. There are various ways in which this can be done. First, the rhythm may exhibit a frequency that is not exactly synchronous with any known environmental periodic factor, such as light, temperature, or other geophysical variable (Weihaupt, 1964). Second, the period of the endogenous rhythm usually deviates from the natural rhythm when studied under constant laboratory conditions. Third, the rhythm may persist when the animal is moved from one part of the world to another. Only when some such criteria apply can we conclude that the rhythm is endogenous.

Since endogenous rhythms tend to gradually deviate from that of the exogenous cycle (such as day-night light or temperature changes), an organism must have a means of synchronizing its endogenous rhythm with the cycle of external events. Jürgen Aschoff (1960) coined the term

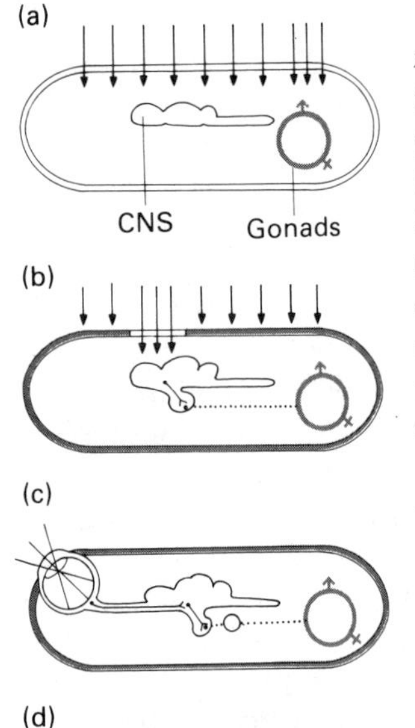

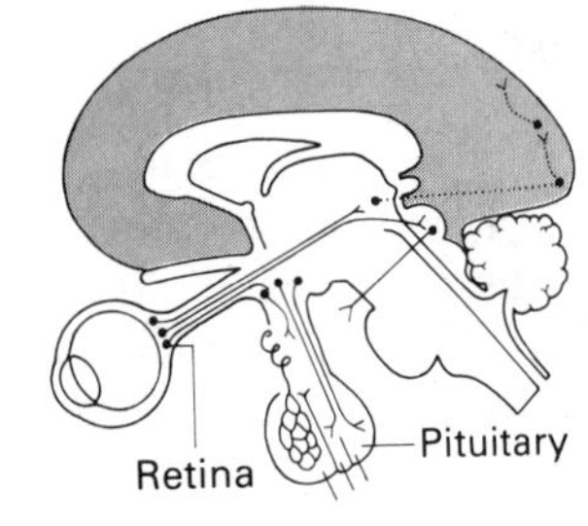

Fig. 16.7 Scharrer's (1964) representation of the influence of light on the gonads. (*a*) In transparent animals, light can exert a direct effect on the internal organs. (*b*) Some opaque animals have a transparent window which admits light to photosensitive parts of the brain, which then stimulates the gonads via hormones released by the pituitary gland. (*c*) A system in which light acts on the retina which sends nervous messages to the hypothalamus. This part of the brain stimulates the pituitary gland to release gonadotrophic hormones. (*d*) Diagram of scheme (*c*) applied to the human brain (After Bligh, 1976)

zeitgeber (meaning "time-giver") for the environmental agent that entrains the organism's behavior to the external environment. For example, when lizards (*Lacerta sicula*) are hatched in an incubator with temperature and light regimens designed to simulate a 16-hour or a 36-hour day, they develop normally and exhibit normal circadian rhythms of activity when tested under constant conditions. Thus, the circadian rhythm of these animals is endogenous and does not depend upon individual experience of the day-night cycle. A 24-hour temperature cycle with an amplitude of 0.6°C is sufficient to act as a zeitgeber and to entrain the lizards' activity rhythm.

In general, it is apparent that many species of animals, ranging from the single-celled (Sweeney, 1969) to the complex multicellular animals (Aschoff, 1965; Bunning, 1967; Pengelley, 1974), have evolved a time sense combining endogenous clocks and exogenous entrainment.

16.4 Reproductive behavior and physiology

Seasonal climatic changes in temperature and precipitation exert powerful influences upon the reproductive success of many species. Among birds, the availability of food for the young appears to be the main determinant of breeding success (Lack, 1968). Birds at middle and high latitudes have a seasonal reproductive pattern in which the eggs are usually laid during the spring, a pattern that enables the young to mature sufficiently to withstand the winter conditions or to endure a long migratory flight. Sandpipers, for example, breed in Arctic regions and nest and incubate in spring when the snow is still on the ground. The eggs usually hatch when the snow melts and there is an abundance of insect life to provide food for the young (West and Norton, 1975).

The reproductive physiology of seasonal breeders is geared to the annual cycle of environmental change so as to anticipate either peak abundance of food or adverse climatic conditions. There are two main ways in which this is done. First, changes in ambient temperature, day length, or other environmental clues induce physiological change at a particular time of year. Second, the physiological changes are programmed on a seasonal basis by means of an endogenous circannual clock.

The most regular and predictable seasonal change in the environment is the fluctuation of day length. In transparent organisms, light can have a direct effect on the gonads, bringing them into reproductive condition at the appropriate time (Scharrer, 1964). Some other animals may have a translucent window that admits light to the brain. The photic stimuli are converted to chemical messages by neurosecretory cells. In some mammals the pineal body, on the dorsal surface of the brain, may act as a light transducer (Wurtman et al., 1968). However, the influence of light on mammalian reproduction appears to be primarily via the retina to the hypothalamus, as shown in Scharrer's scheme (Fig. 16.7).

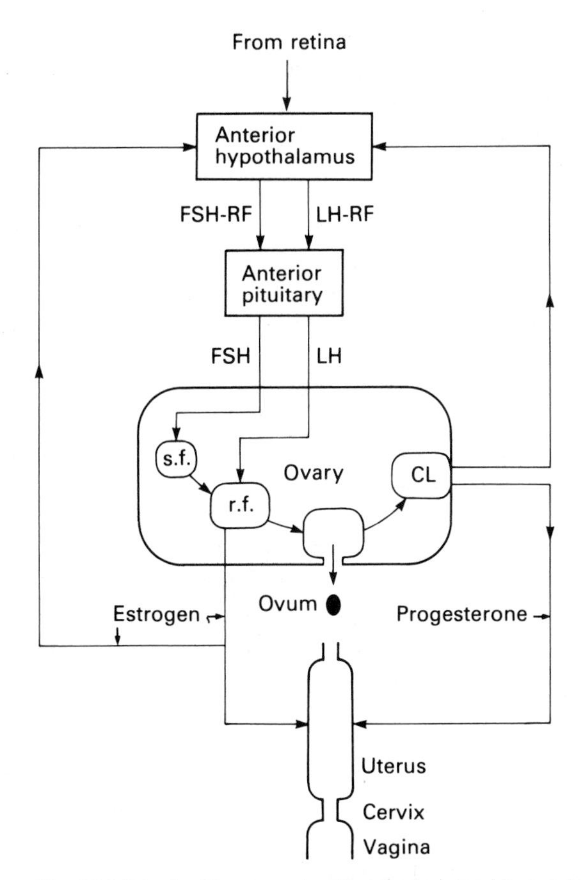

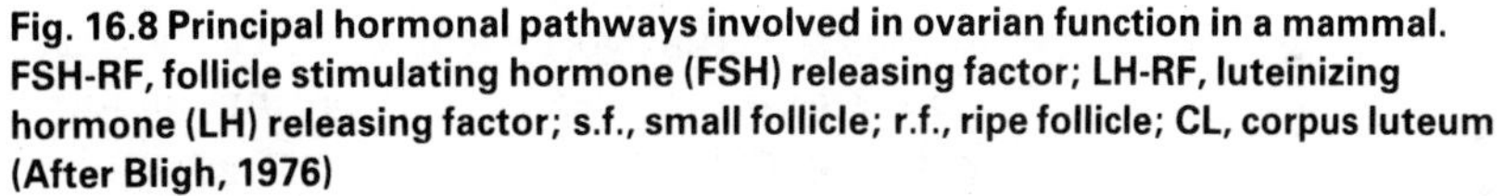

Fig. 16.8 Principal hormonal pathways involved in ovarian function in a mammal. FSH-RF, follicle stimulating hormone (FSH) releasing factor; LH-RF, luteinizing hormone (LH) releasing factor; s.f., small follicle; r.f., ripe follicle; CL, corpus luteum (After Bligh, 1976)

The control of reproductive physiology involves the complex interaction of a number of hormones, as illustrated in Figure 16.8. In most vertebrates the environmental factors stimulate the production of gonadotrophic hormones by the pituitary gland. These hormones stimulate growth and activity in the testes and ovaries, which in turn produce their characteristic sex hormones. At the end of the breeding season, pituitary activity diminishes, the gonads become inactive, and reproductive behavior no longer occurs. In addition to seasonal cycles of sexual activity, many mammals have a much shorter estrous cycle, or "heat." Some, like the red fox (*Vulpes vulpes*), have only one period of heat per year. Others, such as domestic dogs, have two; others have more. Among birds the number of clutches reared in a season may depend upon the food supply (Lack, 1968).

The reproductive pattern of many species is influenced both by photoperiodic factors and by the endogenous circannual clock. Indeed, Brian Follett (1973) has suggested that the two mechanisms may interact. Species living at high latitudes tend to be most strongly influenced by photoperiodicity. At low (equatorial) latitudes there is less annual photoperiodic change, but an annual reproductive cycle may, nonetheless, be advantageous. Animals occupying arid desert environments, for example, may be dependent upon the occurrence of rainfall to trigger the process of reproduction (Marshall, 1970). Some equatorial species show a marked breeding rhythm that appears to be unconnected with seasonal changes. Several sea birds, including the brown booby (*Sula leucogaster*), the sooty tern (*Sterna fuscata*), and the lesser noddy tern (*Anaus tenuirostris*) breed at intervals of 8–10 months. The progressive shift between such breeding cycles and the annual cycle suggests that no one season is preferable under equatorial conditions. Why then do the birds not breed continuously? Perhaps this would require too much energy, so a periodic rest during which moult can occur becomes the most beneficial arrangement.

The annual reproductive cycle is similar to the process of acclimatization in that it involves slow-acting physiological phenomena that change the animal's physiological state in a fundamental way. Reproductive activities, including territorial defense, courtship, mating, and parental care, put extra energy demands and physiological burdens on the animal, which have to be catered to by the system as a whole. If the animal is unable to maintain physiological stability while carrying these burdens, then the reproduction must be abandoned for that year.

16.5 Hibernation

In parts of the world where climate is seasonal, animals may have to adjust to prolonged periods of unfavorable weather. Some animals are able to avoid such conditions by migrating, but others are able to survive unfavorable periods by entering a prolonged phase of dormancy, called *estivation* at high temperatures and *hibernation* at low temperatures. Some desert rodents, such as ground squirrels (*Citellus*), become inactive during the summer and enter a torpid state in which the body temperature falls and in which a general reduction in physiological activity occurs. This means that energy expenditure is lowered and that the animal can stay alive for long periods without feeding. Water losses are reduced as a result of the lowered food intake, which is helped by the habit of retreating into burrows. Some rodents store food in their burrows and maintain high humidity inside by stopping up the entrance (Schmidt-Nielsen, 1964). Estivation is most common in desert-living species, but it is not confined to them. Thus, many species of European earthworm become dormant during the summer. Each animal hollows out a small chamber deep in the soil and curls up into a ball. The onset of such estivation is associated with low humidity, and it is possible to prevent it by keeping the worms in a humid atmosphere.

Hibernation occurs in many species in northern latitudes, enabling them to avoid winter conditions that otherwise would make excessive demands on their energy requirements. True hibernation usually is distinguished from partial dormancy, as shown by the European brown bear (*Ursus arctos*) and the American black bear (*Ursus americanus*). During partial dormancy, the bear's body temperature may fall from about 38°C to about 30°C, although body temperatures below 15°C are lethal. True hibernators allow their body temperatures to fall as low as 2°C. True hibernation is characteristic of small mammals, although similar types of torpor do occur in raccoons and badgers and in some birds. Many species of hummingbird show periods of torpidity during which their body temperature falls to that of the environment, although temperatures below 8°C are lethal. Torpor usually lasts only a few hours in birds, and seasonal dormancy is known in only a single family of birds, the goatsuckers (Caprimulgidae).

A period of dormancy in response to unfavorable climate also occurs among insects and is given the name *diapause*. Diapause is often associated with a particular stage of life cycle. Insects that avoid freezing by supercooling can survive very cold conditions in diapause, in which the body fluids freeze at temperatures well below 0°C. The extent of supercooling may depend upon the degree of acclimatization, which progressively alters the chemical constitution of the body fluids. *Bracon cephi*, a Canadian wasp, may lower its freezing point to −46°C by increasing the concentration of glycerol in the blood.

True hibernation occurs only in small mammals, which cool more readily than large ones due to their proportionately larger body surface. They also warm up more quickly because of their small thermal capacity. Hibernation is characterized by a sleeplike state, with lowered rates of respiration and heartbeat. Hibernating animals often choose a typical sleep site in which to hibernate and adopt a sleeping posture. During hibernation, body temperature is lowered and energy expenditure falls below that of normal sleep. Hibernators often have deposits of a special brown fat that have the primary function of heat production rather than the nutrient energy production when ordinary fat reserves are mobilized. The brown fat is especially important during periods of arousal from hibernation, when body temperature has to be raised quickly.

Some mammals, including the golden-mantled ground squirrel (*Citellus lateralis*) and the woodchuck (*Marmota monax*), have a marked circannual rhythm underlying their seasonal hibernation. The ground squirrels are found in western North America at altitudes of 1,500 to 3,600 meters (4,900 to 11,800 feet), ranging from northern British Columbia to southern California. They normally hibernate for three to four months, during which time their body weight falls considerably. After hibernation their food consumption increases rapidly, and their body weight increases up to the onset of hibernation in October. When isolated in the laboratory, the rhythm of hibernation and associated

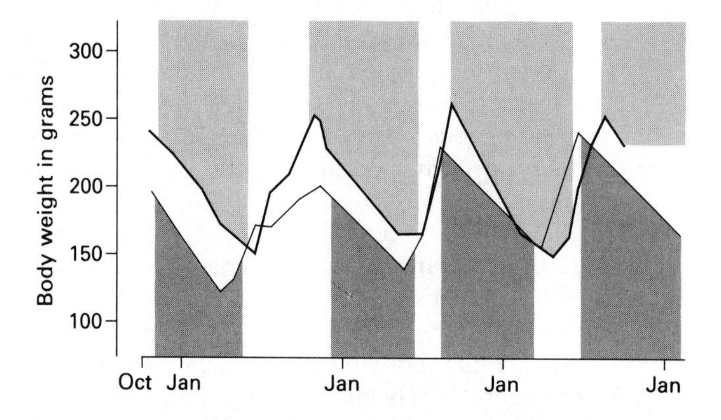

Fig. 16.9 Circannual rhythms in bodyweight and hibernation periods (shaded portions) of two representative animals (*Citellus lateralis*), maintained under constant laboratory conditions (After Pengelley and Asmundson, 1974)

change in body weight may persist for a number of years, even under constant lighting and temperature conditions (see Fig. 16.9). If blood serum from a hibernating ground squirrel is injected into a non-hibernating animal, then hibernation is induced (Pengelley and Asmundson, 1974). The evidence strongly suggests that the basic physiological mechanism responsible for controlling hibernation is driven by an endogenous biological clock entrained to the rhythm of external events by a particular external trigger, or zeitgeber. This has been demonstrated clearly in the woodchuck, which is known also to exhibit a circannual rhythm of hibernation. Woodchucks from the eastern United States were transported to Australia and exposed to the prevailing light and temperature conditions. The woodchucks reversed their normal rhythm within two years, bringing it into line with local conditions, even though the animals were provided freely with food and water and had no need to hibernate.

Hibernation is similar to acclimatization in that it is a long-term change in physiological state enabling the animal to cope with seasonal changes in environmental conditions. The circannual changes that occur in the animal's physiological state can be represented in a physiological space. Both the reproductive and hibernation annual cycles are largely preprogrammed, though they are, in reality, part of a single seasonal program.

16.6 Migration

Migration between two different habitats on a periodic or seasonal basis is shown by many species of animals, including butterflies, locusts, salmon, birds, bats, and antelopes. This is not the only form of migration, since some species undertake migrations as part of explora-

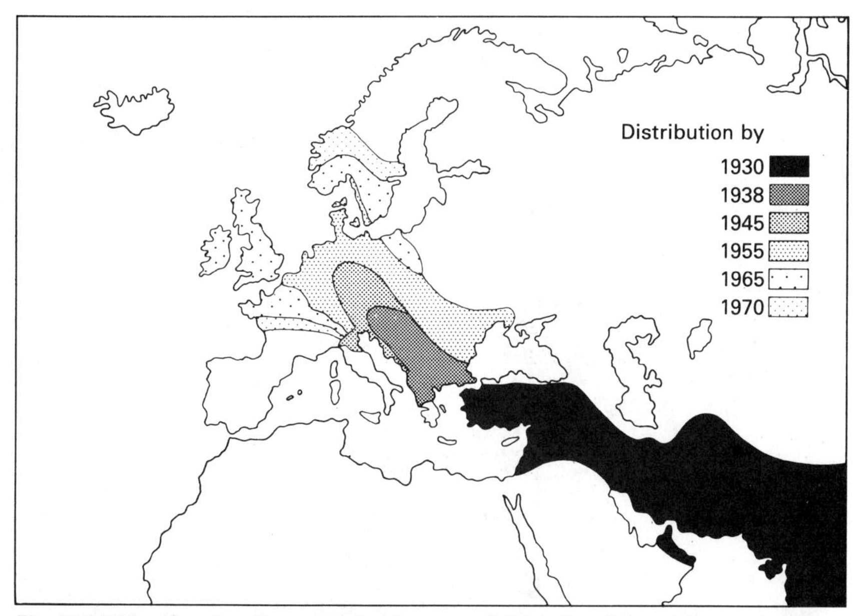

Fig. 16.10 Map showing the expansion in range of the collared dove (*Streptopelia decaocto*) this century (From *The Oxford Companion to Animal Behaviour*, 1981).

tion and colonization of new habitats. For example, population explosion among Norway lemmings (*Lemmus lemmus*) leads to dispersal migration that may involve thousands of individuals, usually immature males. Streams of lemmings run down the mountainsides and into the valleys. Many drown in their attempts to cross expanses of water. Lemmings are good swimmers and usually will not enter water unless they can see land on the other side. However, the pressure of numbers on the shoreline is sometimes such that individuals are induced to swim out into the open sea.

Colonization of new areas is attained sometimes by progressive migration among succeeding generations, as in the European collared dove (*Streptopelia decaocto*) (Fig. 16.10). Exploratory migration is common in young vertebrate animals and may be responsible for successful colonization when ecological conditions are favorable.

Periodic migrations occur in a variety of species in response to changes in environmental conditions. For example, the locust (*Schistocerca gregaria*) inhabits seasonally arid areas. Depending upon their degree of crowding, the insects may develop into one of three adult types. Of these, the *gregaria* individuals aggregate into dense swarms that migrate downwind into areas of low barometric pressure, where they are most likely to encounter rain. They fly by day and stop migrating when they encounter wet conditions. Sexual maturation, copulation, and deposition of eggs then occur. Locust swarms describe seasonal circuits, but the generation time is too short for individuals to complete the entire circuit.

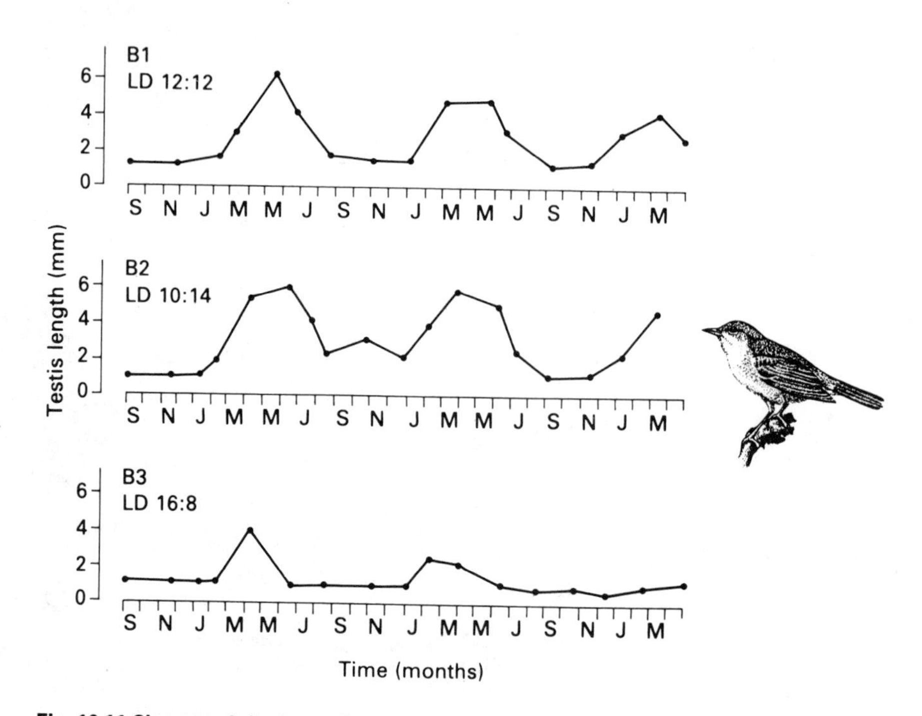

Fig. 16.11 Circannual rhythms of testis length of warblers (*Sylvia borin*) kept under constant photoperiodic conditions for three years. LD is the light-dark ratio (After Berthold, 1974).

Some seasonal migrants, in contrast to periodic migrants, initiate their migration on the basis of a circannual rhythm rather than in response to changes in environmental conditions. Studies have been made of songbird species within the genera *Phylloscopus* (Gwinner, 1971) and *Sylvia* (Berthold, 1973) that exhibit a range of migratory habits. It appears that, in the typical long-distance migrants such as the garden warbler (*S. borin*), the subalpine warbler (*S. cantillans*), and the willow warbler (*P. trochilus*), there are marked seasonal changes in body weight, moult, testis size, nocturnal restlessness, and food preferences (Fig. 16.11). The European populations of these species winter in Africa and migrate across the Sahara Desert (Fig. 16.12). When they are maintained under constant laboratory conditions in Europe, having been hand raised from a few days after hatching, the events that are known to be seasonal in free-living birds appear as seasonal events in caged members of the same species, although the period of oscillation is somewhat shorter than the calendar year.

Middle-distance migrants such as the blackcap (*S. atricapilla*) and the chiffchaff (*P. collybita*) winter in Europe and Africa and show moderate changes in body weight and other migratory indices. These species also show seasonal changes under constant laboratory conditions. The Sardinian warbler (*S. melanocephala*) and the Dartford warbler (*S. undata*)

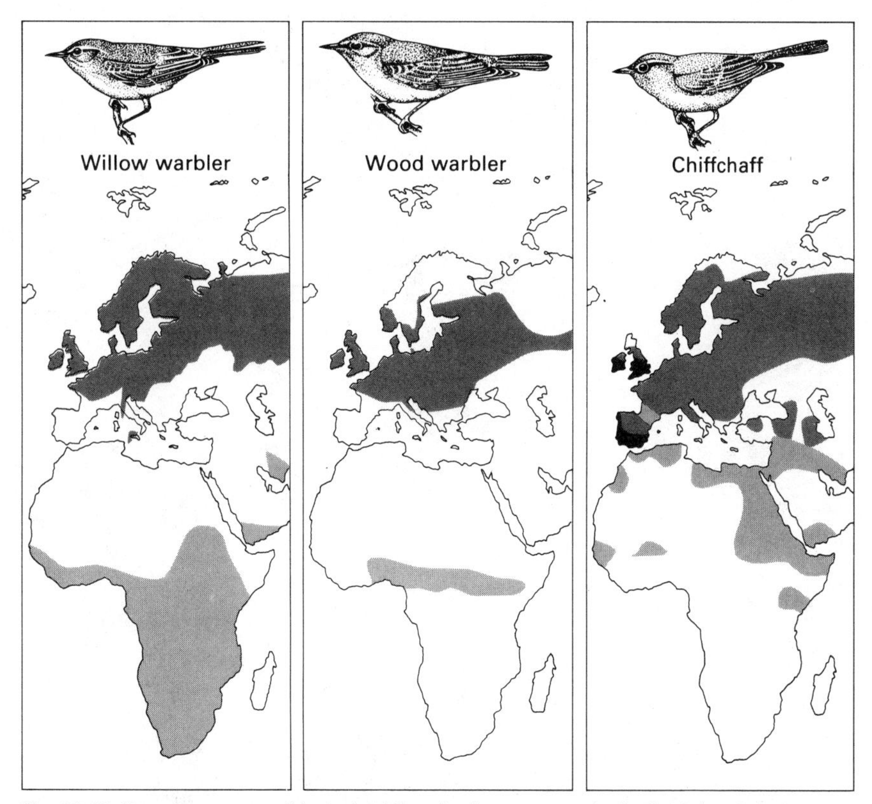

Fig. 16.12 Breeding ranges (dark shade) and winter ranges (pale shade) and their overlap (black) of some migratory warblers (After Schmidt-Koenig, 1979)

are partial migrants that winter within the Mediterranean breeding area, while Marmosa's warbler (*S. sarda balearica*) is resident the year round and endemic on the Balearic Isles and the Pithyuses in the Mediterranean sea. These species have a fairly constant body weight throughout the year, and they have a moult of body feathers of long duration, which alternates with periods of nocturnal restlessness. When tested in the laboratory, some evidence of seasonal changes can be obtained, although there are marked individual differences (Berthold, 1974).

It appears that the species of these two genera show seasonal changes in migratory indices, the intensity of which are correlated with the migratory habits of the species and the timing of which suggest that the onset of migration may be under the partial control of an endogenous circannual rhythm. Similar seasonal rhythms have been documented in several avian species with regard to the physiological processes, including moult, fat deposition, migratory restlessness, and reproduction (Rutledge, 1974).

There is also evidence that endogenous factors control not only the onset of migratory activity but also its pattern. Many species have migratory routes characteristic of particular breeding populations. For

Fig. 16.13 Migratory routes of the eastern and western European populations of the white stork (*Ciconia ciconia*) (After Schüz, 1971). Dark shading indicates intensity of birds

example, white storks (*Ciconia ciconia*) that breed in western Europe fly to their wintering grounds in Africa by a westerly route over Spain and Gibraltar, whereas those that breed in eastern Europe take an easterly route, as illustrated in Figure 16.13. Naive young storks raised in captivity in eastern Europe, but released in western Europe, fly in the southeasterly direction characteristic of birds from eastern Europe (Schüz, 1963). Similar experiments with other species indicate that the direction of migration is influenced genetically.

To reach its winter quarters, a bird must travel not only in the right direction but also the correct distance. Eberhard Gwinner (1972) obtained evidence on the distances traveled on the first migration by juvenile warblers unaccompanied by adults. The evidence suggests that the birds have an endogenous clock that induces as many hours of flight as are necessary to take the birds along each leg of their migratory

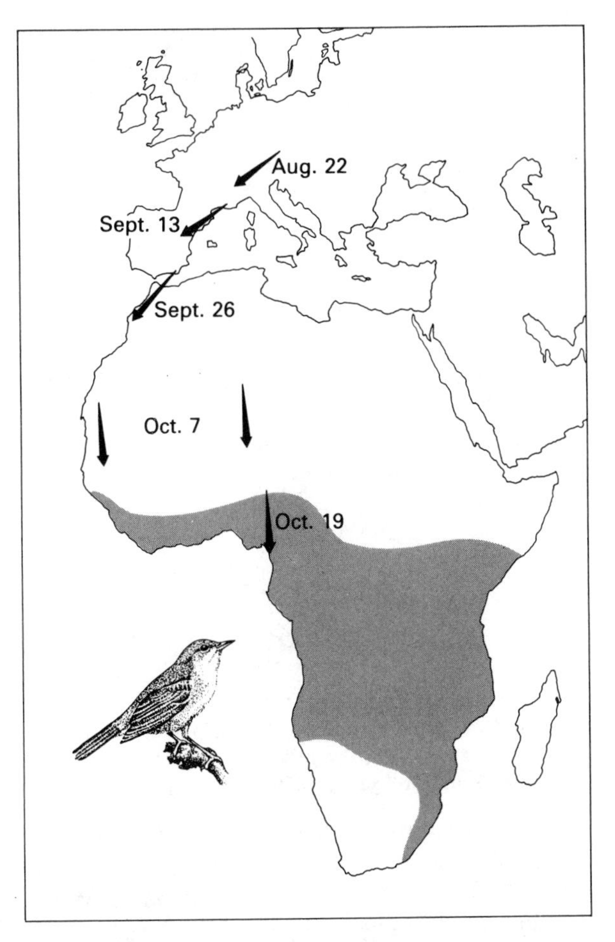

Fig. 16.14 Directional preferences obtained during laboratory tests of migratory restlessness in garden warblers. The direction of the arrows indicates the directional preference obtained on the dates shown. The positions of the arrows on the map indicate the places that migrating warblers would have reached (by the dates shown) during their migration to the wintering grounds (shaded area) in southern Africa (After Gwinner and Wiltschko, 1978)

journey. There is a good correlation between the number of hours of migratory restlessness of caged birds and the distance normally traveled during migration (e.g., Berthold, 1973). The distance traveled by migrating birds that is equivalent to an hour of restlessness can be estimated by comparing the behavior of caged and migrating birds from the same population (Gwinner, 1972). From knowledge of the speed of flight of migrating birds, Gwinner calculated the distance his caged birds would have covered had they been going in the right direction. His calculations gave results close to those obtained by observation of freely migrating birds from the same population. Garden warblers (*Sylvia borin*), tested in cages during their first autumnal migration, headed to the southwest

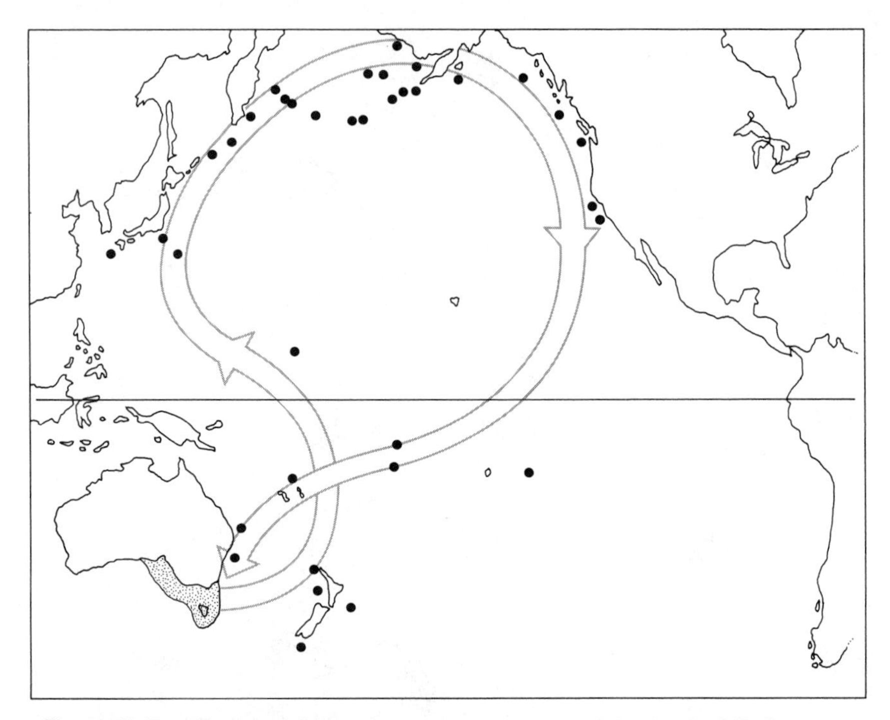

Fig. 16.15 Breeding range (dotted) and migratory route of the slender-billed shearwater (*Puffinus tenuirostris*) from recoveries of banded birds (black dots) (After Schmidt-Koenig, 1979)

during August and September and shifted to south-southeast during October, November, and December (Gwinner and Wiltschko, 1978). These results agree with the timetable of natural outdoor migration, as illustrated in Figure 16.14.

Some birds migrate many hundreds of miles over the ocean, where there are few landmarks. For example, arctic terns (*Sterna paradisaea*) breed in the Arctic and migrate in the autumn to the Antarctic pack ice. In the spring they reverse the migration, though probably by a different route. Some individuals are known to have traveled 9,000 miles (14,500 kilometers). Similarly, the slender-billed shearwater (*Puffinus tenuirostris*) breeds in southeast Australia and migrates to Alaska via Japan. Its return journey takes it down the western coast of North America (Fig. 16.15).

16.7 Lunar and tidal rhythms

As seen from the earth, the moon moves along a path similar to that of the sun and rises 50 minutes later each day, so is seen sometimes in the daytime and at other times only at night. The lunar cycle of 29.5 days is known to influence a variety of aspects of animal behavior. In the palolo

Fig. 16.16 The mating routine of the palolo worm (After Cloudsley-Thompson, 1980)

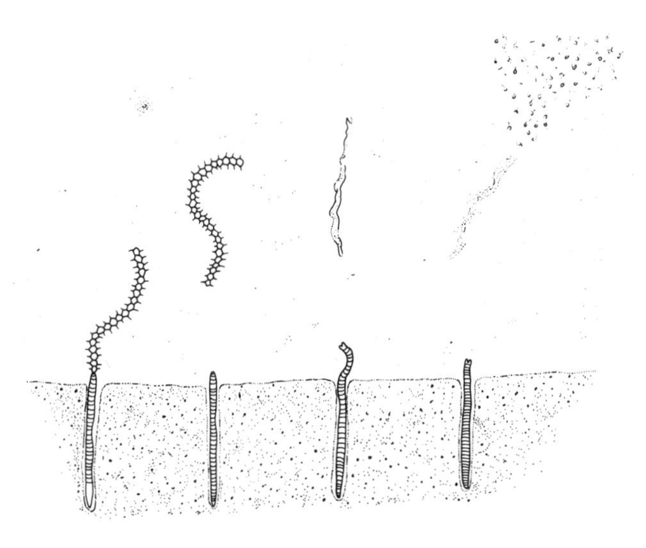

worm (*Eunice*) of the Pacific Ocean, reproductive activity occurs only during the neap tides of the last quarter moon in October and November. As illustrated in Figure 16.16, the posterior part of the worm, containing the genital organs, becomes detached from the anterior part and swims to the surface of the water, where the eggs and sperms are shed (Korringa, 1947). The worms provide abundant food for sharks and other fish, but by synchronizing their reproductive activity the worms are able to swamp these predators and a proportion of the gametes are always able to survive. Worms isolated in the laboratory produce their gametes at the correct time, showing that this activity is probably controlled by an endogenous clock.

Lunar rhythms are also known in terrestrial animals. Thus, Jamaican fruit bats have a pattern of feeding that involves leaving the day roost in the evenings and feeding throughout the dark nights during the period of the new moon. During the full moon, however, they depart from their day roost at the normal time in the evening, but return there when the moon is high, even when obscured by clouds, suggesting that they make use of an endogenous lunar clock to time their foraging and feeding behavior.

The tides result from changes in the combined gravitational pull of the sun and moon. The tidal cycle is thus repeated twice during the lunar cycle. Many marine animals show rhythms of behavior that coincide with the tidal cycle and have been shown to be governed by an endogenous clock. For example, the shore crab (*Carcinus maenas*) has a daily rhythm of activity that is superimposed upon a tidal rhythm. The tidal rhythm ensures that activity is synchronized with high water. This rhythm persists under constant laboratory conditions for about a week before it fades away, and can be restored by cooling the crab to near freezing for about six hours. This cold shock appears to restart the tidal

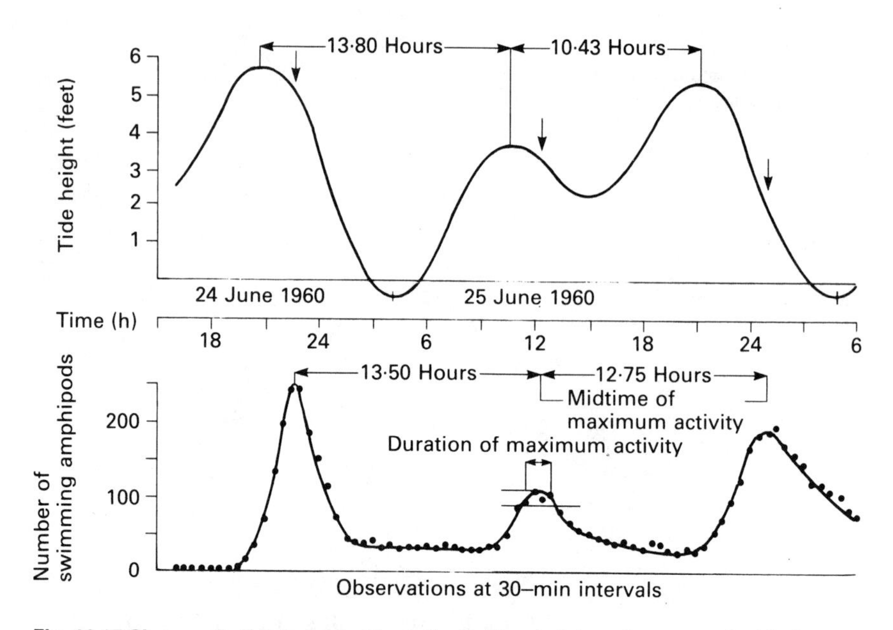

Fig. 16.17 Changes in tide height at La Jolla, California (above) compared with the swimming activity of *Synchelidium* under laboratory conditions (From *The Oxford Companion to Animal Behaviour*, 1981)

clock, perhaps because it resembles the experience of a beach-stranded crab that is eventually reached by a high tide. In one experiment, shore crabs were raised in the laboratory from eggs to adulthood under a day-night regime without tidal influences. These crabs showed only a circadian rhythm of activity. However, a tidal rhythm appeared after the crabs were given a single cold-shock treatment. It seems that the endogenous tidal clock had been dormant until started by the cold shock (Palmer, 1973).

Similar phenomena occur in the tidal rhythms of other crabs. Fiddler crabs (*Uca*) show tidal rhythms of activity that may persist for up to five weeks under constant laboratory conditions. The crabs emerge from their burrows at low tide and actively forage, court, etc. As the tide floods, they retreat back into their burrows. Shore crabs (*Carcinus*) have the opposite rhythm, becoming active at high water. Under constant conditions in the laboratory, the activity rhythm fades away after about a week. It can be restored, however, by cooling the crab to near feezing for about six hours. This cold shock appears to restart the tidal clock. In many crab species the drop in temperature brought about by the flood tides acts as a zeitgeber and sets the phase of the activity rhythm (Palmer, 1973).

Superposition of circadian and tidal rhythms can enable an animal to adapt to the irregular changes of high and low tide that occur in some parts of the world. On the coast of California, for example, a period of 13.80 hours between high tides is followed by one of 10.43 hours, as illustrated in Figure 16.17. The intertidal crustacean *Synchelidium* has a pattern of swimming activity that closely follows this tidal pattern (Fig. 16.17). This swimming pattern continues for several days under constant laboratory conditions.

16.8 Circadian rhythms and daily routines

Most animals experience changes in environmental conditions between night and day. Such changes affect animals both directly and indirectly. Thus there may be changes in food availability, and in numbers of predators, which are brought about by changes in temperature, light intensity, etc.

In adjusting to the differences between night and day the animal adopts a daily routine made up of many different aspects of behavior fitted together, to form a pattern that tends to be repeated day after day. When a single activity is studied it tends to follow a typical daily rhythm. The daily routines of animals have been subjected to relatively little study (Daan, 1981), but daily rhythms have received considerable attention from those primarily interested in circadian clocks (Rusak, 1981).

The most important daily changes in the external environment are those of light intensity and temperature. Animals specialized for daytime vision may be disadvantaged at night, because they are vulnerable to predators, or because they cannot forage efficiently. In cold climates it can benefit small mammals to be active at night, when temperatures are low. Their period of greatest heat production then occurs during the coldest part of the 24-hour cycle, the activity being harnessed as a means of thermoregulation. Small birds, on the other hand, save energy on cold nights by becoming inactive and allowing their body temperature to fall. In hot climates it is advantageous for small mammals to be nocturnal and so avoid the heat of the day.

It is not surprising that rhythms of rest and activity are widespread in the animal kingdom. When it is disadvantageous to be active at night, the best policy is to sit tight in a safe place and save as much energy as possible. This has been suggested as one of the prime functions of sleep (Meddis, 1965). Nocturnal species may hide during the day if they are likely to be preyed upon. If they are themselves nocturnal predators, they may remain hidden and inactive during the day to avoid scaring their prey. Thus, the daily rhythms of the physical environment make some activities advantageous at one time and disadvantageous at another. Much depends upon the overall ecology of the species concerned.

An animal adapted to its environment will settle into a daily routine

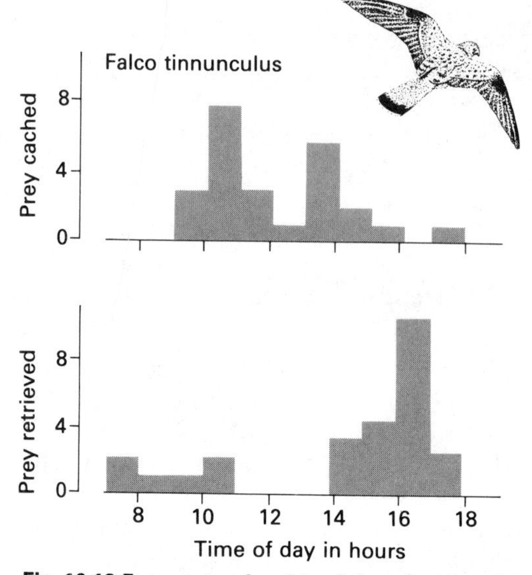

Fig. 16.18 Frequency of caching (above) and retrieval of cached prey (below) by the European Kestrel (*Falco tinnunculus*) (After Daan, 1981)

designed to maximize the survival value of its various activities. This is partly a matter of making the most of opportunities. For example, the European kestrel is a diurnal predator that specializes in preying upon small mammals. The bird relies on vision to detect and catch its prey and hunts most successfully when the light is good. Field studies reveal that kestrels catch prey throughout the day but do not always immediately

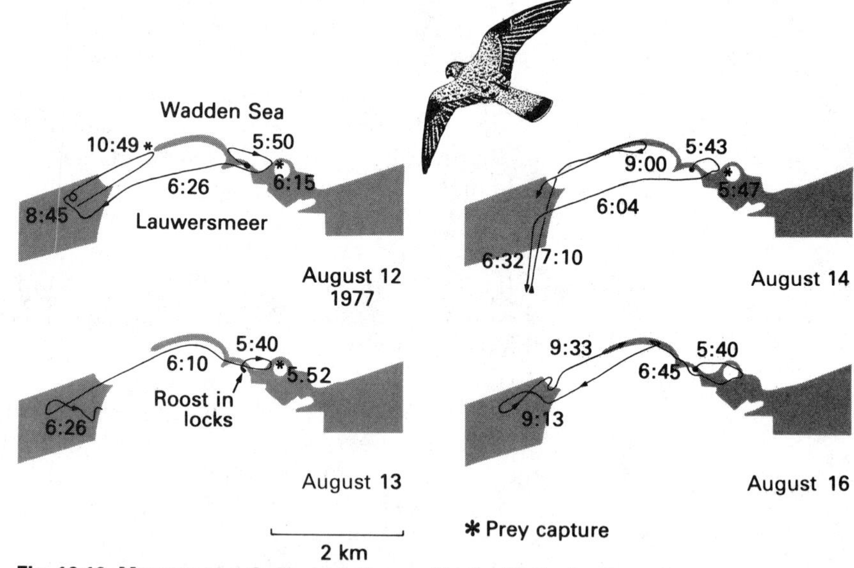

Fig. 16.19 Movements of a kestrel observed in the Netherlands on August 12, 13, 14 and 16, 1977. The numbers along the flight paths refer to the time of day. Land area is shaded, sea is white (After Daan, 1981)

eat what they catch. They tend to cache surplus prey in randomly chosen locations in their hunting area. The caching occurs throughout the day (see Fig. 16.18) and the cached items are typically retrieved at dusk, enabling the kestrel to make the most of the available prey during the day, without spending too much time eating. Moreover, if the bird ate all its surplus prey immediately it would become unnecessarily heavy and its hunting efficiency would probably decline.

In addition to the daily routines typical of their own species, individual animals may develop their own daily habits. Thus, kestrels that have found food at a particular time and place tend to repeat the same search pattern the next day, as shown in Figure 16.19. Such a strategy is appropriate in circumstances where the prey also has its own typical daily routine. We return to the topic of daily routines in Chapter 24, when we consider how they might enable animals to make the best use of their time and energy.

Points to remember

- The range of environmental conditions that an animal can tolerate is limited. It is characteristic of the species but can be somewhat altered by acclimatization.

- Acclimatization involves slow-acting physiological changes which have profound effects on the animal's homeostatic and motivational processes.

- Most animals possess endogenous biological clocks which are entrained to rhythmic aspects of the environment. The clocks may be circannual, lunar, tidal or circadian.

- Many aspects of reproductive physiology and behavior occur on a circannual basis, as do hibernation and migration. These all involve slow-acting physiological changes, somewhat akin to acclimatization.

- Lunar and tidal rhythms may occur in marine animals in addition to circannual and circadian rhythms.

- Circadian rhythms enable animals to adjust to the day–night cycle in an anticipatory manner and to develop daily routines which make the most of the prevailing opportunities.

Further reading

Pengelley, E. T. (1974) *Circannual Clocks–Annual Biological Rhythms*, Academic Press, New York.

2.3 Animal Learning

This section is devoted to the psychology of animal learning. Chapter 17 provides an outline of classical and instrumental conditioning from the traditional psychological viewpoint. Chapter 18 introduces evolutionary aspects and compares the biological and psychological perspectives. Chapter 19 discusses various issues that suggest cognitive explanations of animal learning. This chapter is a prelude to the discussion of the mentality of animals that occurs in Part III.

Ivan Pavlov (1849–1936)

BBC Hulton Picture Library

Ivan Pavlov was born in the Russian town of Ryszan, where he attended the religious school and seminary. He entered St. Petersburg University in 1870 and graduated in natural science in 1875. After obtaining his doctorate at the Military Medical Academy in 1883, he studied in Germany. He became Professor of Pharmacology at the Military Medical Academy in 1890, and Professor of Physiology in 1895. He received the Nobel Prize for Medicine in 1904 for his work on the physiology of digestion. In 1925 Pavlov resigned his professorship in protest against the explusion of sons of priests from the Academy. Himself the son of a priest, he nevertheless was not expelled. Up to the end of his life Pavlov continued to conduct research at various laboratories in the Soviet Union.

Pavlov's early work was on the physiology of circulation, especially the mechanisms that regulate blood pressure. He discovered that the vagus nerve controls blood pressure, and investigated nervous control of the rhythm and depth of the heartbeat. In 1879 Pavlov began work on the physiology of digestion, which culminated in his book *The Work of the Digestive Glands*, published in 1879. He investigated the mechanisms involved in the secretion of various digestive glands, and came to the conclusion that they were controlled exclusively by nervous mechanisms. It is now known that control by hormones also occurs. It is this work that led to the award of the Nobel Prize in 1904.

During the course of his work on the physiology of digestion, Pavlov noticed that salivation could be induced by the sight of food, or other stimuli that normally preceded feeding. This led him to the discovery of the conditional reflex, now regarded as a fundamental aspect of learning.

From 1902 until his death, Pavlov concentrated his researches on the phenomena of conditioning. He is responsible for many of the basic concepts still current in this field of study, and can be regarded as the founder of the experimental study of animal learning.

In a typical experiment, Pavlov showed that if the presentation of food to a dog was repeatedly accompanied by the sound of a bell, then the dog would come to respond to the bell as if it were food. Pavlov measured the dog's salivary response to paired presentations of food and bell and then measured salivation in response to presentation of the bell alone. He regarded the salivation to the food as an unconditional response, and the subsequent salivation to the bell alone as a conditional response, because it is conditional upon prior pairing between food and bell.

Pavlov's aim was to discover the universal laws of learning and to account for these in terms of brain mechanisms. He suggested that the cells of the central nervous system changed structurally and chemically during conditioning. While this notion is not far removed from the modern view, many of Pavlov's ideas about the role of the cortex in learning were premature. Pavlov's major work, translated into English as *Conditioned Reflexes: an investigation of the physiological activity of the cerebral cortex* (1927), had a major impact on the development of psychology.

17 Conditioning and Learning

Ivan Pavlov's work on conditioning became known in the west at a time when considerable interest already existed in mechanistic explanations of behavior, especially in reflexes. The attempts by Jacques Loeb (1859–1924) to account for animal behavior in terms of simple tropisms and taxes (see Chapter 14.2) had considerable influence in Germany and the United States. In physiology, Sir Charles Sherrington published *Integrative Action of the Nervous System* in 1906. He showed how simple reflexes could combine to produce coordinated behavior.

In psychology, John B. Watson (1913) launched the behaviorist school of psychology, which was to become very influential during the first part of the century. The behaviorists would allow only observable stimuli, muscular movements, and glandular secretions to enter into explanations of behavior. In order to account for complex behavior in stimulus-response terms, they postulated covert or implicit stimulus-response relationships. Watson (1907) had already proposed that somesthetic stimuli aroused by the animal's movements served such a purpose. The unobservable processes mediating stimulus and response were said to be comprised of incipient movements and movement-produced stimuli. Thus, Watson (1914) postulated that human thought processes were implicit speech (i.e., talking to oneself), one tongue twitch providing the stimulus for the next response in the chain.

Pavlov gave a lecture in Madrid in 1903, and he delivered the Huxley lecture in London in 1906, of which a report was printed in *Science*. A review of Pavlov's work was published in 1909, and Watson published another in 1916. A translation of Pavlov's book *Conditioned Reflexes* was published in 1927. There was a strong climate of opinion in favor of a purely mechanistic and objective science of behavior. Pavlov's work added fuel to the extreme environmentalist approach espoused by Watson in his behaviorist psychology. Watson (1916) realized that reflex conditioning might serve as a paradigm for learning in general. Behaviorists such as Watson and, later, B. F. Skinner claimed that all animal and human behavior could be accounted for in terms of conditioning. Pavlov's work endowed behavorism with a certain amount of physiological respectability, and the psychology of animal learning came to dominate psychology in the United States up to the end of the 1950s.

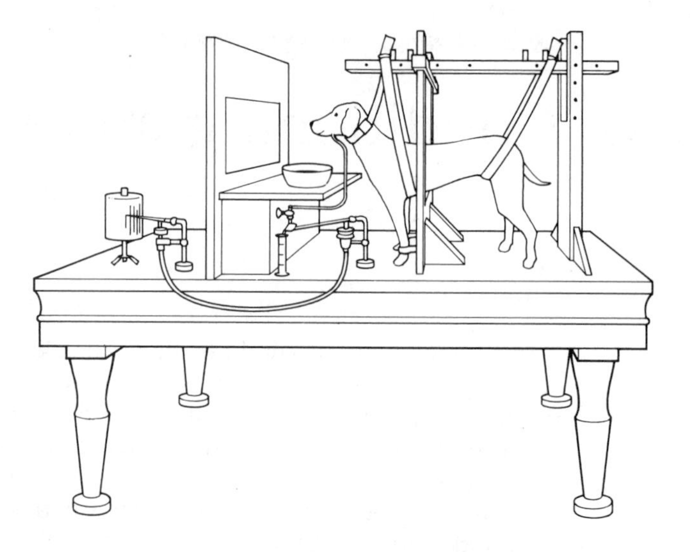

Fig. 17.1 Pavlov's arrangement for the study of salivary conditioning

17.1 Classical conditioning

In his original conditioning experiments, Pavlov restrained a hungry dog in a harness (see Fig. 17.1) and presented small portions of food at regular intervals. When he signaled the delivery of food by preceding it with an external stimulus like the sound of a bell, the behavior of the dog toward the stimulus gradually changed. The animal began by orienting to the bell, licking its lips and salivating. When Pavlov recorded the salivation systematically by placing a small tube in the salivary duct and collecting the saliva, he found that the amount of saliva collected increased as the animal experienced more pairings between the sound of the bell and food presentation. It appeared that the dog had learned to associate the bell with the food.

Pavlov referred to the bell as the *conditional stimulus* (CS) and to the food as the *unconditional stimulus* (UCS). Salivation in response to presentation of food was called the *unconditional response* (UCR), while salivation in response to the bell was called the *conditional response* (CR).

Although Pavlov originally used the words "conditional" and "unconditional", this terminology was mistranslated at an early stage, and the terms conditioned reflex, conditioned response, unconditioned response, and unconditioned reflex became established in the English-language literature. However, it is modern practice to return to Pavlov's original terminology, and we follow his usage in this book. The rationale behind the terminology is that the food unconditionally elicits a set of consummatory responses (such as salivation), one of which is recorded by the experimenter and designated the unconditional response. Conditioning occurs as a result of a contingency, arranged by the experimenter, between the unconditional stimulus (food) and an external stimulus previously unconnected with the feeding situation, like the sound of a bell. After a number of pairings, the bell alone is sufficient to elicit salivation. The bell is then known as the conditional stimulus because as a result of its training, the dog salivates if and when this stimulus is presented. Similarly, the salivation in response to the bell is known as the conditional response, even though it may appear to be the same response as the unconditional response. During the process of conditioning, the presentation of the UCS (food) following the CS (bell) is said to reinforce the conditional reflex of salivation to the CS. The UCS, therefore, is regarded as a *reinforcer.*

Conditioning experiments that employ motivationally beneficial, or positive, reinforcers like the UCS are examples of *positive conditioning.* Conditional reflexes also can be established in experiments that employ negative reinforcers, such as electric shock, which the animal tries to avoid. For example, the onset of a sound stimulus (tone) preceding the delivery of a puff of air to the eye of a rabbit will come to elicit a blink of the eyelid (strictly, the nictitating membrane) as a conditional response. Initially, the air puff (UCS) alone elicits the eye blink (UCR), but after conditioning, the tone (CS) will elicit the eye blink (CR) in the absence of a puff of air. This is an example of negative conditioning.

A reinforcer is characterized not so much by its intrinsic properties as a stimulus but by its motivational significance to the animal. Thus, food acts as a positive reinforcer only if the dog is hungry, and an air puff acts as a negative reinforcer only if it is noxious or unpleasant for the animal. In many cases the reinforcer is innate in the sense that its motivational significance and ability to support conditioning is an integral part of the animal's normal makeup. However, this does not have to be the case, and Pavlov showed that a CS could act as a reinforcer. For example, if a bell is established as a CS by the normal conditioning procedure, it will reliably elicit a CR, like salivation. If a second CS, such as a light, then is paired repeatedly with the bell, in the absence of food, the animal will come to give the CR in response to the light alone, even though food has never been associated directly with the light. This procedure is known as *second-order conditioning.*

Pavlovian, or classical, conditioning is very widely observed in the animal kingdom, and it pervades every aspect of normal life in higher animals, including humans. Pavlov demonstrated that conditioning could occur in monkeys and in mice, and claims have been made for a wide variety of invertebrate animals. In evaluating such claims, however, we must take care to distinguish true classical conditioning from other forms of learning and quasi-learning.

Although classical-conditioning procedures are relatively straightforward, the phenomena they reveal are not so clear-cut and have given rise to considerable discussion in the psychological literature, from the time of Pavlov to the present day. Every student of animal behavior should be familiar with the basic characteristics of classical conditioning because it is hardly possible to carry out an experiment without some conditioning occurring. It may be simply that the animal becomes conditioned to the time of day at which the experimenter appears or that clandestine conditioning phenomena subtly invalidate the experimenter's conclusions. In any event, as a universal feature of higher animals, conditioning is not only of practical importance but also must be incorporated into any coherent understanding of the way animals work. We now briefly discuss some of the major characteristics of conditioning. A more detailed account is given in the excellent books by Nick Mackintosh (1974, 1983).

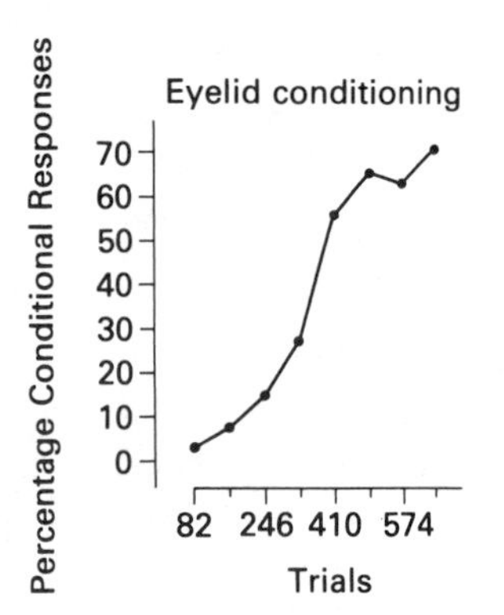

Fig. 17.2 Acquisition of eyeblink response during eyelid conditioning (After Schneidermann et al., 1962)

17.2 Acquisition

We can measure the acquisition of a conditional response in various ways. Pavlov used the amount of saliva collected during presentation of the CS, for example. In the case of eyelid conditioning, the probability of occurrence of the response is measured (see Fig. 17.2). The rate of acquisition may vary considerably, with species, circumstances, and age (Fig. 17.3).

Pavlov held the view that the pairing of CS and UCS leads to the formation of an association between them. The CS becomes a substitute for the UCS and becomes capable of eliciting the responses normally elicited by the UCS. This is usually called the *stimulus substitution theory*. An alternative theory maintains that CRs occur because they are followed by rewards. In other words the CR is reinforced by its consequences. This is usually called the *stimulus-response* theory.

The two theories differ in two main empirical respects. First, on the

Fig. 17.3 The effect of age upon eyelid conditioning (After Braun and Geiselhart, 1959)

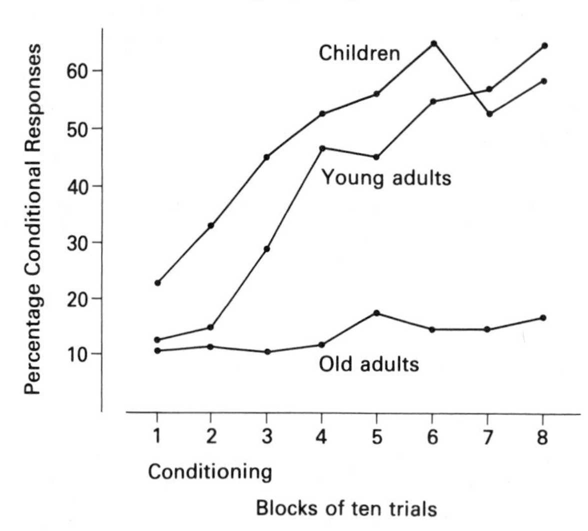

basis of stimulus substitution theory, we would expect the CR to be very similar to the UCR, whereas it should be somewhat different according to stimulus-response theory. Second, Pavlov maintained that an association occurs between the CS and the UCS, and that this association itself constitutes a reinforcement. The stimulus-response theory maintains that learning is dependent upon reinforcing consequences of the CR. Although this topic is controversial, the evidence seems to favor Pavlov's view. Thus, both the CS and the UCS elicit a similar variety of responses. Pavlov deliberately ignored the skeletal components of the CRs because of the temptation to interpret these as signs of expectation or preparation. To avoid such anthropomorphic interpretations, Pavlov concentrated upon the salivary component of the CR. Moreover, as Mackintosh (1974) points out, if responses (such as closure of the nictitating membrane in anticipation of a puff of air, movements of the jaw or licking in anticipation of water, or pecking or salivation in anticipation of food) are relatively unaffected by their consequences when these are explicitly programmed, it is hard to see how they could be established because of their consequences when no such consequences are explicitly programmed. The only alternative is to assume that they are established, as Pavlov argued, because they are elicited by the UCS.

As we shall see, stimulus substitution theory has far-reaching implications for other types of learning and for our understanding of the modifiability of animal behavior in general.

17.3 Extinction and habituation

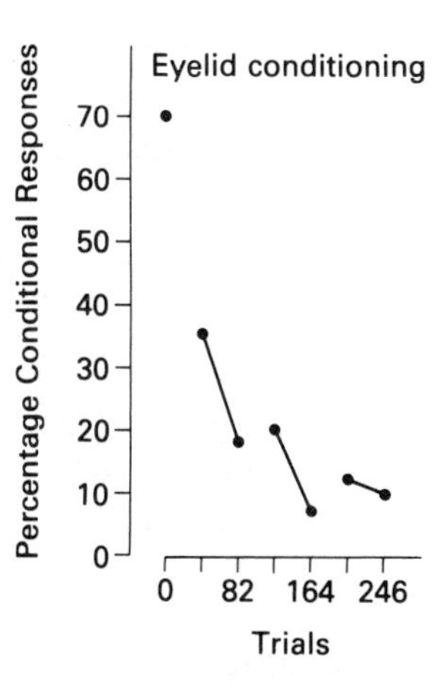

Fig. 17.4 Extinction of eyelid conditional responses in rabbits (After Schneidermann et al., 1962)

We have seen that presentation of an UCS increases the strength of the CR. Pavlov discovered that withholding such reinforcement led to the gradual disappearance of the CR. The process by which learned behavior patterns cease to be performed when no longer appropriate is called *extinction*.

In a classical-conditioning experiment, the dog learns that the bell (CS) signals the presentation of food. Its salivation (CR) is therefore an appropriate response to make in anticipation of the availability of food. If food no longer is made available, then the dog should no longer treat the bell as a signal for food. This is precisely what happens: Omission of food results in a decline in the salivary response to the CS. The behavior of the animal then appears to be the same as before conditioning. Another example of extinction is illustrated in Figure 17.4.

If, after extinction, the CS again is paired with the reinforcer, the CR reappears much more rapidly than during the original conditioning, suggesting that the process of extinction does not abolish the original learning but somehow suppresses it. Further evidence for this conclusion comes from the phenomenon of *spontaneous recovery*, by which a response that has been extinguished recovers its strength with rest. For example, Pavlov (1927) reported an experiment in which the number of drops of saliva secreted to the CS was reduced from 10 to 3 in a series of seven extinction trials. The *latency* (time delay) of the response also increased from 3 to 13 seconds. After a rest period of 23 minutes, the salivation on the first trial with the CS alone amounted to six drops with a latency of only five seconds.

Pavlov argued that the decline in CR observed during extinction must be due to an accumulation of internal inhibition. He showed that if an extraneous, novel stimulus was presented at the same time as the CS in a conditioned trial, the CR was disrupted. According to Pavlov (1927): "The appearance of any new stimulus immediately invokes the investigatory reflex and the animal fixes all its appropriate receptor organs upon the source of disturbance. . . . The investigatory reflex is excited and the conditioned reflex is in consequence inhibited." This phenomenon is called *external inhibition*. If the extraneous stimulus is presented during the course of extinction, the CR increases in strength on that trial. This type of effect, called *Pavlovian disinhibition*, provides further evidence of the inhibitory nature of extinction. Unlike external inhibition, it is not due to competition between reflexes but it is thought to be caused by an increase in arousal. It is a widespread phenomenon and can be shown to occur when a CR is declining, for whatever reason, including habituation.

The evidence suggests that, during extinction, the animal learns that the CS no longer is followed by reinforcement. The CS now is associated with the absence of a reinforcer, and the CR is inhibited accordingly. As we shall see, the idea that animals can learn that certain stimuli predict no consequences plays an important part in modern learning theory. Scientists have found that an important requirement for the development of inhibition is that the CS should be paired with non-reinforcement in a context in which stimuli already are associated with reinforcement. In the animal's ordinary life, numerous stimuli are not associated with reinforcement, but the animal ignores these and learns nothing about them. Only when the animal is surprised by the absence of reinforcement does it learn that particular stimuli signal non-reinforcement (Mackintosh, 1974).

Repeated applications of stimulus often result in decreased responsiveness. This phenomenon, called *habituation*, is a form of non-associative learning that has some similarities to extinction. For example, the escape response of fish to a shadow passing overhead diminishes progressively if the stimulus is repeated every few minutes, until the fish cease to react at all. Similarly, the orientation response of the toad (*Bufo bufo*) toward potential prey progressively declines if non-edible preylike objects are presented repeatedly. It is well known to fruit growers that scarecrows erected to deter birds are effective for a short time, but the birds soon become habituated to them. Attempts to scare

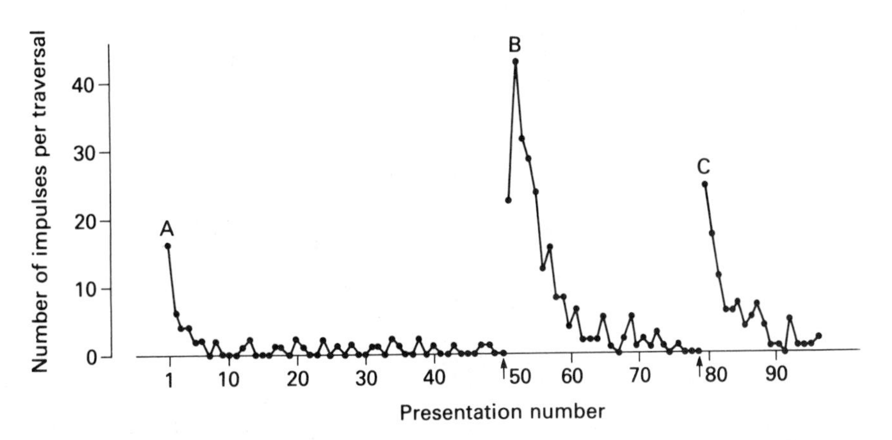

Fig. 17.5 Habituation and stimulus induced dishabituation in a neuron responding to movement stimuli. A is the original habituation, B and C are dishabituations induced by extraneous stimuli (arrows) (After Rowell and Horn, 1968)

birds from airfields by broadcasting alarm calls have run into similar problems of habituation.

Habituated responses show spontaneous recovery when stimulation is withheld. If habituation and subsequent recovery of the response are repeated a number of times, then the habituation becomes progressively more rapid. In these respects, habituation is similar to extinction. If a novel stimulus is presented during the process of habituation, then there is an increase in responsiveness, as illustrated in Figure 17.5. This *dishabituation* is thought to be due to changes in the animal's level of arousal and is very similar to Pavlovian disinhibition.

Habituation is usually regarded as a form of learning and can be distinguished experimentally from the waning of a response due to sensory adaptation or to fatigue. Habituation is similar to extinction in that the animal learns to inhibit responses that are not followed by reinforcing consequences. Both extinction and habituation show spontaneous recovery and disinhibition by extraneous stimuli. Extinction differs from habituation in occurring in relation to previously learned responses, whereas the responses that typically show habituation are innate responses that have not been established through any process of conditioning.

17.4 Generalization

When an animal has learned a particular response to a particular stimulus, it may also show the response to other similar stimuli. Thus, Pavlov (1927) noted that "if a tactile stimulation of a definite circumscribed area of skin is made into a conditional stimulus, tactile stimulation of other skin areas will also elicit some conditional reaction, the

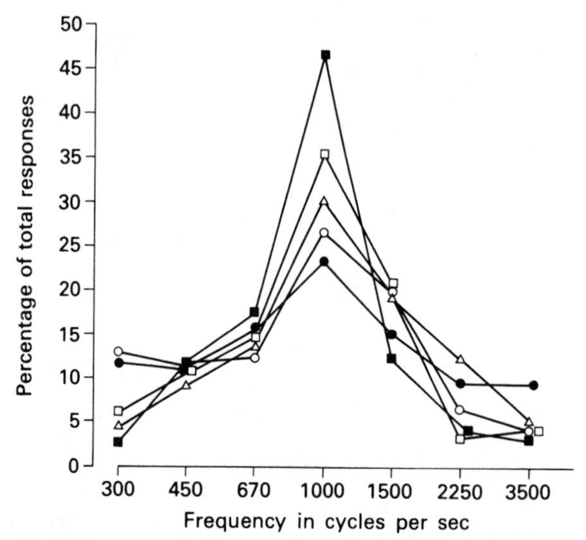

Fig. 17.6 Generalization gradients for individual pigeons responding to tones. The training tone had a frequency of 1000 cycles per second (After Jenkins and Harrison, 1958)

effect diminishing with increasing distance of these areas from the one for which the conditioned reflex was originally established.'' This type of phenomenon is called *stimulus generalization*. Pavlov assumed that it was due to a spreading wave of excitation that traveled across the cerebral cortex from the focus of the CS. This explanation, however, does not accord with modern views on the neuronal structure of the brain (Thompson, 1965).

Current explanations of stimulus generalization concentrate on the stimuli involved. A stimulus used in a conditioning experiment consists of a collection of separate elements. For example, a tone has a given frequency, intensity, and duration. These dimensions of the stimulus may become conditioned during acquisition of a CR. A new stimulus that has some elements in common with the CS may be capable of eliciting the CR to some extent. For example, while a tone of 1,000 hertz is discriminated readily from one of 300 hertz by humans, pigeons trained to respond to the former generalize to the latter. However, the tones have other qualities in common, especially the fact that they are unlike natural sounds in being characterized by a single frequency. It is perhaps not surprising, therefore, that pigeons treat them as similar. If the pigeon mentioned above is rewarded in the presence of a 300-hertz tone, then it has less tendency to generalize to tones of frequencies other than 1,000 hertz. The generalization gradient has been sharpened by discrimination training, as illustrated in Figure 17.6. The discrimination training reduces the number of elements associated consistently with reward. The pigeon is encouraged to pay attention to the frequency of the tone by the fact that the other dimensions of the sound are not associated consistently with reward. This is called the common elements theory of generalization.

Stimulus generalization has important implications for many aspects of animal learning. For example, during extinction of a CR, omission of reinforcement changes the stimulus environment in which the conditioning was established. There is some evidence that the reduction in responsiveness seen during extinction is due in part to a generalization decrement. Similarly, a response habituated to one stimulus will show generalization to another, similar stimulus. To some extent, the animal treats the new stimulus as if it had been presented previously. The degree of habituation to a new stimulus depends upon how similar it is to the stimulus to which the animal previously has been habituated. If the stimulus is both novel and unfamiliar, then dishabituation will occur. Thus, generalization tends to counteract the effects of novel stimuli on habituated responses.

Pavlov held the view that discriminations could be learned when some aspects of the CS are reinforced while others are not. Initially, all aspects of the CS elicit a CR, but if the experimenter reinforces some aspects and not others, then we may designate these CS^+ and CS^-, respectively. Discriminations may be found among aspects of a single physical dimension, such as auditory stimuli of different frequencies or lights of differing brightness. Discrimination also may be formed between combinations of qualitatively different stimuli, called *compound stimuli* by Pavlov (1927). For instance, the CS^+ might be the combination of a tone and a tactile stimulus, and the CS^- might be the tactile stimulus alone. Once the animal has learned the discrimination, the tactile stimulus alone no longer will elicit a CR. As mentioned, every conditioning experiment involves compound stimuli in that the CS inevitably is presented in a particular stimulus situation and can be distinguished from background stimuli only by the process of discrimination learning on the part of the animal. During the interval between trials, some stimuli that the animal associates with reinforcement will be present during the early part of training, prior to the development of discrimination between CS^+ and CS^-. Therefore, we might expect to find that dogs salivate during the intertrial interval of a classical-conditioning experiment, at least during the early stages. Sheffield (1965) reported that his dogs did indeed salivate during the intertrial interval to an extent that declined progressively during training.

17.5 Instrumental learning

While research on classical conditioning was initiated in Russia, the principles of *instrumental conditioning* were discovered and developed in the United States. However, the writings of Conway Lloyd Morgan (1852–1936), of the University of Bristol in England, seem to have provided the initial impetus.

Lloyd Morgan was critical of much of the contemporary research in animal psychology on the grounds of its poor methodology and sloppy reasoning. In his *Introduction to Comparative Psychology* (1894), he enunciated his famous canon: "In no case may we interpret an action as the outcome of the exercise of a higher psychical faculty, if it can be interpreted as the outcome of one which stands lower in the psychological scale." Later (1900) he added the following rider: "To this it may be added—lest the range of the principle he misunderstood—that the canon by no means excludes the interpretation of a particular act as the outcome of the higher mental processes if we already have independent evidence of their occurrence in the agent."

Lloyd Morgan had considerable influence upon the development of behaviorism, particularly upon John Watson and Edward Thorndike. In 1896, he delivered the Lowell lectures at Harvard University, where he provided the stimulus for Thorndike's pioneering research on animal intelligence. Lloyd Morgan recounted how his dog, Toby, had learned to open the latch on the garden gate by putting its head through the railings. Thorndike started his research soon after Lloyd Morgan's visit and devised ways of repeating this observation under controlled conditions in the laboratory.

Thorndike became a very important figure in American psychology, and for half a century his theories dominated both animal and educational psychology. An eminent contemporary wrote: "The psychology of animal learning—not to mention that of child learning—has been and still is primarily a matter of agreeing or disagreeing with Thorndike, or trying in minor ways to improve upon him. Gestalt psychologists—all of us here in America seem to have taken Thorndike, overtly or covertly, as our starting point" (Tolman, 1938).

Thorndike carried out a series of experiments in which cats were required to press a latch or pull a string to open a door and escape from a box to obtain food outside. The boxes were constructed with vertical slats so the food was visible to the cat (Fig. 17.7). A hungry cat, when first placed in the box, shows a number of activities, including reaching through the slats toward the food and scratching at objects within the box. Eventually the cat accidentally hits the release mechanism and escapes from the box. On subsequent trials, the cat's activity becomes concentrated progressively in the region of the release mechanism, and other activities gradually cease. Finally the cat is able to perform the correct behavior as soon as it is placed in the box.

Thorndike (1898) designated this type of learning as "trial, error, and accidental success." It is nowadays called *instrumental learning*, the correct response being "instrumental" in providing access to reward. This type of learning had been known to circus trainers for centuries, but Thorndike first studied it systematically and developed a coherent theory of learning based upon his observations.

To explain the change in the animal's behavior observed during learning experiments, Thorndike (1913) proposed his *law of effect*. This

Fig. 17.7 A cat in one of Thorndike's puzzleboxes

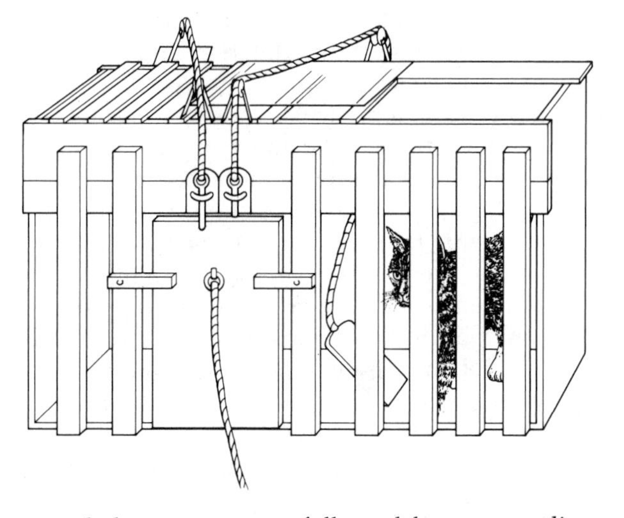

stated that a response followed by a rewarding or satisfying state of affairs would increase in probability of occurrence, while a response followed by an aversive or annoying consequence would decrease the probability of occurrence. Thus, the success of instrumental learning is attributed to the fact that learned behavior can be directly modified by its consequences. Thorndike (1911) assumed that a reinforcer increases the probability of the response upon which it is contingent because it strengthens the learned connection between the response and the prevailing stimulus situation. This became known as the "stimulus-response theory of learning," and versions of this theory were predominant for many years. While recognizing the validity of the law of effect as an empirical statement, most present-day psychologists doubt that behavior is modified by its consequences in the direct way that Thorndike and his followers supposed. To understand this, we first have to consider the nature of reinforcement.

17.6 Reinforcement

There is a fundamental difference between the way a classical-conditioning experiment is conducted and the procedure used in an instrumental-learning experiment. In classical-conditioning experiments, a contingency is arranged between the CS (e.g., a bell) and the UCS (e.g., food). The reinforcer is applied regardless of the behavior of the animal. In instrumental-learning experiments, the reinforcement (e.g., food) is contingent upon some particular behavior of the animal (e.g., pressing a latch). Thus, in classical-conditioning experiments a contingency is arranged between a stimulus and an outcome, while in instrumental-learning experiments the contingency is arranged between a response and an outcome. While these differences do not necessarily imply that different kinds of learning are involved in the two types of experiments, they do suggest that different processes of reinforcement are involved.

We have seen that, in Pavlov's view, the occurrence of a reinforcing stimulus in a particular context will cause the responses elicited by that reinforcer to occur in anticipation of the delivery of the reinforcer. There have been some remarkable recent demonstrations of this effect, as we shall see. It is evident, however, that reinforcement is not always necessary for the occurrence of learned associations among stimuli. This can be seen most clearly by examining the phenomenon called *sensory preconditioning*. The procedure is to present two conditional stimuli (CS_1 and CS_2) together for a number of trials before introducing the UCS. Thus, joint presentations of CS_1 and CS_2 are followed by pairing of CS_1 with the UCS. In the final test phase, the strength of the CR in response to the CS_2 is measured.

The first clear demonstration of sensory preconditioning was that of William Brogden (1939), who gave dogs 200 trials in which a light and a buzzer were presented simultaneously. One of these stimuli was then paired with electric shock to the paw to establish leg flexion as a CR. Test trials to the other CS produced an average of 9.4 CRs compared with only 0.5 CRs obtained from control experiments that did not involve prior pairing of buzzer and light. More recent experiments show that better results can be obtained if fewer pretraining trials are given and if the two CSs are presented not simultaneously but within a few seconds of each other (Mackintosh, 1974).

The results of sensory-preconditioning experiments clearly show that pairing two neutral stimuli is sufficient to establish some learned association between them. It appears that if the stimuli are presented too often, some habituation occurs and learning does not improve. Quite apart from the fact that these results cannot be explained satisfactorily by a stimulus-response theory of learning, it is apparent that reinforcement is not necessary for associations to occur between two neutral stimuli. Pavlovian reinforcement is, therefore, not a necessary condition for the formation of associations, but it does facilitate their formation and render them resistant to habituation. As we see later, animals are quite capable of learning that certain stimuli are irrelevant to their current motivational requirements, and it is not, therefore, surprising that the associations formed between two neutral stimuli should be temporary.

We now turn to the question of *instrumental reinforcement*. Thorndike's law of effect became the mainstay of the behaviorist approach to animal learning. An extreme position was reached by Harvard behaviorist Burrhus Frederic Skinner, who turned the law of effect into a definition of reinforcement. For Skinner, a reinforcer was any event that, if made contingent upon some aspect of behavior, would cause the behavior to increase in frequency. Skinner (1938) also assumed that any reinforcer could strengthen any response in the presence of any stimulus, provided the stimulus could be sensed by the animal and that the response was within its capacity. Thus, the response and reinforcer were regarded as being essentially arbitrary. This became a widespread view among learning theorists up to the end of the 1950s.

A reinforcer that encourages an animal to approach the stimuli that it associates with it is generally called a "positive reinforcer." If the reinforcer leads the animal to avoid the situation in future, then it is a "negative reinforcer."

Animals may learn to fear certain situations as a result of experiencing pain or stress there. They subsequently may show avoidance behavior when they next encounter the situation. Similarly, situations in which the animal encounters natural fear-inducing stimuli may induce fear by association, without the animal's experiencing pain. Repeated presentation of such stimuli, however, may lead to habituation so that the animal no longer avoids them.

As we have seen, aversive stimuli can be used in classical-conditioning experiments. One of the first studies of this type was by Vladimir Bekhterev (1913), who applied a mild electric shock (UCS) to the forepaw of a dog, following an auditory stimulus such as a tone (CS). The dog initially flexed its leg in response to the shock (UCR), and after a number of paired presentations of CS and UCR, it flexed its leg in response to the tone (CR), in the absence of shock. This type of classically conditioned defense reaction remained a paradigm of avoidance learning for a number of years.

For Thorndike (1913), the effects of punishment were symmetrical to those of rewards. The positive law of effect maintained that responses followed by a satisfactory state of affairs increased in probability, while the negative law stated that responses followed by an annoying state of affairs decreased in probability. However, Thorndike (1932) eventually concluded, on the basis of several experiments, that the law of effect did not apply to the effects of punishment. In this view, he was followed by Skinner (1938; 1953) and Estes (1944). The consensus seemed to be that punishment was not effective in weakening stimulus-response connections, though it sometimes had the effect of temporarily suppressing the punished behavior. More recent evidence suggests that punishment can be effective in modifying behavior when it is made contingent upon a particular response (Church, 1963; 1969).

17.7 Operant behavior

Skinner (1937) introduced a distinction between operant behavior and respondent behavior. *Operant behavior* he defined as spontaneous action without any obvious stimulus. *Respondent behavior* included all behavior performed in response to an identifiable stimulus. Skinner believed that all operant behavior was modified and effectively controlled by the

Fig. 17. 8 A pigeon pecking a key in a Skinner box

prevailing reinforcement contingencies. The idea that animal behavior could be manipulated entirely by appropriate schedules of reinforcement represented the extreme behaviorist view (Skinner, 1938). Skinner's behaviorist philosophy brought about a revolution in experimental technique that has persisted to this day.

In place of the trial-by-trial procedure characteristic of classical conditioning and experiments using puzzle boxes or mazes, Skinner devised the free-operant procedure in which the animal is allowed to indulge freely in various activities while the experimenter attempts to manipulate the consequences. Rats and pigeons were chosen most frequently for this type of experiment, although many other animals, were also used including humans. Operant conditioning consists essentially of training an animal to perform a task to obtain a reward. A rat may be required to press a bar, or a pigeon to peck an illuminated disk, called a "key." The typical method of training is called *shaping*.

Let us consider the training of a pigeon to peck a key to obtain food reward. A hungry pigeon is placed in a small cage equipped with a mechanism for delivering grain and a key at head height, as shown in Figure 17.8. This type of apparatus nowadays is called a "Skinner box." Delivery of food normally is signaled by a small light that illuminates the grain. Pigeons soon learn to associate the switching on of the light with the delivery of food, and they approach the food mechanism and eat the grain whenever the light comes on. The next stage of shaping is to make food delivery contingent upon some aspect of the animal's behavior. It is usual to require the pigeon to peck the key, but Skinner claimed that any response could be shaped and that pigeons could be taught to preen or turn in small circles to obtain reward. Key pecking may be shaped by limiting rewards to movements that become progressively more similar to a peck at the key. Thus, when it has learned to approach a key for reward, the pigeon is rewarded only if it stands upright with its head near the key. At this stage, the pigeon usually pecks at the key of its own accord, but it can be encouraged by temporarily gluing a grain of wheat to the key. When the pigeon pecks the key, it closes a sensitive switch in an electronic circuit that causes food to be delivered automatically. From this point on, the pigeon is rewarded only when it pecks the key, the manual control of reward no longer being required. The animal is now ready for use in an experiment.

Many types of experiment utilize this operant-conditioning procedure. For example, discrimination learning can be studied by rewarding the animals for responding only when a certain color or pattern is presented, or by allowing the animal to choose between two keys that are differentiated visually. The technique has proved particularly useful for studying the effect of different types of reward patterns. Thus, instead of rewarding a pigeon for every peck, it can be rewarded for every *n*th peck so there is a fixed ratio between number of pecks and number of rewards. This procedure is called a *fixed-ratio reward schedule*.

Other commonly used schedules include *variable ratio, fixed interval*, and *variable interval*. On an interval schedule, reward is given at time intervals specified by the experimenter. The animal is rewarded for the first response after a given interval has elapsed. Different schedules of rewards have been found to have different effects on the animal's performance. For example, a variable interval schedule produces a very uniform rate of responding and provides a good benchmark against which to test the effects upon behavior of various factors like reward size.

Skinner is regarded as having treated behaviorism as a philosophy of the science of behavior, not as the science itself. His approach was operationalist and his psychology was antitheoretical. Although Skinner maintains that all behavior is produced by reinforcement, he recognized (1975) that "natural selection is responsible for the fact that men respond to stimuli, act upon the environment and change their behavior under contingencies of reinforcement." Similarly, "the fact that operant conditioning, like all physiological processes, is a product of natural selection throws light on the question of what kind of consequences are reinforcing and why."

Skinner's approach depends upon the efficacy of reinforcement in modifying behavior. His claim that any activity can be modified is illustrated by the various games that pigeons can be trained to play. Thus, Skinner (1958) reports that "the pigeon was to send a wooden ball down a miniature alley towards a set of toy pins by swiping the ball with a sharp sideward movement of the beak. The result amazed us. . . . The spectacle so impressed Keller Breland that he gave up a promising career in psychology and went into the commercial production of behavior."

Ironically, it was Keller Breland and Marian Breland (1961) who first cast doubt upon the assumption that any activity could be modified by reinforcement. They found that, in attempting to train animals to perform various tricks, some activities appeared to be resistant to modification by reinforcement. For example, they attempted to train a pig to insert a coin into a piggy bank. The pig would pick up a wooden token, but instead of dropping it into the container, it would repeatedly drop it on the floor, "root it, drop it again, root it along the way, pick it up, toss it in the air, drop it, root it some more, and so on" (Breland and Breland, 1961). Similarly, they encountered chickens that insisted on scratching at the ground when they were supposed to stand on a platform for 10 to 12 seconds to receive a food reward. Subsequently, there have been numerous reports of this type. Thus, Peit Sevenster (1968; 1973) successfully trained male three-spined sticklebacks (*Gasterosteus aculeatus*) to swim through a small ring to obtain access to a female. He was unable, however, to train males to bite a glass rod to gain access to a female because the male insisted on directing courtship behavior toward the rod. These and other studies are reviewed by Sarah Shettleworth (1972).

Breland and Breland (1961) interpret their findings as evidence of instinctive drift by which "learned behavior drifts towards instinctive behavior" whenever the animal has strong instinctive behavior similar to the conditioned response. They point out that their results violate Skinner's (1938) principle of least effort, by which animals are supposed to obtain their rewards in the quickest and most comfortable manner. In all cases, reward is delayed considerably by the misbehavior of the animals in their studies. In Chapter 18 we discuss further examples of this type and try to assess the extent to which they can be accounted for by a universal theory of learning.

Points to remember

● Classical, or Pavlovian, conditioning leads to the establishment of an association between a previously meaningful stimulus (the unconditional stimulus) and a previously neutral stimulus (the conditional stimulus) such that the animal comes to respond to the latter (the conditional response) in the way that it previously responded (with the unconditional response) to the former.

● The process of establishing the association, called acquisition, usually takes a number of trials during which some reinforcement is given.

● A decline in responsiveness can occur as a result of habituation or extinction. Habituation results from repeated presentation of a stimulus without immediate consequences. Extinction results from repeated presentation of a conditional stimulus without reinforcement.

● Generalization occurs when an animal responds to a stimulus which is similar to a conditional stimulus.

● Instrumental learning appears to differ from classical conditioning because the animal is required to make a response before receiving reinforcement. However, it is not possible to conduct an instrumental-learning experiment without also setting up the conditions for classical conditioning. It is possible that reinforcement is not really instrumental in the formation of learned responses.

Further reading

Breland, K. and Breland, M. (1961) 'The misbehavior of organisms,' *American Psychologist*, **16,** 661–4.

Mackintosh, N. J. (1983) *Conditioning and Associative Learning*, Clarendon, Oxford.

18 Biological Aspects of Learning

The mechanisms of learning in animals are complex partly because simple forms of learning like habituation coexist with more complicated processes. In people, for example, we can find examples of habituation, conditional responses, instrumental learning, reasoning, and cognition. While other animals may not be so adept at learning by experience and reasoning, many undoubtedly possess these abilities to some degree. However, animal species vary considerably in their ecological circumstances and evolutionary history. Should we not expect natural selection to have tailored each animal's learning abilities to suit its ecological niche? Many biologists are inclined to take this view. Psychologists tend to assume that some learning mechanisms are common to many species. Is this view justified? In this chapter we attempt to answer this question.

18.1 Evolutionary aspects of learning

When an animal learns, its behavioral repertoire is changed forever. Although its learned behavior may be extinguished or forgotten, the animal can never return to its previous state. As we saw in Chapter 17, most apparent unlearning is actually learned inhibition of previous learning. Learning changes the animal's makeup and is, therefore, likely to alter its fitness.

The processes of learning have long been subject to natural selection. Therefore, we can expect the consequences of learning to be adaptive and to increase the animal's fitness. The idea that learning is adaptive has long been recognized by both psychologists (e.g., Thorndike, 1911; Seligman, 1970) and ethologists (e.g., Lorenz, 1965). However, there has been little systematic attempt to look at learning in terms of evolutionary theory.

If the environment never changed, animals would gain no benefit by learning. A set of simple rules designed to produce appropriate behavior in a given situation could be established by natural selection and incorporated as a permanent feature of the animal's psychological makeup. There are some features of the environment that do not change, and we usually find that animals respond to these in a stereotyped, innate manner. For example, gravity is a universal feature of the environment, and the antigravity reflexes tend to be consistent and stereotyped (Mittelstaedt, 1964; Delius and Vollrath, 1973).

Some features of the environment change on a cyclic basis, either diurnal, lunar, or seasonal. Here again, there is no real need for the animal to learn to adapt to such changes. As we saw in Chapter 16, fundamental aspects of the animal's physiological and motivational makeup can change on a periodic basis as a result of acclimatization or as dictated by a biological clock. In general, environmental changes that are predictable from an evolutionary point of view can be handled on the basis of preprogrammed changes in the animal's makeup (McFarland and Houston, 1981).

Change throughout an animal's life history may be evolutionarily predictable also. In some cases, the animal's behavior may change through maturation, without learning being involved. In other cases, the animal's behavior may change as a result of learning, but learning that is preprogrammed, that is, occurring at a particular stage of the animal's life, more or less independently of variation in individual behavior. For example, human children have an immense facility for language learning between the ages of two and seven, learning whatever language they hear around them. Similarly, some birds have distinct sensitive periods of song learning during which they normally learn by hearing their parents or other conspecifics. Such learning is preprogrammed; that is, it occurs during a particular period of the life history, independent of the individual's circumstances. A child in one environment may learn German, another English. A bird in one environment may learn one type of song, in another environment a different song.

In Chapter 20, we see how young animals may become imprinted upon certain features of the environment at particular stages of development, and thereby learn about certain characteristics of their parents, siblings, or habitat. Such imprinting may influence future habitat selection, courtship and social behavior of the individual animal. Why do such animals have to learn the characteristics of their habitat, parents, or future mates, when other species show no such imprinting? This is a difficult question to answer. In the case of habitat selection, imprinting may enable animals to choose habitats similar to those of their kin and thus to enhance their inclusive fitness (see Chapter 6).

Pat Bateson (1979) has suggested that some young birds may need to discriminate between the parent that cares for it and other members of the species, since parents discriminate between their own and other young and may attack young not their own. He also suggests that imprinting enables an animal to learn about certain characteristics of close kin so it can subsequently choose a slightly different mate and thus promote a balance between inbreeding and outbreeding. Whenever the problems facing the juvenile animal differ slightly from generation to generation, there may be an advantage for individuals that are predisposed to learn certain types of things, such as the nature of the habitat, of close relatives, and of food sources. This type of learning is a form of *contingent maturation*; that is, the course of development is contingent upon certain experiences.

Unpredictable environmental changes that occur within an individual animal's lifetime cannot be anticipated by preprogrammed forms of learning. The individual must rely upon its own resources in adapting to such changes. The ability to modify behavior appropriately in the face of unpredictable environmental changes usually is taken as a sign of intelligence. Yet we should not forget that much can be explained in terms of conventional learning mechanisms.

Historically, learning has been investigated in artificial laboratory conditions, without much regard for the animal's special requirements and preoccupations. As we saw at the end of Chapter 17, however, doubts about the efficacy of reinforcement first came from within psychology. A common reaction to the discovery that reinforcement is not always effective in altering behavior in the way the experimenter intended has been to deny the existence of any general laws of learning. Instead, the learning abilities of different species are said to be adapted specifically to the ecological constraints typical of the animal's normal way of life (e.g., Rozin and Kalat, 1972; Shettleworth, 1972). This is obviously an attractive idea for ethologists and evolutionary biologists, but does it mean that we necessarily have to abandon all attempts to lay down general principles of learning? The issue is controversial, and before coming to any conclusion, we should look again at the role of reinforcement in instrumental learning.

18.2 Constraints on Learning

In an instrumental-learning situation, the animal is reinforced for performing some act. The reinforcement must occur in some stimulus situation. Particular features of the stimulus situation will be associated regularly with reinforcement, thus providing the essential conditions for classical conditioning. The reinforcer used in an instrumental-learning experiment is also a potential classical UCS that will automatically elicit an innate set of responses. In chickens, these responses include scratching for food, and in pigs, rooting—responses that will become classically conditioned to the stimuli that regularly accompany food (Breland and Breland, 1961) (Fig. 18.1).

What, then, of the arbitrary responses selected for reinforcement by Skinner and his followers? Bruce Moore (1973) has pointed out that in almost every case the responses are not really arbitrary but part of the repertoire of instinctive behavior normally associated with the reward. Thus, the swiping movement employed by the pigeon in Skinner's bowling alley, mentioned previously, is a normal part of pigeon feeding

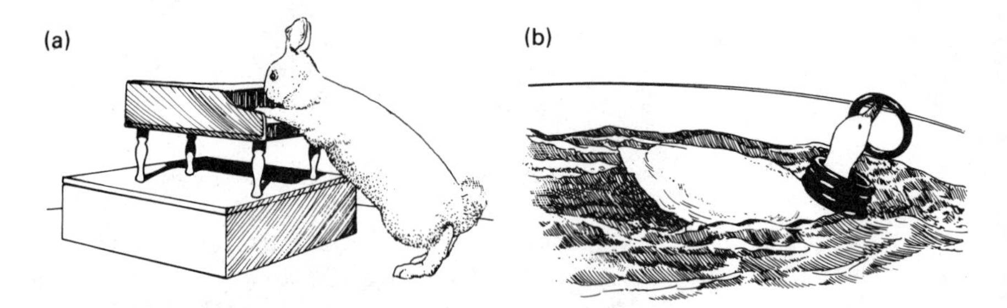

Fig. 18.1 Examples of tricks taught to animals by Keller Breland, using operant conditioning methods. (*a*) The rabbit plays up and down the keyboard several times to obtain a reward. (*b*) The duck must collect a number of rings before obtaining a reward

behavior—flicking loose earth aside to uncover seeds. If this interpretation is correct, then we might expect that pigeons pecking a key for food reward would do so differently than when working for water reward. Examination of a film of pigeons pecking at keys shows that this is in fact the case (see Fig. 18.2). When students were asked to distinguish between filmed sessions, of food and water tests, without knowing which reinforcer was being used, they were able to judge with 87 per cent accuracy (Moore, 1973). Birds pecking for food strike the key with an open beak, making sharp, vigorous pecks. When pecking for water, the bill remains closed and contact with the key is more sustained. Water pecking is often accompanied by the pumping (swallowing) movements typical of pigeon drinking behavior (unlike other birds, pigeons drink by sucking up the water).

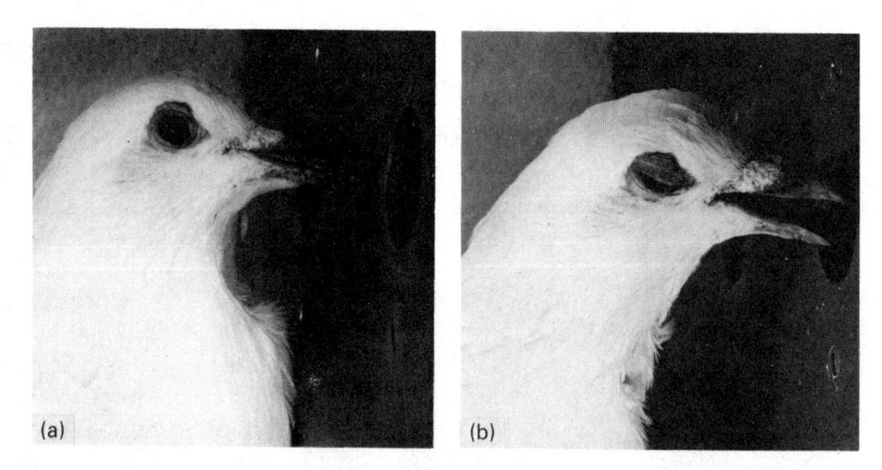

Fig. 18.2 Pecking for food and water rewards by pigeons; (*a*) shows pecking for water and (*b*) shows pecking for food in an auto-shaping experiment (*Photograph: Courtesy of Bruce Moore*)

Further evidence comes from the phenomenon of *autoshaping.* If pigeons are exposed to repeated pairings of key illumination and food presentation, they begin to peck the key without any of the previously described type of training or shaping (Brown and Jenkins, 1968). The key-pecking behavior typical of innumerable operant-conditioning experiments is established not as a result of instrumental reinforcement but by a straightforward Pavlovian procedure: The lighted key is paired with grain, and the pigeons peck the key as if it were grain. The point can be demonstrated even more clearly if the key illumination is paired with a reinforcement to which pecking is *not* the pigeons' natural response, for instance, by offering the pigeons a sexual reward. Paired male and female pigeons are housed in adjacent chambers separated by a sliding door. Once a day, the stimulus light is turned on and the sliding door removed so the male can begin courtship display. Within 5 to 10 trials, the males begin to make conditioned responses toward the stimulus light. The pigeons direct their courtship toward the light, behaving as if it were a female (Moore, 1973).

To prevent the pigeon from obtaining any instrumental reinforcement during the autoshaping experiment, the food reward can be withheld whenever the pigeon pecks the key. Such is the strength of Pavlovian conditioning that autoshaped pigeons persist in pecking under these conditions (Williams and Williams, 1969). These and other autoshaping procedures have established beyond doubt that complex behavior patterns such as feeding, drinking, and courtship can be established by means of classical conditioning. They also suggest that much of the so-called operant behavior established by instrumental-conditioning techniques, such as shaping, is in fact the result of classical conditioning. This does not mean, however, that the law of effect necessarily has to be abandoned.

18.3 Learning to avoid enemies

Avoidance reactions are a form of defensive behavior by means of which animals minimize their exposure to situations that appear to be dangerous. Fear-provoking stimuli may be sign stimuli (see Chapter 20) responded to without any prior experience, or they may be stimuli to which a fear or avoidance response has become attached through the process of conditioning. Natural stimuli that induce innate avoidance behavior include those associated with predators, such as the hawklike silhouettes avoided by young birds, and with poisonous plants and animals, such as snakes and fungi. Innate avoidance behavior varies considerably from species to species. It includes freezing posture, especially in animals that normally are well camouflaged; running-for-cover and thigmotaxic behavior, which involves contact with objects and avoidance of open areas; and warning signals, either to deter attack or to

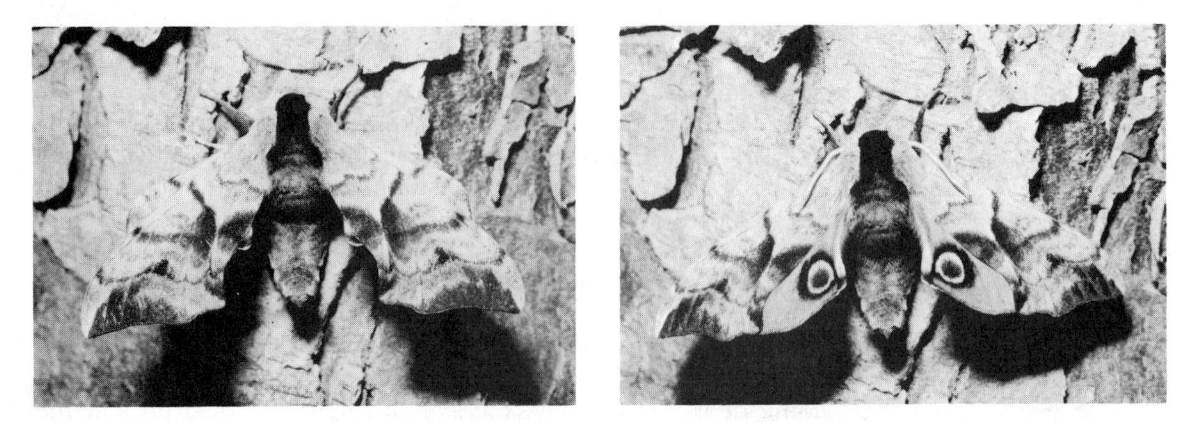

Fig. 18.3 Warning display of a hawk-moth disturbed by a potential predator. The sudden appearance of the eye spots often scares away small birds

Fig. 18.4 An example of a warning device

warn other animals. Figures 18.3 and 18.4 provide some examples.

In its natural habitat, an animal that learned how to escape from predators by trial and error probably would survive for a few trials only. Animals tend to have innate defense reactions that are essentially reflex in nature though modifiable by learning. Robert Bolles (1970) pointed out that the experimenter in an avoidance-learning experiment selects—either deliberately or by virtue of the design of the apparatus—the avoidance response that will be effective, which may or may not coincide with the animal's innate response. The degree of compatibility between the innate and to-be-learned behavior greatly influences the ease with which the avoidance response is learned.

For a number of years, psychologists had reported considerable variation in the facility with which animals learn avoidance responses (Mackintosh, 1974). Thus, rats learn within about five trials to run from one box to another to avoid shock, but require hundreds of trials to learn to press a lever to avoid shock. Pigeons have great difficulty learning to peck a key to avoid shock but less difficulty learning to depress a treadle. Another complicating factor is that innate avoidance responses may vary with the nature of the situation. Thus, in response to shock, rats tend to run when escape is possible and to freeze when escape is not possible. If the situation includes a target such as another member of the species, then shock is likely to induce aggressive behavior (Logan and Boice, 1969). If the situation requires the animal to approach the stimulus associated with shock, then the training of avoidance responses is very difficult. Thus, if a rat is required to press a lever directly below the light that serves as a warning signal for shock, avoidance learning is difficult to establish. If the warning light is located far away from the lever, then conditioning is much easier (Biederman et al., 1964). Exposure to shock in an open-field apparatus increases the amount of time rats spend in close contact with the wall (Grossen and

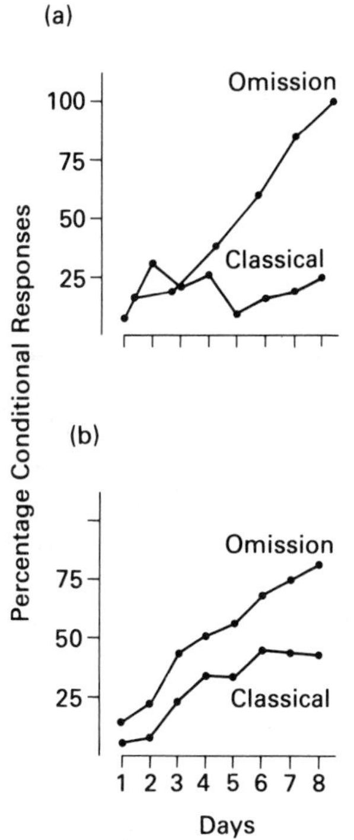

Fig. 18.5 Comparisons of classical and omission schedules of reinforcement; (*a*) wheel running in guinea pigs, (*b*) leg flexion conditioning in dogs ((*a*) After Brogden et al, 1938 and (*b*) After Wahlsten and Cole, 1972)

Kelley, 1972). Rats learn to jump onto a ledge to avoid shock more rapidly when the ledge is next to the wall than when it is in the center of the apparatus.

In terms of conventional learning theory, innate defense reactions are UCRs. A Pavlovian analysis of avoidance learning involves two cardinal propositions (Mackintosh, 1974). First, successful avoidance responses are closely related to the responses normally elicited by the aversive stimulus. As we have seen, avoidance learning is much more rapid if the response required by the experimenter is compatible with the animal's natural reaction to the punishing stimulus. We will shortly see this principle illustrated in another context.

The second proposition of the Pavlovian view of avoidance learning is that establishing an avoidance response should not directly depend upon the avoidance contingency as such. It is the presentation of a negative reinforcer that strengthens the response elicited by that reinforcer. Omission of reinforcement, whether or not a result of avoidance behavior, should not improve the classicial conditioning process. At this point, however, it is worth noting that omission of reinforcement sometimes has no effect and sometimes profound effect on avoidance learning (Mackintosh, 1974). For example, goldfish swim from one compartment to another when this results in the omission of shock but they will swim just as readily when shocks are unavoidable (Woodward and Bitterman, 1973). Conversely, omission of shock is an effective reinforcer to induce guinea pigs to run in a wheel (Brogden et al., 1938) or rats to run from one compartment of a box to another (Miller, 1948). This procedure is superior to the classical-conditioning procedure (see Fig. 18.5).

Mackintosh (1974) points out that the distinction between classical and instrumental analysis of avoidance learning may hinge on the question of whether or not external stimuli are established as a signal for shock or for safety and whether or not the responses to these stimuli provide the animal with adequate feedback. In cases where the animal avoids shock by jumping out of the box or by running away from a noxious stimulus, the avoidance response itself is distinguishable from the animal's other behavior, and it also transports the animal into a different external stimulus situation. However, for a continually swimming goldfish, swimming to a particular place (to avoid shock) is not a very distinctive response. For a rat that is motionless, on the other hand, running to a place where shocks are not received is a distinctive response with obvious consequences. It should not be surprising, therefore, that the rat finds this type of avoidance response easier to learn than pressing a bar.

Rats that can press a lever to avoid shock and gain access to a compartment where shock is never delivered learn to press more quickly than rats that can press the lever but can never gain access to the shock-free chamber (Masterson, 1970). Pressing a lever leaves the animal in much the same external stimulus situation as before, whereas escape to another compartment removes the animal from the stimulus associated with shock. Avoidance learning is much more effective in a one-way shuttle box than in a two-way shuttle box, presumably because the two-way box lacks a place associated with safety.

18.4 Learning to avoid sickness

We have seen that interoceptive cues can act as Pavlovian conditional stimuli (Chapter 17) and that learning plays an important role in homeostasis (Chapter 15), especially in the regulation of food intake. Evidence for the importance of learning was available at an early stage of research into specific hungers. Classic studies by Harris et al. (1933) and Richter et al. (1937) clearly demonstrated that thiamine-deficient rats developed a preference for foods containing thiamine. Scott and Verney (1947) offered distinctively flavored thiamine-supplemented food or an unflavored thiamine-deficient food to thiamine-deficient rats. After a preference had developed for the flavored food, the flavor was switched to the other food, but the rats continued to prefer the flavored food, even though it was not deficient in thiamine, suggesting there was no specific recognition of thiamine.

The main problem in recognizing that learning was involved in the development of specific hungers is the long delay between the ingestive behavior and its reinforcing consequences. In the case of sodium appetite, the sodium-deficient rat can learn which food contains salt because it can taste and recognize salt and so obtain rapid reinforcement for selecting that particular food. In the case of a vitamin deficiency, however, the animal cannot directly detect the vitamin's presence in the food. It can learn about the food's properties only after some of it has been digested and there has been an appropriate alteration in the animal's physiological state. Although it was suggested (e.g., Scott and Verney, 1947) that the animals learned to associate the stimulus characteristics of the food with the subsequent beneficial feelings of recovery from sickness, this view was not acceptable to most psychologists at the time, for two main reasons. First, a generalized feeling of well-being is rather vague, with overtones of hedonism (Young, 1961). Second, the delay between the ingestive behavior and its consequences is very much greater than learning theorists thought possible. Attempts were made to bypass this second problem by postulating some mediation process like prolonged aftertaste from feeding. However, such possibilities were disproved.

John Garcia and his co-workers (1955) fed rats a harmless substance and then produced a physiological after-effect by an independent means. After gamma radiation the rats developed an aversion to previously preferred saccharin-flavored water. Garcia et al. (1961)

Fig. 18.6 A rat feeding in its natural environment

pointed out the importance of this type of effect for the psychology of animal learning and cited nine confirmatory studies. It was another 10 years, however, before the full implications of the phenomenon were appreciated and before researchers realized that specific hungers and poison avoidance were different aspects of the same phenomenon.

For centuries, people have attempted to eradicate local rat populations by various means, including poisoning. A wide variety of poisons have been tried (Chitty, 1954), with limited success. A perennial problem has been the rat's proverbial bait shyness (Rzoska, 1953). As we saw in Chapter 15, rats tend to avoid unfamiliar foods and to sample them tentatively. A rat that takes a small amount of poisoned food and survives will never touch the same type of food again (Fig. 18.6). The following account of the laboratory equivalent of this occurrence is given by Sam Revusky and John Garcia (1970):

> An animal is made to consume a flavored substance, such as saccharin solution, and is later subjected to toxic after-effects produced by such independent means as injection of poison or X-irradiation. After it has recovered from the toxicosis the animal will avoid consuming the flavored substance. The animal behaves as though it thinks that consumption of the substance made it sick. This specific aversion will not develop if toxicosis occurs in the absence of previous consumption, or if consumption occurs without being followed by toxicosis. It differs from the types of learning usually investigated in that it can occur after a single pairing even when the interval between ingestion and toxicosis is a number of hours.

Little is known about the nature of the internal cues by which animals assess their relative sickness or well-being. Michel Treisman (1977) has suggested that the nausea and vomiting typical of motion sickness in humans, which arise from abnormal visual-vestibular stimulation (Howard, 1982), occur because one of the first indications an animal has that it has been poisoned is a disturbance in the vestibular system. The vomiting, therefore, serves to expel any poison from the system, and the ensuing depression serves to keep the animal quiet while it recovers. Thus, Treisman suggests, the vestibular system—which is particularly sensitive and susceptible to penetration by small molecules—has evolved a secondary function as a poison detector. This hypothesis has yet to be confirmed.

Sickness can be regarded as a state function, that is, an index of the animal's overall physiological state. It can be argued that it must be the rate of change of this function that the animal associates with the feeding stimuli (McFarland, 1973). This line of argument leads to the following conclusion:

> It is impossible for an animal both to learn to prefer flavoured substances on the basis of improved physiological state, and to avoid them on the opposite basis. The two processes will simply cancel each other out. For example, an animal that learns to avoid a flavour because of subsequent sickness must also learn to

approach the same flavour when it recovers from the sickness (McFarland, 1973).

Paul Rozin and James Kalat (1971) note that much of the evidence that appears to demonstrate positive preferences can be reinterpreted in terms of learned aversions. The problem is that many of the experiments designed to test for preferences (e.g., Garcia et al., 1967; Campbell, 1969; Zahorik and Maier, 1969) do not give the rats a fair choice. Zahorik and Maier gave a choice among a taste associated with recovery from dietary deficiency, a taste associated with deficiency, and a novel taste. However, they did not offer the rats food with a familiar safe taste. Thus, it is possible that the apparent preference for the food associated with recovery was simply a contrast to the rats aversive behavior toward the other foods (on account of the novelty of one and the association of the other with dietary deficiency).

A number of studies have attempted to demonstrate a clear positive preference (e.g., Revusky, 1967; Zahorik et al., 1974), but the question remains controversial (McFarland, 1973; Rozin, 1976a). The situation is complicated because rats have a natural aversion to novel foods, although this is modified in deficient animals. For instance, thiamine-deficient rats show an immediate, marked preference for new foods, even when the new food is thiamine deficient and the old food has a thiamine supplement (the preference reverses within a few days) (Rodgers and Rozin, 1966). Rozin (1968) found that a rat suffering from effects of poisoning or from a dietary deficiency, when faced with a choice among a familiar-safe food, a familiar-aversive food, and a new food, shows a preference for the familiar-safe food. It seems that the rat learns to avoid the deficient food when it is the only food available (Rozin and Kalat, 1971). The rat tends to become anorexic, refusing to eat. As a result of this learned aversion, the rat shows an immediate preference for a novel food, thus enabling it to learn about its consequences.

Rozin (1976) has suggested that rats categorize food into four classes: novel, familiar-safe, familiar-dangerous, and familiar-beneficial. The last category remains a subject of controversy. In any event, usually it is agreed that rats "are strongly biased toward learning effectively and rapidly what makes them sick, and rather poor at learning what makes them well" (Rozin, 1976). We now turn to the question of how this learning occurs.

18.5 Stimulus relevance

How does an animal learn to associate eating a particular food with the delayed physiological consequences? During the interval between eating a poisoned food and suffering toxic effects, numerous events with which the toxicosis could be associated are likely to occur, including other meals. The problem can be met in part by appealing to the principle of *stimulus relevance* (Capretta, 1961). According to this principle, the associated strength of a cue with some consequences (reinforcer) depends partly upon the nature of the consequences. Similar notions are those of *preparedness* (Seligman, 1970) and *belongingness* (Garcia and Koelling, 1966). Revusky and Garcia (1970) take the view that:

> The relevance principle responsible for association of delayed physiological consequences with flavors is that flavor has a high associative strength relative to physiological consequence, while an exteroceptive stimulus has low associative strength (at least in the mammal). If the consequence is an event which normally emanates from the environment, such as shock or receipt of a pellet food, the converse is true.

Various experiments seem to substantiate this view. For example, Garcia and Koelling (1966) labeled food with light, sound, and taste cues simultaneously and paired them with either electric shock or poisoning. Different groups of rats were presented with different stimulus-consequence pairs. They found that the rats would associate the taste of food with sickness but not with electric shock. They would associate the visual and auditory stimuli with shock but not with sickness. Similarly, Garcia et al. (1968) used two sizes of food pellets, which were either uncoated or coated with flour or powdered sugar. The experimental design is illustrated in Table 18.1. Rats ate pellets differing either in size or in flavor. They were punished with electric shock or with irradiation. As predicted, the eating of pellets distinguishable by flavor was suppressed markedly by sickness but not by shock. Conversely, eating pellets that differed in size was suppressed by shock but not by sickness.

Some scientists (e.g., Garcia et al., 1970; Rozin and Kalat, 1971) claim that the rat's ability to associate flavors selectively with sickness over long intervals represents a specialized learning system that does not obey the conventional law of learning. Others (e.g., Mackintosh, 1974; Revusky, 1977), however, maintain that almost every property of conventional laboratory conditioning is also present in the learning of taste aversions.

Birds quickly learn aversions to the sight of food (Brower, 1969). Thus,

Table 18.1 Plan of the experiment by Garcia *et al*, (1968) and the results expected on the basis of stimulus relevance

Variable	*Punishment*	*Expected result*
Size of pellet	Shock	Aversion
Size of pellet	Toxicosis	No aversion
Flavor	Toxicosis	Aversion
Flavor	Shock	No aversion

Japanese quail, in contrast to rats, learn poison-based aversions more rapidly to the color than to the taste of drinking water (Wilcoxon et al., 1971). Similar results have been obtained with chickens (Moore and Capretta, 1968). If pigeons are trained to press a treadle in the presence of a combined visual and auditory stimulus (tone), the relative importance of the visual and auditory cues depends upon the nature of the reinforcement. Pigeons responding for food do not respond to the tone presented alone but do respond to the light alone. Conversely, pigeons responding to avoid electric shock respond in the presence of the tone but not the light (Foree and Lolordo, 1973). Thus, it appears that the cues to which the animal normally attends while feeding are those that are effective in learned aversions to the physiological consequences of feeding. Rats normally attend to olfactory and taste cues, while birds normally attend to the visual characteristics of their food. In responding to electric shock, rats attend to visual stimuli and pigeons to auditory stimuli.

Some evidence indicates that animals perceive electric shock as an external stimulus similar to an attack from an opponent or predator (Ulrich and Azrin, 1962). For a bird, auditory stimuli like alarm calls may be most effective in situations where there is an external threat. Nick Mackintosh (1974) points out that, during their lifetime, rats may learn that changes in visual or auditory stimulation are uncorrelated with changes in their internal state, whereas changes in taste stimuli do predict such changes. The tendency of rats, as adults, to associate flavor rather than visual or auditory stimuli with poisoning may result from the rats' ability to learn about, and generalize from, the correlations to which they are exposed during their lifetime.

The long delay that can occur between stimulus and reinforcement in taste-aversion learning poses a more serious problem for conventional learning theory. In the conventional learning experiment, the reinforcer starts to lose its effectiveness after only a few seconds of delay. An association between flavor and sickness can occur even with an interval of several hours between the two (Revusky and Garcia, 1970; Rozin and Kalat, 1971). It has been suggested that the flavor may persist for a long time after ingestion of the food, but a number of studies rule out this possibility. For example, rats can associate the temperature of their drinking water with subsequent delayed sickness, a factor that could not possibly persist for more than a few seconds (Nachman, 1970). Moreover, as we have seen, birds are able to associate the color of water with subsequent sickness. Paul Rozin (1969) found that rats could learn aversions to particular concentrations of substances, and it is difficult to see how different concentrations could remain discriminably different half an hour after ingestion. Several studies show that exposure to a second flavor during the interval between ingestion of a novel test substance and subsequent induced sickness does not abolish the association between the test substance and sickness. If the second flavor is similar, it will not become associated with sickness and will not interfere with the aversion to the test substance (Revusky, 1971).

Normally, when a rat is poisoned after exposure to a novel flavor, few, if any events occur during the interval between ingestion and sickness that would have sufficient stimulus relevance to be associated with the sickness. However, this is not the case with normal exteroceptive conditioning. Sam Revusky (1971) argues that the possible delay of reinforcement is very short in such cases because a longer interval would permit interference from other stimuli, which could become established as signals for the UCS. Long delays are possible because only flavors are associated readily with sickness (in rats), whereas a variety of stimuli can be associated with electric shock. When a novel flavor (high stimulus relevance) is introduced into the interval between ingestion of a test substance and subsequent sickness, then interference does occur (Revusky, 1971; Kalat and Rozin, 1971). Revusky's suggestion goes some way toward reconciling the phenomena of taste-aversion learning with conventional learning theory, but difficulties remain. For example, birds tend to associate the visual characteristics of food with subsequent sickness, and there must be numerous visual stimuli that could interfere with the formation of such associations.

18.6 The biological and psychological perspectives

Psychologists often are accused of thinking that all animals are alike and of placing too much reliance on experiments with laboratory rats and pigeons. Many ethologists and some psychologists (e.g., Rozin and Kalat, 1972) argue that the learning capacities of a species are tailored to its particular niche. This view might suggest that there is no general learning process common to many species and that the learning capacities of a given species are an amalgam of specific learning processes. For example, we have seen that the processes involved in learning to avoid noxious food seem different in some respects from the learning processes traditionally studied in laboratories.

There are some types of learning, such as the song learning of birds and imprinting of juvenile animals (see Chapter 20), that are obviously different from the associative learning discussed in Chapter 17. The animal may be evolutionarily preprogrammed to learn certain things at a certain age.

Other aspects of learning, such as learning to avoid noxious foods, have also been seen as learning that is biologically tailored to the animal's way of life. This view is open to dispute, because almost every properly conventional laboratory conditioning can be demonstrated in food-aversion learning (Revusky, 1977). Knowledge of which foods cause illness and which are nutritious, which route is dangerous and which is safe, requires some integration of the predictive relationships among events in the environment. Such knowledge can be gained either through genetic programming or by learning.

Some psychologists working on animal learning have concluded that all types of learning have basic similarities since they involve the common problem of learning about causal relationships. Learning enables animals to associate cause with effect, and so to predict significant events. Predictive relationships are based upon events in a causal chain that has universal features. Thus, events do not occur without a cause and do not occur before a cause. We take these relationships for granted, and it is reasonable to suppose that animals do the same. The best predictors of events are the causes of the events, and animals that can detect and learn about such events will be well equipped to respond to an important and universal feature of the natural world.

On the basis of such considerations, some psychologists (e.g., Dickinson, 1980) believe in "the existence of a basic associative learning mechanism, which is common to a variety of species and designed to detect and store information about causal relationships in the animal's environment." This does not mean that there are not also ways of learning unique to particular species or that some animals may not have a range of learning capacities. It simply means that most animals have one aspect of learning in common. However, this approach differs from previous psychological approaches to learning because it has a biological perspective. As we see in the next chapter, it is possible to ask what features of causal relationships are likely to be important to animals. This is a question about design, an essentially biological way of thinking. An interesting, and perhaps surprising, result of this approach is that it leads to an essentially cognitive view of simple associative learning.

Points to remember

- Learning does not necessarily confer an evolutionary advantage. Some learning is evolutionarily preprogrammed as part of normal development, some occurs only in certain circumstances, and only some is innovative.

- Learning can be constrained by the nature of the relationship between the type of reinforcer and the type of response to be learned. Such constraints can be seen in learning to avoid enemies and in learning to avoid sickness.

- Constraints on learning also occur as a result of stimulus relevance. In other words, the associative strength of a cue which has a particular consequence depends partly upon the nature of the consequence.

- The biological view of learning is that the constraints are primarily innate, while the psychological view is that the same laws of learning apply to all animals. In some cases these two views provide different explanations of the same phenomena.

Further reading

Bolles, R. C. (1970) 'Species-specific defense reactions and avoidance learning, *Psychological Review*, **77**, 32–48.
Shettleworth, S. J. (1972) 'Constraints on learning,' *Advances in the Study of Behavior*, **4**, 1–68.

19 Cognitive Aspects of Learning

Most present-day psychologists recognize that there is a spectrum of learning ability ranging from the simple learning of primitive animals to the cognitive abilities of humans. The problem is to assess the extent to which these abilities occur in particular animal species. Cognition involves learning and thinking processes not directly observable, but for which there is often indirect evidence. We start by considering some aspects of learning that cannot easily be explained in terms of conditioning.

19.1 Hidden aspects of conditioning

Pavlovian conditioning traditionally has been viewed as the most mechanistic and least cognitive aspect of learning. Indeed, as we have seen, a major objective of the behaviorist school of psychology was to account for the behavior of animals without reference to unobserved processes, whether physiological or cognitive. Thus, classical conditioning has often been used to explain apparently cognitive aspects of complex behavior (Rescorla, 1978), but the reverse perspective has not found favor with behaviorists. It has been evident for a long time, however, that some aspects of classical conditioning do not conform easily to the behaviorist view.

When a CS is present together with a UCS, we observe an increase in conditional responses to the CS. How can we know that these responses are a consequence of the experimentally arranged relationship between the CS and the UCS? It is possible that the CS may elicit certain reflexes that are enhanced by exposure to the UCS. In some cases these reflexes may resemble the CR that is to be conditioned. In human eyelid conditioning, for example, a visual CS may elicit reflex eyelid closure. In conditioning emotional reactions, any novel stimulus may elicit an emotional reflex, like the galvanic skin response (a change in electrical conductivity of the skin due to the action of the sweat glands). Such potentiation of the CR by the CS is called *sensitization* (Gormezano, 1966). For example, the common octopus (*Octopus vulgaris*) can be trained in a laboratory aquarium to emerge from its home to attack a crab or a neutral stimulus associated with food (see Fig. 19.1). The probability

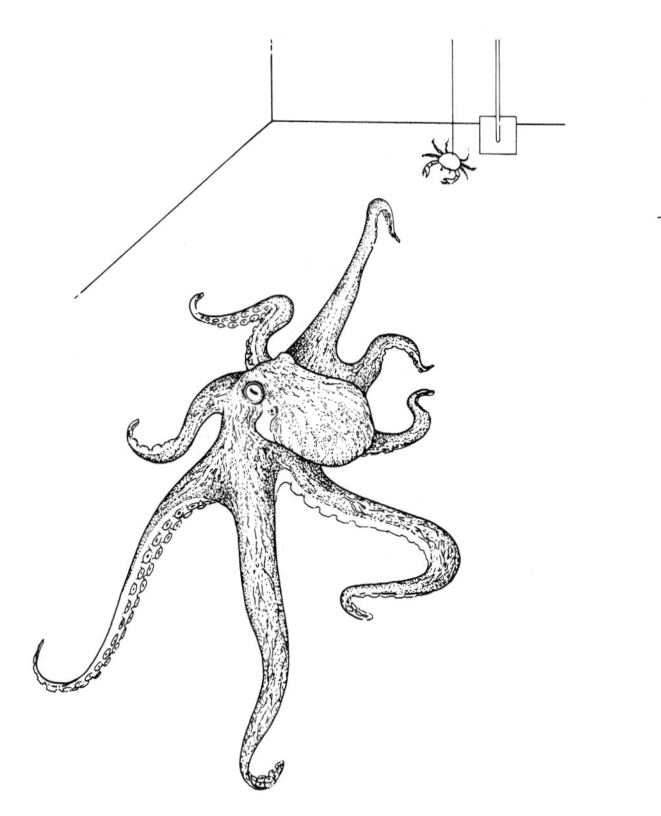

Fig. 19.1 Octopus learning to associate two neutral stimuli with food

that the octopus will attack a neutral stimulus like a white disk suspended on a rod is increased if the octopus has received food recently and decreased if it has recently received a mild electric shock. The effect occurs even if the octopus is fed or shocked in its home, so it cannot be due to positive or negative reinforcement of attack behavior.

It is possible that the UCR may come to be elicited by stimuli other than the UCS even though there is no contingent relationship between them. This is usually called *pseudoconditioning* (Grether, 1938). A possible explanation is that there is generalization to stimuli similar to the UCS. It can also happen that exposure to electric shock alters the animal's internal state so the animal comes to avoid any external stimulus. This has been shown to occur in polychaete worms (Evans, 1966) and octopus (Young, 1960) and appears to be an adaptive form of behavior (Wells, 1968).

Experimental psychologists normally take precautions against sensitization and pseudoconditioning (Mackintosh, 1974), but other aspects of conditioning are not easily observed and may be difficult to account for in behaviorist terms. These include interoceptive and temporal conditioning.

Secretions of internal organs can be conditioned readily to external stimuli. They also can be conditioned to internal stimuli such as changes

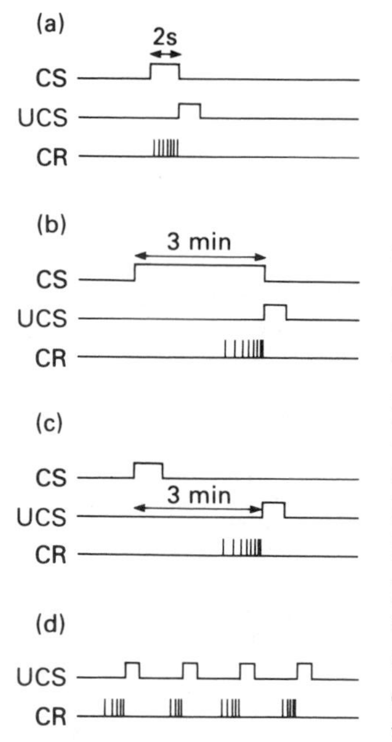

Fig. 19.2 Schematic representation of Pavlovian conditioning; (*a*) usual procedure with the UCS occurring immediately after the CS; (*b*) delayed conditioning; (*c*) trace conditioning; (*d*) temporal conditioning

in body temperature, blood sugar and the carbon dioxide content of inhaled air (Bykov, 1957). Such *interoceptive conditioning* may be no different in principle than ordinary classical conditioning, even though it does not always involve observable behavior.

In simple *temporal-conditioning* experiments, there is no CS. The UCS is presented at fixed intervals of time, and the CR develops as a correspondingly periodic response (see Fig. 19.2). Such experiments were conducted in Pavlov's laboratory, and there have been many variations (Church, 1978; Richelle and Lejeune, 1980). The most natural explanation of temporal conditioning is that the sense of time, based upon some internal clock, can act as a stimulus just like any external stimulus. However, time lacks the physical attributes normally associated with a stimulus, and entities like the internal clock are not directly observable. For these reasons the simple explanation has not found favor with the behaviorists, and they have made many attempts to account for temporal conditioning in terms of mediating behavior. However, it is now accepted that animals do have internal clocks and are capable of learning on the basis of temporal cues. There is good evidence that the internal clock advances at a fixed rate and that the animal can read the clock when necessary. There is also evidence that animals can exercise some control over the internal clock and can make it stop or run (Roberts, 1981).

In the traditional description of Pavlovian conditioning, a UCS that normally evokes a UCR is arranged to follow a CS that initially fails to evoke the response. After repeated pairings of UCS and CS, the CS comes to evoke the response, which is now called the CR. This description implies the incorporation of a new stimulus into an existing reflex system. An alternative view, anticipated to some extent by the American psychologist Edward Tolman (1932), is that the animal learns about the relationships among events outside its control and that it then generates appropriate behavior (Rescorla, 1978). Rather than emphasizing the animal's responses to stimuli, the cognitive approach asks whether or not the animal's behavior changes as a result of exposing it to particular relationships among events. Whether or not the events initially evoke responses is largely irrelevant.

The cognitive approach emphasizes the distinction between learning and performance. Animals may form associations among events without altering their behavior at the time. On the one hand, the reflex tradition places Pavlovian conditioning in a compartment of its own, as a particular type of learning. The cognitive approach, on the other hand, sees Pavlovian conditioning as an example of associative learning. Animals can associate many types of stimuli—exteroceptive, interoceptive, those arising from an internal clock, and those emanating as feedback from the animal's own behavior. In some cases the animal may choose to reflect its learning directly in its behavior, and in some cases it may not.

19.2 The nature of cognitive processes

Cognition refers to the mental processes that cannot be observed directly in animals but for which there is, nevertheless, scientific evidence. Suppose we allow hungry pigeons to observe food presentations accompanied by the illumination of a small electric light. During the observation period, the pigeons are not allowed to approach the light or the food. Other (control) pigeons observe light and food presentations unrelated in time. At the end of the initial observation period, the pigeons are allowed to approach the light and food stimuli. All the pigeons tend to peck at the food delivery mechanism. However, the experimental pigeons also peck at the light, while the control pigeons do not. This shows that the experimental pigeons must have formed a mental association between the food and the light, even though their behavior during the initial phase of the experiment was the same as that of the control pigeons.

Of course, many aspects of cognition exist apart from those involved in associative learning. Some of these seem to be more in the nature of innate skills rather than forms of learning. For example, pigeons are capable of feats of navigation and time perception that are beyond the ability of human beings (see Chapter 14.4). Pigeons also seem to be extraordinarily good at the formation of natural concepts. They can be trained to discriminate between photographs showing water versus non-water, tree versus non-tree, or human versus non-human (Herrnstein et al., 1976; Malott and Siddall, 1972; Siegel and Honig, 1970). They can make these discriminations even though the relevant cue is presented in a variety of ways. For example, the pigeon can recognize water in the form of droplets, a turbulent river, or a placid lake. Humans can be picked out whether they are clothed or naked, alone or in a crowd, etc.

Although the pigeon can form a concept of water, trees, or people, the process need not require much in the way of mental abstraction. A more revealing type of test is the "matching" experiment. The animal first is shown a single object, the sample, and then is shown an array of objects that includes the sample. The correct response is to choose the sample object from the array. Primates, including children, learn to solve this type of problem within a few trials, but pigeons require hundreds of trials. If the original objects are replaced by a new array of objects, pigeons usually have to learn the task all over again, but dolphins and chimpanzees master the second task within a few trials (Premack, 1978). It seems that pigeons categorize objects in the external world, while primates, and some other animals, are capable of relational concepts. Indeed, it is probable that chimpanzees use same/different categorization in their normal lives.

As we see in Chapter 26, there is some evidence that chimpanzees are able to form a cognitive map, or model, of parts of their external environment. Free-ranging animals may take roundabout routes that suggest they have a mental picture of the spatial layout of their environment (Menzel, 1978). The social relationships of chimpanzees are particularly complex (Passingham, 1982), and it has been suggested that their cognitive abilities have evolved in relation to the essentially political nature of their social life. Evidence for mental abstraction in chimpanzees comes partly from studies of their capacities for language and partly from observations of social behavior.

For example, a chimpanzee named Sarah was trained to construct sentences by selecting from an array of colored plastic tokens (Premack, 1976). On one occasion, when asked to provide a token to complete the sentence "—— is the color of chocolate," Sarah selected the token meaning brown. The tokens for chocolate and for brown were not brown in color, so it seems that Sarah may have possessed an abstract notion of the color brown. The token meaning brown and that signifying chocolate are connected by a purely symbolic relationship, which Sarah appeared to associate with the token for the color brown.

To take another example, when a subordinate chimpanzee sees food near a dominant member of the group, it usually does not approach the food until the dominant animal has departed. If the dominant animal has not noticed the food, however, the subordinate animal may try to obtain it by stealth. The subordinate animal appears to realize that the dominant animal is an obstacle to obtaining the food only if the dominant animal knows that the subordinate wants it.

Although these examples seem persuasive, one must beware of accepting cognitive interpretations of behavior at face value. While it is not necessary to adopt the extreme skepticism of the behaviorists, a certain amount of caution is advisable. In the case of Sarah, for example, we would want to know whether she had ever been in a position to form a simple association between the tokens for chocolate and for the color brown. Were the instructions properly understood by Sarah or did she simply choose the color token she most associated with chocolate?

19.3 Insight learning

The idea that animal learning involves cognitive processes has a long history. It gained clear expression in the work of the *Gestalt* school of psychology, which believed that animals gained insight into problems through an innate tendency to perceive the situation as a whole.

A classic series of experiments was carried out by Wolfgang Köhler between 1913 and 1917. Throughout World War I, Köhler was interned on the island of Tenerife. He devoted his energies to studying the chimpanzee at the Anthropoid Station there and reported his work in *The Mentality of Apes*, published in 1925. Köhler's experiments required chimpanzees to use tools to obtain food rewards. For example, in one

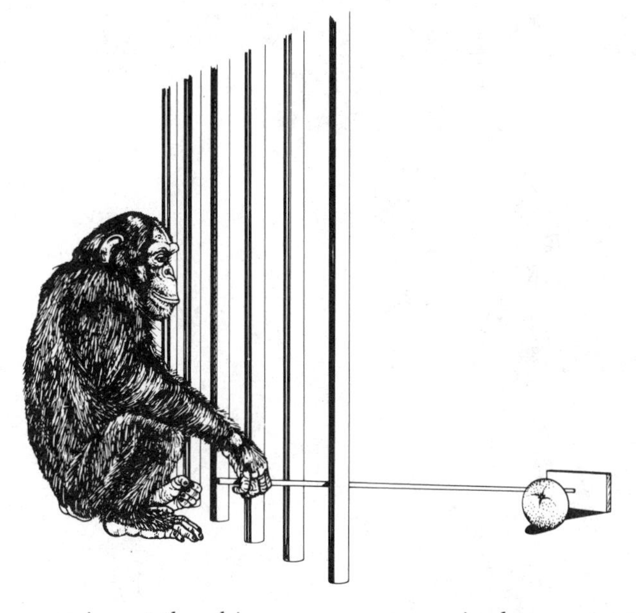

Fig. 19.3 A chimpanzee using a stick to obtain an apple

experiment the chimpanzee was required to use a stick to rake in food from outside its cage, as shown in Figure 19.3. Köhler claimed that his problems differed from those set by Thorndike (see Chapter 17) in an important respect. He noted that Thorndike's animals could not develop an understanding of the latch mechanism that opened their cage door because it was on the outside of the cage and hidden from view. They had little alternative but to solve the problem by trial and error. In Köhler's experiments, all the ingredients necessary for the solution of the problem are visible.

One of Köhler's chimpanzees was given two bamboo poles, neither of which was long enough to reach the fruit placed outside the cage. However, the poles could be fitted together to make a longer pole. After many unsuccessful attempts to reach the fruit with one of the short poles, the chimpanzee gave up, started playing with the poles, and accidentally joined them together by pushing the narrower pole inside the hollow end of the other. The chimpanzee then jumped up and immediately ran to the bars of the cage to retrieve the fruit with the long pole. Köhler interpreted this as an example of insightful behavior. In another experiment, fruit was suspended high up in the roof of the chimpanzee cage. It could be reached by stacking boxes on top of each other, as shown in Figure 19.4. Some chimpanzees learned to solve this problem.

In accounting for the results of his experiments, Köhler claimed that his animals exhibited *insight*, a term used to denote the apprehension of relationships among stimuli or events. Insight learning differs from trial-and-error learning because it involves the sudden production of a new response. The philosopher Bertrand Russell was amused by the

Fig. 19.4 One of Köhler's chimps standing on stacked boxes to obtain food suspended from above (From Köhler, 1925)

difference between the views of Thorndike and Köhler. He noted:

> All the animals that have been carefully observed . . . have all displayed the national characteristics of the observer. Animals studied by Americans rush about frantically, with an incredible display of hustle and pep, and at last achieve the desired result by chance. Animals observed by Germans sit still and think, and at last evolve the solution out of their inner consciousness.

The Gestalt interpretation of animal problem-solving has been criticized on various grounds. The experiments are devised to determine whether or not animals behave with insight under conditions that are supposed to require such behavior. There is no independent evidence that the task requires insight, but if the animal succeeds in the task, it is said to have demonstrated insight. This line of reasoning is not universally accepted. Another difficulty is that it is very hard to know whether the insightful response is genuinely new. The chimpanzees engage in a considerable amount of irrelevant behavior, play, and abortive attempts to obtain the reward. Is it possible that they arrive at the solution to the problem through a cumulative process of trial and error?

This question was examined by Paul Schiller (1952), who systematically investigated the innate components of the problem-solving behavior of chimpanzees. In one study, for example, he provided pairs of sticks that would fit together to 48 new chimpanzees without any problem to solve. Of these, 32 fitted the sticks together within an hour, and 19 of the 20 adults in the group fitted them together within five minutes. On the basis of these and other investigations, doubt has been cast on the Gestalt interpretation of problem-solving behavior (e.g., Chance, 1960). It appears that the previous experience of the animal is very important. Familiarity with sticks and boxes makes an enormous difference to the way in which the problems are tackled. Chimpanzees allowed to play with these objects learn about their properties. The ability to manipulate objects in ways that are relevant to problem-solving is largely a matter of maturation. Once the animal discovers it can achieve a particular manipulation, it tends to repeat it over and over again. Some manipulations are simply too difficult for the younger chimps, but once the behavior becomes established in the animals' repertoire, it can be deployed in a variety of contexts.

The main differences between insight learning and other forms of learning appears to lie in the ability of the more intelligent animals to draw on experience gained in other contexts. However, this does not necessarily mean that insight must be ruled out as an aspect of learning. The problem-solving abilities of animals are difficult to investigate because the human investigator may have little idea as to how the animal sees the situation. It is obviously unfair to set a problem that is beyond the animal's manipulative ability or against its natural inclination. On the other hand, each species is well suited by nature to perform

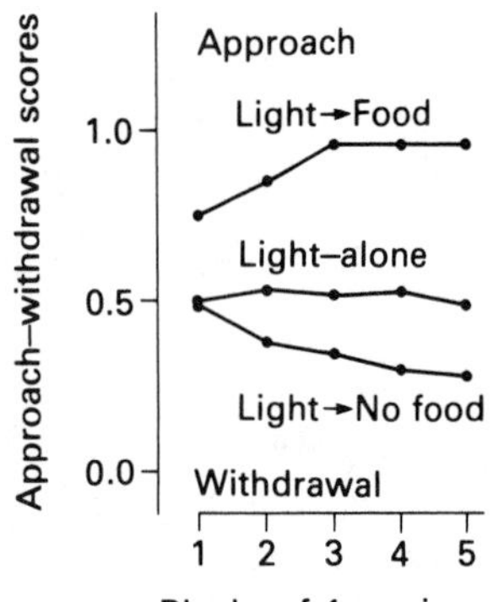

Fig. 19.5 The acquisition of a tendency to either approach or withdraw from a lighted key by pigeons exposed to various correlations between illumination of the key and delivery of food. In the light → food condition illumination of the key was paired with food, and the pigeons developed an approach response. In the light alone condition the pigeons were never presented with food and developed no approach or withdrawal response. In the light → no food condition they were presented with food, but never during, or shortly after, illumination of the key. These pigeons developed a withdrawal response (After Dickinson, 1980)

some apparently clever feats, and we must not be misled into thinking that these are evidence of "insight" or "intelligence." Indeed, these words are really only labels for phenomena that still require explanation.

19.4 Associative learning

To investigate the hypothesis that animals possess mechanisms to detect and learn about causal realtionships, we must specify the nature of such relationships. There are basically two types of causal relationships, and there is little doubt that animals can learn about both (Dickinson, 1980). One event (the cause) can cause another event to happen (the effect) or not to happen (non-effect). The first event need not be an immediate cause of the effect or non-effect, but it may be an identifiable link in a chain of cause and effect. Indeed, the event noticed by the animal may not be part of the causal chain but merely a sign that the causal event has occurred. It is the apparent cause that is important to the animal.

That animals can learn about both types of causal relationships can be demonstrated by a simple experiment. Hungry pigeons can be placed in a Skinner box fitted with two illuminated keys and a food delivery mechanism. One group of pigeons is exposed to a light-food (cause-effect) relationship, and another group is exposed to a light–no-food (cause–non-effect) relationship. A third group is exposed to the light alone. In the first case, one of the disks is illuminated for 10 seconds, at irregular intervals, and food is presented as soon as the light goes off. In the second case, the light and the food are presented the same number of times as in the first case, but care is taken to ensure that food is never presented soon after the light. In the third case, no food is presented. In the first case the light signals the presentation of food, and in the second case it signals the non-presentation of food. Either of the lights can be illuminated on a particular trial, and a record can be kept of whether the bird tends to approach or withdraw from the light. The results of this experiment (see Fig. 19.5) show clearly that the pigeons exposed to the light-food relationship tend to approach the light, as might be expected on the basis of normal classical conditioning. However, the pigeons exposed to the light–no-food relationship were not indifferent to the light but definitely avoided it (Wasserman et al., 1974). Only pigeons that were exposed to only the light apparently were indifferent to it (Fig. 19.5).

In considering the results of this experiment, we must bear in mind a number of important points. First, we cannot claim that the pigeon learned that the light caused food or non-food. What animals learn when exposed to causal relationships can be discussed only in terms of theories about the internal changes that occur as a result of the learning experience (see following discussion). Second, the fact that the pigeons approached the light in one case and withdrew from it in another is

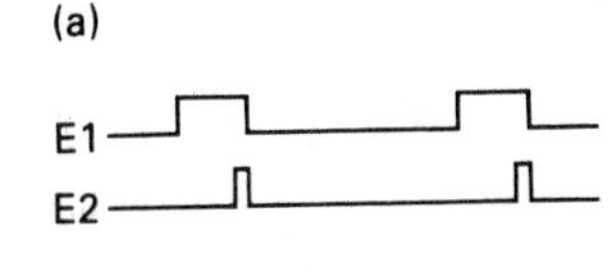

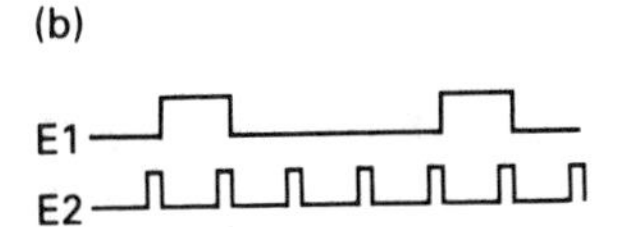

Fig. 19.6 The pattern of events when there are different correlations between E1 and E2. In (*a*) E2 only occurs during or shortly after E1, and the two events are positively correlated. In (*b*) E2 is just as likely to occur when E1 is absent as when it is present, and the two events are uncorrelated. (After Dickinson, 1980)

interesting but not directly relevant to the matter at hand. Many differences in behavior caused by the presentations could be used as an indication that learning had taken place. The particular behavior shown may become important when we come to consider what the animals learned. Third, it makes sense to refer to the light as the apparent cause of the non-presentation of food only if the food would have been presented had the light not occurred. We cannot expect the pigeon to learn that the light is a cause of the non-presentation of food if the food had never been presented in the situation, as was the case for the third test group. An animal has the opportunity to learn about a cause–non-effect association only if there is some reason to expect the effect to occur (Dickinson, 1980).

Although we can sometimes conclude that a change in behavior shows that learning has occurred, we cannot assume that nothing is learned in the absence of behavioral change. Anthony Dickinson (1980) calls this the problem of *behavioral silence*. In the experiment described previously, for example, we cannot assume that the pigeons in the light-alone condition learned nothing. Indeed, there is evidence that rats can learn to ignore stimuli that predict no change in the consequences of their behavior (Mackintosh, 1973; Baker and Mackintosh, 1977). If rats are first exposed to a schedule in which two stimuli occur independently, the rate at which they subsequently learn to associate the two stimuli is retarded compared with animals that had no prior exposure to the stimuli. The implication is that animals can learn that particular stimuli are irrelevant and that the learned irrelevance interferes with subsequent learning about the stimuli.

Other forms of behaviorally silent learning also can occur. Animals can learn that two events are unrelated, either in the sense that an effect is unrelated to a particular cause or that an effect is unrelated to a whole class of causal events. If the class of causal events is the animal's behavioral repertoire, this form of learning is called *learned helplessness* (Maier and Seligman, 1976), meaning that animals can learn that there is nothing they can do to improve a situation. Such learned helplessness retards future learning in that situation.

For an animal to learn about a simple causal relationship, there must be an overall positive correlation between the two events (Dickinson, 1980). From the animal's viewpoint, there is always a variety of possible causes of an event, apart from that provided by the experimenter. As illustrated in Figure 19.6, there must be a fairly close relationship between two events for one to be seen as the cause of the other. To test the importance of such background, or contextual, cues, Nick Mackintosh (1976) trained rats to press a lever for food and then introduced a variety of stimuli. One group was given a light stimulus on each trial, while the other groups were given a compound stimulus consisting of light plus noise. The noise was soft (50 decibels) for some groups and loud (85 decibels) for others. On all trials the animals were given a mild

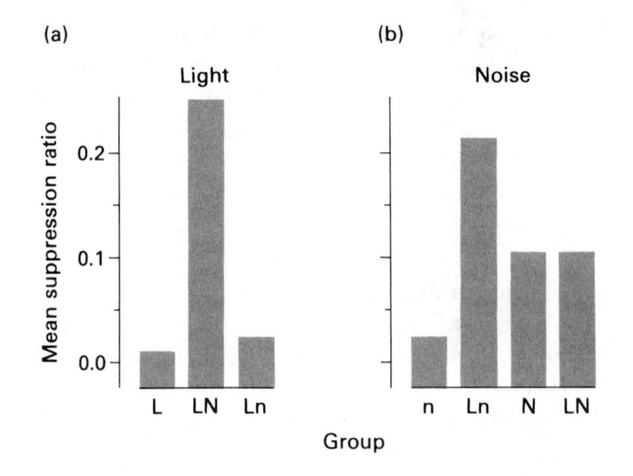

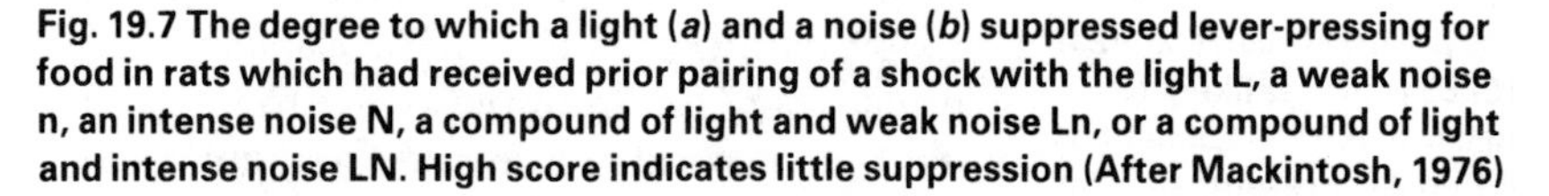

Fig. 19.7 The degree to which a light (*a*) and a noise (*b*) suppressed lever-pressing for food in rats which had received prior pairing of a shock with the light L, a weak noise n, an intense noise N, a compound of light and weak noise Ln, or a compound of light and intense noise LN. High score indicates little suppression (After Mackintosh, 1976)

electric shock just after the presentation of each stimulus. At the end of the experiment, the light was presented alone to all groups to see how much the animals had learned about the light-shock relationship.

The results (Fig. 19.7) show that presentation of the light suppressed the bar-pressing behavior to different extents in different groups. The rats given light alone or light plus soft noise showed considerable suppression, but those that had experienced the light in the presence of a loud noise showed much less suppression. Thus, the presence of a powerful second stimulus decreased the amount by which the light was associated with the shock, even though there was perfect correlation between light and shock. This phenomenon is called *overshadowing*. The degree of overshadowing depends upon the relative salience of the overshadowed and overshadowing stimuli. This is why the soft noise had little overshadowing effect.

Animals will learn to associate two events only if they are initially accompanied by an unexpected or surprising occurrence (Mackintosh, 1974). In a normal conditioning experiment, the surprise is provided by the reinforcer. Thus, if a stimulus is paired with shock, and if neither the stimulus nor the contextual cues initially predict its occurrence, the shock will be surprising. Suppose, however, that the animal already has experienced shock in the presence of stimulus A, then if both A and B are correlated with shock, the presence of A will block learning about B. This phenomenon, first discovered by Kamin (1969), is known as *blocking*. An experiment by Rescorla (1971) demonstrates this effect and also shows that the more surprising the reinforcer, the more the animal learns (see Fig. 19.8).

In the natural environment, certain types of causes are more likely to

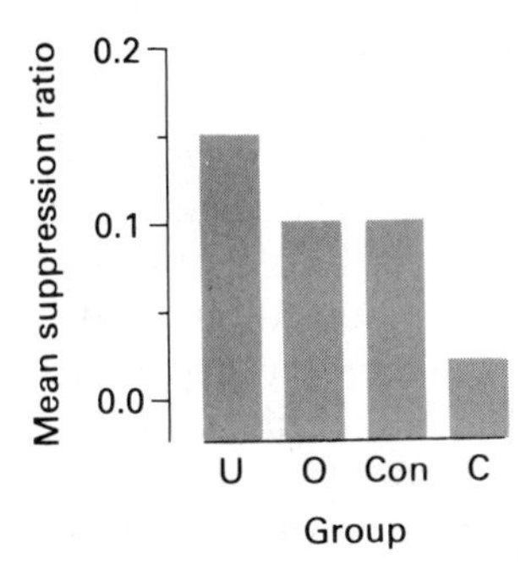

Fig. 19.8 The degree to which stimulus A suppressed lever-pressing for food in rats previously exposed to pairings of an AB compound with shock. The uncorrelated group U had received prior pairings of B with shock. The simple overshadowing group O was pre-exposed to the shock alone. The control group Con had experienced random presentations of B with shock. The correlated group C had experienced a negative correlation of B with shock. High score indicates little suppression (After Rescoria, 1971)

produce certain effects than other types of causes. For example, if a cat jumps into an apple tree at the same time a dog barks and then an apple falls to the ground, we are more likely to think that the cat rather than the dog's bark caused the apple to fall. Both the cat's jumping and the dog's barking bear the same temporal relationship to the fall of the apple, but other aspects of these events lead us to assume that the cat was the cause of the apple's falling. Similarly, we can show that animals are more likely to form associations among certain types of stimuli than others. For example, rats readily associate taste with subsequent illness but do not easily learn to associate a tone or light with illness (Domjan and Wilson, 1972; see also Chapter 18). Rats have been shown to associate two events more quickly when they are in the same sensory modality (Rescorla and Furrow, 1977) and when they are in the same spatial location (Testa, 1975; Rescorla and Cunningham, 1979).

Finally, for an animal to form an association between two events, it is usually necessary for the events to occur close together in time. The temporal relationships between two events have been the subject of numerous experiments in animal learning (see Dickinson, 1980), and it appears that learning is most effective when the onset of one event (the cause) occurs shortly before the onset of the other (the effect). However, some evidence supports the view that this relationship is not due to a direct effect of the time interval on the learning process but that changes in the time interval increase the extent to which the first event is overshadowed by contextual cues (Dickinson, 1980), implying that the extent to which the time interval influences learning should depend upon the relevance of the contextual cues. In particular, we would expect contextual cues to be much less important in the case of taste–illness learning than in the case of tone–shock learning, because the background stimuli provided by the experimental situation are not normally relevant to taste-aversion learning.

Sam Revusky (1971) was the first to suggest that lack of overshadowing might be responsible for taste-aversion learning that occurs in spite of the very long time intervals between food ingestion and subsequent sickness. He showed that the effective time interval could be reduced by interposing a background of relevant (taste) stimuli.

In summary, we have seen that animals can learn to associate two events if the relationship between them conforms to what we normally call a causal relationship. Thus, animals can learn that one event (the cause) predicts another event (the effect) or that one event predicts that another event (the non-effect) will not occur. They also can learn that certain stimuli predict no consequences in a given situation or that a class of stimuli (including the animal's own behavior) is causally irrelevant. The conditions under which these types of associative learning occur are those we would expect on the hypothesis that animals are designed to acquire knowledge about the causal relationships in their environment. Thus, the animal must be able to distinguish potential causes from contextual cues, and for this to occur there must be some surprising occurrence that draws the animal's attention to particular events, or the events must be (innately) relevant to particular consequences. If these conditions are not fulfilled, contextual cues may overshadow potential causal events, or learning may be blocked by prior association with a now irrelevant cue. Thus, the conditions under which associative learning occurs are consistent with our commonsense views about the nature of causality. They are not consistent with the traditional view that animal learning is an automaton-like association of stimulus and response. The phenomenon of behaviorally silent learning suggests that some kind of cognitive interpretation of animal learning is required. However, this does not mean that we can jump to conclusions about the cognitive abilities of animals, or about the nature of the animal mind.

19.5 Representations

Representation is an issue that usually is considered to be central to the question of animal mentality. Do animals have internal representations, or mental images, of objects they are searching for or of complex spatial or social situations (Kummer, 1982)? This issue has attracted considerable attention from philosophers (e.g., Dennett, 1978) and has aroused interest in various branches of the behavioral sciences.

We have seen that animals can learn to associate two events if the relationship between them conforms to what we normally call a causal relationship. Some of the conditions under which associative learning occurs are not consistent with the traditional view that animal learning is an automaton-like association of stimulus and response. They are more consistent with the view that animals are designed to acquire knowledge about the causal relationships in their environment.

In considering the nature of the internal representation that encodes the learning experience, Anthony Dickinson (1980) makes a distinction between declarative and procedural representations. A *declarative representation* is a mental image of the object or goal desired. If a rat uses declarative representation to find food in a familiar maze, it has an image of food in mind and knows that it must choose, say, the left-hand turn to find it. A *procedural representation* is a set of instructions leading automatically to the object desired without envisioning it. Thus if a rat uses a procedural representation to find food in a similiar maze, it takes that left-hand turn not because it "knows" there is food there, but because it associates the left turn with the attainment of food (Fig. 19.9).

In a declarative system, knowledge is represented in a form that corresponds to a statement or proposition describing a relationship among events in the animal's world, a form of representation that does not commit the animal to use the information in any particular way. In a procedural system, however, the form of representation directly reflects

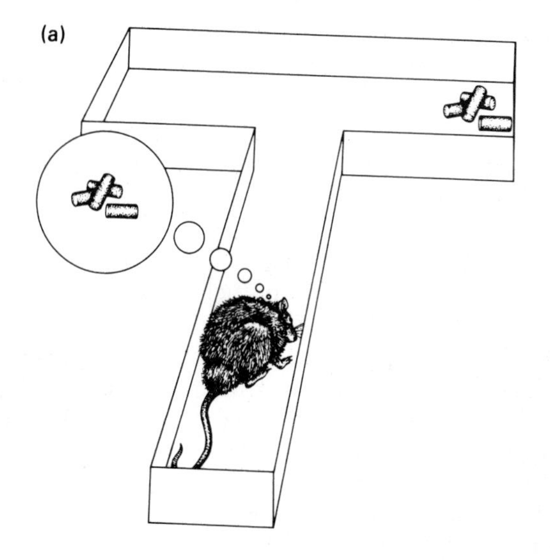

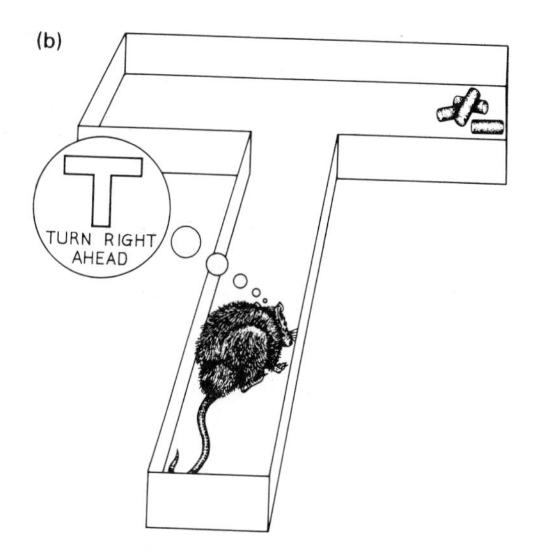

Fig. 19.9 A simple example of declarative (left) and procedural (right) representation. In one case the rat has a mental image of the goal, in the other it follows a simple rule or procedure

the use to which the knowledge will be put. For example, P. C. Holland (1977) exposed rats to a tone–food relationship by occasionally presenting an eight-second tone and then delivering food pellets into a receptacle. Holland noticed that the rats developed a tendency to approach the food receptacle during the tone. This observation might suggest that during learning, the procedure of approaching the food receptacle is established, so that the learned information is stored in a form related closely to its use. The alternative possibility is that the rat learns that the tone causes the food, and thus establishes a declarative representation. That the rat tends to approach the food receptacle during the tone then would have to be accounted for in some other way, because a declarative representation is passive, that is, it does not control the animal's behavior. The procedural representation thus provides a less complex explanation of the rat's behavior.

Suppose, however, that after the tone–food association is established, the rats are exposed to a food–illness relationship to the point where they refuse to eat food when it is presented. The rats now will have formed two separate associations, tone–food and food–illness. The question is whether they are capable of integrating the two. On the one hand, a procedural account of learning would imply that the rats should not be able to integrate the two procedures, which have no factors in common. A declarative system, on the other hand, provides a basis for integration because both representations have the "food" term in common. Holland and Straub (1979) showed that rats can integrate information from such associations learned at different times. Rats exposed to a food–illness following tone–food showed a disinclination to approach the food receptacle when the tone was presented again.

There is little doubt that animals can integrate separately formed associations, explained most easily in terms of a declarative system. However, under certain circumstances there are failures of integration that point to the possibility of an underlying procedural representation (Dickinson, 1980).

In a declarative system, there must be some way of translating the stored representation into overt behavior. Various mechanisms have been suggested (Dickinson, 1980), but these need not concern us here. The important point is that, although a procedural theory offers a less complex explanation of simple learning situations, a more complex theory may be necessary to account for the observed phenomena, and this may require some form of declarative representation. Once we accept that declarative representation is a necessary ingredient of our explanation of behavior, then we have to admit to some form of animal mentality. However, we have to be careful to distinguish between evidence for declarative representation and the facility that this concept affords scientists attempting to explain behavior. It may be that the notion of declarative representation is merely a convenient crutch upon which to support current learning theory.

Points to remember

- Some aspects of conditioning are not apparent to an observer, and this suggests that cognitive processes may be involved.

- Animals that appear to suddenly arrive at the solution to a problem are sometimes said to have shown insight. However, it is not always clear exactly how this differs from ordinary learning.

- Some aspects of associative learning are said to require cognitive explanation, because it seems that they must involve a mental image of the goal to be achieved. The alternative possibility is that the animals are simply following complex procedural rules.

Further reading

Dickinson, A. (1980) *Contemporary Animal Learning Theory*, Cambridge University Press, Cambridge.

PART 3 *Understanding Complex Behavior*

In the third part of this book we look at complex behavior patterns. To understand these we need to think in terms of both the design and mechanisms of behavior.

The combination of design and mechanism has traditionally been the province of ethology, as distinct from the related scientific disciplines of evolutionary biology and psychology. Evolutionary biologists (including sociobiologists) look at behavior in terms of its past evolution and do not address problems of present-day causation. Psychologists are primarily concerned with the proximate causes of behavior and rarely make use of rigorous argument based upon the theory of natural selection. Ethologists, on the other hand, have sought to combine the mechanistic and evolutionary approaches to behavior, asking not only how behavior is controlled but how the mechanisms evolved and why particular mechanisms appear in particular circumstances.

In this third part of the book we look first at some areas of classical ethology in which this dual approach has been employed. We then look at more recent attempts to combine design and mechanism. Finally, we consider the difficult areas of animal language and cognition, in which it seems that the dual approach is likely to prove useful in the future.

3.1 Instinct

In this section we discuss issues from classical ethology which involve consideration of both design and mechanism in behavior. Chapter 20 deals with the problem of instinct, starting from early ideas and progressing to interactions with learning as exemplified by imprinting. Chapter 21 reviews displacement activities, which still pose problems of causal explanation but which have provided fertile ground for consideration of evolutionary aspects of communication. Chapter 22 discusses animal communication in both evolutionary and motivational terms. It also provides a basis for subsequent discussion of animal language.

Konrad Lorenz (born 1903) and Niko Tinbergen (born 1907)

Konrad Lorenz and Niko Tinbergen are generally regarded as the founders of modern ethology. Although their approach was anticipated by the work of Charles Whitman (1842–1910) and Wallace Craig (1876–1954) in the USA and Oskar Heinroth (1871–1945) in Germany, their work provided the basis for the future development of ethology and their approach offered an alternative to the then dominant American behaviorism.

Konrad Zacharius Lorenz was born in Austria. He studied medicine in Vienna and also studied comparative anatomy, philosophy, and psychology. He became demonstrator and then lecturer in comparative anatomy and animal psychology. At the same time he studied animal behavior at his family home in Altenberg. In 1940 he was appointed Professor of Philosophy at the University of Königsberg, but in 1943 he was drafted into the army medical service. In 1944 he was taken prisoner of war by the Russians. He was released in 1948, became attached to the University of Münster, and then moved to Seewiesen with the founding of the Max Planck Institute for Behavioural Physiology, where he remained until his retirement in 1973.

Nikolaus Tinbergen was born in The Hague, the Netherlands, and studied biology at the University of Leiden. In 1930 he went on an expedition to Greenland, and in 1938 he visited Lorenz at Altenberg. During World War II he was interned in a hostage camp in the Netherlands, afterwards to become Professor of Zoology at the University of Leiden. In 1949 he was invited to become a Lecturer in Zoology at the University of Oxford, where he founded the Animal Behaviour Research Group. He retired in 1974.

In 1973 Konrad Lorenz and Niko Tinbergen, together with Karl von Frisch, were awarded the Nobel Prize for Medicine. Both Lorenz and Tinbergen emphasized the importance of straightforward observation of animal behavior under natural conditions. Lorenz's approach was somewhat more philosophical and his numerous theories became quite influential. Tinbergen was a gifted field biologist who carried out many elegant experiments in the natural environment. A significant feature of the work of Lorenz and Tinbergen was their attempt to combine evolutionary, or functional, explanations of behavior with causal, or mechanistic, explanations. For example, in his 1963 paper *On aims and methods in ethology*, Tinbergen posed four questions he thought necessary to a full account of any aspect of animal behavior. The ethologist, he believed, should aim to answer the questions of the causation, development, survival value, and evolution of any behavior pattern under study.

Konrad Lorenz
Photograph: Hermann Kacher

Niko Tinbergen
Photograph: B. Tschanz

This aim is probably the most notable characteristic of ethology. While evolutionary biologists seek functional accounts of behavior, and psychologists seek explanation in terms of proximate causes or mechanisms, ethologists have followed the example of Lorenz and Tinbergen in retaining interest in, and enthusiasm for, all four approaches to animal behavior.

20 Instinct and Learning

The history of the concept of *instinct* is bound up inextricably with notions of voluntary behavior and of our responsibility for our actions. Plato and most of the ancient Greek, philosophers regarded human behavior as the result of rational and voluntary processes, with individuals being free to choose whatever course of action their reason dictates. This view, called *rationalism,* persists to this day.

In the thirteenth century, the philosopher Thomas Aquinas wrote:

> Man has sensuous desire, and rational desire or will. He is not absolutely determined in his desires and actions by sense impressions as is the brute, but possesses a faculty of self-determination, whereby he is able to act or not to act. . . . The will is determined by what intelligence conceives to be the good, by a rational purpose. This, however, is not compulsion: compulsion exists where a being is inevitably determined by an external cause. Man is free because he is rational, because he is not driven into action by an external cause without his consent, and because he can choose between the means of realising the good or the purpose which his reason conceives.

Today we recognize that an individual neither should be held responsible for his behavior when acting under compulsion nor should be rewarded or punished under such circumstances. Thomas Aquinas clearly regarded animal behavior as being determined by sensuous desire, though he appeared to recognize some elementary process of judgment in animals: "Others act from some kind of choice, such as irrational animals, for the sheep flies from the wolf by a kind of judgment whereby it considers it to be hurtful to itself; such a judgment is not a free one but implanted by nature".

A few of the ancient Greek philosophers—notably, Democritus—dissented from the common rationalist view and held that events in the mental world are caused in a similar way to events in the physical world. This view is called *materialism.* Such explanations of behavior had little influence up to the time of René Descartes. In his *Passions of the Soul* written in 1649, Descartes held that animals were mechanical automatons, while human behavior was under the dual influence of a mechanical body and a rational mind. A pinnacle of materialism was reached by Thomas Hobbes in 1651, for whom the explanation of all things was to be found in their physical motions. For Hobbes, the will is simply an idea that man has about himself. In accounting for mental events in material terms, in seeking mechanistic explanations of purposive behavior, and in regarding the will as an epiphenomenon, Hobbes anticipated much modern scientific thought. However, such thoroughgoing materialism was not to become accepted for many decades.

The *associationists*, like the materialists, denied any freedom of the will, but they did not necessarily attempt to account for behavior in physical or physiological terms. Both John Locke in 1700 and David Hume in 1739 took the view that human behavior develops entirely through experience, according to laws of association, a view very influential in the early days of psychology.

In this chapter we discuss the relationship between instinct and learning. You will find that the dichotomy is not nearly so wide today as it was in the past. To understand how this shift in opinion has taken place, it is helpful to start with an historical sketch.

20.1 The concept of instinct

Early writers regarded instinct as the natural origin of the biologically important motives. Thus, Thomas Aquinas believed that animal judgment is not free but implanted by nature. Descartes regarded instinct as the source of the forces that govern behavior, being designed by God in such a way as to make the behavior adaptable. The associationists appeared to reject all notions of instinct, although Locke did write of "an uneasiness of the mind for want of some absent good . . . God has put into man the uneasiness of hunger and thirst, and other natural desires . . . to move and determine their wills for the preservation of themselves and the continuation of the species."

While the associationists believed that human behavior is maintained by the knowledge of and desire for the consequences of behavior, others like Hutcheson in 1728 argued that instinct produces action prior to any thought of the consequences. Whereas instinct previously had been regarded as the source of motivational forces, Hutcheson made instinct the force itself. This concept of instinct was seized upon by the new rationalists (e.g., Reid in 1785, Hamilton in 1858, and James in 1890) as a convenient vehicle for the non-rational elements of behavior. Thus, human nature was seen as a combination of blind instinct and rational thought.

The idea of instinct as a prime mover was taken up by psychologists such as Freud (1915) and McDougall (1908). Sigmund Freud developed a motivational theory of neurosis and psychosis that emphasized the irrational forces in human nature. He saw behavior as the outcome of two basic energies: a life force underlying life-maintaining and life-continuing human activities and a death force underlying aggressive and destructive human activities. Freud thought of the life and death forces as instincts whose energy required expression or discharge.

McDougall thought of the instincts as irrational and compelling sources of conduct that oriented the organism toward its goals. He postulated a number of instincts, most of which had a corresponding emotion—for example, flight and the emotion of fear, repulsion and the emotion of disgust, curiosity and the emotion of wonder, and pugnacity and the emotion of anger.

These various conceptions of instinct were derived from the subjective human emotional experiences. This essentially unscientific practice involves difficulties of interpretation, of agreement among different psychologists, and of determining the number of instincts that should be allowed or recognized. Darwin (1859) was the first to propose an objective definition of instinct in terms of animal behavior. He treated instincts as complex reflexes made up of units compatible with the mechanisms of inheritance, and, thus, a product of natural selection that had evolved together with the other aspects of the animal's life. Darwin's concept of instinct is then similar to that of Descartes, with evolution replacing the role of God.

Darwin laid the foundations of the classical ethological view propounded by Lorenz and Tinbergen. Lorenz (1937) maintained that much of animal behavior was made up of a number of fixed-action patterns characteristic of the species and largely genetically determined. He subsequently postulated that each *fixed-action pattern*, or instinct, was motivated by *action-specific energy* (Lorenz, 1950). This was likened to a liquid in a reservoir; each instinct corresponded to a separate reservoir, and when an appropriate releasing stimulus was presented, the liquid was discharged in the form of an instinctive drive that gave rise to the appropriate behavior. Tinbergen (1951) proposed that the reservoirs, or instinct centers, were arranged in a hierarchy so that the energy responsible for one type of activity, like reproduction, would drive a number of subordinate activities, such as nest-building, courtship, and parental behavior. Lorenz and Tinbergen provided numerous examples of what they regarded as instinctive behavior patterns.

The classical ethological concept of instinct is currently regarded as unsatisfactory, for two main reasons. The first reason is connected with the suggestion that instinctive forces or drives energize certain aspects of behavior. For reasons that are discussed fully in Chapter 15.5, motivation is seen no longer in terms of drives, super-reflexes, or instinctive urges. The second reason concerns the implication that certain aspects of behavior are innate, that is, developing independent of environmental influences. As we saw in Chapter 3, genetic influences upon behavior are no longer thought of as being independent of environmental influences. The term *innate* has come to be used for activities characteristic of the species, bearing in mind that early environmental influences are also characteristic of the species in that the circumstances in which the different members of the species are born and raised are often very similar.

Fig. 20.1 A robin attacking a dummy. The red breast of the dummy is a sign stimulus (Drawn from a photograph in Sparks, 1982)

The primitive idea of instinctive behavior was that detailed instructions relating to the performance of the behavior and the stimuli that elicit the behavior are encoded in the genes. The ontogeny of the behavior is fixed in that, within limits, the developmental circumstances make no difference to the form of the behavior. Instinctive behavior thus is characteristic of the species and is made up of fixed-action patterns released by specific sign stimuli. Instinctive behavior is adaptive because natural selection acts upon it as it does on other genetically determined traits. There is also a tendency in the early ethological literature to imply that where behavior is clearly adaptive, it must therefore be instinctive, as opposed to learned behavior, which is not acted on by natural selection.

20.2 The innate releasing mechanism

The early ethologists (e.g., Uexkull, 1934; Lorenz, 1935) thought that animals sometimes respond instinctively to specific, though often complex, stimuli. Such stimuli came to be called *sign stimuli*, an example of which is illustrated in Figure 20.1. A sign stimulus is part of a stimulus configuration, and may be a relatively simple part. For instance, a male three-spined stickleback has a characteristic red belly when in breeding condition. This is a sign stimulus that elicits aggression in other territorial males. As we can see from Figure 20.2, crude models suffice to elicit aggression, provided they have a red underside. In contrast, a freshly killed male stickleback without a red belly is ineffective in provoking attack from other males. Thus, many of the details of the structure and texture of a male stickleback apparently are ignored by other males. The red coloration is much more effective if it is on the underside of the model.

Such configurational relationships are a common feature of sign stimuli. In the case of sticklebacks it does not seem to be an essential feature, however. Tinbergen (1953) describes how he was studying the behavior of territorial male sticklebacks in aquariums placed in a window. Whenever a red post office van passed along the road outside the window, the sticklebacks immediately attempted to attack it as if it were a rival male. Recently the BBC was able to repeat these observations while making a historical documentary film for television (Sparks, 1982).

The selective responses to stimuli suggested to early ethologists that there must be some built-in mechanism by which such sign stimuli were recognized. This supposed mechanism came to be called the *innate releasing mechanism* (*IRM*) (Lorenz, 1950; Tinbergen, 1950). There are three important aspects of this concept. First, the mechanism is envisaged as being innate, that is, both the recognition of the sign stimulus and the resulting response to it are inborn and characteristic of the

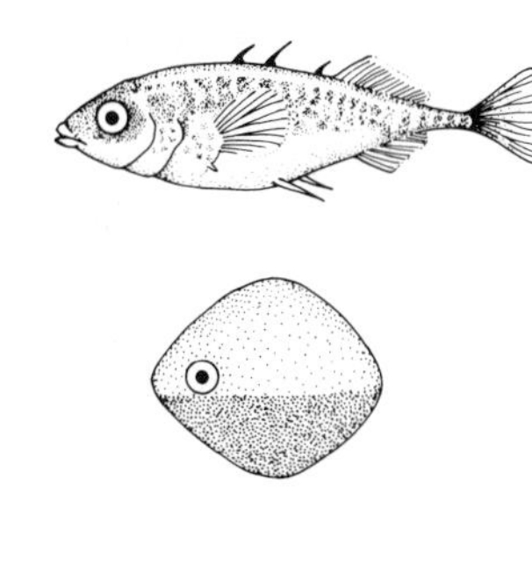

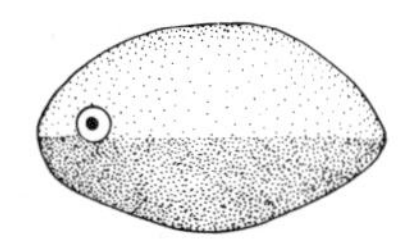

Fig. 20.2 Models of male sticklebacks that elicit attack by territorial males. At the top is a dead male lacking in nuptial colors. The four lower models are very crude, but their red underside provides a sufficient sign stimulus to elicit attack (After Tinbergen, 1951)

species. As we saw in Chapter 3, however, the early ethologists' notion of innateness is somewhat different than that current today. Second, the IRM has the role of releasing the response to the sign stimulus, implying that the IRM holds back the pent-up action-specific energy, or drive, until the appropriate sign stimulus is recognized, upon which the energy is released in the form of appropriate behavior. So central was this aspect of the IRM that sign stimuli often were referred to as "releasers." Third, the response released by the IRM was stereotyped and part of the animal's innate repertoire of fixed-action patterns. Fixed-action patterns, as originally conceived by Lorenz (1932), were activities with a relatively fixed pattern of coordination, somewhat akin to reflexes. They were innate and typical of the species.

Lorenz drew several distinctions between fixed-action patterns and reflexes. First, fixed-action patterns can be released by a variety of stimuli, whereas reflexes are elicited by specific stimuli. Second, whereas animals are motivated to perform fixed-action patterns, this is not true of reflexes. Third, fixed-action patterns can appear in the absence of external stimulation and are then called *vacuum activities*. Some of these points would be disputed now. For example, the startle reflex occurs in response to a variety of stimuli. Many fixed-action patterns now are known to be generated in a predictable manner, without feedback control, which accounts for their stereotyped appearance.

A typical example of the IRM concept is provided by Baerends (1950):

> There is important evidence to support the conception that every releasing mechanism has its own sign stimuli. For example, the digger wasp *Ammophila adriaansei*, that catches caterpillars and drags them to its nest as food for its larva, may respond to the perception of a caterpillar in different ways, all depending on which instinct is activated. When it is hunting, a caterpillar is caught and stung; when it is found near the nest opening, just after the wasp has opened the nest, it is drawn in; but when it lies close to the nest when the wasp is filling the nest entrance, it may be used as filling material. Finally, when we put it into the nest shaft when the wasp is digging out the nest, then it brings the caterpillar away, exactly as she would deal with another obstacle—for instance, a piece of plant root. It is, therefore, the same object that, with different conditions of the animal, releases different responses. Still, in every situation the caterpillar is always sending visual as well as chemical stimuli to the sense organs of the wasp where they will always be transformed into impulses. But then it depends on the instinct activated in the wasp which of these impulses will be intercepted somewhere and which can pass along a still unknown way in the nervous system finally to stimulate the principal motoric centre of the reaction. There are indications that in each case different stimuli are working. When hunting, *Ammophila* very likely become aware of the presence of a caterpillar by its odour, but when it loses the caterpillar during the transport to the nest in the first place optical stimuli are used to find it.

Here we have the elements of many of the features of stimulus filtering discussed in Chapter 12.5. In demonstrating that animals are

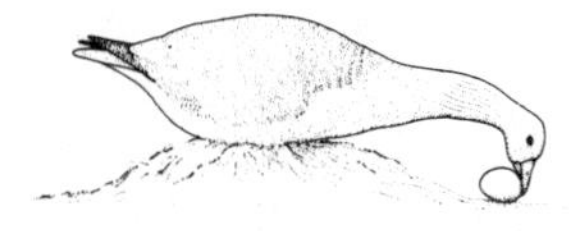

Fig. 20.3 A greylag goose retrieving an egg into its nest (From *The Oxford Companion to Animal Behaviour*, 1981)

selective in their responses to complex stimuli, the ethologists made a major contribution to our understanding of perception in animals. However, their concept of the IRM is open to a number of criticisms. Hinde (1966), for example, pointed out that the term IRM often implies that the mechanism is specific to a particular response when the evidence for its existence is based solely on the study of that response. Yet, it may be that a particular sign stimulus is relevant to more than one aspect of behavior. Second, the selectivity in responsiveness need not be confined to stimuli that release responses. It may also occur with respect to those aspects of the stimulus important in the orientation of the behavior. Third, the process of selectivity need not be purely innate but may be influenced by learning. As we saw in Chapter 12.5, the notion of a sign stimulus is not far removed from that of selective attention, a concept that originated in the study of animal learning.

Some early ethologists were well aware of these problems. Thus, Tinbergen (1951) specifically distinguishes between releasing and directing aspects of the stimulus and illustrates this distinction by reference to the studies of egg retrieval in the greylag goose (see Fig. 20.3), conducted by Lorenz and Tinbergen in the 1930s.

20.3 The discovery of imprinting

One of Lorenz's main contributions to ethology is his work on the development of social relationships, especially the phenomenon of imprinting. In this he was influenced greatly by his observations on jackdaws and by the previous work of Heinroth. Although Heinroth is often given the credit for being the first to use the term *imprinting* (*Prägung*), Spalding had conducted extensive studies on imprinting many years earlier. Between 1872 and 1875, Spalding published six papers reporting his extensive observations of the hatching of domestic chicks and their behavior during the first few days of life, work that anticipated much of the early ethological work on instinct (see Chapter 3) and included the observation that chicks only two and three days old would follow any moving object and develop an attachment to it. Spalding died in 1877 at the age of 37, his work forgotten until discovered and republished by Haldane in 1954. Had Spalding lived longer, he probably would be regarded as the founder of ethology (Thorpe, 1979).

Heinroth wrote his major papers in 1910 and 1911 on the ethology of ducks and geese. He made detailed studies of various species and was a pioneer of the comparative method (see Chapter 5.4). He observed the behavior of goslings hatched in an incubator and then handled by a human prior to being introduced to a goose family. Although the parent geese regarded the goslings as their own, the goslings showed no inclination to regard the geese as their parents. Each gosling ran off, piping, and attached itself to the first human being that came past; it regarded the human being as its parent (Heinroth, 1910). Heinroth explains that to introduce such goslings successfully into a goose family, they must be removed from the incubator and immediately placed in a sack so they do not catch sight of a human being.

Lorenz (1935), in extending Heinroth's observations, argued that imprinting, unlike ordinary learning, took place at a particular stage of development and was irreversible. Lorenz confirmed Heinroth's observations on goslings (see Fig. 20.4) and also studied imprinting in mallard ducklings, pigeons, jackdaws, and many other birds. He confirmed Heinroth's observation that birds imprinted on humans would often direct their subsequent sexual behavior toward them. Thus, Lorenz (1935) notes how a Barbary dove (*Streptopelia risoria*) imprinted on humans would direct its courtship behavior toward his hand and would attempt to copulate if the hand was held in a certain position. Lorenz (1935) emphasizes that the behavior shown as a result of imprinting is innate but that the recognition of the object of imprinting is not innate. He maintains that the young animal becomes imprinted upon whatever moving object is encountered during a particular phase of development and that it subsequently directs its filial, sexual, and social behavior toward that object.

As Heinroth (1910) and Lorenz (1935) observed, the young of many precocial species (see Chapter 3), which can run around soon after birth, show a fairly indiscriminate attachment to moving objects. Newly hatched goslings and ducklings separated from their mother will follow a slowly walking person (Fig. 20.4), a crude model duck, or even a cardboard box. A lamb will follow the person who feeds it on a bottle, even when not hungry. Even when the lamb has been weaned and has joined a flock, it will approach and follow its former keeper. Thus, as a juvenile the lamb follows the person as if it were its parent, and as an adult it retains some attachment to the person, illustrating that imprinting can have both long- and short-term aspects.

Although the "following response" is elicited by a wide range of stimuli, some are more effective than others. Up to a point, the effectiveness of a stimulus increases with its conspicuousness (Bateson, 1964), although if an object is too startling, it elicits fleeing rather than approach. Thus, ducklings approach a human who sways from side to side but flee if the same person moves vigorously. Some species have particular preferences. Domestic chicks, for example, most readily follow blue or orange objects; mallard ducklings prefer yellow-green objects, and their following is enhanced if the object emits appropriate sounds; wood ducks (*Aix sponsa*) nest in holes in trees and the young normally are called out of the nest by the mother from some distance away, yet these ducklings will approach a source of intermittent sound in the absence of any visual stimuli (Gottlieb, 1963).

In general, the more an animal follows and becomes familiar with one

Fig. 20.4 Konrad Lorenz followed by goslings (*Photograph: Dmitri Kasterine; courtesy of Radio Times*)

object, the less it is attracted to others. The "following response" can be enhanced with food rewards, and in nature it is rewarded by contact with the mother and the warmth she provides. Some researchers have claimed (e.g., Hess, 1958; 1959a) that the degree of imprinting is determined partly by the effort the young animal exerts in following the parent object. Hess (1958) arranged for ducklings to follow a moving object in a runway. Some ducklings were forced to clamber over hurdles in order to keep up with the parent, while others could run freely. Hess claimed that the ducklings obliged to negotiate hurdles attained higher imprinting scores than those that had no such obstacles. Similarly, he claimed, ducklings that followed the parent model up an inclined plane imprinted more strongly than those that could run on the flat. On the basis of such results, Hess formulated a law of effort, relating the strength of imprinting to the energy expended in following.

Attempts to replicate the phenomenon have resulted in many failures and many demonstrations of the contrary (see Sluckin, 1964). Indeed, there are considerable differences among species in the tendency to follow the parent model. Gottlieb (1961) reported that Peking ducklings initially follow more vigorously than mallard ducklings but that their degree of imprinting is no greater. As we shall see, many species show types of imprinting that do not involve following at all.

20.4 Sensitive periods in learning

Lorenz (1935) recognized that imprinting was confined to a particular period of development, and he thought that this was due entirely to endogenous factors, similar to those involved in embryological induction. However, we know now that the period during which imprinting can occur is affected considerably by experience. Ducklings and domestic chicks tend to stay close together, even in the absence of a parent. Guiton (1959) found that chicks kept in groups cease to follow moving objects three days after hatching, whereas those reared in isolation retain the "following response" for much longer. He was able to show that socially reared chicks become imprinted upon each other.

If developmental age is measured from the beginning of embryonic development, then the onset of the sensitive period of mallard ducklings is much more marked than if age is measured from hatching (Gottlieb, 1961; 1971), which suggests that post-hatching experience is relatively unimportant and that the onset of the sensitive period is due to maturation. More recent studies, however, indicate that post-hatching experience is also important (Landsberg, 1976). Changes in mobility (Hess, 1959a) and maturation of the visual system and parts of the brain have been suggested (see Bateson, 1979).

Newly hatched birds of many species do not avoid novel objects initially but tend to approach and explore them. After a few days they

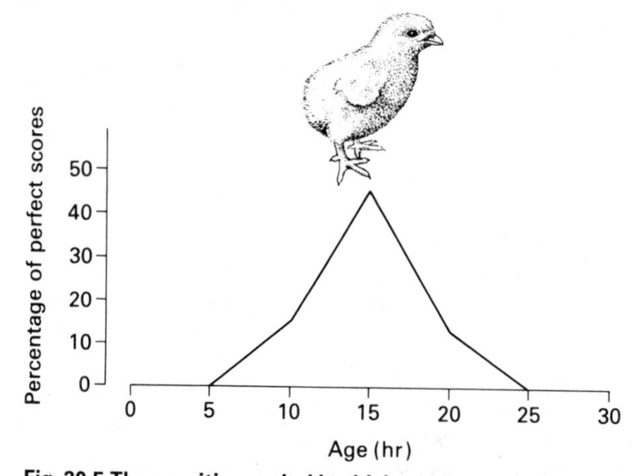

Fig. 20.5 The sensitive period in chicks: scores obtained by chicks of different ages in a laboratory test of the following response (After Hess, 1959a)

become more timid and show signs of fear of unfamiliar objects. The time at which this occurs is influenced by rearing conditions (Bateson, 1966). For the newly hatched bird, nothing is familiar and nothing is strange, but as it becomes familiar with some stimuli, it is able to differentiate others. Bateson (1964) found that chicks avoid moving objects less if the objects have the same color pattern as the walls of the pen in which the chicks were reared. This shows that within a few days of hatching the chicks can learn the characteristics of their immediate environment and discriminate them from novel stimuli. They avoid objects they detect as being unfamiliar. In the natural environment the mother and siblings soon would become familiar, and most other objects would be initially unfamiliar. Hess (1959b) found that the graph of increase in speed of locomotion with age corresponded closely to the graph showing the onset of the sensitive period in chicks (see Fig. 20.5). He also showed that the graph of fear responses with age (measured as the proportion of birds giving distress calls in a standard situation) corresponded with the end of the critical period, as shown in Figure 20.6.

An alternative working model has been suggested by Bateson (1978; 1979), who draws an analogy between the process of development and a train traveling one way from a place called Conception. All windows are closed for the first part of the journey so that the passengers cannot see out. At a later stage of the journey through life, some windows open, exposing the passengers to the outside world. Each compartment of the train, with its occupants, represents a particular behavioral system, sensitive to the environment at different stages of development. Once the windows of a compartment are open, the occupants can learn about the environment. The windows may then close again, or they may remain open. It is also possible that the occupants could change during the journey. Bateson's developmental train is attractive because it represents some of the complexities of the real situation. The external

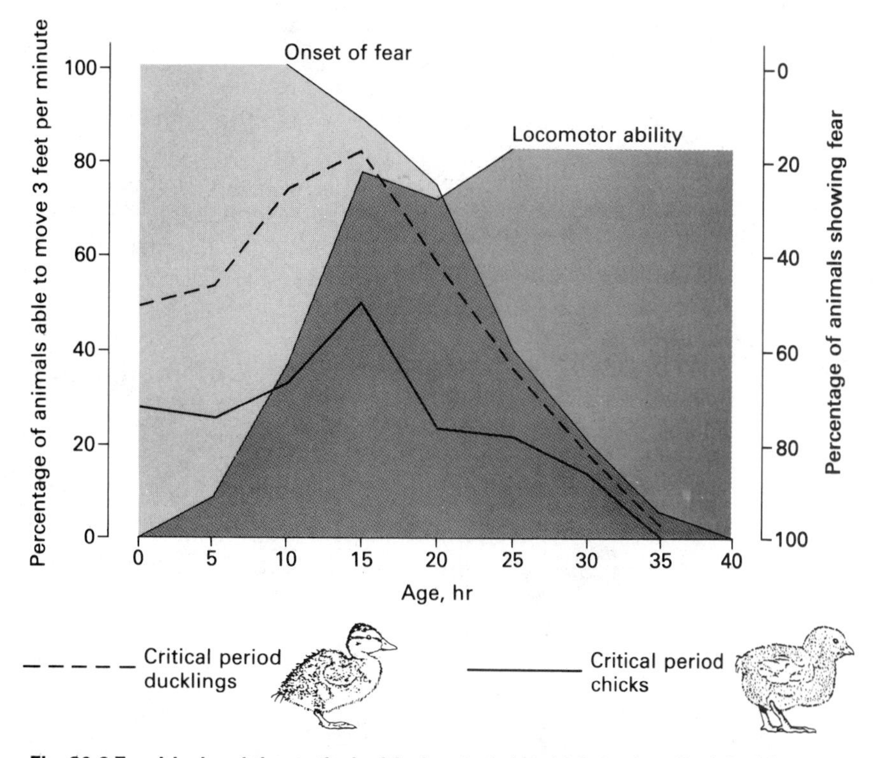

Fig. 20.6 Empirical and theoretical critical periods for chicks and mallard ducklings. The empirical critical periods for chicks and ducklings are shown by continuous and dashed lines, respectively. The pale shaded area indicates the proportion of birds showing fear responses. The darker shaded area shows the proportion of birds passing a locomotion test. The dark area (where the two shades overlap) indicates the theoretical critical period, on the assumption that it is determined by these two factors (After Hess, 1959b)

environment (landscape) is continually changing, and the different developmental systems (compartments) may change their nature (occupants) and may be programmed to respond to outside influences (via the windows) at different stages of development.

The evidence suggests that there may be more than one type of sensitive period. Early studies of imprinting show that both filial and subsequent sexual behavior can be affected by early experience. More recent studies (e.g., Schutz, 1965; Vidal, 1976; Gallagher, 1977) suggest that the sensitive period for sexual imprinting occurs later than that for filial imprinting. For example, Vidal exposed cockerels to a moving model for one of three periods: 0 to 15 days, 16 to 30 days, or 31 to 45 days after hatching. Subsequently, they were reared with a female or isolated up to the age of 150 days, at which time they were given a series of choice tests. He found that the chicks exposed to the model from 31 to 45 days showed the strongest sexual preference for the model even though they had shown the least amount of filial behavior to the model during the period of exposure. The sensitive period for filial imprinting normally would be from 1 to 30 days (see Fig. 20.6). Sensitive periods of learning are known to occur also in many other contexts. Thus, starvation has stunting effects on the growth of rats if it occurs early in life (Dobbing, 1976; Smart, 1977). Handling influences the subsequent behavior of rats, provided it is done while they are still being reared by the mother (Denenberg, 1962). Sensitive periods occur in the learning of song in birds and language in humans. Many other examples are reviewed by Bateson (1979).

20.5 Long-term aspects of imprinting

As a result of filial imprinting, an attachment develops between offspring and parent or foster parent, and ceases to be important once the juvenile reaches adulthood. However, such early experience may have long-term effects upon subsequent social behavior. Among dogs, for example, there is a sensitive period, from 3 to 10 weeks, during which normal social contacts develop. If a puppy is reared in isolation beyond about 14 weeks of age, its social behavior will not develop normally. Like some birds, dogs readily accept humans as social partners, and a puppy will form a lasting relationship with its owner if the attachment is formed at the height of the sensitive period. In primates also, contact between mother and infant is of great importance for the development of normal social relationships (Hinde, 1974).

In birds, as noted earlier, imprinting can have profound effects upon subsequent sexual preferences, and this is often called *sexual imprinting*. Cross-fostering experiments with domestic breeds of fowl, ducks, and pigeons can be carried out readily. If individuals of one distinctively colored breed are reared by parents of a different breed and color, the offspring usually prefer to mate with birds of their foster parents' color rather than of their own. For example, Warriner et al. (1963) used black and white varieties of domestic pigeon (*Columba livia*). Sixteen previously unmated pigeons were placed together in a large cage and allowed to form pairs. These pigeons had been raised exclusively by black or white pigeons, as shown in Table 20.1. Their courtship and pairing behavior in the large cage was observed and recorded. This experiment was repeated four times, each time with fresh birds so that 64 pigeons were involved in the experiment as a whole. The results showed that males paired with females of the same color as the male's foster parents in 26 of the 32 cases. In 5 of the 6 remaining cases, the female paired with a male of the same color as her foster parents. Thus, in most pairs the male's preference predominated over that of the female. The results show a clear effect of early experience on choice of male pigeons. In the case of females the evidence is inconclusive because the female preference is so often masked by the dominance of the male.

Table 20.1 Characteristics of pigeons used in each replication of the sexual imprinting experiments of Warriner et al. (1963). Sixteen previously unmated pigeons of the sex, color, and rearing-parent type indicated were placed together in a large cage and allowed to form pairs.

Color of pigeon	*Sex*	*Rearing parents*	
		White	*Black*
White	male	2	2
	female	2	2
Black	male	2	2
	female	2	2

Similar cross-fostering experiments can be done with birds of different, though closely related, species. Experiments have been carried out with ducks and geese, pigeons and doves, jungle fowl and domestic fowl, house and tree sparrows, and various species of gulls and finches. The results usually show a sexual attachment to the species of foster parent. For example, if male zebra finches (*Taeniopygia guttata*) are raised by Bengalese finches (*Lonchura striata*), then they court Bengalese finches when adult. Even if given a choice between an enthusiastic female zebra finch and an unenthusiastic female Bengalese finch, a cross-fostered male zebra finch will prefer the female Bengalese finch (Immelmann, 1972). The sexual preference for the foster-parent species is not restricted to particular individuals, but generalizes to all members of the species. Among male zebra finches, a preference for a particular individual may develop after pair formation, but the initial preference is for the species of female that played a parental role. In filial imprinting, by contrast, an attachment to particular aspects of the parent or parent substitute is much more likely to develop.

Schutz (1965; 1971) carried out a series of experiments with various species of duck. He found that males tend to prefer sexual partners similar to the female that reared them, whereas females prefer to mate with males of their own species, irrespective of their early experience. Of 34 male mallard ducks raised with other species or domestic varieties, 22 mated with their foster-parent species and 12 mated with their own species. However, among 18 mallard females raised by another species, all but three mated with their own species. This type of result was obtained with the other duck species studied by Schutz, with the exception of the Chilean teal (*Anas flavirostris*). Of seven Chilean teal females fostered by mallards, all subsequently paired with mallard males. The other species studied by Schutz are sexually dimorphic, the female coloration differing from that of the males (see Fig. 20.7). Chilean teal are monomorphic, both sexes having inconspicuous female-type plumage. Ducklings typically are cared for by the female parent and normally would not be greatly exposed to the male color pattern.

Fig. 20.7 Sexually dimorphic mallard ducks (left) and monomorphic Chilean teal (right) (*Photograph: F Schutz*)

Different species of duck are often found together on a lake, and the full-colored females look more alike than the conspicuous males. Thus, the females of dimorphic species can discriminate easily among males, but the males have a more difficult task. It has been suggested that the males of dimorphic species have to rely more on early experience to learn to identify their own species (Schutz, 1971), as we see later in this chapter.

Sexual imprinting occurs most readily to conspecifics, less readily to inappropriate species like human beings. In the absence of any alternative, however, reliable sexual imprinting may occur and may be long lasting. Thus, birds that are hand reared can become sexually imprinted on people; this has been reported for more than 25 species (Immelmann, 1972). Sexual preferences based upon imprinting often persist for a number of years. Thus, mallards fostered by other species of ducks and geese continue to court members of the foster species, even though they obtain little cooperation. Immelmann (1972) cross fostered Bengalese and zebra finches and then isolated them from their foster species for a number of years. Most bred successfully with members of their own species, but when they eventually were given a choice between their own and their foster species, they strongly preferred to court the species they had been raised by years before.

20.6 Imprinting as learning

Lorenz (1935) believed that imprinting is fundamentally different from other forms of learning; but this is not a popular view today (Bateson, 1966; Hinde, 1970). Imprinting involves a narrowing of pre-existing preferences, a process that has much in common with other forms of perceptual learning. Imprinting may also involve learning to respond in particular ways to the imprinting situation.

Perceptual learning is a rather controversial topic because it is often difficult to determine to what extent animals have to learn to organize their perceptual world and to what extent their recognition of external stimuli is innate. Numerous experiments show that animals deprived of perceptual experience during early life do not discriminate stimuli as well as normal animals (see Hinde, 1970). However, the interpretation of sensory deprivation experiments in terms of perceptual learning is open to criticism on a number of grounds. For one thing, deprived animals may perform poorly in discrimination tests because they have learned to rely on other senses, because exposure to novel stimuli is emotionally traumatic or because of deterioration of sensory mechanisms. For another, experiments that provide specific experience with particular stimuli can provide good evidence for perceptual learning. As we saw in Chapter 3, the development of full song in the chaffinch requires both exposure to the song during a particular phase of early life and the opportunity to practice singing it at a later stage. The phase during which the bird stores a description of the complete song can be described as perceptual learning. Similarly, the zebra finch raised by a Bengalese finch to which it responds sexually after years of separation provides a clear example of perceptual learning.

In a classic study, Gibson reared rats in cages with shapes cut out of metal fixed to the walls. In her tests, the ability of the rats to approach one of two shapes to obtain a food reward was evaluated. The observers found that rats previously exposed to one of the shapes learned more quickly than rats not so exposed (Gibson and Walk, 1956; Gibson et al., 1959). Bateson's (1964) finding that chicks imprint more readily to patterns with which they are familiar from their home environment is comparable and illustrates the importance of perceptual learning in imprinting.

Imprinting also appears to involve instrumental learning. Ducklings exposed to a toy train one day after hatching can be trained to peck at a pole by running the train past them just after they had pecked (Hoffman et al., 1966). The behavior of ducklings exposed to the train at a later stage of development could not be shaped in this way. Bateson and Reese (1969) found that mallard ducklings and chicks would learn to depress a treadle to gain exposure to a rotating light to which they had not been imprinted previously, but only if trained during the imprinting-sensitive period. Thus, it appears that the chick may learn about its mother through perceptual learning and also that through instrumental conditioning it learns responses that bring it into her presence (see Hinde, 1974). Bateson and Wainwright (1972) showed that as the chick becomes familiar with an imprinting stimulus, it begins to prefer slightly different stimuli. They tested chicks in the apparatus shown in Figure 20.8, by means of which they obtained a quantitative measure of the chick's preference between familiar and unfamiliar stimuli. The results showed that as familiarity with the initial imprinting stimulus increased,

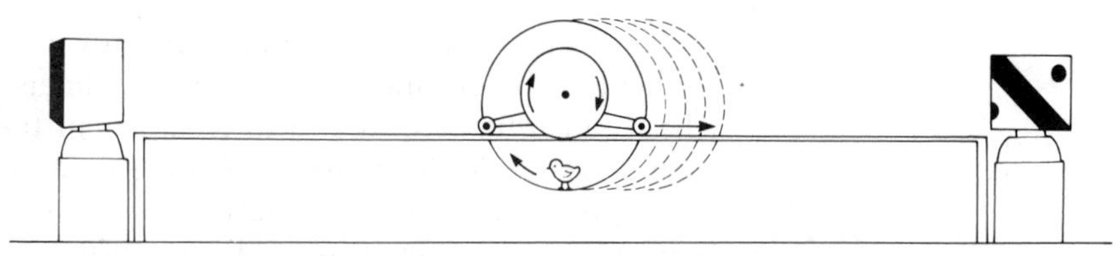

Fig. 20.8 Apparatus used by Bateson and Wainwright (1972) to test chick's preferences among various imprinting stimuli. As the chick attempts to approach one stimulus, it is propelled away from it (by a special arrangement of gears, not shown here). As the chick nears the other stimulus, it reverses its preference and attempts to approach it. The relative preference between two stimuli can be gauged from the positions of the chick when its preference changes.

the chicks developed an increasingly strong preference for stimuli differing slightly from the familiar one. They suggested that, in the natural situation, this tendency would have the effect of familiarizing the chick with different aspects of the mother. The ability to recognize a parent from many angles could only develop as the young bird built up a composite picture of its parent's characteristics (Bateson, 1973).

As we see in Chapter 17, some types of learning can occur in the absence of reinforcement and yet can be modified by reinforcement. Moreover, animals will perform responses that are instrumental in providing access to a familiar stimulus such as the song of a conspecific or the sight of a mate. The role of reinforcement in imprinting clearly falls into this same category, and the claim that imprinting is a special form of learning cannot be upheld on these grounds. In Chapter 18 we see that imprinting can be classified, on functional grounds, as a form of *preprogrammed* learning. However, this term also applies to many other aspects of learning such as song learning, place learning, and latent learning. The features that once made imprinting seem to be different than other forms of learning have been shown by research to be commonplace, partly as a result of our changed view of learning.

20.7 Functional aspects of imprinting

Lorenz (1935) suggested that imprinting was important in species recognition, but later studies indicate that it does not play an essential role. A bird of a species that imprints readily, nevertheless, can respond to conspecifics even if it has no relevant experience (Schutz, 1965; Immelmann, 1969; Gottlieb, 1971). Thus, although imprinting may provide an alternative to innate recognition of members of one's own species, it probably serves other functions.

From an evolutionary viewpoint, it is often important that mating occur only between members of the same species and that parents care only for their own offspring. Imprinting tends to occur in species in which attachment to parents, to the family group, or to a member of the opposite sex is an important aspect of the social organization. For example, among flocks of goats or sheep there is always a possibility of kids and lambs losing contact with the mother and approaching other females. Shortly after she gives birth, the mother goat labels her offspring by licking them and is sensitive to the smell of her kid for about an hour. During this period, a 5-minute contact with any kid is sufficient for it to be accepted as her own. If no such contact occurs, the kid will not be allowed to suckle (Klopfer and Gamble, 1966). Similarly, chicks of colonial gulls may be abandoned by their parents during a disturbance in the colony, and they usually hide under nearby vegetation. When the parents return they call the chicks from hiding, but some chicks may have strayed outside their parents' territory. Many gull chicks develop a specific recognition of their parents' call, and vice versa. These contact calls, learned at a particular stage of development, are used after periods of separation and help to maintain the integrity of the family units (Beer, 1970).

The sensitive period for sexual imprinting in ducks and geese corresponds closely to the period of parental care and is more prolonged in geese than in ducks. Normally, the young bird is susceptible to imprinting while a member of a family group. The sensitive period usually ends before the juvenile is likely to mix with birds other than its immediate kin. Bateson (1979) suggests that sexual imprinting is more important for kin recognition than for species recognition. He suggests that sexual imprinting enables an animal to learn the characteristics of its close kin and subsequently to choose a mate that appears slightly different, but not too different, than its parents and siblings. This would make it possible for the animal to strike a balance between the advantages of inbreeding and outbreeding. The advantages of outbreeding are commonly held to be that it introduces beneficial genetic variety and reduces the potency of lethal recessive genes. Although the validity of these suggestions is disputed (see Maynard Smith, 1978a), the existence of some selection pressure from outbreeding is widely recognized (Bischof, 1975). Similarly, there are thought to be certain advantages to inbreeding, particularly in maintaining the integrity of co-adapted complexes of genes. To strike a balance between these opposing pressures, the animal should choose a mate with a particular degree of relatedness, like a first cousin. But how are such kin to be recognized?

Bateson (1979) suggests that in order for a bird to recognize its kin it should delay learning about its siblings until they are old enough for their juvenile characteristics to provide a reliable indication of their adult appearance. As Figure 20.9 shows, the appearance of the mallard duckling, Japanese quail chick, and domestic fowl chick changes with age at a rate that is different for each species. Sexual imprinting in the mallard starts at about week 4 and lasts about a month (Schutz, 1965), which coincides with the time when the duckling begins to take an adultlike appearance. In the quail, sexual imprinting occurs in the first

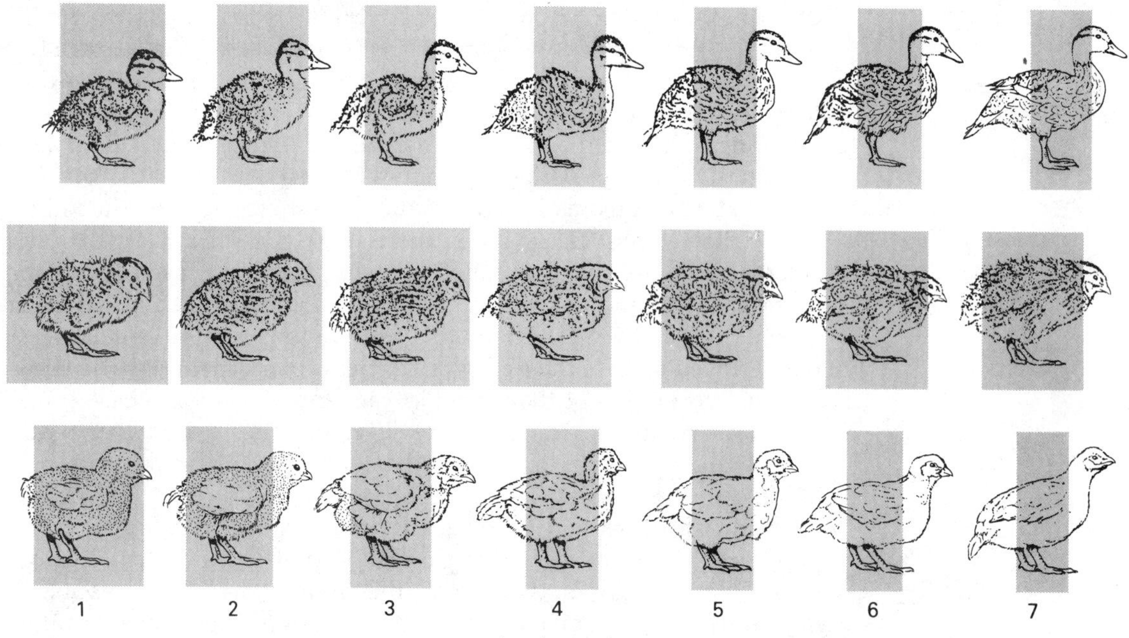

Fig. 20.9 Changes in external appearance with age in a mallard duckling (above), Japanese quail (middle), and domestic fowl chick (below). The 10 cm scale (shaded) shrinks as the birds increase in age, thus presenting a sibling's eye view of each bird (After Bateson, 1979)

few weeks after hatching (Gallagher, 1977), and by week three the chick's plumage is already fairly adultlike (see Fig. 20.9). By contrast, the domestic chick takes much longer to develop adult plumage, and the sensitive period for sexual imprinting is about five to six weeks (Vidal, 1976).

Bateson (1979) suggests that the timing of sexual imprinting is associated with the development of adult plumage, thus giving the juvenile bird a good opportunity to learn about the appearance of its siblings. Bateson (1980) found experimentally that Japanese quail prefer to mate with birds that differ slightly in plumage color from their parents. Normal brown males reared in groups preferred to mate with a brown female when given a choice between a brown and a mutant white female. However, when given a choice between a strange brown female and a familiar one with which they were reared, they preferred the former. Similarly, there is some evidence that Bewick's swans (*Cygnus columbianus bewickii*) avoid mating with close kin. These swans have distinctive facial markings (see Fig. 20.10), and members of the same family tend to have similar faces. Mated pairs have facial patterns that differ more than would be expected by chance (Bateson et al., 1980), suggesting that the young birds avoid inbreeding by choosing mates with facial markings that differ from the family pattern.

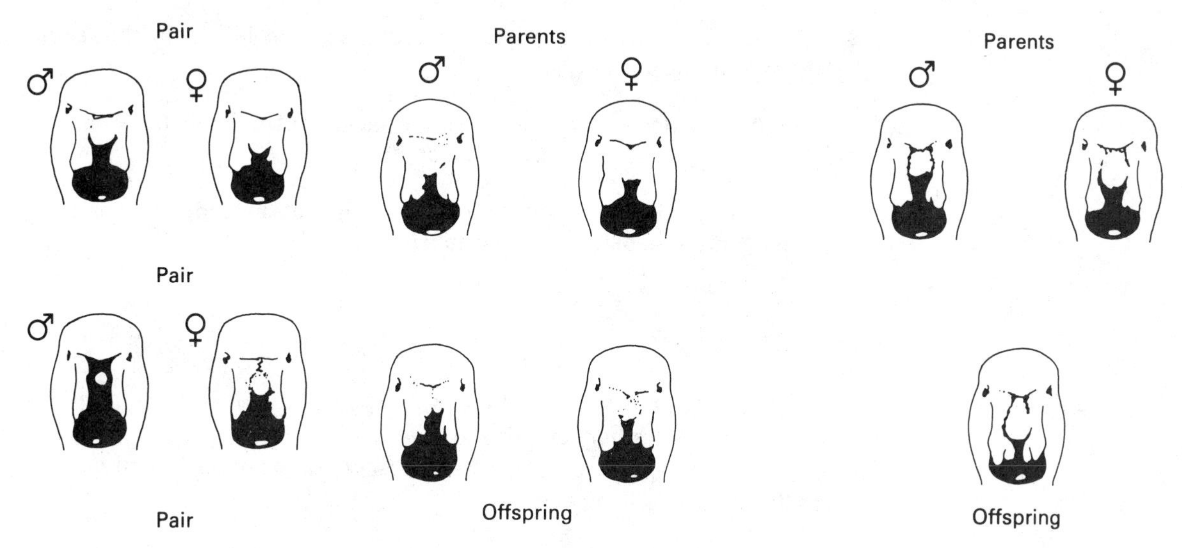

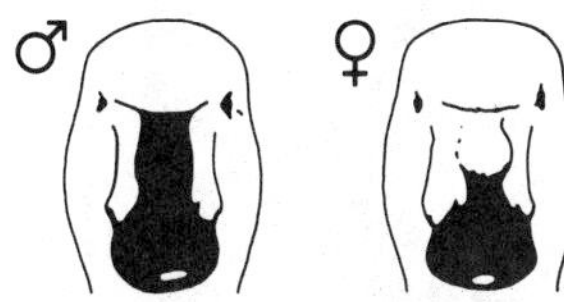

Fig. 20.10 Facial pattern in Bewick's swans. Offspring tend to resemble their parents (right), suggesting that the pattern is inherited. Mates are often different in appearance (above), suggesting that the swans actively outbreed (From Bateson, Lotwick and Scott, 1980)

Some researchers (Westermark, 1891) have suggested that humans tend to choose mates socially, psychologically, and physically similar to themselves (Lewis, 1975; see also Chapter 8.4). In contrast, there is evidence that satisfactory marriages are not formed between people who spend their early childhood together. Studies of Taiwanese arranged marriages (Wolf, 1966; 1970) and Israeli kibbutzim (Shepher, 1971) indicate a lack of sexual attraction between people who spend their childhood together. In Taiwanese arranged marriages the bride is adopted into the family of the husband as a small child. In many kibbutzim the children are raised from birth in peer groups consisting only of children of the same age. In both cases there is some social pressure in favor of ultimate marriage, yet it is rarely successful. Although such human relationships are complicated by social convention and taboo, the evidence for some biologically based negative imprinting is considerable.

Points to remember

- The concept of instinct has changed over the years with the growing realization that all behavior is the result of genetic and environmental influences.

- The early ethologists thought that the recognition of sign stimuli required a special mechanism, which they called the innate releasing mechanism. This notion has much in common with selective attention.

- The young of many bird species develop an attachment to moving objects soon after hatching. This imprinting process occurs during a particular sensitive period of development.

- Imprinting may have long-term consequences, particularly upon the sexual behavior of animals.

- Imprinting used to be thought of as a special form of learning, but it has much in common with ordinary conditioning.

- Imprinting may be important in the development of kin recognition and in preventing interbreeding among close relatives.

Further reading

Bateson, P. P. G. (1979) 'How do sensitive periods arise and what are they for?' *Animal Behaviour*, **27**, 470–86.
Gottlieb, G. (1971) *Development of Species Identification in Birds*, University of Chicago Press, Chicago.

21 Displacement Activities

Fig. 21.1 A male stickleback fanning its nest

A displacement activity is characterized by its apparent irrelevance to the situation in which it appears. For example, while courting a female, a male three-spined stickleback may swim to its nest and perform fanning movements, as illustrated in Fig. 21.1. In the normal course of events, the courtship serves to entice the female into the nest, where she lays her eggs. The male then fertilizes the eggs and chases the female out of his territory. The male alone takes care of the eggs and guards them against predators. His parental activities include fanning the nest to induce a current of water through it so that the waste products of respiration are removed and a supply of oxygenated water is maintained.

Sometimes the male will court a second female and add her eggs to his clutch. Thus, a courting male already may have eggs in his nest, and it would make sense for him to visit the nest periodically to ventilate the eggs. Often, however, there are no eggs in the nest and it is difficult to see why a male should break off his courtship to perform fanning movements at the nest. Such fanning behavior appears to be irrelevant in two main respects. First, it occurs in the absence of normal stimulus, which is carbon dioxide emanating from the respiring eggs. Second, it does not serve its normal function of ventilating the eggs, and it is not obvious what other function it might have.

The frequent occurrence of such apparently irrelevant activities greatly puzzled the early ethologists, and they continue to demand explanation. In this chapter we will examine the various explanations offered for displacement behavior, an area of ethology in which I have been personally involved. It provides us with a case history of changing attitudes to the mechanisms of behavior. Finally, displacement activities provided an impetus for a very significant development in the understanding of animal communication, the ritualization of behavior during its evolution.

21.1 The causation of displacement activities

Displacement activities were described by a number of early ethologists, (e.g. Huxley, 1914; Makkink, 1936; Kortlandt, 1940; and Tinbergen,

Fig. 21.2 Examples of displacement activities. 1, displacement nesting behavior during aggressive encounters in herring gulls. 2, displacement grass-pulling (nest-making) during a territorial dispute. 3, displacement sleep during a fight between oystercatchers, and 4, between avocets. 5, displacement sand-digging (nest-making) by a male three-spined stickleback during a territorial dispute. 6, displacement preening during courtship in the sheldrake, and 7, in the garganey, and 8, in the mandarin, and 9, in the mallard, and 10, in the avocet. 11, displacement food-catching during courtship in the European blue heron. 12, 13 and 14, displacement sexual behavior in the European cormorant during aggressive encounters. 15, displacement food-begging in the herring gull during courtship. 16, displacement feeding in domestic cocks during fighting (From Tinbergen, 1951)

1940). The word *displacement* was used first in this context by Edward Armstrong (1947) and by Niko Tinbergen and Jan van Iersel (1947). Tinbergen (1952) emphasized the following characteristics of displacement activity:

Displacement activities are recognizably similar to, or derived from, motor patterns normal to the species.

The movements shown appear to be irrelevant, entirely out of context with the behavior immediately preceding or following them.

A displacement activity seems to appear when an activated drive is denied discharge through its own consummatory act(s).

Fig. 21.3 Threat posture of the male three-spined stickleback, derived from the sand-digging behavior that precedes nest-building

Figure 21.2 shows some of the classical examples of displacement activities.

Kortlandt's (1940) and Tinbergen's (1940) proposed explanations of displacement activities were based on the notion that each instinct is energized by its own "action-specific energy." Kortlandt distinguished between *autochthonous* behavior, which is energized by its own drive, and *allochthonous* behavior, which is energized by the drive built up by other activities. Displacement activities were considered allochthonous. In Tinbergen's (1952) formulation: "Displacement activities are outlets through which the thwarted drives can express themselves in motion." This concept was expressed by the word *Ubersprung*, or sparking-over.

Tinbergen (1951) noted that displacement activities occurred when "there is a surplus of motivation, the discharge of which through the normal paths is somehow prevented." He identified the causes of such prevention as conflict of two strongly activated antagonistic drives and as strong motivation in situations where there is a lack of the external stimulation required to release the relevant consummatory behavior. A conflict between two antagonistic drives is found in animals fighting at the boundary line between their territories. Thus, male sticklebacks meeting at the boundary between their territories adopt a head-down attitude, as illustrated in Figure 21.3. Tinbergen (1951) identified this posture as belonging to the digging behavior at the start of nest building:

It is a striking fact that displacement activities often occur in a situation in which the fighting drive and the drive to escape are both activated. Within its own territory, a male invariably attacks every other male. Outside its territory the same male does not fight but flees before a stranger. In between the two situations, that is, at the territory's boundary, opposing males perform displacement activities. The natural conclusion, viz, that displacement activities, in this situation, are an outlet of the conflicting drives of attack and escape (which of course cannot discharge themselves simultaneously, because their motor patterns are antagonistic), has been tested experimentally (Tinbergen, 1940). A red dummy was offered to a male stickleback in its territory and was duly attacked. Instead of withdrawing the dummy, it was made to "resist" the attack by hitting the attacking male with it. When this "counterattack" is carried out vigorously enough, the territory-holding male can be defeated in its own territory. It withdraws and hides in the vegetation. If the dummy is now held motionless in the territory, it will continuously stimulate the male's fighting drive. The tendency to flee, however, diminishes with time. Gradually the fighting drive regains its superiority over the tendency to flee, and after a few minutes the male will attack the dummy again. Just before this happens, however, the male performs displacement digging. This shows, therefore, that displacement digging occurs when the two drives involved are in exact equilibrium. There is little doubt that the various displacement activities occurring during territorial fights must be explained in the same way.

That displacement activities tend to occur at points of equilibrium between two different motivational tendencies has been amply confirmed. Notably, this does not, of itself, provide evidence that displacement activities are due to a surplus of motivational energy. As we shall see, alternative explanations exist.

Reviews of field studies (e.g., Tinbergen, 1952; Bastock et al., 1953) indicate that displacement activities tend to occur in three types of situations: (1) physical thwarting of appetitive behavior, (2) thwarting of consummatory behavior by removal of its objective or goal, and (3) simultaneous activation of incompatible tendencies. These observations are confirmed by laboratory studies (e.g., McFarland, 1965). The feature that these situations have in common is that the ongoing behavior is thwarted, either physically or by non-availability of an expected consequence of the behavior or by an incompatible activity.

The classical ethological explanation of displacement activity is that they are allochthonous, caused not by their normal drive but by an overflow, or sparking-over, from the thwarted drive. Tinbergen (1951) also implied that displacement activities could release tension and act as an outlet for nervous energy. This view of displacement activity seems to make intuitive sense. We all occasionally experience subjective feelings of tension in social situations and we may suppose that fidgeting with cigarettes or jewelry relieves the tension. Tinbergen (1951) identified this type of human behavior as displacement activity:

> Another phenomenon suggesting an instinctive organization in man basically similar to that found in animals is displacement activity. Displacement activities are by no means rare in man. They are not easily recognized as in animals because in man learned patterns, like lighting a cigarette, handling keys or handkerchief, etc., often act as displacement activities. However, innate patterns may function as outlets in man too. The general occurrence of scratching behind one's ear in conflict situations almost certainly has an innate basis. It is striking how often activities belonging to instinct of comfort (care of the skin) are shown in conflict situations: in women it mostly takes the form of adjusting non-existing disorders of the coiffure, in man it consists of handling a beard or moustache, not only in the days when men still had them but also in this "clean-shaven" era.

The early ethologists prided themselves on their objective study of behavior (e.g., Tinbergen, 1942), and it is perhaps unfortunate that they allowed themselves to lapse into subjective speculation about displacement activities. As we shall see, the situation is by no means as straightforward as they thought.

The first doubts about the classical interpretation of displacement activity came with the discovery that it need not be entirely allochthonous. Turkey cocks will break off in the middle of a fight and show bouts of displacement feeding or drinking (Raber, 1948). Which displacement activity appeared depended upon whether it was food or water that was available to the birds. Richard Andrew (1956b) suggested that displace-

Fig. 21.4 Apparatus for inducing approach-avoidance conflict in chaffinches. As the bird approaches the food, by hopping from perch to perch, the experimenter flashes the light bulb in the food dish, which induces a weak avoidance tendency

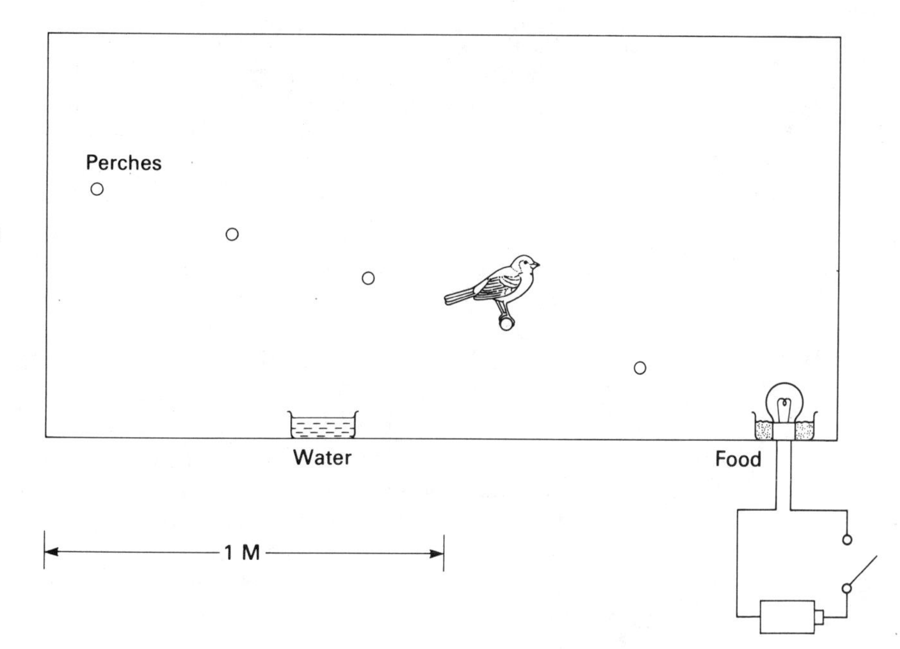

ment grooming in finches was controlled by the peripheral stimuli that normally elicit grooming. He showed that grooming is suppressed easily by other activities, and he postulated that the stimuli for grooming were continually present. Andrew suggested that, in conflict situations, when neither of the primary competing activities can be expressed fully, the suppression of grooming is reduced sufficiently to allow displacement grooming to occur.

This *disinhibition* theory of displacement activity was stated more explicitly as a result of a study of nesting terns (van Iersel and Bol, 1958). When the birds were in a conflict between the tendency to stay on the nest and the tendency to escape, they often indulged in bouts of displacement preening. These ethologists found that the amount of displacement preening was greater after rain, when the wet feathers normally would provide a stimulus for preening. They postulated that equilibrium between two conflicting drives lowered their power to inhibit a third drive, which then could be expressed as overt behavior. Similarly, studies of displacement fanning in sticklebacks showed that it occurred when the aggressive and sexual tendencies were in equilibrium (Sevenster, 1961). The displacement fanning was facilitated by mimicking the presence of developing eggs by irrigating the nest with water containing dissolved carbon dioxide.

Hugh Rowell (1961) studied approach-avoidance conflict in chaffinches. He induced the conflict either by presenting a stuffed owl that induced a mobbing response or by frightening hungry birds by means of a flashing light inside a food dish. The aviaries in which these experiments were carried out were equipped with rows of perches that enabled the experimenter to tell at which point the bird was in equilibrium between approach and avoidance, as shown in Figure 21.4. Rowell was able to show that displacement preening and bill wiping occurred at such equilibrium points and that these displacement activities were influenced by their normal causal factors. Thus, more displacement preening was seen when the birds had been showered with a mixture of soot and water. Rowell considered that three conditions must be fulfilled for a displacement activity to occur: (1) There must be a state of equilibrium between two conflicting drives or tendencies; (2) the equilibrium must persist long enough for a displacement activity to be performed; and (3) there must be adequate external stimuli relevant to the normal (non-displacement) performance of the activity.

In my own work (McFarland, 1965), I showed that a motivational conflict was not a necessary condition for displacement activity. Working with thirsty Barbary doves, I found that they would show displacement feeding and preening in an approach-avoidance situation similar to that used by Rowell. However, they also would show these displacement activities if prevented from reaching their habitual water dish by a glass partition or if there was no water in the dish when the thirsty bird arrived to drink there. Displacement feeding was enhanced by the presence of grain on the floor of the cage and by making the birds slightly hungry. Moreover, doves that had been trained to peck an illuminated disk to obtain food rewards would perform this learned feeding behavior as a displacement activity in situations where they were thirsty but thwarted in obtaining water. Thus, displacement activities need not be innate.

The original disinhibition theory of Andrew (1956b), van Iersel and Bol (1958), Rowell (1961), and Sevenster (1961) agreed with the classical surplus drive theory in confirming that displacement activities occurred at the equilibrium points in conflict situations. This theory differed radically, however, in its account of the motivation of the displacement activity. Instead of discharging the surplus energy from a different activity, the displacement activity occurred in response to its normal stimuli; that is, the displacement activity was autochthonous and not allochthonous. The significance of motivational conflict was that it generated short periods of equilibrium, or motivational stalemate, during which the displacement activity was disinhibited. Thus, displacement activities were low-priority activities that showed through gaps created by conflict among the major competing drives.

This early version of disinhibition theory was criticized (McFarland, 1966b) for being couched in terms of drives, or neural centers, for being imprecisely formulated, and for accounting for displacement activities only in cases of motivational conflict. Margaret Bastock et al. (1953) had noted that displacement activities could occur in both conflict and thwarting situations. They also suggested that a failure in some sort of negative feedback might provide a factor common to all types of situations in which displacement activities were observed.

21.2 Attention and displacement activity

Bastock et al. (1953) postulated a mechanism along the lines of the reafference theory (see Chapter 14). They proposed that, at the initiation of an activity, a center is charged in an output copy of the normally expected stimuli. Failure in sensory feedback (that is, in confirmation of the expectation) was supposed to lead to an accumulation of nervous energy that could be responsible for displacement activity. I adapted this basic idea, replacing accumulation of energy with a mechanism for disinhibition (McFarland, 1966b).

Different activities, I noted, in addition to being incompatible in motor terms, also were incompatible in terms of attention. The evidence suggested that animals could not attend to two different sets of stimuli simultaneously. I suggested that attention was controlled partly by the consequences of behavior: The animal learned what consequences to expect in a given situation, and these expectations, in the form of an *efference copy* (see Fig. 21.5), were compared with the actual consequences of behavior on a given occasion. In a stable environment the animal's expectation is approximately correct, and the comparison with the actual consequences (in which one value is subtracted from the other) will give zero discrepancy. However, if the behavior is disrupted in any way, there will be a discrepancy in the feedback from the consequences of behavior, and this will cause the animal to switch attention away from the stimuli controlling its behavior at the time.

Disruption of the consequences of behavior can occur as a result of physical thwarting, conflict with another motivational system, or non-confirmation of an expected consequence, situations that give rise to displacement activity and produce a feedback discrepancy (Fig. 21.5) that causes a switch of attention. The ongoing behavior is terminated and some other (displacement) activity takes over. However, a displace-

Fig. 21.5 Diagram of a reafference system

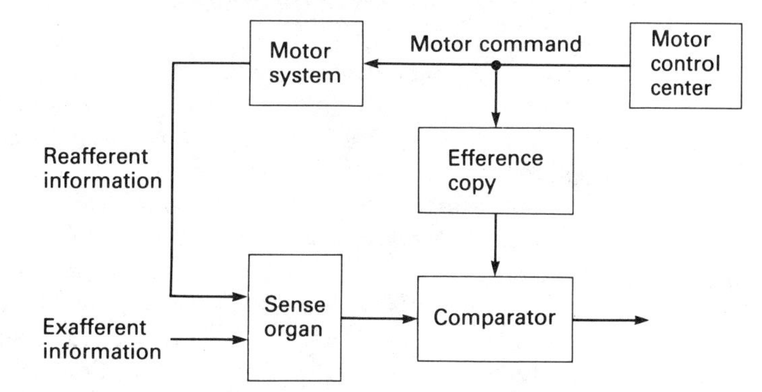

ment activity will have its own output copy, and the consequences of the activity usually will not match with this. For example, displacement feeding in birds normally does not result in food being eaten (McFarland, 1965). The displacement activity occurs in non-typical situations, which are in various ways discrepant from what the animal normally expects from such behavior. Attention, therefore, soon will be switched back to the original activity. In this way it is possible to account for the fact that displacement activities are typically of short duration and incomplete compared with normal activities.

This theory hinges on the idea that frustration diverts attention and thus causes a disinhibition of behavior, which is seen as a displacement activity. As we saw in Chapter 14.3, evidence for the existence of attention mechanisms in animals comes from the study of animal discrimination learning. Sutherland (1964) suggested that animals not rewarded periodically during training learn more about incidental aspects of the situation than animals rewarded consistently. With such partially rewarded animals, therefore, it might take longer to extinguish their learned behavior (see Chapter 17) because they have attended to and learned about more aspects of the situation and have more to unlearn during extinction.

I extended Sutherland's theory and suggested that any disruption of ongoing behavior would cause attention to be diverted from the stimuli controlling that behavior. My experiments showed that partially rewarded doves learn more about incidental cues, and respond more to novel stimuli, than consistently rewarded doves. Non-reward of this type is one of the situations in which displacement activities are common, and the suggestion is that a mechanism by which frustration diverts attention also acts as a mechanism for the disinhibition of displacement activities. This theory is supported by the finding that doves that take longest to unlearn a simple approach response (and are therefore more attentive to incidental stimuli) also spend more time in displacement activities during extinction.

21.3 The concept of disinhibition

While there is considerable evidence for the idea that frustration diverts attention (e.g., McFarland, 1966a; McFarland and McGonigle, 1967; McGonigle et al., 1967), the case for claiming that shifts in attention are responsible for displacement activities is less strong. The problem is that it is very difficult to tell what an animal is attending to from one moment to the next. A possible way around this problem is to devise tests for the occurrence of disinhibition, leaving aside for the present the question of whether or not attention shifts are responsible for the disinhibition. As we have seen, behavioral disinhibition was suggested (e.g., Andrew, 1956b; van Iersel and Bol, 1958; Rowell, 1961; Sevenster, 1961) as a means by which behavior B can become temporarily interpolated between occurrences of another activity (A, B, A), or between conflicting activities (A, B, C). I suggested that, instead of concentrating on the conflict or thwarting circumstances of displacement activity, we should consider how one activity normally might come to follow another (McFarland, 1969a). If A and B are incompatible activities, then the occurrence of one precludes the occurrence of the other. Ethologists (e.g., Hinde, 1966; 1970) use the term *inhibition* to express this relationship. Thus, if an animal is doing A but would be doing B if A were not possible, then the occurrence of B is inhibited by the occurrence of A.

As we saw in Chapter 15, the strength of the tendency to perform an activity depends upon a combination of internal and external causal factors. Suppose that the tendency for activity A is initially higher than the tendency for B, as illustrated in Figure 21.6. The animal is observed to be engaged in activity A. If B tendency is rising while A tendency remains the same, then B tendency eventually will become stronger than A tendency, and we might expect to observe the animal's behavior change from activity A to activity B. Moreover, if B tendency could be manipulated in such a way that it became greater than A tendency at an earlier point in time, then we would expect to see the animal's behavior change from activity A to activity B at an earlier time. This may seem a trivial point, but it leads to the following operational definition of *motivational competition*: "A changeover due to competition can in practice be recognized when a change in the level of causal factors for a second-in-priority activity results in an alteration in the temporal position of the occurrence of that activity" (McFarland, 1969a). Thus, it is envisaged that the systems controlling activities A and B compete with each other, as we see in Chapter 25. The system with the higher tendency is the one that wins the competition, and it is that activity that is observed. This may all seem very obvious, but it has interesting implications in a situation in which manipulation of second-priority causal factors does not alter the time of occurrence of the relevant behavior.

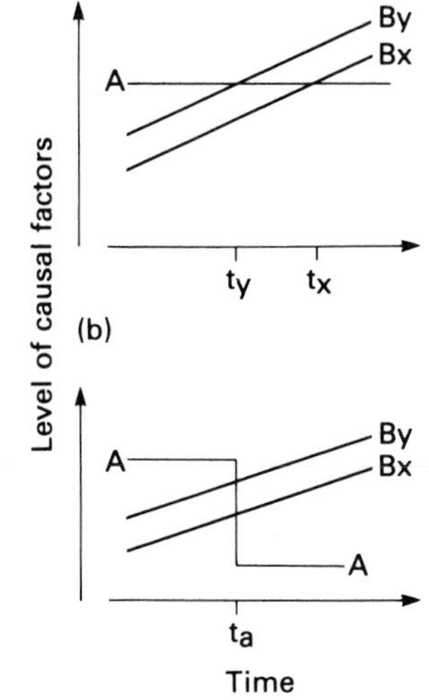

Fig. 21.6 (*a*) situation in which activity B ousts activity A by motivational competition. (*b*) situation in which activity A disinhibits activity B. If the tendency to perform activity B is raised from B_x to B_y then B is performed earlier under model (*a*) but timing is unaffected in model (*b*)

The occurrence of a second-priority activity is inhibited by the occurrence of a top-priority activity. If this inhibition were removed, then the second-priority activity would be disinhibited. This is the essence of the disinhibition theories of displacement activity that we have discussed in this chapter. McFarland (1969a) operationally defined *disinhibition* as follows: "The time of occurrence of a disinhibited activity is independent of the level of causal factors relevant to that activity." A possible representation of this situation is illustrated in Figure 21.6. The important point is that B tendency is not instrumental in the sudden change in A tendency, and therefore the strength of B tendency does not determine the time of occurrence of activity B. Of course, if B tendency were raised sufficiently, it could become greater than A tendency and an appropriate change in behavior would be observed. Activity B then would not be disinhibited.

It is important to realize that these definitions are purely operational and imply nothing about the mechanisms that may be responsible for disinhibition. Operational definitions simply identify those situations in which an empirical test gives a particular result. If the result is not obtained, then by definition the phenomenon does not occur. Unfortunately, the status of operational definitions is not always properly understood, and some researchers have argued that circumstances that ought to have given rise to a particular phenomenon and that have failed to do so therefore invalidate the concept of disinhibition. The exact motivational implications of disinhibition remain controversial (e.g., Roper and Crossland, 1982; Houston, 1982; McFarland, 1983). However, the operational definition of disinhibition cannot be disproved by theoretical argument. All that can be determined is whether or not the phenomenon occurs, and whether the definition is useful.

Disinhibition has been shown to occur in a variety of situations (e.g., McFarland and L'Angellier, 1966; McFarland, 1970b; Halliday and Sweatman, 1976). For example, during feeding behavior, doves (*Streptopelia*) typically pause for a few seconds and then resume feeding. If water is made available, the birds may drink during the pause but the time of occurrence of the pause is not affected by the presence or absence of water. Similarly, if paper clips are fixed to the primary wing feathers, the birds will try to remove them during the pause, but again the timing of the pause is not affected by the presence or absence of the paper clips (McFarland, 1970b). The water dish and the paper clips are examples of stimuli that alter the strength of the second-priority activities of drinking and preening, respectively. However, manipulation of these stimuli

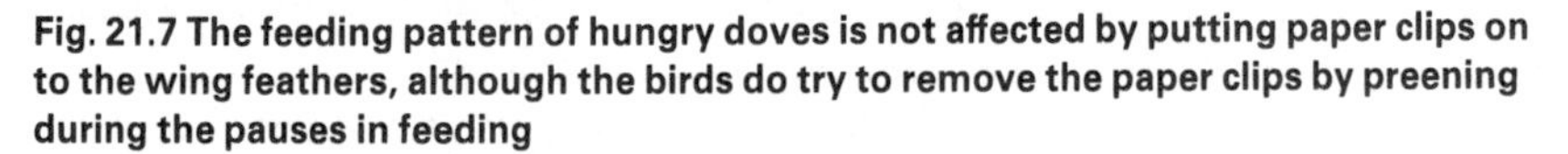

Fig. 21.7 The feeding pattern of hungry doves is not affected by putting paper clips on to the wing feathers, although the birds do try to remove the paper clips by preening during the pauses in feeding

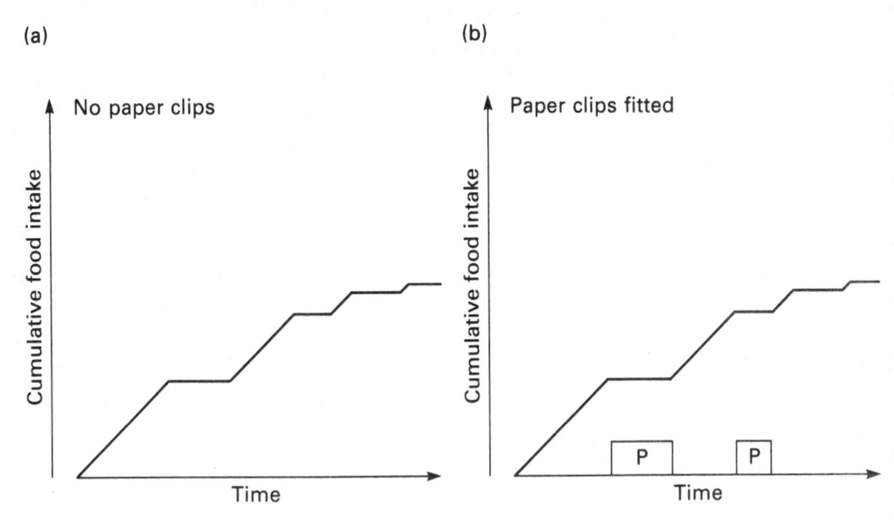

does not affect the time at which these activities occur, as illustrated in Figure 21.7. According to the operational definition of disinhibition (above) these activities have been disinhibited. The importance of this type of result is that a disinhibited activity can be identified in a situation in which there is no motivational conflict or thwarting of the ongoing behavior. It raises the possibility that disinhibited activities may occur as part of ordinary behavior sequences. Motivational disinhibition has been found to occur in a variety of species and in behavior sequences including feeding, drinking, copulation and courtship. Such instances cannot be accounted for in terms of theories of conflict or frustration but must form part of the normal organization of behavior sequences.

Displacement activities were noticed first in conflict and thwarting situations. They were shown later to be due to disinhibition. This does not mean, however, that such activities can only occur in situations of conflict or frustration. The early ethologists had no means of identifying disinhibited activities in ordinary behavior sequences. Indeed, disinhibited activities are impossible to identify by observation alone. Only by experiment can the two cases be distinguished. As we see in Chapter 25, various methods may be used to gain evidence that disinhibition is occurring. All involve some form of experimental manipulation, and all ultimately must relate to the way in which disinhibition is operationally defined.

What are we now to conclude about displacement activities? Are they, on the one hand, merely instances of disinhibited activities that happen to occur in situations of conflict and frustration? Are they, on the other hand, special activities that relieve tension or alleviate frustration? These questions remain unresolved, but the story of displacement activities illustrates how thinking has changed since the days of the early ethologists.

21.4 Functional aspects of displacement activity

Displacement activities were seen originally as being functionally irrelevant as well as allochthonous (caused by an irrelevant drive). Normally, animal behavior appears to result in the attainment of a goal with an obvious biological function. Behavior that has no obvious goal may seem to be functionally irrelevant. For example, we might say that the goal of feeding behavior is to ingest food. It is characteristic of displacement feeding in birds, however, that the food is picked up in the bill and then dropped rather than swallowed. Displacement feeding thus would seem to be functionally irrelevant in that it is feeding behavior that does not fulfill its normal function. It may be, however, that the displacement activity fulfills some other function such as relieving motivational tension or communicating with other animals.

Displacement activities often occur in sexual or aggressive encounters, where motivational conflict is likely to arise. As displacement activities usually occur at the equilibrium point in a conflict, they provide potential information for the other participant in an encounter. For example, we saw how male three-spined sticklebacks tend to indulge in bouts of displacement digging in disputes with rivals near their territorial boundary where the tendencies to attack and escape are equal. It is possible that sticklebacks could learn to recognize territorial boundaries by the displacement digging done by neighbors. In situations where this type of communication is advantageous, natural selection will tend to make the behavior more efficient and reliable as a source of information. The evolutionary process by which this comes about is known as *ritualization*.

Ritualization is an evolutionary process by which behavior patterns become modified to serve a communication function. The concept was instigated by Julian Huxley (1923) but was developed into a full-fledged theory by Niko Tinbergen (1952). The main evidence that ritualization has occurred during evolution comes from comparative studies. For example, the beak wiping performed by male zebra finches (*Taeniopygia guttata*) during courtship is thought to be a displacement activity. Beak wiping normally occurs after feeding, but in courtship it appears to occur at the equilibrium points in approach-avoidance conflicts. In the related spice finch (*Lonchura punctulata*) and Bengalese finch (*Lonchura striata*), no beak wiping occurs during courtship, but the male performs a bow, remaining with its head lowered for a few seconds. The similarity of this movement to beak wiping and the context in which it occurs suggest that the bow is a ritualized form of beak wiping.

Behavior patterns change in many ways during the process of ritualization. They often become stereotyped and incomplete. Ritualized grooming, for instance, often is restricted to certain parts of the body. The movements are curtailed and may become merely token movements. In other cases, movements become frozen into postures. For example, the threat posture of the male three-spined stickleback is very similar to part of the sand-digging movement (see Fig. 21.3) and is probably a ritualized form of displacement sand digging (Tinbergen, 1951).

An activity that is normally variable in intensity may come to have a typical intensity (Morris, 1957) once it has become ritualized. For example, when the black woodpecker chips out a nest cavity, it does so with a variable rhythm. The males also drum against dry wood to attract females and to ward off other males. In this case the drumming is highly stereotyped and follows a precise rhythm. During excavation of the nest cavity, the bird doing the work flies to the entrance of the nest and drums with a very slow rhythm at the edge of the entrance hole. This is a signal to the other bird to come and take over the work of excavation. Here we have examples of normal activity (drumming to excavate) and of two forms of ritualized activity (drumming to advertise and drum-

(a)

(b)

(c)

(d)

Fig. 21.8 Ritualized displacement preening in (*a*) the shelduck *Tadorna tadorna*, (*b*) the mallard *Anas platyrhynchos*, (*c*) the garganey *Anas querquedula*, (*d*) the mandarin *Aix galericulata* (From *The Oxford Companion to Animal Behaviour*, 1981)

ming to change shifts). The ritualized signals are unambiguously different from each other and from the normal behavior.

Ritualization of behavior often involves evolution of special structures or markings serving to enhance the display by making it more conspicuous. Lorenz (1941) noted that preening is a common feature of courtship in ducks. In the common shelduck (*Tadorna tadorna*) this seems to be an unritualized displacement activity. It is similar to normal preening and occurs during motivational conflict. The northern mallard (*Anas platyrhynchos*) lifts one wing and restricts its preening to the brightly marked feathers thus revealed (see Fig. 21.8). This behavior is ritualized in two respects: (1) It is restricted to certain parts of the body, and (2) it is enhanced by the specially conspicuous markings revealed during the display. In the garganey duck (*Anas querquedula*), the preening movements are even more incomplete and are restricted to the light blue feathers on the outside of the wing. The preening movements serve no grooming function and are merely sham preening. In the mandarin drake (*Aix galericulata*), the courtship preening is highly ritualized, the bill merely touching an enlarged rust-red feather that is raised up in the air (see Fig. 21.8). The preening movement is enhanced by the crest raised on the back of the head. Comparison of these duck species clearly shows a progressive ritualization of the original displacement and development of ways of making the behavior more conspicuous and effective as a signal.

Points to remember

● Displacement activities require special explanation because they seem to occur out of their normal motivational context.

● In some way the animal diverts its attention from the ongoing behavior and indulges in bouts of seemingly irrelevant activity.

● In order for this to occur, the top priority behavior is switched off, thus disinhibiting the displacement activity.

● Some displacement activities are thought to have become ritualized during evolution, so that they now serve a communication function.

Further reading

Tinbergen, N. (1951, 1965) *The Study of Instinct*, Clarendon Press, Oxford.

22 Ritualization and Communication

The early ethologists made considerable progress in understanding animal communication. Charles Darwin (1872) had set the stage in emphasizing the role of communication in the emotional expressions of animals. For many years, however, the subject remained curiously neglected. Konrad Lorenz (1932; 1935) used the term *releaser* for "those characters exhibited by an individual of a given animal species which activate existing releasing mechanisms in conspecifics and elicit certain chains of instinctive behavior patterns." Thus did Lorenz lay the foundations of the classical ethological view of communication, further developed by Niko Tinbergen (1951; 1953).

Various specific features of an animal's morphology may be ritualized and act as sign stimuli to which other members of the species respond instinctively. In the social context these sign stimuli were often known

Fig. 22.1 A herring gull pecking at the red spot on its parent's bill (see back cover) (*Photograph: Jim Shaffery*)

Fig. 22.2 Models used in experiments on the food-begging behavior of herring gull chicks. The length of the grey bar indicates the number of pecks delivered to the model. (*a*) Models differing in bill color, (*b*) models differing in the color of the spot on a yellow bill, (*c*) models with grey bills and spots differing in contrast (After Tinbergen and Perdeck, 1950)

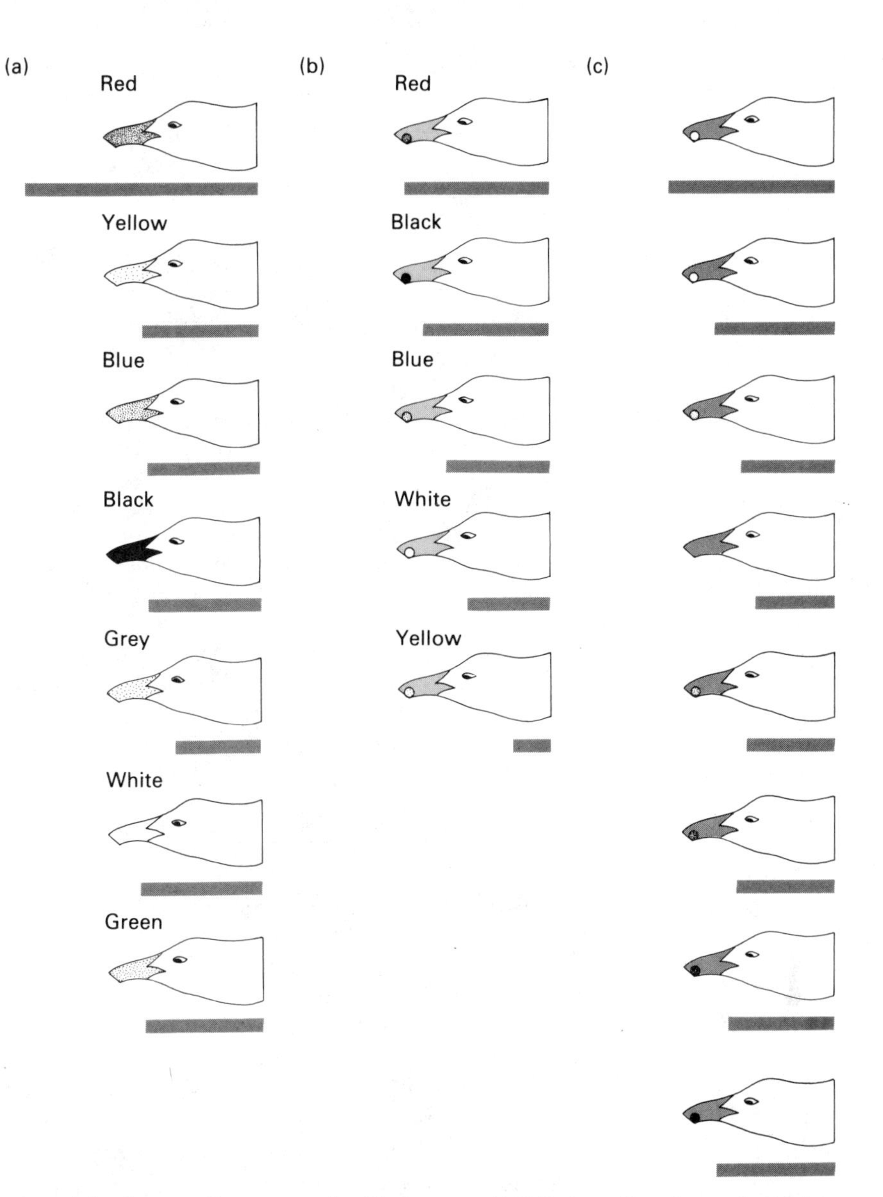

as *social releasers*. For example, Tinbergen (1951) describes how the chick of the herring gull is fed by a parent. When the parent arrives at the nest after foraging, it calls the chicks out of hiding. The chick approaches the parent and pecks at the red spot at the tip of the parent's beak. This stimulates the parent to regurgitate food, which it then picks up from the ground and holds at the tip of its bill, as illustrated in Figure 22.1.

The stimuli that elicit the begging response of the herring gull were subjected to extensive analysis by Tinbergen and his co-workers (e.g., Tinbergen, 1949; Tinbergen and Perdeck, 1950). Using a series of cardboard models of a gull's head, they measured the responsiveness of chicks in terms of the number of pecks aimed at the model in a given period of time. The models were systematically varied in shape, coloration, and patterning, as illustrated in Figure 22.2. The experiments showed that the greatest response was to models held fairly close to the ground, moving slightly, and with a long, thin, downward-pointing protrusion. The chicks normally aimed their pecks at the tip of the bill and were most responsive if this was marked with a red spot on a contrasting background. The color of the bill and of the head were found not to affect the begging response.

The red spot on the herring gull's bill has all the characteristics of a sign stimulus. In the terminology of classical ethology, it "releases the begging response" of the chick. This type of social releaser was seen as the basis of social communication in animals:

> So far as our present knowledge goes, social cooperation seems to depend mainly on a system of releasers. The tendency of the actor to give these signals is innate, and the reactor's responses are likewise innate. Releasers seem always to be conspicuous, and relatively simple. This is significant, because we know from other work that the stimuli releasing innate behavior are always simple "sign stimuli." It seems, therefore, as if the structures and behavior elements acting as releasers are adapted to the task of providing sign stimuli (Tinbergen, 1953).

Classically, social releasers were considered to be characteristic of each species and to have evolved through a process of ritualization. Their recognition by means of the innate releasing mechanism also was considered a species characteristic, and the communication system was believed to have evolved in such a way that the mechanisms that send and receive the signal are kept in tune with each other. In this chapter we see how subsequent research has modified this basic picture.

22.1 Ritualization

In Chapter 21 we saw that displacement activities may become ritualized during the course of evolution so they come to serve a communication function. Not only displacement activities but any activity already a potential source of information to other animals may become ritualized. Darwin (1872) noted that the protective facial expressions of mammals play a role in communication. The protective reflexes, which include narrowing of the eyes, flattening of the ears, and raising of the hair around the neck, serve to protect the sense organs at moments of danger. Such responses signal information to other animals, who can interpret them as signs of fear or anger. Thus, primitive facial expressions provide good material for the selection of an efficient communication system. The expressions can be made more effective by exaggeration, by accompanying vocalization, and by distinctive markings that draw attention to the face or that emphasize a change in facial express-

Fig. 22.3 Plate from Darwin's book *The Expressions of the Emotions in Man and Animals*

Fig. 22.4 Darwin's principle of antithesis illustrated by the posture of a dog when friendly (left) and angry (right)

ion. The absence of hair on parts of the human face draws attention to the main features used in communication as seen in Figure 22.3.

Darwin noted that signals opposite in meaning often are conveyed by expressions or postures that are opposites. Human facial expressions indicating pleasure and anger use opposing sets of muscles, and the posture of an angry dog is in many ways the opposite of its posture when friendly (see Fig. 22.4). Darwin (1872) called this the *principle of antithesis.* Protective responses and their antitheses are thought to have been particularly important in the evolution of facial expressions among primates (Andrew, 1963).

Another aspect of behavior that is thought to have provided an important origin for ritualized displays is *intention movement;* which is an incomplete behavior pattern that provides potential information that an animal is about to perform a particular activity. For example, when a bird is about to take off in flight, it first crouches, raises its tail, and withdraws its head, as illustrated in Figure 22.5. The crouching may occur a number of times before the bird takes off, or it may not precede flight at all.

The importance of intention movements as signals to other animals can be seen from the study of Michael Davis (1975) on the flight reactions of pigeons. He found that a pigeon, on leaving a flock, does not disturb the others, provided the normal flight-intention movements occur. If a pigeon flies away suddenly, without any prior intention signals, then all the birds fly off. It appears that flight not preceded by intention movements is a type of alarm signal.

Fig. 22.5 Many birds crouch and raise the tail as a prelude to take-off

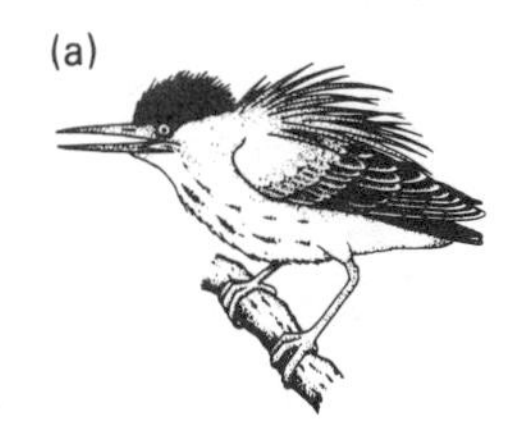

(b)

Fig. 22.6 Displays of the green heron *Butorides virescens*. (*a*) Forward threat display, (*b*) snap display, (*c*) stretch display (From *The Oxford Companion to Animal Behaviour*, 1981)

A number of ethologists have studied ritualized intention movements (e.g., Daanje, 1950; Tinbergen, 1953; Andrew, 1956a). Such movements occur, for example, in the courtship of the American green heron and the golden-eye duck. The American green heron, studied by Andrew Meyerriecks (1960), nests in dead trees in salt marshes. The males arrive in spring and defend a nest tree against rival males. Intruders are challenged by a forward threat display (Fig. 22.6) that is thought to have evolved from intention attack behavior. The male adopts a horizontal posture, points its beak toward the opponent, erects its feathers, and vibrates the tail. Females are attracted by the male advertising call but initially are threatened by the male. As the females persist, the behavior of the male changes, and his readiness to accept the female is signaled by the snap display (Fig. 22.6). The beak is pointed diagonally downward and the mandibles are snapped together. This display is similar to the behavior in which the male breaks twigs from trees to build its nest, and it may be a ritualized form of displacement nest building.

After accepting the female, the male performs the stretch display, which appears to be a ritualized form of flight intention. It is the antithesis of the forward display in many respects. Whereas the forward display is accompanied by a harsh call and ruffled feathers that increase the bird's apparent size, the stretch display is accompanied by a soft call and sleeked feathers. In the forward display the beak, the bird's main weapon, is directed toward the opponent, thus exhibiting the bright red lining of the mouth. In the stretch display the beak is directed away from the female. Whereas the forward display signifies threat, the stretch display symbolizes appeasement and is followed by mutual displays on the part of male and female, including mutual billing and preening. Copulation occurs shortly after this contact between male and female.

22.2 Conflict

Motivational conflict is a common starting point for ritualization. It occurs when two tendencies compete for dominance in the control of behavior (see Chapter 25). Because conflicting tendencies cannot be expressed simultaneously in behavior, the behavior seen during conflict is very different from the normally smooth run of activity.

There are three logical main types of conflict, though only one is of practical importance. (1) *Approach-approach* conflict occurs when two simultaneous tendencies are directed toward different goals. Although it is possible for an animal to reach a point where the tendencies are equal, such a situation is usually transitory because any departure from the point of balance will result in an increased tendency to approach the other. This instability occurs because of the *goal gradient* (Fig. 22.7), by which the tendency to approach a goal increases with proximity to the

Fig. 22.7 The goal-gradients of approach and avoidance

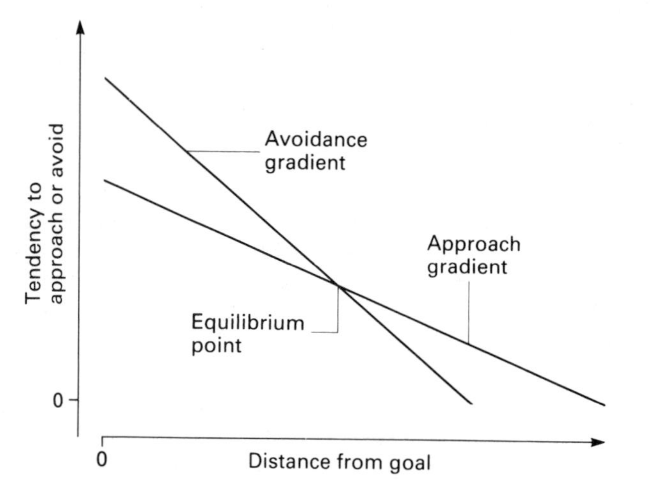

goal. (2) *Avoidance-avoidance* conflict occurs when two avoidance tendencies occur simultaneously. Since the tendency to avoid objects usually increases with proximity to the object (see Fig. 22.7), the animal will tend to move to a position where the avoidance tendencies are equal and then to escape from the situation by moving away at right angles to the line between the two objects. An avoidance-avoidance situation is therefore unstable. (3) *Approach-avoidance* conflict occurs when an animal has simultaneous tendencies to approach and to avoid the same object. As illustrated in Figure 22.7, the gradients of approach and avoidance normally cross in this type of situation; that is, near the object the avoidance tendency is stronger than the approach tendency, while the reverse is true when the animal is far from the object. At some intermediate distance is a point of equilibrium where the tendencies to approach and avoid are the same. Thus, approach-avoidance conflict tends to be stable because the animal moves toward the equilibrium point whichever side it starts from.

Approach-avoidance conflict is very common in animal behavior. Thus, the male three-spined stickleback tends to find itself in such a conflict (between fear of and aggression towards a rival male) when near the boundary of its territory. Courtship also involves approach-avoidance conflict, since each animal is initially wary of the other, yet sexually attracted. In theory, the animal reaches an impasse at the equilibrium point in a conflict situation because whichever way it moves, its approach or avoidance tendency brings it back to the equilibrium point. In practice, one tendency eventually dominates, either because the animal's fear of the object declines or its tendency to approach increases or because some other motivational factor intervenes. However, animals may be readily observed to oscillate backward and forward during approach-avoidance conflict or to remain stationary in an ambivalent posture. Behavior that consists of separate components

Fig. 22.8 The upright threat posture of a herring gull (After a photograph by Niko Tinbergen)

of the conflicting tendencies is called *compromise behavior*. Alternatively, the animal may take up an *ambivalent posture* that compounds elements of the conflicting tendencies (Andrew, 1956c). For example, when a duck is offered bread by a person in a park, it may approach and then stop, craning its neck forward to reach the bread while turning its body away. As we saw in Chapter 21, displacement activities are typical of conflict situations and are thought to occur at the point of equilibrium.

The motivational state of an animal in a conflict situation usually is readily apparent. It is therefore ideal material for ritualization. Many displays seem to consist of ritualized aspects of conflict behavior, and many have been analyzed from this point of view. For example, the upright threat posture of the herring gull (Fig. 22.8) contains elements of both fear and aggression (Tinbergen, 1959). On the one hand, the sleeked feathers, stretched upright neck and sideways position of the body with respect to the opponent are signs of fear. On the other hand, the downward-pointing bill and raised wing carpels (elbows) are signs of aggression.

The interpretation of displays, in terms of their evolutionary origin and present meaning, is fraught with difficulties. Nevertheless, ethologists have been fairly successful in analyzing the diversity of behavior shown in threat and courtship in terms of ambivalence among a relatively small number of behavioral tendencies. The usual approach is to postulate three basic tendencies, one of which, if acting alone, would lead to sexual behavior, one to attack, and one to fleeing. These activities rarely are observed in unadulterated form. The observed behavior is interpreted in terms of a mixture of the three basic tendencies. Thus, a threat posture usually will be due to a particular combination of the tendencies to attack and to flee. This approach has been applied to a wide variety of activities, including courtship, nest-relief ceremonies, and mobbing behavior. Recognition that the behavior involves conflict and analysis in terms of the underlying incompatible tendencies is the essence of the approach, which has come to be known as the "conflict theory of display" (Baerends, 1975), and is typified by the work of Tinbergen (1959; 1962) and his co-workers.

The interpretation of displays in terms of conflict involves four lines of evidence (Tinbergen, 1962; Hinde, 1966):

(1) *The situation*. For instance, near a territory boundary the animal is likely to be both fearful and aggressive, while in the presence of a potential mate, sexual motivation may be involved as well.

(2) *The behavior accompanying a display*. There may be obvious conflict behavior like alternately approaching and retreating from a rival. There may be color changes correlated with the animal's motivation, such as the color patterns of male guppies that give an indication of the extent of the fish's sexual motivation.

(3) *The behavior immediately preceding or following a display*. Martin Moynihan (1955) used this method to assess the relative strengths of the

attack and flight tendencies involved in various displays of the black-headed gull (see Fig. 22.9). However, there are a number of difficulties with this method. First, the behavior shown immediately after a display may be a reaction to the other animal's response to the display. This difficulty can be circumvented by including in the analysis only that behavior shown when the behavior of the rival is unchanging or by testing theories about displays by using motionless dummies of rival animals. A second problem is that the behavior that precedes or follows a display may be due to a complex of motivational factors rather than an expression of a simple tendency. Third, the behavior may be motivationally unrelated to the display, as in the cases of displacement activities and time sharing (see Chapter 25.4).

(4) *The nature of the display.* Each posture can be analyzed in terms of its components (angle of the head, limbs, etc.). For example, Tinbergen (1959) distinguished between the herring gull's aggressive threat posture (Fig. 22.8) and its anxious threat posture, in which the beak is held more horizontally. Edwina Baher and I showed that the positions of the features in different regions of the body of the aggressive or defensive dove were typical of these states and different than feather movements due to temperature changes (see Fig. 22.10) (McFarland and Baher,

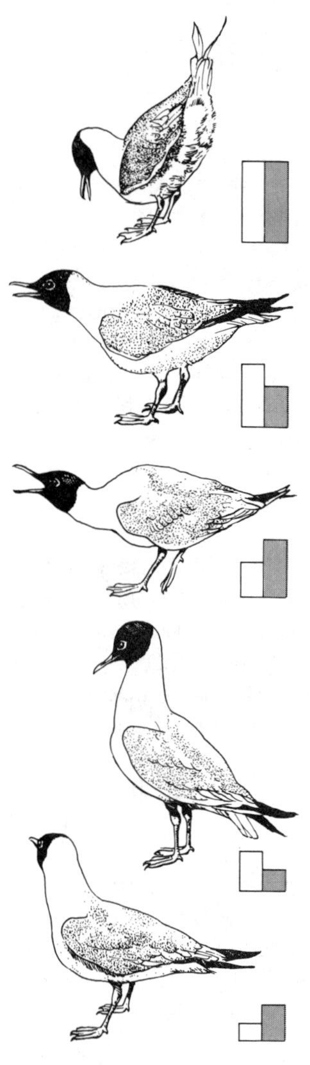

Fig. 22.9 Threat postures of the black-headed gull. The different postures are associated with differences in the balance between the tendencies to attack (dark column) and escape (pale column) (After Moynihan, 1955)

Fig. 22.10 Feather postures in the dove *Streptopelia risoria*. To obtain an index of feather position the body was divided into regions, and the feather posture in each region was noted once per minute. Different postures were given different scores, and the scores are summed over ten tests to give a minimum of zero (fully sleeked) and a maximum of 20 (fully raised). The birds were tested, while sitting on the nest, by moving a wooden rod 2 inches closer to the bird every 2 minutes. The birds reacted with a defensive posture which was characterized by systematic changes in the posture of the feathers in the different body regions. c = crown, n = neck, d = dorsal region, v = ventral region, w = wing, b = breast (After McFarland and Baher, 1968)

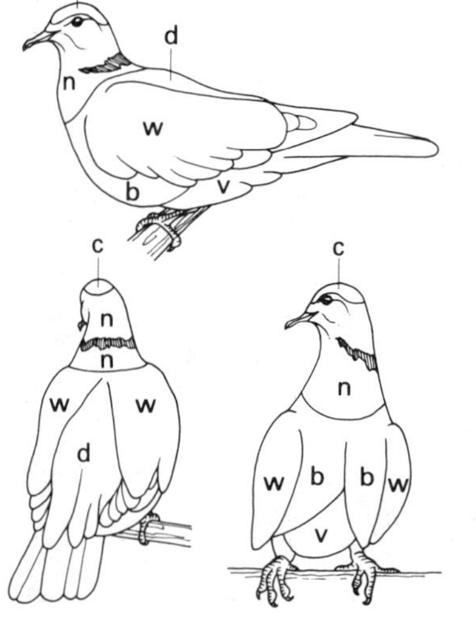

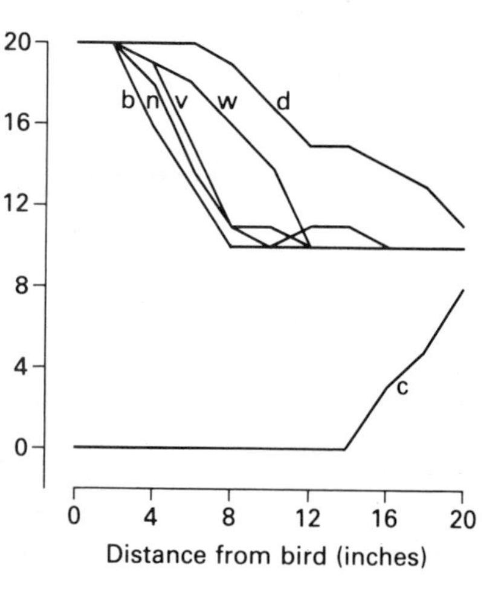

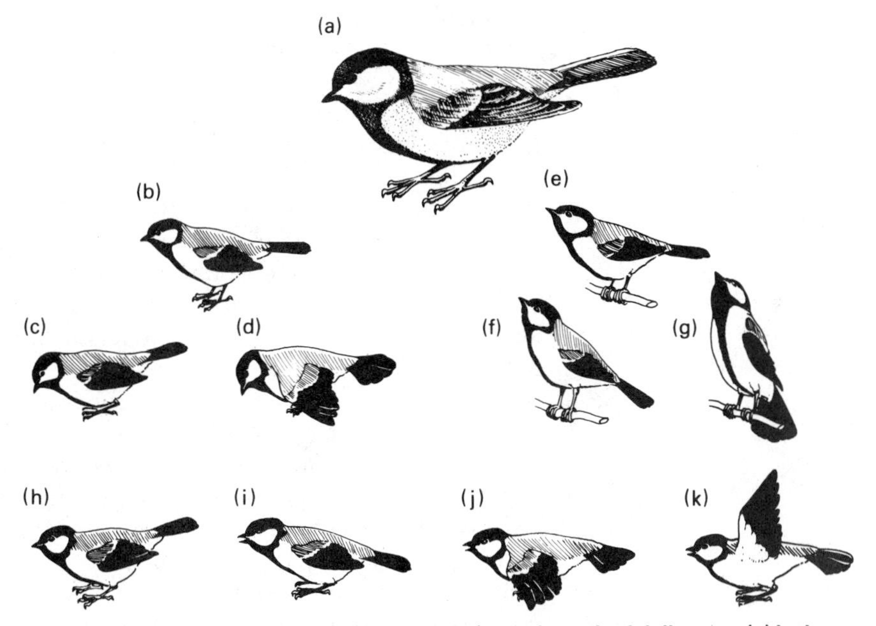

Fig. 22.11 Postures of the great tit (*Parus major*) seen in territorial disputes. (*a*) is the normal relaxed posture, which grades into a head down threat display (*b–d*), or into a head-up threat display (*e–g*). The horizontal display (*h, i*) is sometimes accompanied by a wings-out display (*j, k*). (After Blurton-Jones, 1968)

1968). Comparison of species sometimes can be used to substantiate this type of evidence. For example, the herring gull uses wing beating during fights, and its upright threat posture includes raising of the wing carpels as if in preparation to fight. Skuas do not use wing beating in fights, and their upright threat posture does not involve exposure of the carpels.

While these lines of evidence often can be used to obtain plausible accounts of the motivation of displays, there is sometimes an element of circularity in the procedure. The circumstantial evidence obtained from field observations has been substantiated in some cases by experiment. Robert Hinde (1952) described various threat postures in the great tit (*Parus major*), which he observed in territorial boundary disputes. The postures are summarized in Figure 22.11. Hinde formed the hypothesis that the threat displays were caused by simultaneous arousal of attack and fleeing behavior. Nick Blurton-Jones (1968) set out to test this hypothesis in experiments in which the tendencies to attack and to flee were manipulated independently.

In the initial part of his study, Blurton-Jones made observations of birds in uncontrolled situations using the methods of analysis outlined earlier. As a result of very thorough analysis, he concluded that the threat displays resulted from the same causal factors as attack and that the factors normally causing fleeing contributed little to the display. He postulated that threat displays occurred when causal factors for attack were prevalent but when attack was prevented from occurring by the tendency to flee or some other factor. A tendency to stay in one place, resulting from a tendency to feed there, was related to the proportion of horizontal displays, as opposed to head-down or head-up displays. Thus, the type of threat display seemed to be influenced by the circumstances that prevented attack.

In the experimental analysis, the tendencies to attack, flee, and feed were manipulated independently. Attack provoked by poking at the bird with a pencil was accompanied by the head-down, horizontal, head-up, and wings-out posture. Fleeing was evoked by a small lamp bulb and was accompanied by crest raising and feather fluffing. Presentation of food elicited hopping toward, reaching for, and taking the food. By presenting stimuli simultaneously, Blurton-Jones was able to create artificial conflicts. Simultaneous presentation of attack and flight stimuli created an attack-flee conflict in which attack was reduced greatly and the head-up and other threat postures increased. Thus, a prolonged bout of attack behavior could be changed to threat display by a presentation of the flee stimulus outside the cage where the bird could not reach it. This confirmed the conclusion from the observational part of the study that anything that stopped attack in the presence of the attack stimulus would evoke threat. Blurton-Jones reasoned that this result would not be expected if the threat displays were unritualized combinations of components of attack and escape. However, the displays might be ritualized to the extent that they were no longer influenced by the fleeing tendency but only by the attack tendency. It is possible for a display to become completely emancipated from its original causal factors and to acquire causal factors of its own. However, this does not seem to have occurred in the threat displays of the great tit. Both the observational and experimental methods indicate that threat displays have essentially the same motivation as attack behavior, though the exact form of the display depends upon the circumstances.

22.3 Communication among species

Animals living in close association with humans often behave as if humans were members of their own species. Pet owners have plenty of opportunity to see this kind of behavior. It took the owner of a pet tortoise some time to realize that the animal was making repeated attempts to court his shoes. In zoos the male kangaroos may behave as if the upright posture of the keeper were a challenge to fight. If the keeper adopts the bowed posture characteristic of peaceful kangaroos, then the confrontation can be avoided (Hediger, 1964). Similarly, many people treat animals as if they were human beings. They talk to their pets and may even provide them with human adornment like nail polish. The

Fig. 22.12 Common characteristics of the juveniles illustrated here include short faces, rounded head shape, and high foreheads

tendency to attribute human characteristics to animals, called *anthropomorphism*, probably results from an instinctive recognition of sign stimuli important in human social behavior. For example, infant head shape is an important factor in evoking parental responses from human adults, and it is often said that people also respond to similar characteristics in young animals, as shown in Figure 22.12. Such appealing features often are exaggerated in cartoons and advertisements.

In addition to the purely fortuitous similarity between the features of one species and the sign stimuli of another species, there are many instances where natural selection has favored communication among species. This is a particularly common feature of antipredator devices. When discovered by a predator, many animals adopt a posture designed to intimidate the predator. In some cases the display is pure bluff. For example, many species of moth and butterfly suddenly expose eyelike spots on their hind wings when they are disturbed while resting. Such eye spots also are found in cuttlefish, toads, and caterpillars (see Fig. 22.13). Some researchers have shown experimentally (e.g., Blest, 1957; Coppinger, 1969; 1970) that the sudden appearance of bright colors may startle predatory birds, giving the moth a chance to escape.

Eye spots, whether permanent or suddenly revealed, also have a deterrent effect, presumably on account of their resemblance to the bird's predators. David Blest (1957) placed dead mealworms on a box and encouraged birds such as chaffinches, buntings, and tits to feed on

Fig. 22.13 Hawkmoth (*a*) in resting posture (*b*) exposes its eye-spots when disturbed by a predator

(a)

(b)

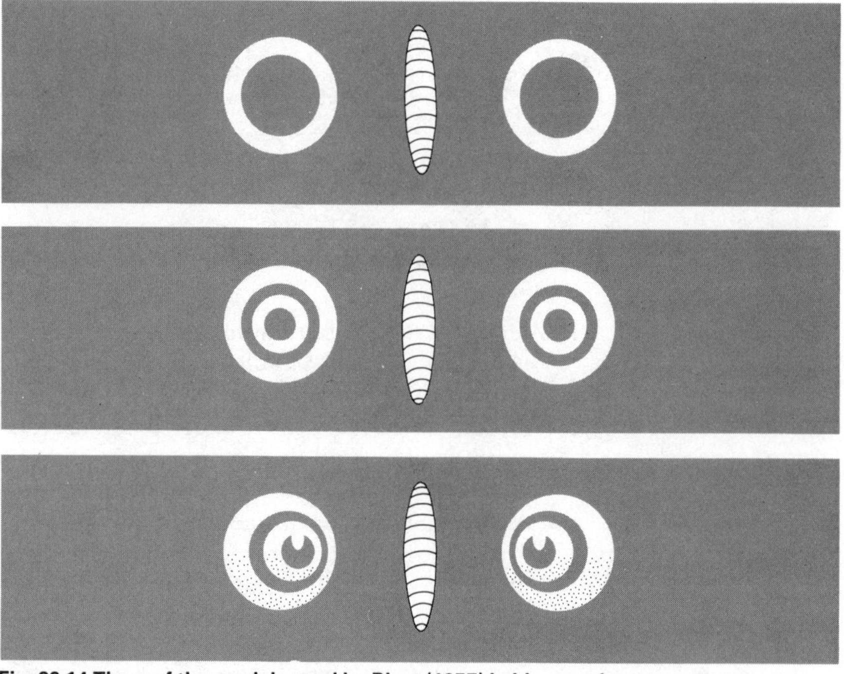

Fig. 22.14 Three of the models used by Blest (1957) in his experiments on eye-spot patterns. When a bird landed on the apparatus to obtain a mealworm (center) it activated a switch that lit up the circles, or eye-spots, on either side. The birds were least frightened by the model shown at the top, and most by that at the bottom

them. When the birds were accustomed to the situation, he tested various eyelike patterns. As soon as the bird alighted on the box, it completed an electric circuit that lit up two patterns each side of the mealworm. Blest found that circular patterns were more effective than crosses in frightening the birds and that the more eyelike the pattern, the more effective it was in eliciting escape behavior (see Fig. 22.14). Blest found that birds soon habituate to the eye spots, and it therefore would seem advantageous for insects to keep their eye spots hidden until needed.

The display of eye spots is a form of mimicry of other species' sign stimuli. Many types of display involve mimicry of other animals' markings or behavior. The saber-toothed blenny mimics the distinctive color patterning of the cleaner wrasse (see front cover) and is thus able to deceive large fish into permitting it to approach. Instead of removing parasites like the cleaner wrasse, however, the blenny bites a piece out of the large fish and then escapes. Some snakes mimic the color patterning and warning displays of poisonous species. Thus, the harmless coral snake *Lampropeltis elapsoides* mimics the distinctive bands of red, yellow, and black of the poisonous coral snake *Micrurus fulvius* (see Fig. 22.15). The African carpet viper (*Echis carinata*) has a warning

Fig. 22.15 The poisonous coral snake *Micrurus fulvius* (left), and its non-poisonous mimic *Lampropeltis elapsoides* (right) (After Edmunds, 1974)

display in which it folds its body upon itself and produces a rasping or hissing noise by rubbing the folded parts against each other (see Fig. 22.16). This display is mimicked by some of the harmless snakes of the genus *Dasypeltis*. Some hole-nesting birds hiss like a snake if disturbed on the nest. Since it is dark in the hole, predatory mammals might be intimated by the display, even though the bird does not visually resemble a snake (Hinde, 1952). Some hawkmoth caterpillars have markings on the head that closely resemble the head of a snake when inflated, as shown in Figure 1.3. When the caterpillar is disturbed it inflates its snake-head and moves it around. It may even strike at the predator (Wickler, 1968; Edmunds, 1974).

Mimicry is a form of deceit. The display of eye spots or snakelike stimuli give protection to the extent they induce in the predator behavior appropriate to dangerous stimuli. If the potential prey is not really dangerous, then the predator is deceived. Another example of deceit can be seen in the angler fish (*Lophius piscatorius*) that dangles a wormlike bait on the end of a rodlike appendage (see Fig. 22.17). When a prey fish approaches the lure, it is captured by the angler fish. Evolutionary deceit exists in situations in which natural selection favors developments in members of one species that trick members of another species into behavior that is deleterious. Natural selection will, of course, tend to sharpen the powers of discrimination of the victim, but

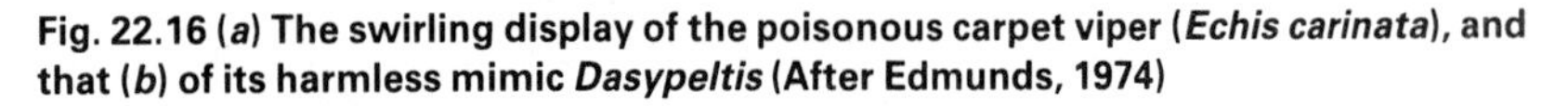

Fig. 22.16 (*a*) The swirling display of the poisonous carpet viper (*Echis carinata*), and that (*b*) of its harmless mimic *Dasypeltis* (After Edmunds, 1974)

(a)

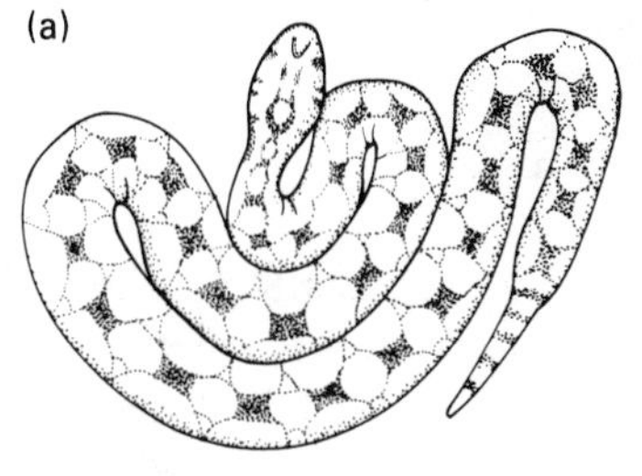

(b)

Fig. 22.17 The angler fish *Lophius piscatorius*

this can be counteracted by the evolution of more effective mimicry. If the model is sufficiently common in relation to the mimic, it is very difficult for the victim species to avoid being deceived. Thus, because wormlike objects are a common form of prey, the angler fish easily can exploit the prey-recognition system of its victims. To be able to distinguish true prey from the mimic, the victim would have to spend more time inspecting each prey item, at the expense of feeding efficiency. Provided the dangerous prey are not too common relative to real prey, natural selection will result in a compromise; that is, if the risk of encountering the poisonous prey is low, it can be offset by the benefits of efficient feeding. Nevertheless, it is important to recognize that in cases of communication among species that involve deceit, the forces of natural selection acting on each species tend to work in opposition.

Situations in which communication among species is mutually beneficial usually involve *symbiosis*. In one form of symbiosis, known as *commensalism*, one species benefits from the relationship while the other remains unaffected. For example, the trumpet fish (*Aulostomus*) sometimes joins schools of yellow sturgeon and takes advantage of this camouflage to approach smaller prey fish. It darts out from among the sturgeon and seizes its prey. The sturgeon fish remains unaffected by the association, and there appears to be no communication between the two species. Similarly, cattle egrets (*Bubulcus ibis*) live in close association with cattle and feed on insects disturbed by them. The birds do not remove parasites from the cattle as do tick birds. There is no communication between the cattle and the cattle egrets, and it seems that the cattle do not benefit from the relationship.

In true symbiosis, or *mutualism*, both species benefit and communication usually occurs between them. For example, the honey badger (*Mellivora capensis*) lives in symbiosis with a small bird called the honey guide (*Indicator indicator*). When the bird discovers a hive of wild bees, it searches for a badger and guides it to the hive by means of a special display. Protected by its thick skin, the badger opens the hive with its large claws and feeds on the honeycombs. The birds feed upon the wax and bee larvae, to which it could not gain access unaided. If the honey guide cannot find a badger, it may attempt to attract people. The natives understand the bird's behavior and follow it to the hive. It is an unwritten law that the bird be allowed to take the bee larvae.

22.4 Manipulation

In recent years there have been attempts to interpret animal communication in terms of evolutionary theory. For example, Richard Dawkins and John Krebs (1978, p. 286) argue that "communication is said to occur when an animal, the actor, does something which appears to be the result of selection to influence the sense organs of another animal, the reactor, so that the reactor's behavior changes to the advantage of the actor." In other words, the behavior of the actor is selected to manipulate the behavior of the reactor. The signals given by the actor are the tools designed by natural selection to exploit rivals or members of other species. Thus, the eye spot display of a butterfly exploits the fact that small birds are frightened by a pair of staring eyes, and the wormlike lure of the angler fish exploits the sign stimuli to which smaller fish respond while foraging. In the case of communication among members of different species it is easy to see how one animal may deceive or manipulate another. However, Dawkins and Krebs suggest that the same reasoning applies to communication between two members of the same species. For example, they note that:

> If a dog can cause rivals to flee simply by baring his teeth, selection will favour dogs who exploit this power. Tooth-baring will become ritualised, exaggerated for increased power to frighten, and the lips may be pulled back further than is strictly necessary merely to get them out of the way. Over evolutionary time teeth may get larger, even if this makes them less significant for eating (p. 288).

This view may not seem very different from the traditional ethological view. However, Dawkins and Krebs maintain that the recipient of the signal is deceived into responding by the similarity between the ritualized signal and other sign stimuli that the reactor normally would respond to, just as we saw in our discussion of communication among members of different species. This is where they depart from the traditional ethological view, which has always maintained that evolutionary developments in communication must occur in both actor and reactor (e.g., Blest, 1961); that is, during the ritualization at displays, or the evolution of any specialized mode of communication, any change in the signal must be matched by a corresponding change in the signal recognition mechanism.

While Dawkins and Krebs can argue that in the case of the snarling dog mentioned earlier, the exaggerated or ritualized signal is still meaningful to the recipient, this can hardly be the case with the majority of ritualized communication systems. For example, the ritualized displacement preening of the mandarin drake (see Fig. 21.8) or the displays of the male green heron (see Fig. 22.6) hardly can have evolved without some corresponding evolution of female recognition mechanisms. Similarly, the elaborate fernlike antennae (Fig. 12.1) by which the male silkmoth detects the female pheromone hardly can have evolved to serve some other function. Thus, communication systems often involve complex coevolution by actor and reactor. This is a situation in which evolutionary theory offers more than one possible explanation.

In evolutionary terms, it is hard to understand how an individual animal can benefit by informing another individual about its true motivational state or about what it is likely to do next. It seems more plausible that an animal would attempt to deceive others to gain an advantage. The main reason for coming to this theoretical conclusion is that a population of honest animals is open to invasion by dishonest cheaters, and it is difficult to see how the invasion could be stopped. Thus, blushing in humans is an example of non-verbal communication that is probably ritualized. Blushing is not usually under voluntary control, and it occurs in people who are embarrassed or mildly frightened. It is usually confined to the face and neck, the focal area of communication, and it is the opposite of what we would expect on simple physiological grounds. Normally, mild fear results in sympathetic arousal that causes blood to move away from the skin and other peripheral parts of the body. If blushing is a ritualized activity that informs other people that the actor is embarrassed, it is difficult to see how it has evolved. How does the actor benefit by providing this information? Why are populations of blushers not invaded by non-blushing cheats?

Another interesting example is provided by a study of the Harris sparrow (*Zonotrichia querula*) (Rohwer and Rohwer, 1978). These birds form flocks in winter, and dominant birds exert priority over subordinates in gaining access to food. The dominant males have more black plumage than subordinates, although all males are capable of growing black feathers, which is what they do when they adopt their breeding plumage in the spring. Why the paler males do not simply grow dark plumage and gain the benefits of apparent dominance is something of a puzzle. The Rohwers painted subordinates to look like dominant sparrows, but these birds did not rise in status. Pale birds treated with the hormone testosterone became more aggressive, but their opponents did not retreat during disputes. However, birds that were both painted and treated with testosterone were able to assert their dominance. Thus, it looks as though a combination of dark coloration and dominant behavior is necessary to attain a dominant status that is recognized by other birds. If a subordinate male attempts to cheat by growing dark feathers, it would not be successful. However, it is difficult to see why a cheat could not also secrete more testosterone. Perhaps high levels of testosterone carry costs that subordinate birds cannot afford.

Points to remember

- Ritualization of an activity usually results in increased conspicuousness and exaggeration of particular details. It usually occurs in activities that have a potential communication function, such as intention movements, conflict behavior and displacement activity.

- Conflict behavior occurs in many types of social situations and is characterized by stereotyped displays which have an important role in communication.

● Communication among species often involves deceit, in the sense that the recipient is tricked into responding to stimuli that mimic some aspect of their normal life. How far this principle can be extended to communication in general is a matter of debate.

Further reading

Sebeok, T. (ed.) (1977) *How Animals Communicate*, Indiana University Press, Bloomington, Indiana.

3.2 Decision-making in Animals

In this section we look at complex behavior of animals, taking as an initial example the behavior of honeybees. This example, which takes up the whole of Chapter 23, shows how complex behavior can occur in a relatively simple animal. In Chapter 24 we look at decision-making in animals from an evolutionary viewpoint and discover many affinities with consumer economics. In Chapter 25 we look at mechanisms of decision-making, including motivational competition, time-sharing and optimal decision-making.

Karl von Frisch (1886–1983)

Karl von Frisch was born in Vienna and educated at the Universities of Munich and Vienna. He became Professor of Zoology at the University of Rostock in 1921, at Breslau in 1923, and at Munich in 1925.

As a child, Karl von Frisch was interested in natural history, and while still at school he published some of his observations, including experiments on the light sensitivity of sea anemones. Most of his subsequent research on animal behavior was concerned with how animals obtain information about their environment. For much of his life von Frisch spent winters in his laboratory, studying fish, and summers at his family home at Brunnwinkl, studying honeybees. He discovered that fish are capable of color vision and of discriminating underwater sound waves. Both these discoveries were contrary to the prevailing scientific opinion, and thus aroused opposition. Von Frisch also discovered that when the skin of a minnow is damaged, a pheromone is released that causes other minnows to flee from the area. In these studies von Frisch's success was based upon careful behavioral observation and a profound understanding of biological function. This was also shown in his work on honeybees.

It was widely thought that bees were color blind, but von Frisch reasoned that the color of flowers must function to attract bees and other insects. He demonstrated that, although bees ignore the wavelength of light when escaping from a box, they are responsive to color when foraging for food. During these experiments he noticed that a single "scout" would appear at colored food dishes set out in the open, but once the scout had departed it was only a short time before many bees arrived. This observation led von Frisch to the discovery of the bee language system.

In 1973 Karl von Frisch shared the Nobel Prize for Medicine with Konrad Lorenz and Niko Tinbergen.

Photograph by courtesy of Maximilian Renner

Although von Frisch's greatest material contribution was his work on honeybee communication, he also stands out as a pioneer of the argument from design. Time and again he made important discoveries on the basis of his understanding of biological function.

23 The Complex Behavior of Honeybees

In this chapter we examine the behavior of a particular species in detail: the honeybee (*Apis mellifera*). The honeybee has been the subject of considerable scientific research, and quite a lot is known about its behavior. Because the honeybee is an insect, we tend to think of it as a mere automaton. However, its behavior is surprisingly complex. By understanding the nature of this complexity we hope to gain some insights into the organization of complex behavior in general and, perhaps, the complex behavior of other species, especially those more closely related to humankind.

23.1 The honeybee life cycle

Fig. 23.1 Comb of the tropical dwarf honey bee *Apis florea* (After a photograph in Gould, 1982)

Honeybees evolved in the tropics and wild species are still to be found there, one of which is the dwarf honeybee *Apis florea*. This bee builds a sheet of comb that hangs on a tree branch, exposed to the open air, as illustrated in Figure 23.1. The comb is made up of individual cells used to store honey and pollen, and to raise bee larvae. In the warm, equitable climate these bees can live their lives in public all year round.

It is thought that the tropical bees are primitive, and that the familiar honeybee (*Apis mellifera*) was able to colonize the temperate zones of the world by evolving the practice of building its comb within hollow trees, thus enabling them to gain protection from the changeable weather and to withstand very cold temperatures by massing together and generating metabolic heat in the insulated cavity. During the spring, when food becomes abundant, the queen bee lays many thousands of eggs. She is constantly attended by young female "nurses," which groom and feed her. They are attracted by a special pheromone and as they groom the queen they ingest chemicals that suppress the activity of their ovaries, and prevent them from becoming possible rivals to the queen. The queen is capable of laying up to 3,000 eggs per day. Each egg hatches two days after deposition and the resulting larva is fed by the workers, receiving food roughly once per minute for an entire week. The queen lays a few eggs in especially large cells prepared by the workers. The larvae in these cells are given special nourishment and develop into fertile queens rather than sterile workers. Of these virgin queens, one becomes the new queen of the hive; but before this happens the old queen prepares for departure.

Once the new queen cells have been started, the workers tending the queen stop feeding her. Her egg production rate declines and she loses weight. When the queen has lost enough weight to be capable of flight, she starts to communicate with the virgin queens: She produces a pulsed tone by vibrating her thorax, and any mature virgin queens respond from their cells with a similar pulsed tone of higher frequency. In this way the queen indicates that she is ready to depart, and the virgin queens signal that they are ready to take her place.

The old queen departs from the hive together with about one half of the total population. The departing bees swarm together and form a compact cluster on a nearby tree, where they remain for several days. Scout bees fly off and investigate possible sites for a new hive. They return and perform a dance on the vertical surface of the swarm, a dance that provides information about the direction, distance, and quality of a new hive site. The dance also stimulates other workers to fly off and inspect the site, who, when they return, also dance in a way that indicates the quality of the site. A long and vigorous dance indicates an attractive site. Usually more than one possible site is initially reported, and each report recruits further scouts. Differences in the quality of the potential sites become apparent from the number of dancing bees and the vigor of their dances. Eventually the bees from the inferior sites cease to advertise their discoveries and universal agreement is reached. The swarm then moves to the chosen site and begins to build a new hive. The bees prefer sites that are of a suitable size and are protected from climatic extremes, such as holes in trees, or in the ground. The new hive should not be too far for the queen to fly from her temporary resting place, nor too near the old hive.

When the old queen has left, the new ones emerge from their cells. The first to emerge usually kills those still in the cells. If two emerge simultaneously they fight until one is killed. A few days later the new queen flies out of the hive to mate with several drones. She then returns to begin her egg-laying career.

The drones are few in number, and spend most of their time in the hive doing nothing. On sunny afternoons they may fly out to visit the traditional mating places in the locality, which are usually about 20–30 meters above ground and are simply air spaces about 20 meters in diameter. The same localities are used year after year, even though the drones do not survive the winter and the new queens have not visited the mating area before. How the bees find these places is a mystery. The drones chase the arriving queens and the successful one mates with her, and dies soon afterwards. The vast majority of drones never mate and they are evicted from the hive when fall comes.

In midsummer the bees start preparing for the winter. The colony size increases slowly. The workers devote some time to raising young, but most of their time and energy is spent collecting and storing food for the winter. Each worker takes on a series of tasks. Initially she cleans the cells, but as her mandibular glands develop she feeds the queen and her brood. When the worker's wax glands become functional, she helps to cap cells and build the comb. At about three weeks old she begins to forage, and by about six weeks she is dead, usually from wear and tear.

23.2 Foraging by honeybees

The foraging honeybee, scouting for food, is faced with a formidable task. She must leave the hive and search for food. She must recognize suitable sources of food. She must then register her whereabouts in relation to the hive. She must make her way back to the hive, and upon arrival must communicate her findings to other workers and persuade them to fly out and collect food from the newly discovered source. As with the bumblebees (discussed in Chapter 24), this must all be accomplished in the most economical manner, taking account of the quality of the food and its distance from the hive.

Honeybees forage primarily upon flowers that produce copious pollen and secrete a sugary nectar. Many plant species have flowers designed to attract bees and other insects, which act as vehicles for transporting pollen from one flower to another, thus cross-fertilizing the plants. Some flowers are open only at specific times of day, so the bees have to learn not only which kinds of flowers are producing nectar, but at what time and what place it is available. The first question we have to consider is: How do bees recognize suitable flowers?

When a foraging honeybee finds a source of food it may be a long way (up to 10 kilometers) from the hive. The discoverer must return to the hive to inform other workers about the find. To do this she must use information about the direction of the hive, even though she may have travelled by a circuitous route. The second question that we shall consider is: How do bees navigate?

Upon arrival at the hive the forager must communicate to the other workers the direction and distance of the food source, and must give some indication of the quality of the food. To do this she must attract the attention of the other workers, who may already be engaged in other tasks or may have already received messages from other foragers. Our third question is: How do bees communicate?

23.3 Flower recognition by bees

In 1912 Karl von Frisch started his experiments on honeybees. Going contrary to the prevailing view, he reasoned that honeybees were likely to have color vision. Why else would flowers be so colorful? He found

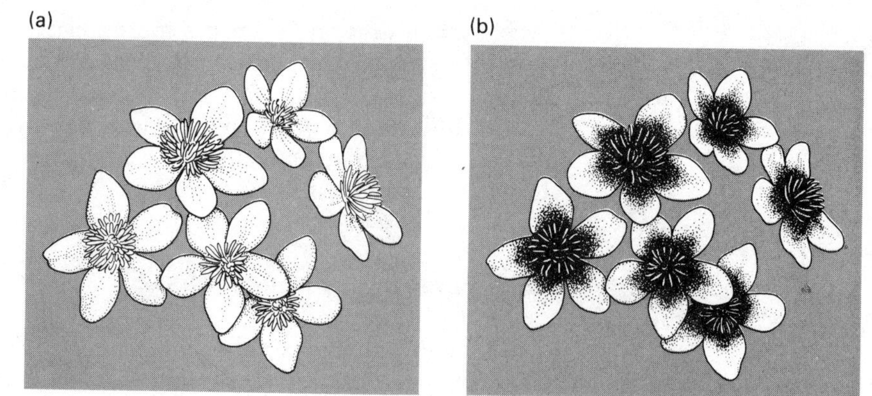

Fig. 23.2 Flowers that appear to us to be white in daylight (left) may have distinct patterns (called honey guides) when viewed in ultraviolet light (right). Bees are sensitive to ultraviolet light and respond to the honey guides

that the bees quickly learned to visit a dish of sugar solution placed near the hive, collect some of the food, and take it to the hive. Von Frisch then placed the dish on a colored piece of paper. After a number of further visits by the bees, he set out many pieces of paper, some colored and some various shades of grey. The bees searched for the food only on papers of the appropriate color, thus demonstrating that they could distinguish the color from shades of grey. Von Frisch also discovered that the bees were unable to tell red from grey, but that they could distinguish between grey paper made by different manufacturers. Further investigation showed that some sheets of paper reflected more ultraviolet light than others and that the bees were sensitive to this.

We now know that honeybees have well-developed color vision that differs from human vision in being insensitive to red, but that extends into the ultraviolet, where human eyes are totally insensitive. Von Frisch discovered that many flowers have well-developed markings, called *honey guides*. Some of these markings are visible only in ultraviolet light (see Fig. 23.2), so they are normally invisible to humans but visible to bees.

When honeybees are foraging they readily alight upon colored flower-like shapes. They can be trained to alight upon particular shapes to obtain food, but they prefer to alight on shapes with a broken outline and a radial pattern. Experiments with honey guides indicate that small details of flower pattern can influence the behavior of honeybees (Manning, 1956). In one group of orchids, the bee orchids, each species mimics the odor and appearance of a different species of bee. The male pollinates the flower while attempting to copulate with it (Baerends, 1950).

23.4 Honeybee navigation

When a bee flies out from the hive in search of a new food source, it takes a circuitous route as it visits various possible feeding localities. It

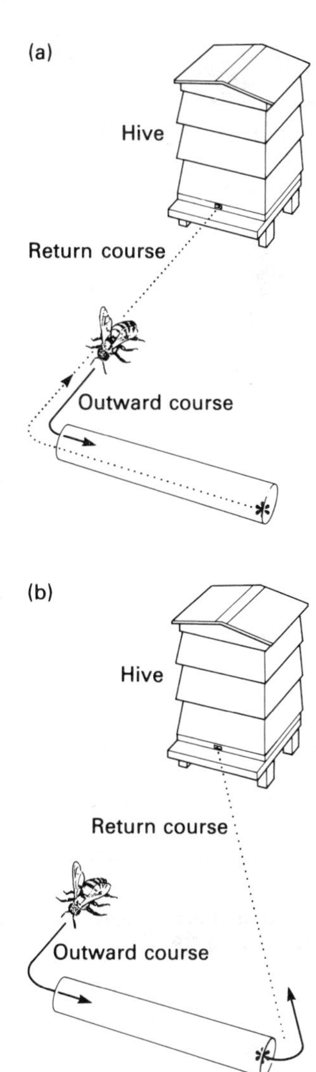

Fig. 23.3 Martin Lindauer's experiment to investigate the energy expended by bees during the flight from the hive to a food source. In (*a*) the bee has a long return course, while in (*b*) it has a short one. In both cases the outward course is the same, and when the bees returned to the hive their dance indicated this

flies home by a direct route without retracing its outbound path. It has been suggested that the bee keeps track of each leg of its outbound path by measuring the distance in terms of energy expended, and direction by reference to the angle with respect to landmarks and to the sun.

By requiring bees to walk to a source of food along a gallery (Fig. 23.3), Martin Lindauer (1963) was able to manipulate the distance of the return flight. He found that the bees correctly gauged the energy expended in the outbound flight after both long and short return flights. Von Frisch showed that bees fitted with 55-milligram lead weights, or with tinfoil flaps that increase drag, overestimate distances because of the extra energy they expend.

In investigating the importance of landmarks, von Frisch and Lindauer (1954) trained bees to obtain a honey reward by flying in a particular direction. In one experiment the flight path was along the north-south edge of a pine forest. When the bees had been trained, they were tested near the edge of a similar forest, the edge of which ran east-west (see Fig. 23.4). Most bees followed the line of the plantation, only a few taking the correct southerly course. Landmarks are most effective when they are linear and lead directly to the food. Trees in the middle of a field may be ignored (see Fig. 23.5), even though they would appear to be useful landmarks.

Once a scouting bee finds a food source it flies directly home. A simple experiment shows that the scouts use the sun as a compass. The bees are trained to a feeder, and then the position of the feeder is moved while some of the bees are feeding. When they depart for home, these bees fly in the direction that would have been correct if the feeder had not been moved. If the bees are trapped in the feeder long enough for the sun to move appreciably, they still fly in the correct direction, showing that they have a time-compensated sun compass (see Fig. 23.6).

Bees may often find themselves in situations where there are no suitable landmarks and the sun is hidden behind clouds. Under such circumstances they can still fly in the homeward direction, so they are obviously capable of using some other navigational cues. As we have seen (Chapter 14), bees are sensitive to the plane of polarization of sunlight in the ultraviolet region of the spectrum.

Von Frisch looked at different parts of the sky through an octagonal filter made from eight pieces of triangular polaroid (Fig. 23.7), and saw different patterns of brightness, even when the sun was behind clouds. The pattern of polarization of light in the sky is symmetrical about the sun due to particles in the atmosphere scattering the light. Thus, bees can sense the direction of the sun even when it is obscured by clouds. However, such information can be ambiguous, especially if the animal has a restricted view of the sky, as do primitive bees in the African forest.

Only a small patch of sky has to be visible to the bee, but there will

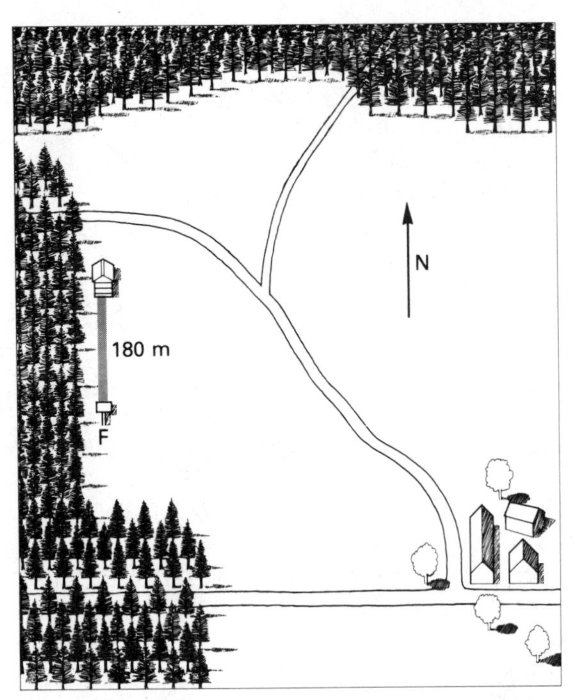

(a)

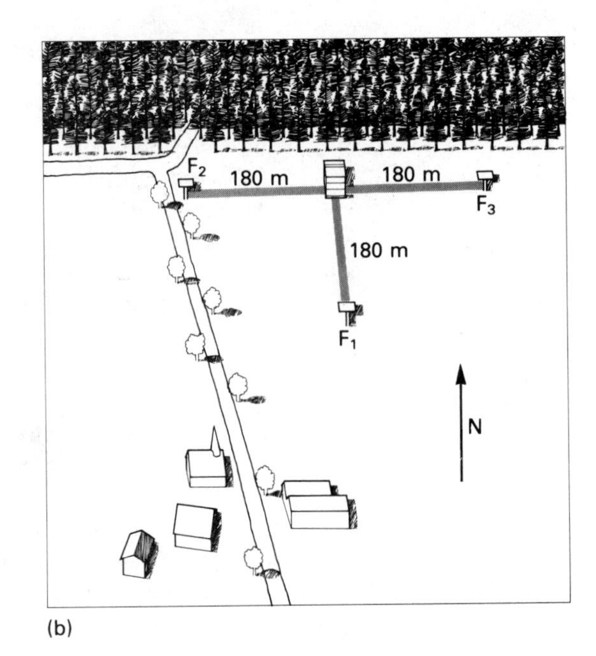

(b)

Fig. 23.4 Experiments to determine the role of linear landmarks in the orientation of bees. (*a*) The bees are trained to visit a food table F along the north-south edge of a pine plantation. (*b*) The bees are then tested near an east-west edge, where they were given a choice of three food tables F_1, F_2, and F_3. Most bees chose to fly east-west, along the edge of the wood, even though they had been trained to fly north-south (After Lindauer, 1961)

Fig. 23.5 Bees trained to fly past the tree to the food table F, ignored the tree and flew due south when the hive was placed west of the tree (After Lindauer, 1961)

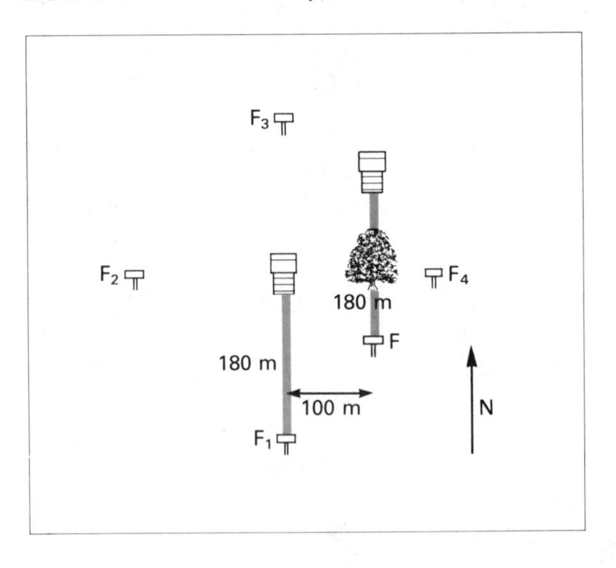

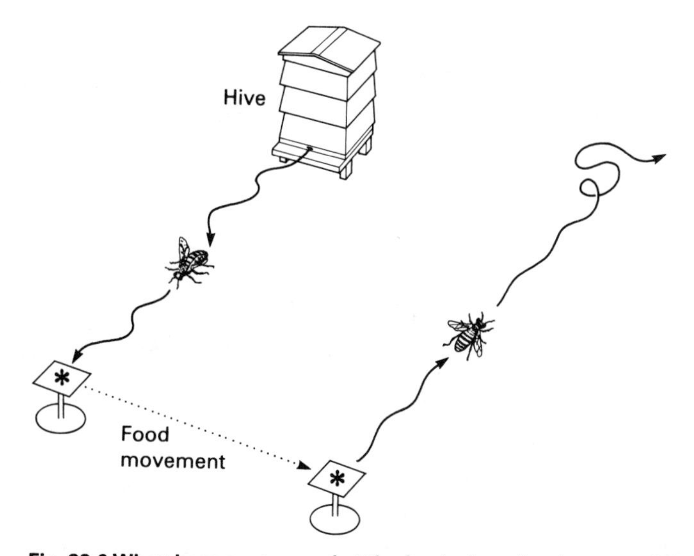

Fig. 23.6 When bees are trapped at the feeder for a few hours, and the feeder is moved during this time, they fly in the correct direction when released. This shows that they can compensate for the change in the sun's position with time

usually be two patches that look the same and are positioned symmetrically with respect to the sun. The bees overcome this problem by acting as if the patch they see is the one to the right of the sun. This convention will sometimes lead to errors, but if all bees consistently use the same convention, then the errors will tend to cancel each other. The problem can also be overcome by knowing where the sun ought to be at each time of day. Like many other insects, bees have an endogenous clock entrained by a zeitgeber (see Chapter 16). It has been suggested that, since worker bees spend much of their time in the darkness of the hive, the zeitgeber is not provided by the time of sunrise or sunset, as it is in some animals, but instead by daily changes in the earth's magnetic field. Bees are known to be sensitive to magnetic fields, and during magnetic storms their time sense is disrupted (Gould, 1980).

Fig. 23.7 Four different brightness patterns seen at the same time of day when different parts of the sky were viewed through an octagonal filter made of eight pieces of polaroid. Depth of shading indicates brightness (From *The Oxford Companion to Animal Behaviour*, 1981)

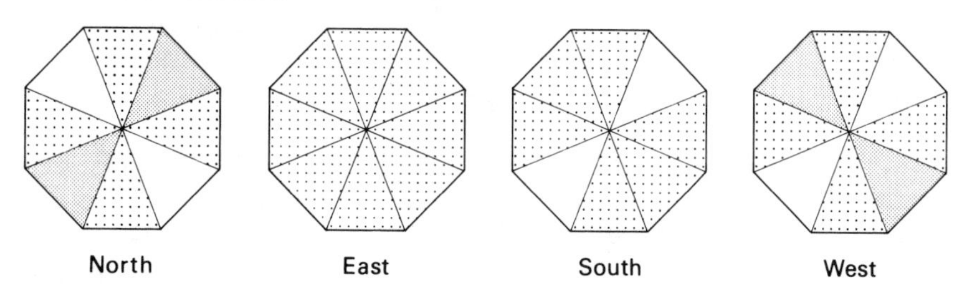

Fig. 23.8 The round dance

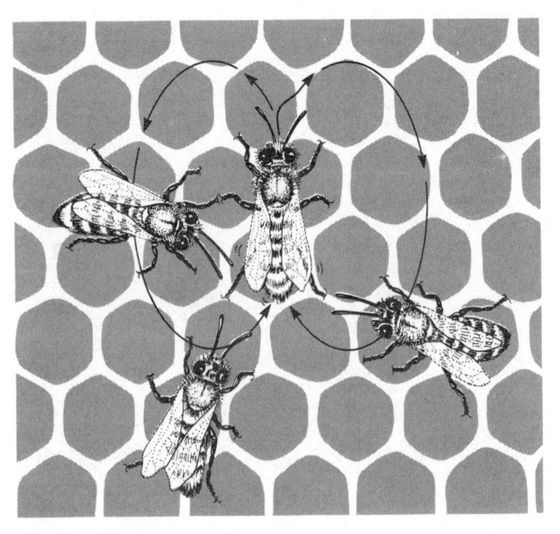

Fig. 23.9 The waggle dance

23.5 Communication among honeybees

Von Frisch observed the behavior of the bees inside a specially constructed hive with a glass wall and noticed that returning foragers performed dances that attracted the attention of other bees. He identified two types of dance: a round dance (Fig. 23.8) and a waggle dance (Fig. 23.9). At first he thought the round dance indicated nectar and the waggle dance pollen, but later he discovered this was incorrect. He found that foragers returning from food sources differing in distance and direction from the hive performed waggle dances differing in their details.

The bees perform the dance on vertical sheets of comb in the darkness of the hive. The angle between the axis of the dance and the vertical (Fig. 23.10) corresponds to the angle between the direction of the food and the direction of the sun. As the sun moves west the dances rotate counterclockwise. The duration of the waggle portion of the dance corresponds to the distance of the food from the hive. The round dance is simply a waggle dance that indicates food so close that no waggles are necessary. As we see in Chapter 26, different geographical races of bees have differing dance dialects. The more primitive tropical honeybees dance on the horizontal surface formed by the top of the comb, the axis of the dance pointing in the direction of the food source, which is also what happens when temperate-zone honeybees are forced to dance on a horizontal surface.

The returning scout attracts other workers by a display (Fig. 23.11) during which she fans her wings and releases a recruitment pheromone. However, this is done only if the food source is of worthwhile quality. The value of the food is judged by the forager in relation to its distance

Fig. 23.10 The waggle dance of the honeybee. The angle between the axis of the dance and the vertical is the same as the angle between the food source and the sun

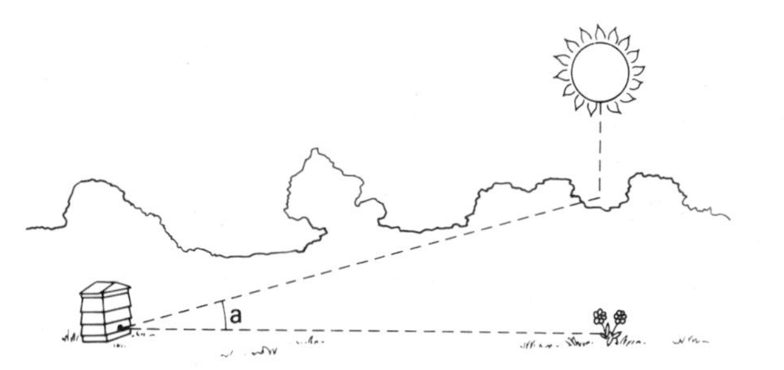

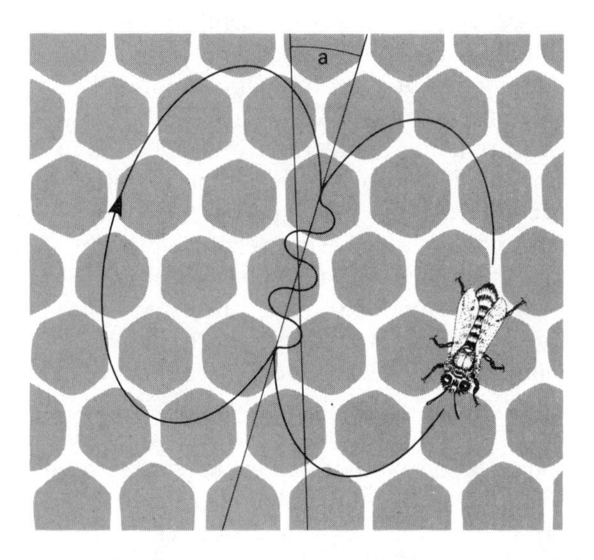

Fig. 23.11 The recruitment display of the honeybee (After a photograph in Gould, 1982)

from the hive, and to the quality of alternative sources of food: The greater the distance of a food source, the sweeter it must be to elicit recruitment and dancing. In spring and early summer, when food is generally plentiful, a food source must be really sweet to elicit recruitment. In late summer and fall, when food becomes scarce, even low-quality food will result in recruitment. How the other workers obtain information from the waggle dance is not entirely understood. The returning forager brings back traces of the scent of the flowers she has visited. The other workers crowd around the dancing bee, pick up the scent, and quickly learn the food odor, making use of their memory of it when they arrive near the feeding place. In addition to reacting to odor cues, the bees probably utilize sound cues emitted by the dancing bee in the darkness of the hive. During the waggle portion of the dance, the bee emits a sound with a pulse rate approximately 2.5 times as fast as the rate of waggling (Wenner, 1962; 1964). This sound—probably in addition to the rate of waggling as felt by the workers crowding around—provides information about the distance of the food source.

It has been suggested by some that the dance is not important in recruiting foragers, which is accomplished primarily by means of olfactory cues, as in some other social insects. Much of von Frisch's early experimental data is consistent with this interpretation. However, although honeybees rely upon odor to locate food under certain circumstances, the more recent experiments both by von Frisch and others (Gould, 1976) provide convincing evidence that the dance is the primary system used to communicate the location of food sources. For example, the foraging bees' sensitivity to light can be reduced eight times by painting their ocelli black. (The *ocelli* are the three simple eyes on top of the bee's head.) If the vertical comb on which the dance takes place is provided with artificial sunlight that the returning ocelli-blackened scouts cannot see, then their dances are oriented with respect to gravity, just as they would normally be. However, the unpainted recruits respond to the artificial sun, giving the experimenter a means of manipulating the recruits' interpretation of the dance. If the position of the artificial sun is altered, the recruits can be sent off in a direction determined by the experimenter. This experiment shows that the recruited workers were relying upon the dance to obtain information about the distance and direction of the food.

James Gould (1976) has suggested that honeybee communication has taken so long to analyze, and has been so controversial, because of its complexity. Not only do the scouts have a sophisticated dance routine, but if deprived of the opportunity to use this type of communication, they can often fall back on the odor system of recruiting foragers. The polarization pattern of the sky can be used instead of direct sunlight. Bees dancing in the darkness of the hive can use the direction of gravity as a substitute for the sun's direction. The bees' endogenous clock can provide a substitute for the movement of the sun when the bees are inside the hive or are entrapped by an experimenter. It seems that the bees have a solution for every contingency. As we shall see, this kind of contingency planning is the key to complex behavior in apparently simple animals.

23.6 The organization of complex behavior

In temperate climates, foraging honeybees are faced with a variable environment. As the bee leaves the hive on a scouting trip, the sun may be visible or obscured by clouds. Flowers previously rich in nectar may be closed or dead or removed by some animal. The odors remembered from the last trip may no longer be present or may be mixed in with a whole galaxy of new odors.

We can imagine a computer program designed to cope with contingencies such as those that face a foraging honeybee. As the bee flies out on a scouting trip the program asks a series of questions, each based upon the answer to the previous question. The final outcome is contingent upon the outcome of the previous stage, and so on. This kind of programming can be incorporated into a simple logic network specifically designed for a particular task. To some extent bee behavior appears to be organized in this way.

However, honeybee behavior is more complex than a simple IF–THEN flow chart. Bees are capable of adapting to new circumstances by learning. Randolf Menzel investigated how a bee learns the color of a food source. He changed the colors of artificial flowers during the approach, landing, feeding, and departure phases of a visit and found that the bee learns about the color of a flower during the final two seconds before landing on it (Menzel, 1979). The bee learns about the location of landmarks only as she flies away from a flower. If the landmarks are removed during feeding and replaced when the bee has departed, the bee will not be able to remember them even though they were present when she initially approached the flower. Thus, it appears, color and landmark learning are closely tied to particular aspects of behavior. Similarly, the bees learn the location of the hive as they depart from it each day. Day-to-day changes in the appearance of the hive or its surrounding vegetation do not trouble them. However, if the hive is moved a few feet while bees are out foraging, they have great difficulty in locating it when they return.

Von Frisch showed that bees are able to learn flower odors extremely rapidly, and to distinguish one from among 700 others. The flowers' color, shape, location, and time of opening are learned progressively less readily. If the scent of a familiar artificial flower is changed, then the bee rapidly learns the new scent, but the color, shape, etc., which remain unchanged, nevertheless have to be learned all over again (Menzel and Erber, 1974). In other words, foraging bees appear to learn about flowers' characteristics (scent, shape, color, location, etc.) as a package. If the scent is changed experimentally then the whole package has to be relearned. Thus, it seems, learning in bees is preprogrammed; that is, specific types of learning take place in particular situations that are characterized by the bees' behavior, such as leaving the hive, approaching a flower, and leaving a flower.

To explain honeybee behavior we do not have to invoke any special mental powers or cognitive abilities. We are tempted to account for it in terms of sets of procedural rules. However, this may prove complicated if we are to encompass the animal's total behavioral repertoire. It is always possible, moreover, that some feature of honeybee behavior will be discovered that will force us to abandon the procedural mode of explanation. In our present state of knowledge, to endow bees with mental capacities would seem fanciful. When we move to more complicated animals the situation is not so straightforward.

Many vertebrates are capable of behavior that appears to be much more complex than that of honeybees. At the same time, much of vertebrate behavior is rather stereotyped. Perhaps, as Gould and Gould

(1982) suggest, the mental abilities of vertebrates are overestimated, and there is not such a big gap between them and honeybees. In the chapters that follow we shall address this kind of question in various ways. In considering the more sophisticated aspects of vertebrate behavior, we will have at the back of our mind the complex behavior of honeybees.

Points to remember

- Honeybees provide many examples of complex behavior in a seemingly primitive animal.
- Their foraging involves sophisticated feats of navigation, including landmark recognition, the use of the sun as a compass and the use of the polarization of light in the sky and of aspects of the earth's magnetic field.
- In recognizing flowers bees make use of their sense of smell, color vision and visual-pattern detection.
- In communicating their findings to others, bees can indicate the distance and direction of food sources.
- The evidence suggests that the complex behavior of bees is the result of the systematic application of simple programming rules.

Further reading

Frisch, K. von (1967) *The Dance Language and Orientation of Bees*, Belknap, Cambridge, Massachusetts.

Lindauer, M. (1961) *Communication Among Social Bees*, Harvard University Press, Cambridge, Massachusetts.

24 Evolutionary Optimality

Many animals and plants appear to be perfectly designed to fulfill their role or purpose. Darwin realized that such apparently perfect adaptations can be accounted for in terms of natural selection. Most biologists now agree that the theory of natural selection can adequately account for even the most intricate adaptations (Cain, 1964), although some (e.g., Gould and Eldredge, 1977; Gould, 1980) still dispute it. In this chapter we consider some of the problems posed by considering behavior to be a result of evolutionary design. In the first part of the chapter we discuss optimal foraging, a subject in which many of these problems are apparent. In the second part we consider how we might circumvent these problems.

In Darwin's day, teleological arguments were often used in attempts to disprove the theory of natural selection. Scientists said that no mechanism as passive and chancey as natural selection could account for the evident purpose inherent in the design of animals and plants. Teleological arguments invoke an end, or purpose, to account for phenomena that seem inexplicable in other terms. Teleological philosophy is no longer regarded as relevant to discussion about evolution or biological function (Woodfield, 1976), though it remains important in discussions of the purpose of intentional behavior of individuals, as we see in Chapter 26.

In recent years the notion of design has become an important part of biological thinking because we now realize that natural selection is a designing agent. Modern evolutionary theory implies that animals tend to assume those characteristics that ensure that, within a specified stable environment, there is no selective disadvantage vis-à-vis the other animals with which they compete. In other words, over a sufficiently long period in a stable environment, animals tend to assume characteristics that are optimal with respect to the prevailing circumstances. This does not mean that every animal is perfectly adapted, since all individuals differ genetically and since many are displaced from the prime habitats to which they would best be suited by competing species.

24.1 Optimal foraging

Many animals are able to select among various kinds of food according to their physiological requirements. However, in order to obtain food, animals have to expend energy. They may also have to spend valuable time, spend physiological commodities such as heat and water, and risk exposing themselves to predation. It seems obvious, therefore, that some forms of foraging will be better than others, in terms of the animal's overall fitness.

In discussing foraging from this viewpoint, we first have to distinguish between various basic evolutionary strategies. In Chapter 7 we saw that it is sometimes useful to think in terms of animals employing a strategy to increase future representation of their genes, even though such evolutionary strategies are really the passive result of natural selection. Just as we can think in terms of strategies for defense and for sexual reproduction, so we can think in terms of foraging strategies. Thus, some species employ a sit-and-wait strategy, ambushing their prey from a prechosen position. Some species hunt by stalking, others by searching. To give a more specific example, it is fairly common for one individual, the *scrounger*, to make use of the foraging investment of another individual, the *producer*, to obtain food. Such relationships can occur between two individuals of different species, or among members of the same species (see, e.g., the review of kleptoparasitism (scrounging) in birds, by Brockmann and Barnard, 1979). Such relationships are little different in principle than some of the sexual strategies discussed in Chapter 10, in which subordinate males steal copulations from dominant males.

Chris Barnard and Richard Sibly (1981) studied the alternative foraging strategies of captive flocks of house sparrows (*Passer domesticus*). Here the producers obtain most of their food by actively foraging for mealworms provided by the experimenters, while the scroungers obtain most of their food by copying actively foraging birds or by following them around without actively searching themselves. Sometimes a scrounger snatches food from a producer. On theoretical grounds, we would expect the payoff to scroungers to increase with the number of producers because of the increased opportunity to parasitize them. The payoff to producers, in contrast, should decrease with the increasing number of scroungers because the producers stand an increasing chance of forfeiting the benefits of their foraging investment.

This type of situation lends itself to analysis in terms of evolutionarily stable strategies (see Chapter 7), and Barnard and Sibly used this approach. They found that individual birds did not opportunistically change their strategy in accordance with the composition of the flock as a whole. Scroungers fared better when more producers were present, but they did less well when greatly outnumbered by producers. This is perhaps because a large number of producers deplete the available food very quickly.

In order to carry out a quantitative analysis of foraging strategy, it is necessary to assume that the animal's behavior is designed to maximize

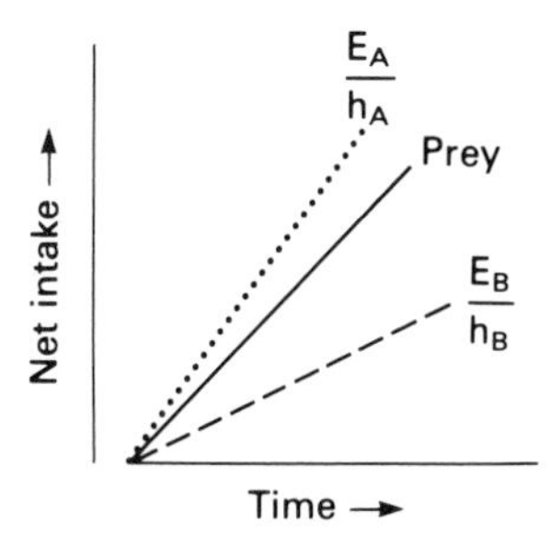

Fig. 24.1 A simple optimal foraging model. Each type of prey (A and B) is characterized by its profitability E/h (net food yield per unit handling time). The graph shows the probabilities of two prey as the slopes of the dotted and dashed lines. If the predator eats only the more profitable prey, its intake is given by the solid line. The slope of this is less than the slope of the dotted line, because of the time taken to search for larger prey. When the slope of the solid line is equal to, or less than, the slope of the dashed line, then it is worth while for the predator to eat smaller prey. Thus as the larger prey becomes rarer the solid line rotates to the right, and the animal switches to the less profitable prey

some entity. In the analysis by Barnard and Sibly, the payoff was the number of mealworms eaten in a 10-minute period. Thus, they assumed that the birds were maximizing the rate of food intake. This assumption is obviously not strictly valid because other factors may have been associated with the contribution to fitness of the two strategies. In particular, the costs of the two strategies may have been different. However, to carry out this type of analysis, it is often convenient to assume a fairly simple index of fitness, like rate of obtaining food, or some measure of foraging efficiency.

When a predator encounters a prey item, it has to pay a cost in time taken to catch and eat the prey, often called *handling time*. This cost is offset by the net energy value of the prey item, which is the gross value minus the energy expended in handling and digesting the food. The *profitability* of the prey is the net energy value divided by the handling time. When an animal is able to choose among prey items, it can be assumed that it will choose the most profitable prey. For example, in one study, bluegill sunfish (*Lepomis macrochirus*) were allowed to hunt for water fleas (*Daphnia*) in a large aquarium (Werner and Hall, 1974). The investigators found that the fish showed no preferences among small, medium, and large *Daphnia* when these were available at low densities. When the *Daphnia* were abundant, however, the fish chose the largest and most profitable *Daphnia* and ignored the small ones, the result that would be expected on a profitability model, as we can see from Figure 24.1. However, another possible explanation is that the fish take whichever *Daphnia* appears largest at each choice (O'Brien et al., 1976). A nearby small *Daphnia* may appear larger than a larger one farther away. As the density of prey increases, the probability that there will be a large *Daphnia* nearby also increases. By selecting whichever prey item appears larger, the fish could behave in the manner predicted by the profitability model. Notice, however, that the profitability model specifies what ought to happen in a given situation, while the take-the-largest model proposes a rule that the fish might actually use.

John Goss-Custard (1977a) studied foraging in the redshank (*Tringa totanus*), a wading bird that hunts for food along the seashore and on mudflats. He found that when these birds are feeding exclusively on polychaete worms (*Nereis diversicolor* and *Nephthys hombergi*), they tend to pass over the smaller worms and to select those over a certain size. Their size preference is influenced by the rate at which they encounter the larger worms but not by their encounter rate with small worms. This finding is consistent with the view that the redshank's foraging strategy is designed to maximize energy profitability; that is, they select worms that provide the greatest amount of energy per unit of energy expended on foraging. The smaller worms are not so profitable because of the relatively low net energy returns on time spent foraging for them.

Taken at face value, these results might suggest that the redshank make decisions about which prey to take on the basis of energy

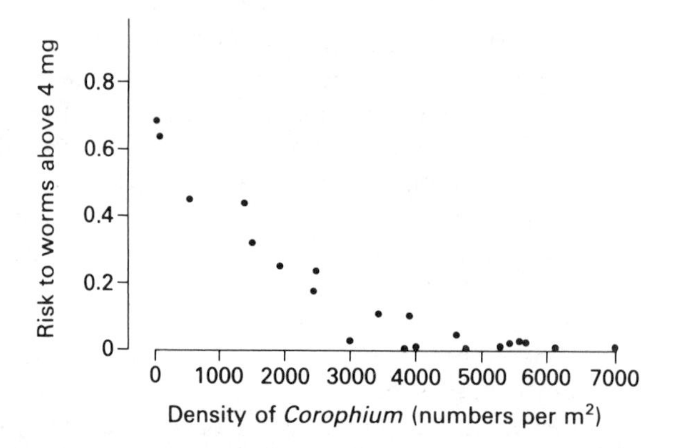

Fig. 24.2 Index of the risk of worms being eaten by redshank, in relation to the local density of *Corophium*. Each point represents the average for a particular study site. (After Goss-Custard, 1977b)

trade-offs. However, Goss-Custard (1977b) also found that when the amphipod crustacean *Corophium* was available in addition to polychaete worms, the birds tended to select *Corophium* (Fig. 24.2). He was able to discount the possibility that the habitat typical of *Corophium* was one in which polychaete worms were harder to find because he observed that some birds concentrated on worms while the majority was feeding on *Corophium*. On the basis of previous work, Goss-Custard had formed the hypothesis that redshank achieved a higher rate of net energy intake by

Table 24.1 Comparison of the rates at which redshank obtained energy in three sites where they mainly took *Corophium* with the rates they would have achieved by taking worms instead (After Goss-Custard, 1971a).

Site	*Rate of ingesting energy* (cal/min)	
	Potential rate from Nereis *alone*	*Actual rate from mainly* Corophium
9	234	88
10	224	70
11	185	93

Table 24.2 The effort expended in three sites on collecting 1 kcal by birds feeding mainly on *Corophium* as they actually did, and on *Nereis* alone (After Goss-Custard, 1971a).

Site	*Distance searched in meters*		*Number of pecks and probes made*		*Time spent swallowing prey*	
	Corophium	*Nereis*	*Corophium*	*Nereis*	*Corophium*	*Nereis*
9	103	42	470	165	62	48
10	150	44	671	167	121	49
11	106	56	543	198	79	48

feeding on *Corophium* than by taking worms. However, when he came to analyze the energy content of the prey (Table 24.1) and the energy costs of obtaining prey (Table 24.2), Goss-Custard found that the birds would have obtained between two and three times more energy per minute by taking worms exclusively compared with what they obtained by feeding on *Corophium*. Clearly, energy was not the only factor relevant to foraging redshank when *Corophium* was available. Presumably, the *Corophium* contain something other than energy that is important to the redshank.

For many animals, opportunities to concentrate on one type of activity to the exclusion of others are rare. This is particularly true of animals that may be in danger from predators. For example, Chris Barnard (1980) studied the foraging behavior of house sparrows (*Passer domesticus*) on an English farm in winter. The sparrows moved around in a flock of variable size and fed in two habitats that differed in the apparent risk of predation from cats and hawks. Inside the cattlesheds the sparrows fed upon barley seed concealed in the bedding straw. Their feeding behavior was influenced primarily by the density of seeds and was relatively unaffected by the number of birds in the flock (Fig. 24.3). In the open fields, however, the sparrows' feeding behavior was inter-

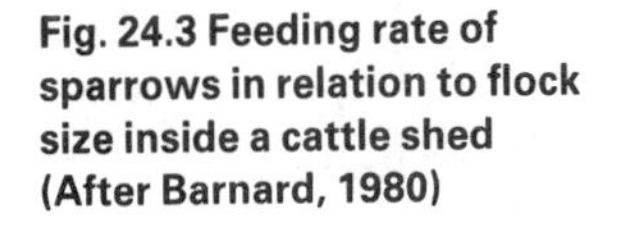
Fig. 24.3 Feeding rate of sparrows in relation to flock size inside a cattle shed (After Barnard, 1980)

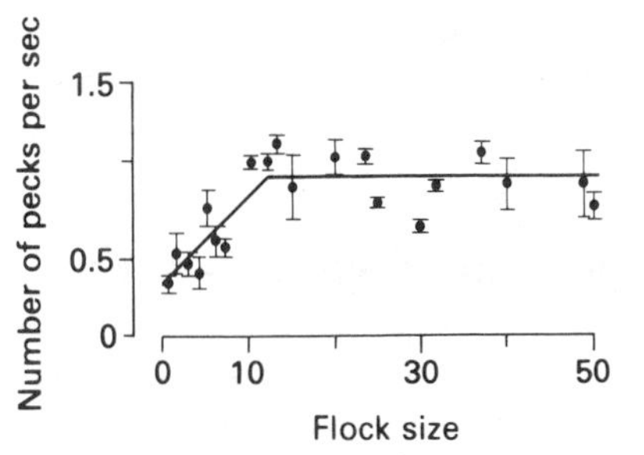

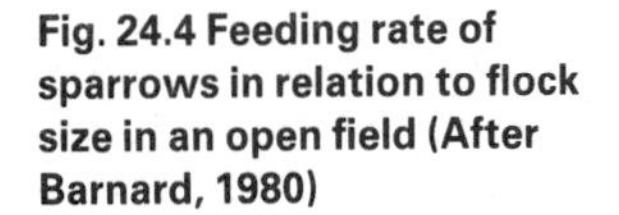
Fig. 24.4 Feeding rate of sparrows in relation to flock size in an open field (After Barnard, 1980)

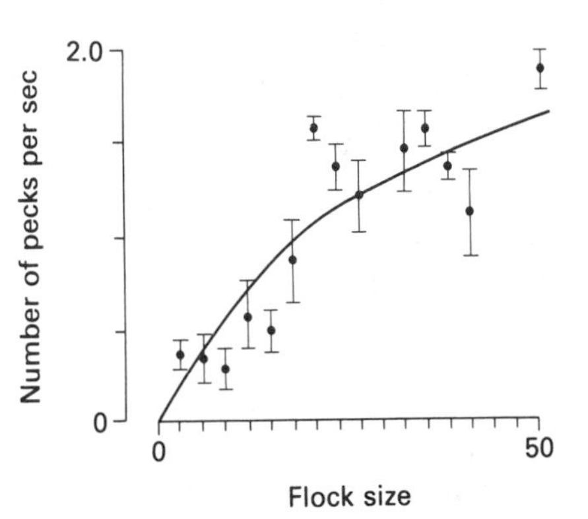

spersed by scanning for predators. Their rate of feeding was more influenced by flock size (Fig. 24.4) and less influenced by food density. When feeding in a large flock, each bird can devote more time to feeding because the need for vigilance is less than in a small flock (Pulliam, 1973). There is evidence that large flocks of birds spot predators sooner than small flocks (Powell, 1974; Siegfried and Underhill, 1975). The sparrows in the open field also spent less time scanning when they were feeding close to a hedge or bush that might offer some cover from predators. In a field of freshly sown spring barley, they would usually stay close to cover even though the density of seeds was higher in the open field.

This study suggests that some kind of balance or trade-off between foraging and antipredator behavior occurs in a number of ways in feeding sparrows. The birds feed at a lower rate when the danger of predation is high, and they somehow choose to feed in a relatively safe, less profitable place rather than a more profitable but riskier location. Barnard (1980) found that the distance at which the birds took flight from a predator was influenced by the density of seeds on which they were feeding at the time of alarm. This may be due to the greater benefit of feeding in relation to the risk of predation, and in terms of the mechanisms involved, it may be due to the fact that the birds were more preoccupied by feeding on the more abundant seeds and did not notice the predator so soon. As we see in Chapter 25, the question of how animals divide their attention among different tasks is important when we come to consider the mechanisms by which animals make decisions.

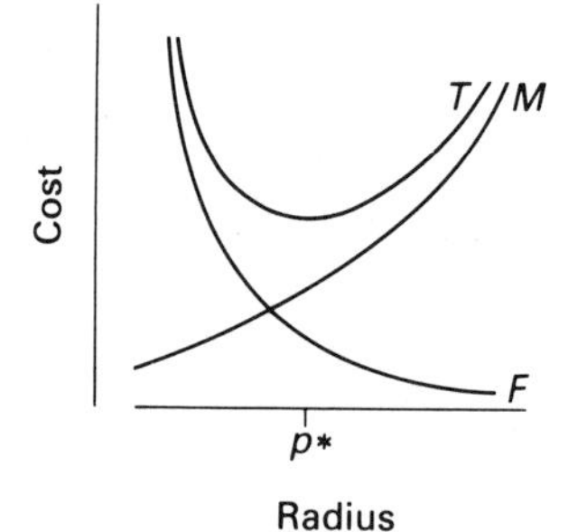

Fig. 24.5 The energy cost of maintaining blood flow in relation to the radius of the blood vessel. F is the cost due to friction, M is the cost due to maintenance of the vessel wall, and T is the total energy cost. p* is the optimal radius at which T is at a minimum (After Milsum and Roberge, 1973)

24.2 Trade-offs in animal behavior

Most design operations require compromise among the merits of incompatible aspects of a given situation. In the design of a simple blood vessel, for example, the costs due to friction decrease with increasing radius while the costs due to maintenance increase, as illustrated in Figure 24.5. This type of direct comparison of costs and benefits is called a *trade-off*. Successful design usually exploits the trade-offs inherent in a situation. Identification of trade-offs in animal behavior is therefore a good start to an analysis of the way in which natural selection has operated.

Reto Zach (1979) discovered a simple trade-off in the foraging behavior of crows that feed on shellfish on the west coast of Canada. The crows hunt for whelks at low tide, usually selecting the largest ones. When they find one they hover over a rock and drop the whelk so it breaks open to expose the edible inside. By dropping whelks of various sizes from different heights, Zach discovered that the number of times a whelk has to be dropped in order to break is related to the height from which it is dropped, as shown in Figure 24.6. The crows have to expend

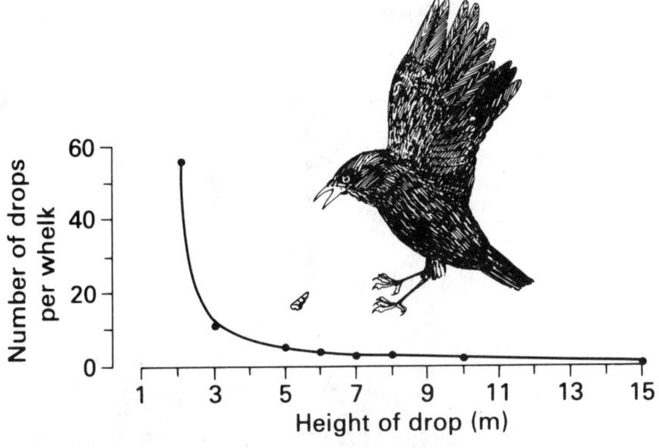

Fig. 24.6 Number of times a whelk shell needs to be dropped to break it, when dropped from different heights (After Zach, 1979)

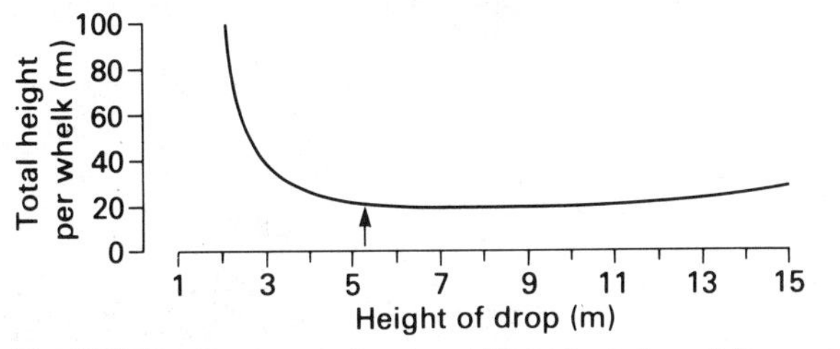

Fig. 24.7 Total amount of upward flight (number of drops × height of each drop) is minimal at the height (arrow) usually used by crows (After Zach, 1979)

energy in flying up to drop a whelk, so Zach calculated the total amount of upward flying a crow would have to do to break a whelk from a given height. He showed that the lowest flying cost is incurred if the whelks are dropped from about 5 meters, as Figure 24.7 shows. Just as in blood vessel design, there is a trade-off, in this instance between the cost in number of drops required to break the whelk and the height of the drop. Calculations based upon this trade-off reveal that there is an optimal drop height (Fig. 24.7) analogous to the optimum radius of a blood vessel (Fig. 24.5). Zach found that the height from which crows usually drop their whelks corresponds closely to the calculated optimal drop height of 5 meters. Thus, it appears, the crows somehow have been programmed to exploit this particular feature of the foraging situation.

The calculations of costs and benefits of crows foraging for whelks, like those for the design of blood vessels, are simplified by the fact that trade-offs occur among factors all measurable in terms of energy. Zach's calculations suggest that the crows choose particular-size whelks and drop them from that particular height that provides the largest net energy profit. However, it is not always possible, or even desirable, to consider optimal trade-offs purely in terms of energy. As we saw in the case of the redshank, energy is not always the prime consideration. Moreover, animals are not always in a position to concentrate exclusively upon foraging but may have to watch out for predators, territorial

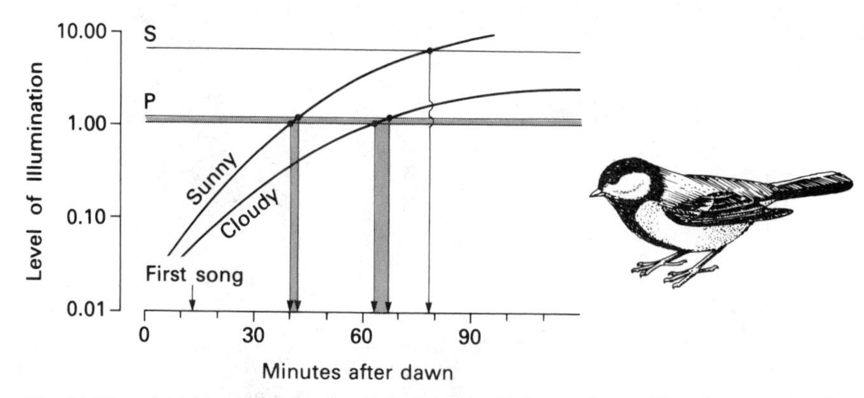

Fig. 24.8 Luminance of great tit feeding sites in relation to time of day, for sunny and overcast weather. P and S are the luminance values at which measures of profitability and searching efficiency, as determined in the laboratory, respectively reach 95% of their maximum value (After Kacelnik, 1979)

rivals, etc. What we have to do is to consider the situation as a whole, bearing in mind that foraging is only one activity out of the animal's repertoire and that loss and gain of energy is not the only consequence of behavior.

In the great tit (*Parus major*), song is an important aspect of territorial defense and advertisement (Krebs, 1971). The major daily episode of singing during early spring occurs at dawn, starting when the sun is just below the horizon. This dawn chorus is typical of many birds, yet is something of a puzzlement because most small birds lose up to 10 per cent of their body weight overnight in winter and therefore ought to be very hungry when they wake up. Why do they sing rather than forage first thing in the morning? There seems to be two main reasons. First, the advantage of territorial defense just after dawn is particularly high because at this time birds without territories probe the defenses to exploit any vacancies that have occurred as a result of predation since the previous day. Second, the benefits of foraging are low because foraging efficiency is limited by the light intensity. Alex Kacelnik (1979) determined by laboratory experiment that great tits cannot forage efficiently below a certain level of illumination and showed that this level normally occurred between 40 and 80 minutes after dawn, depending on the weather (see Fig. 24.8). Thus, the birds probably gain more benefits from singing first and foraging second than they would by foraging first and singing later.

To judge whether one course of action is better than another, we need to have a common currency in terms of which they can be compared. We also need to have some criterion by which we can measure the relative value of different aspects of behavior. Every activity has costs and benefits associated with it. In this chapter we see how costs and benefits can be evaluated precisely, but for the present we confine ourselves to general considerations. In the final analysis, what matters is whether a particular activity is likely to increase or decrease an animal's inclusive fitness. A great tit that sings at dawn may increase its fitness by deterring intruders, but it may decrease its fitness by postponing foraging, especially if it has lost weight during the night. Such increments in fitness are conveniently called *benefits*, while decrements in fitness are called *costs*. Thus, the foraging animal benefits from the energy gained but may incur costs as a result of neglecting to defend its territory or as a result of making itself more vulnerable to predators.

One advantage of thinking in terms of costs and benefits is that it enables us to compare the relative merits of disparate activities. Different aspects of foraging can be compared in terms of the food or energy gained, and different types of territorial defense can be compared in terms of their effectiveness in deterring intruders. In comparing foraging and territorial defense, however, we have to consider how much each contributes to, or detracts from, the animal's inclusive fitness. Fitness is the only common currency by which we can make comparisons among different aspects of behavior. Indeed, it is, in effect, the currency by which nature makes such comparisons because fitness is determined through the agency of natural selection.

It is valuable not only to be able to specify what the consequences of different activities have in common but also to evaluate them in terms of some criterion. In considering applicants for a university post, for example, we may wish to compare candidates in terms of common factors such as teaching ability and research potential. In addition, we need a criterion by which we can judge whether a candidate who is good at teaching but poor at research is preferable to one who is poor at teaching but good at research (see McFarland, 1976; 1977). Similarly, in evaluating animal behavior we need an optimality criterion that enables us to say whether one activity is better than another under certain particular circumstances. In the final analysis we would hope to be able to specify the optimality criterion in terms of inclusive fitness. To do this properly, however, would require a considerable knowledge of the animal's circumstances. In the next section we see how we might set about obtaining such knowledge.

24.3 Cost functions

We have seen that animals often are faced with alternative courses of action among which exists a trade-off between costs and benefits. The best course for an animal to take is the one that provides the least cost or greatest benefit. From evolutionary considerations we expect well-designed animals to behave in a way that will maximize inclusive fitness. Thus, the concepts of cost and benefit are related to the concept of fitness.

Animals incur costs and benefits in a great variety of ways, which correspond to the selective pressures of the natural environment. Natural selection acts upon the animal as a whole, but in analyzing the costs and benefits of behavior, we have to separate out the costs (decrements in fitness) and benefits (increments in fitness) that can be ascribed to each aspect of the animal's internal state and behavior. The total specification of the animal in these terms is called the *cost function*, which can be defined as the specification of the instantaneous level of risk incurred by (and reproductive benefit available to) an animal in a particular internal state, engaged in a particular activity in a particular environment (McFarland, 1977).

For example, suppose we consider a herring gull (*Larus argentatus*) sitting on its nest incubating its eggs. Normally, both parents incubate in turn, and a nest left unattended soon is subject to predation by neighbors. A sitting bird will not leave the nest until relieved by its partner, unless it is flushed from the nest by a predator such as a fox or a human. Usually the partner leaves the nest to forage for food and returns within a few hours. Sometimes, however, the return may be delayed as a result of mishap, injury, or capture by a scientist. What should the sitting bird do? On the one hand, its partner may return at any time, but on the other hand, the sitting bird becomes increasingly hungry as time passes. Eventually, the sitting bird should quit the nest to search for food. Because herring gulls breed in many successive seasons, it is not in the genetic interests of the individual to endanger its life for a single clutch of eggs. The question is, at what point should the bird decide to quit?

If the incubating bird quits too early, it may endanger its clutch unnecessarily because its mate merely may have been delayed. If it quits too late, it may endanger its life unnecessarily because it may not be able to find food as quickly as usual. To determine exactly when the bird should quit, these alternative risks have to be evaluated quantitatively, an evaluation carried out effectively by nature in designing the animal to make the most appropriate decisions.

To investigate this type of problem, it is necessary to measure in the field the risk run by animals engaged in various types of activities. In the case of incubating herring gulls, for example, we would need to know the risk to the eggs of being left in the nest unattended and the risk to the bird of remaining on the nest without food. Various types of experiments have tried to evaluate the costs and benefits of incubation. For example, Rudi Drent (1970) investigated herring gulls nesting on the Dutch Friesian island of Schiermonnikoog. He found that 20 per cent of the eggs were eliminated by predation, while 12 per cent were lost in meeting the physical demands of incubation. When not disturbed by people, the birds keep the eggs covered about 98 per cent of the time. This effectively seals the nest contents from outside air and enables the bird to regulate the temperature of the eggs. The main problem faced by

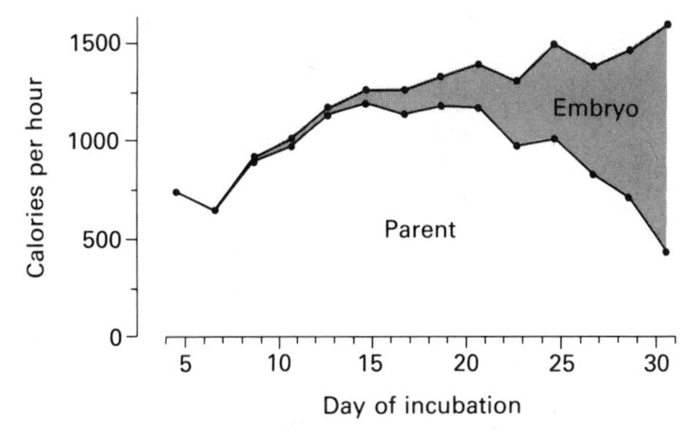

Fig. 24.9 The energy cost of incubation in the herring gull, calculated for the entire clutch. Shaded area represents the heat produced by the embryo. Upper line therefore indicates total energy, lower line cost to parent (After Drent, 1970)

the incubating bird is the regulation of its own body temperature. During the cold night it has to produce enough heat to keep itself and the eggs warm, while in the midday heat it may have to resort to panting to keep itself cool. By attaching monitoring instruments to the eggs and nest, Drent was able to make detailed observations of the factors that affect the developing egg. Initially, the egg is entirely dependent upon the parent for warmth, but as the embryo develops, its metabolism starts to produce heat (Fig. 24.9). However, the embryo is not capable of temperature regulation, and the eggs steadily cool if left exposed. There is no compensatory heat production by the embryo, whereas a newly hatched chick, by contrast, reacts to a drop in air temperature by increasing its heat production. Thus, the chick has a certain degree of thermal homeostasis.

In the absence of the parent, the embryo risks death by either overheating or chilling, depending upon the prevailing weather. However, a more important factor is the total amount of heat supplied by the time the hatching date arrives. The rate of development of the embryo depends upon its metabolism, which in turn is affected by the temperature inside the egg. If the temperature is low the embryo will develop slowly. During incubation, the egg continually loses weight as a result of evaporation of water (Fig. 24.10). This loss of water results in an air space at one end of the egg, necessary for the embryo's breathing during the period shortly before hatching. The embryo must not develop too quickly or the air space will not be formed properly by the time it is ready to hatch. It must not develop too slowly or the egg will have lost so much water that the embryo will become dehydrated. Hatching must occur when only a certain amount of water has been lost from the egg, and the embryo must have developed enough to be ready to hatch at that time. It does not matter much if the eggs are allowed to cool for relatively short periods, as long as the egg temperature does not fall very low. On a hot day there may be a danger of overheating if the

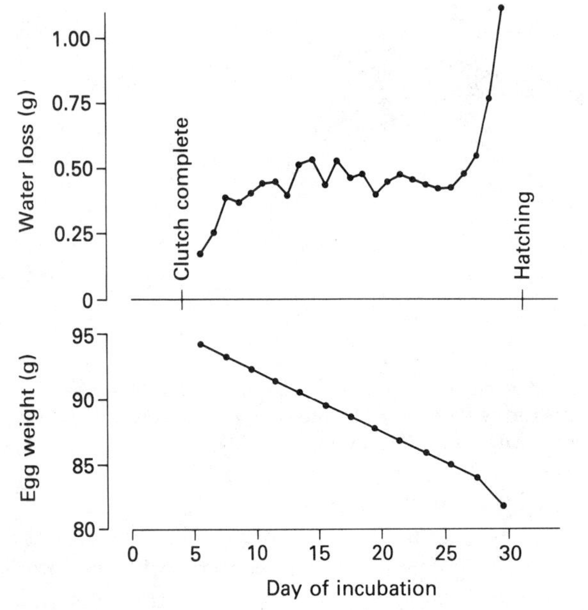

Fig. 24.10 Water loss (above) of herring gull eggs measured from the weight change (below). The air space becomes larger as water is lost, but if too much is lost the embryo becomes dehydrated (After Drent, 1970)

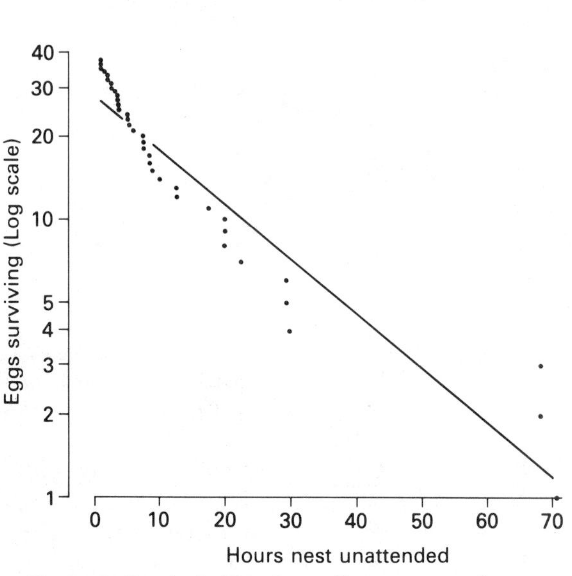

Fig. 24.11 Survival of herring-gull eggs exposed to predation by the absence of the nest owners (Data from Sibly and McCleery, 1985)

eggs are exposed, but normally, eggs exposed for an hour or two do not suffer physiologically. Incubating herring gulls thus can tolerate occasional disturbances that oblige them to leave the nest. If they are disturbed repeatedly, however, their average rate of heat supply to the eggs may fall short of that required for successful hatching.

A herring gull embryo is about twice as likely to be lost through predation as from failure to develop properly. Incubating birds have to leave the nest in response to certain types of predator, such as humans. During their absence the eggs may be taken by neighboring gulls, by cannibalistic gulls, or by rats, hedgehogs, crows, or other predators. The risk to an egg can be calculated from field observations during periods when the eggs are exposed or during experiments in which the parents are removed from the nest (Fig. 24.11).

The incubating gull sitting on its nest awaiting the return of its partner cannot leave the territory without endangering the eggs. As we have seen, the risk to the eggs can be estimated quantitatively as part of the overall cost function. On the one hand, part of the cost of not incubating is the risk that the eggs will fail to hatch as a result of exposure to predators or to climate. On the other hand, there is a cost associated with incubation, made up of the risk of being caught by a predator and the risk of starvation.

These aspects of the cost function also can be investigated experimentally. For example, the body weight of an incubating bird can be

Fig. 24.12 These nest-balances (left) can be placed in a hole dug under the nest of a herring gull. The nest is then placed on the platform (right). The owners continue to behave normally towards their nest, enabling changes in the weight of the nest to be electrically recorded (*Photographs: Robin McCleery*)

measured by means of a specially designed balance placed in a hole in the ground under the nest (Fig. 24.12). The amount of fat a bird is carrying can be calculated from its weight in relation to skeletal size (measured when the bird initially is caught and marked). The amount of food a herring gull obtains from foraging can be estimated by the change in weight measured on the nest balance before and after foraging trips. The quality of the food can be estimated from analysis of the fecal remains gathered from around the nest and from observations made on the bird while foraging. Herring gulls may fly a number of miles to obtain food, so to obtain observations on foraging, particular birds are fitted with radio transmitters (Fig. 24.13) and tracked by means of a directional radio receiver (Fig. 24.14). Thus, by using a variety of techniques it is sometimes possible to arrive at fairly accurate estimates of the costs and benefits incurred by animals living a normal life in a natural environment (Sibly and McCleery, 1985).

In general, the cost function is made up of two main parts: (1) the costs associated with the activity occurring at a particular time and (2) the costs associated with all the other activities not occurring at that time. For example, an incubating gull incurs costs associated with incubation that include the physiological cost involved in maintaining the eggs at a certain temperature and behavior costs like those arising from the risk of being caught by a predator while sitting. At the same time, some costs arise because incubation is incompatible with other activities, usually connected with the animal's state. For example, while incubating, the animal's hunger inevitably rises. Being hungry involves a risk of starvation because food may be unavailable when the animal decides to start foraging. As we shall see, such considerations are

Fig. 24.13 A herring gull fitted with a radio transmitter (*Photograph: Robin McCleery*)

related intimately to the animal's ecological circumstances. In seeking to understand how animals respond to the costs and benefits inherent in their situation, it is sometimes helpful to pursue the analogy between the economic behavior of humans and similar behavior in animals.

24.4 Functional aspects of decision-making

There are two basic principles of economic decision-making. First, it must be rational, and second, it must involve some evaluation of the situational pros and cons. Decision-making is basically rational if it is self-consistent, a property usually called *transitivity* of choice. Suppose a person has to choose between options A, B, and C. If A is preferred to B, and B to C, then it is rational to expect A to be preferred to C. We can now write a consistent order of preference, A-B-C. The relationships among A, B, and C are said to be *transitive*. If A is preferred to B and B to C, while A is not preferred to C, then the decision to choose C above A is irrational, and the relationships among A, B, and C are *intransitive*. Economists base their theory upon the concept of the "rational economic person," which implies that all preference relationships are transitive (Edwards, 1954).

Fig. 24.14 The radio receiving equipment mounted on a Land Rover, and used for tracking herring gulls on Walney Island, Cumbria, UK. (*Photograph: Richard Sibly*)

Rationality does not necessarily imply the use of reason. People can make some decisions via reasoning, but they can also make rational decisions in a purely automatic way, as if designed or programmed to do so. People also make irrational decisions. In fact, it is impossible to prove that human choices are transitive because to do so would involve repeated choice experiments under identical circumstances, which is impossible because circumstances are never exactly the same twice, if only because the memory of having made one choice changes the circumstances for the next (Edwards, 1954; 1961). Economists have to take transitivity of choice as a working hypothesis, an assumption upon which the elementary theory is based.

Transitivity of choice implies that something is maximized in the decision-making process. To see that this must be so, let us consider the following situation. Suppose A, B, and C can be evaluated numerically in some way. If A has a higher score than B, then we write A > B. A will be chosen over B by a person using a maximization principle (like "choose the option with the larger score"). If we know that B is larger than C, we can write B > C. If C is chosen over A, it would appear that C has been allocated a higher score than A, but we know that A > B > C, which implies that C has a lower score than A. Thus, if C is chosen over A, C must be preferred even though it has a lower score than A. A

Fig. 24.15 Parallel concepts in economics and ethology: (above) general scheme showing the most important contributors (with the dates of their major works); (below) detailed scheme as applied to the individual person or animal (After McFarland and Houston, 1981)

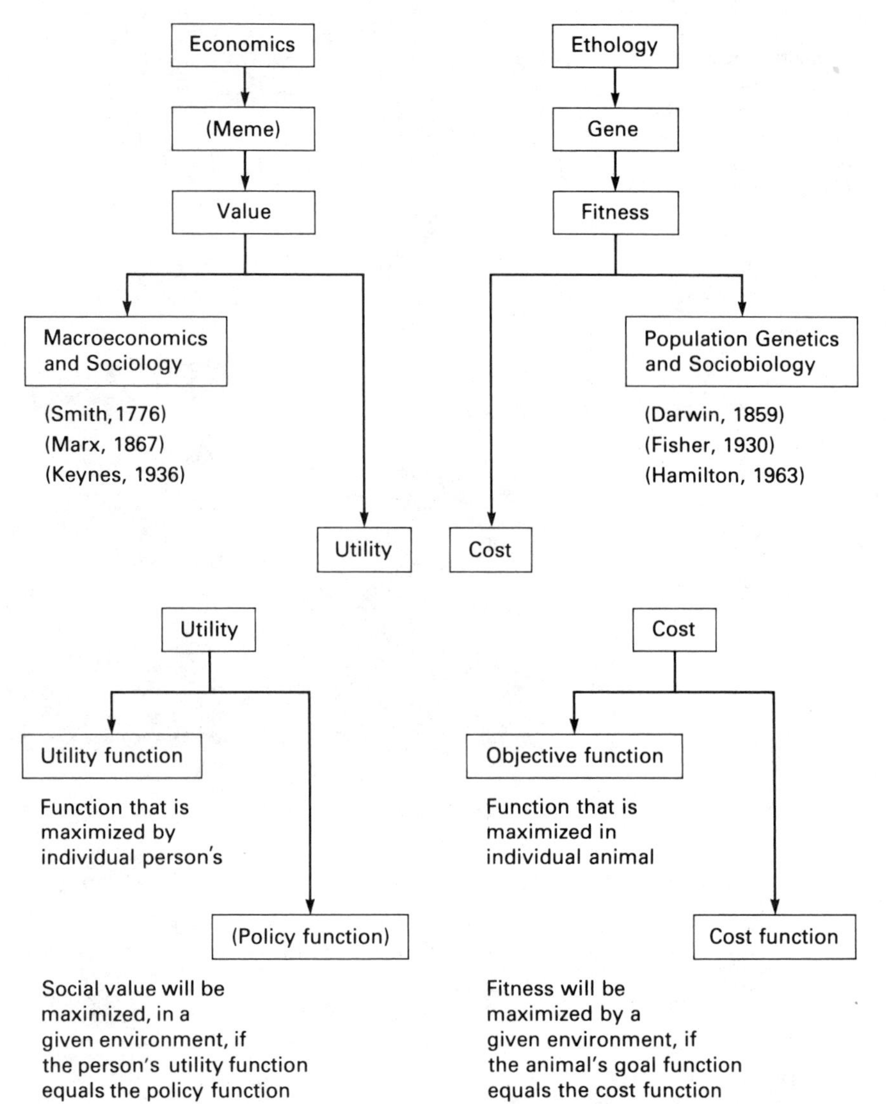

person making choices on this intransitive basis could not be choosing the option with the largest number of points; such a person could not be using a maximization principle. If a person's preferences are transitive, however, then we can deduce that he or she is using a maximizing principle, although the person may not be aware of it.

The name given to the quantity that is maximized in the choice behavior of the rational economic person is *utility*. I may obtain a certain amount of utility from buying china for my collection, a certain amount from sport and a certain amount from reading books. In spending my time and money on these things, I choose what maximizes the amount of return satisfaction or utility. I am not aware of maximizing utility, but (if I am rational) I appear to behave in a way that maximizes utility. Thus, utility is a *notional measure* of the psychological value of goods, leisure, etc., notional because we do not know whether it enters people's choice behavior. We only assume they behave as if utility is maximized.

The equivalent of utility in animal behavior is benefit (negative cost). Just as utility is a notional measure of the value of behavior, so cost is a notional measure of the change in fitness associated with an animal's behavior and its internal state (see Fig. 24.15). We do not know whether cost enters directly into the decision-making processes of animals. We can only test the hypothesis that animals behave in a way that maximizes benefit (minimizes cost). The components of cost include factors associated with the behavior occurring at a particular time and risks associated with the animal's internal state. Thus, the incubating gull incurs physiological cost in keeping the eggs warm, and it incurs costs as a result of its increasing hunger while on the nest. The various costs can be combined to form a cost function, which evaluates every aspect of the animal's state and behavior in terms of its associated cost. In economics, the equivalent (but inverse) function is called a *utility function* (see Fig. 24.15). At this point, it may be helpful to analyze a particular example of animal behavior in economic terms to see how the concept of utility, and other economic concepts, can be useful in the study of animal behavior.

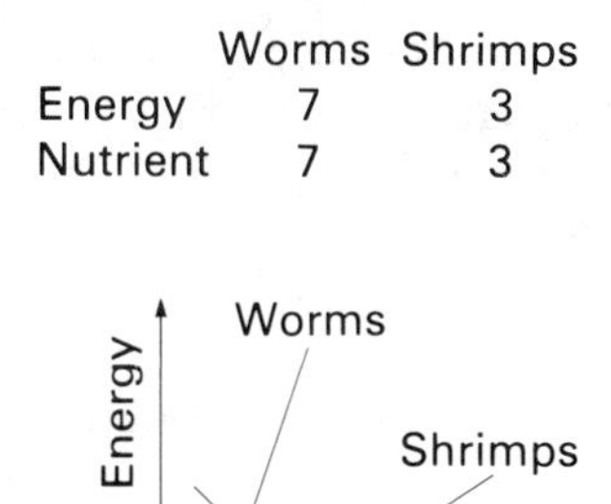

Fig. 24.16 Consequences of eating worms and shrimps in terms of the energy and nutrient gained. The budget line is based upon the energy prices of worms and shrimps used in this model (After McFarland and Houston, 1981)

24.5 The animal as an economic consumer

We saw above that redshank foraging on mudflats prefer *Corophium* to polychaete worms, even though much less energy is obtained from *Corophium*. We conclude that the *Corophium* must provide something other than energy that is of importance to the redshank. What might an economist have to say about this situation?

To simplify matters, we will refer to the redshank prey as worms and shrimps, and we will assume that each prey provides a certain amount of energy and a certain amount of some unknown nutrient. The energy and nutrient are present in different proportions in the two prey, as shown in Figure 24.16, meaning that the consequences of eating two prey are different and can be portrayed as different trajectories in a diagram representing the possible consequences of choice behavior.

The economist would ask next about the relative price of worms and shrimps. The redshank have to expend more energy to obtain shrimps than to obtain the equivalent energy return from worms. The energy price of shrimps is about twice that of worms. Figure 24.16 shows how many worms and shrimps a bird could obtain for a given amount of

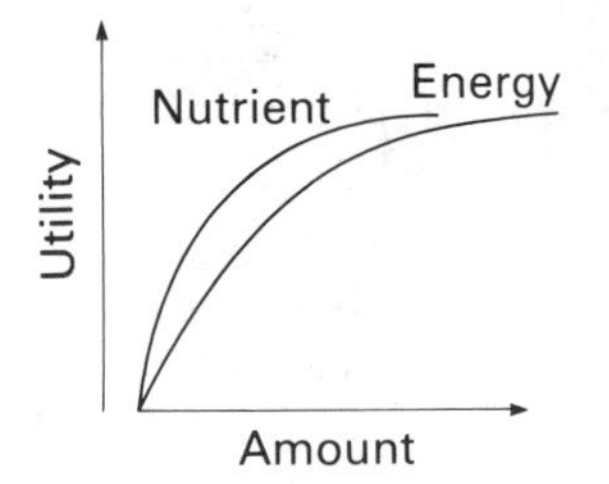

Fig. 24.17 Hypothetical utility functions for nutrient and energy

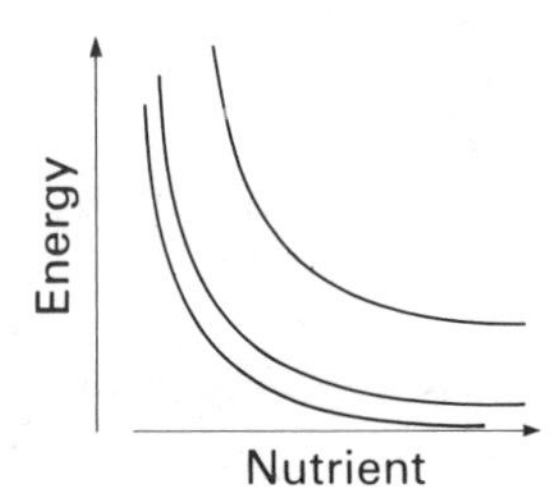

Fig. 24.18 Iso-utility functions for nutrient and energy, based upon Figure 24.17

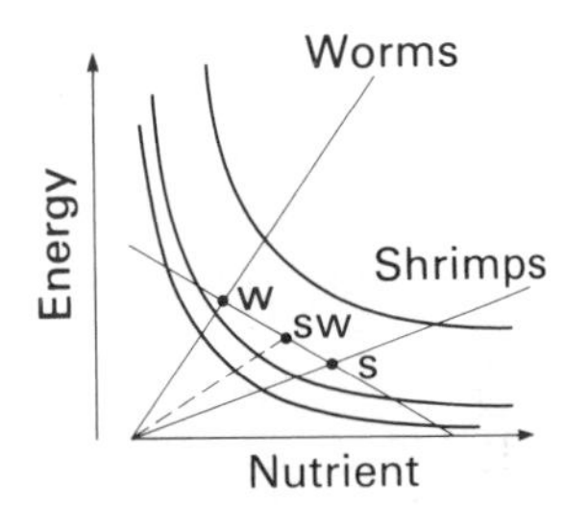

Fig. 24.19 Iso-utility functions from Figure 24.18 superimposed on Figure 24.16. The dotted line indicates the best possible mixture of worms and shrimps, assuming that there is no cost of changing between them (After McFarland and Houston, 1981)

energy spent on foraging, which is equivalent to the amount of goods A and B that a shopper could obtain for a given amount of money. The amount of energy an animal has at the time of foraging acts as a constraint on what it can purchase. The energy obtained by foraging is not immediately available for use because the food has to be digested, just as the money a person earns by working is not immediately available to spend. The *budget line* in Figure 24.16 represents the constraints on foraging imposed by the bird's available energy and means that, within a given period of time, the bird cannot obtain more worms and shrimps than the quantities represented inside the triangle formed by the budget line.

The next question is: What utility does a redshank derive from energy and nutrient? The utility of such commodities usually obeys a law of diminishing returns; that is, if an animal already has a good supply of energy, a little extra does not add much to the utility. If the animal is short of energy, however, then that same small amount of energy will make a large contribution to utility (see Fig. 24.17). The same considerations commonly apply to economics. Thus, I may derive a certain amount of satisfaction or utility from my large china collection. If I add one more piece of china to my collection, then I will increase my satisfaction by a small amount. If I had a smaller collection, however, then adding that same piece would increase my satisfaction by a larger amount.

Figure 24.17 shows hypothetical utility functions for nutrient and energy. Figure 24.18 shows how these functions can be combined to give a set of *iso-utility curves* (joining all points of equal utility), which shows that the redshank can obtain the same utility from a large amount of energy combined with a small amount of nutrient as from a small amount of energy combined with a large amount of nutrient. The shape of the iso-utility curve is determined by the shape of the corresponding utility functions in Figure 24.17. In the economic analogy, I might derive the same total utility from purchasing two pieces of china and one tennis ball as from one piece of china and six tennis balls. For this reason, economists often call iso-utility curves *indifference curves*.

Suppose we now combine Figures 24.16 and 24.18 into Figure 24.19. We can now see that certain combinations of worms and shrimps will yield higher utility than others. The least utility is obtained by foraging for worms alone (w), more is obtained from shrimps alone (s), but even more is obtained by a certain combination of worms and shrimps (sw). Two main factors affect the combination yielding the highest utility: the shape of the iso-utility (determined by the utility functions) and the slope of the budget line (determined by the relative prices of the two commodities).

In consumer economics, a person that has more money to spend can purchase more goods and there is a parallel shift in the budget line (Fig. 24.20). If the price of a particular item is lowered, then more can be purchased for the same amount of money, also making the budget

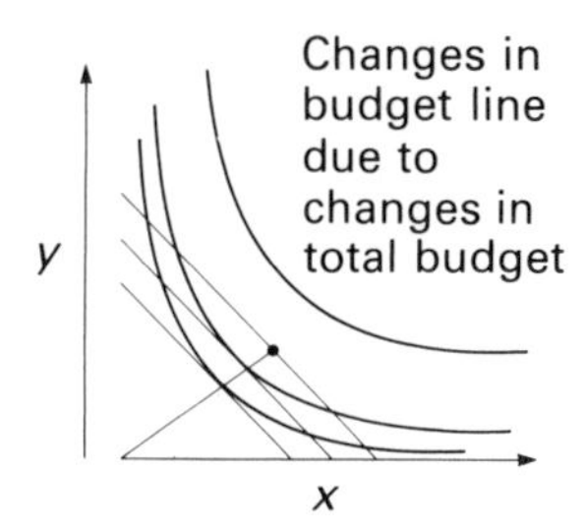

Fig. 24.20 Different budget lines result from differences in the amount of energy initially available to the animal. Note that the optimal preference does not change

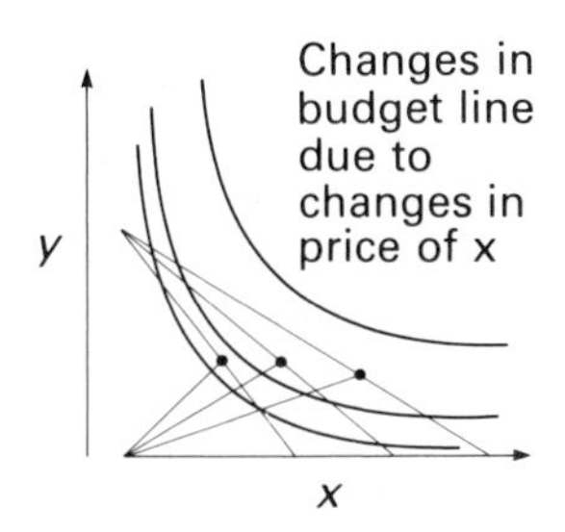

Fig. 24.21 Changes in the budget line due to changes in the price of x. Note the changes in optimal preference

triangle larger (Fig. 24.21). However, a price change for a single item alters the *slope* of the budget line, which does not occur if the overall budget is increased. Comparison of Figures 24.20 and 24.21 shows that there is little change in the optimal preference when the overall budget is altered but that changes in price cause a considerable shift in optimal preferences.

In likening an animal to an economic consumer, we regard cost or utility as the common currency of decision-making. We distinguish between these concepts and the energy cost or price associated with particular activities or goods. Although energy cost is a common factor in the alternative possible activities in the examples we have discussed so far, this is not always the case. It is a mistake, therefore, to regard energy as a common currency for decision-making, even though it may be a common factor in particular cases. Energy will be a factor common to all alternatives only if energy availability acts as a constraint in the circumstances being considered. An economic example may help to make this distinction clear.

Suppose a person enters a supermarket to purchase a variety of goods. As we have seen, utility functions will be associated with each possible item of purchase, and the utility of particular goods will depend upon the shopper's preferences and experience. Factors such as the palatability, visual attractiveness and list of ingredients will influence the utility associated with food items. The concept of utility enables very different items to be compared. Thus, a person may obtain the same utility from a packet of mixed herbs as from a bottle of soda. Utility is thus the currency by which different items are evaluated. However, another currency common to all items is money, important if the shopper has a limited amount to spend since it constrains the amount that can be purchased. However, although every item in the supermarket has a marked price, money is not always the relevant constraint. A shopper in a great hurry may carry more money than can be spent in the time available. Time is thus the constraint that bites, and it may be that the shopper can choose among bottles of soda more quickly than among different packets of mixed herbs. Time taken to choose each item will determine the slope of the time-budget line, and it may turn out that the shopper's choice when pressed for time is different than that of the shopper with limited cash.

The lesson to be learned from this example is that many possible constraints act upon an animal's choice behavior. These constraints may vary with circumstances, but it is only the constraint that bites at the time that is important. For this reason, time and energy should not be confused with utility or with cost (in terms of fitness). In drawing the analogy between an animal and an economic consumer, cost is equivalent to utility, energy is equivalent to money, and time is equivalent to time. The animal can earn energy (money) by foraging (working) and may spend it upon various other activities. Over and above the basic continuous level of metabolic (subsistence) expenditure, the animal can save energy (bank money) by hoarding food or depositing fat, or it can spend it upon various activities, including foraging (working). When the price of activities is high, the animal is subject to a tight energy-budget constraint, and when the price is reduced the animal experiences an increase in real income and the budget constraint is relaxed (McFarland and Houston, 1981).

24.6 Time budgets and energy budgets

If we have a limited amount of money to spend on a variety of goods and activities, we often partition it among the different purposes. We usually review our expenditure over a particular period of time—such as a day, a week, or a year—and call this a budget. Budgeting may occur in advance of any expenditure, or it may occur in retrospect. In the one case we allocate particular sums to particular purposes, while in the other we review the expediture that has already occurred. In either case, the notion of budgeting implies a certain discipline in spending money.

Since, as we have seen, money in consumer economics is analogous to energy in animal behavior, it is natural to ask whether or not animals have energy budgets. However, we have seen also that energy is only one type of constraint that impinges on animal behavior. Another important constraint is time, and we also can ask whether time budgets are relevant to animal behavior. A time or energy budget should not simply be an account of how an animal spent time or energy. An animal whose use of time and energy was completely chaotic would have no budget. However, we can expect natural selection to design animals so their available time and energy is put to maximum use. It seems reasonable, therefore, to expect animals to treat time and energy as valuable resources and to budget accordingly.

Bernd Heinrich (1979) likened the foraging bumblebee (*Bombus*) to a shopper:

> A bee starting to forage in a meadow with many different flowers faces a task not unlike that confronting an illiterate shopper pushing a cart down the aisle of a supermarket. Directly or indirectly, both try to get the most value for their money. Neither knows beforehand the precise contents of the packages on the shelf or in the meadow. But they learn by experience.

The bees are dependent upon flowers for the energy required to rear the young, but they may have to expend considerable amounts of energy in foraging. Bumblebees are able to exist in cold climates by virtue of their remarkable themoregulatory physiology. They can maintain a high body temperature at a low environmental temperature, which enables them to be active even though it involves a high rate of energy expenditure. They can conserve energy by greatly reducing their activity level and conserving heat. When food resources from flowers are scarce, the bees nevertheless manage to make a profit by foraging slowly. When food is abundant, they raise their body temperature and forage rapidly. Thus, they budget their energy expenditure in accordance with the prevailing circumstances.

A foraging bumblebee spends most of its time travelling. In moving from flower to flower, bees try to keep their flight time and distance to a minimum, flying between 11 and 20 kilometers per hour and spending only two to four minutes inside the nest between foraging trips. Simple calculations (Heinrich, 1979) show that the time budget is more important than the energy budget for a foraging bumblebee. Suppose, for example, that one bee has flowers close to the nest and can forage there continuously, while another bee is foraging three kilometers from the nest. If the second bee flies at 15 kilometers per hour, it must travel 24 minutes per trip. Foraging on fireweed, both bees could collect a honeycropful of nectar (about 30 milligrams of sugar) in about 10 minutes. The commuting bee collects 30 milligrams of sugar in 34 minutes but expends about three milligrams in flight metabolism. Thus, commuting takes up about two-thirds of the time per trip but only one-tenth of the energy. The bee foraging close to the nest would obtain 102 milligrams of sugar in a foraging trip of the same duration. Thus, to make commuting worth while, the flowers far away from the nest would have to be 3.4 times more rewarding than those close by.

Bumblebees change their foraging behavior in response to changes in nectar abundance. The more nectar per flower, the more they search other flowers in the vicinity. Heinrich carried out experiments in which he laid screens of bridal veil over some patches of clover and left others unscreened. The bees depleted the nectar in the unscreened clover, but that in the screened clover accumulated. When the screening was

Fig. 24.22 Changes in foraging behavior of bumblebees foraging in two patches of white clover. One (left) was utilized by many bumblebees and had only 0.003 mg sugar per flower top, the other (right) had been screened with bridal veil to allow nectar to accumulate to a level of 0.01 mg sugar per flower. The graphs show that the bees made longer flights between flower heads when nectar supplies were low (After Heinrich, 1979)

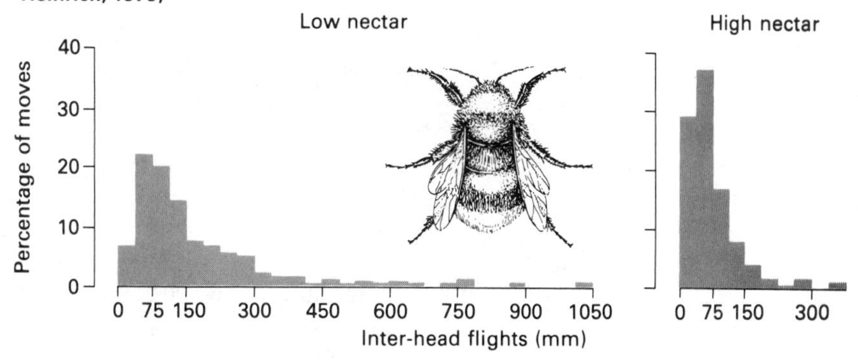

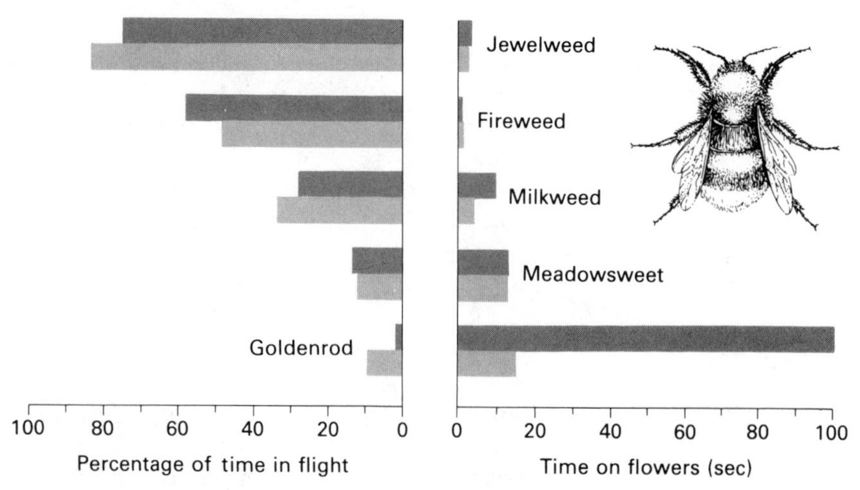

Fig. 24.23 The proportions of time that bumblebees spend in flight and perching are closely related to the kinds of flowers visited and not much affected by air temperature. Pale bars indicate results obtained at 20°C, dark bars show results obtained at 30°C (After Heinrich, 1979)

removed so the bees could visit both rich and poor areas of clover, in nectar-rich areas they probed about 12 florets per head and made short flights between heads. In areas of low nectar the bees probed only about two florets on each flower head and made longer flights between heads (Fig. 24.22). In this way the bumblebees concentrated their foraging in the most profitable areas.

Fig. 24.24 (*a*) Body temperatures of bumblebees at different air temperatures, while foraging from profitable flowers. (*b*) Calculated foraging costs at different air temperatures for a worker bee weighing 0.2 g, regulating its thoracic temperature at 30°C, and spending half its time in flight and half perched on flowers (After Heinrich, 1979)

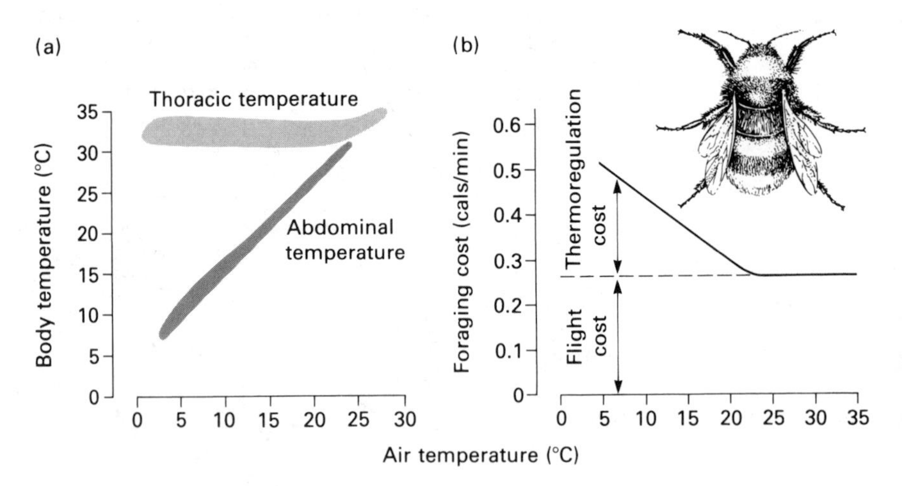

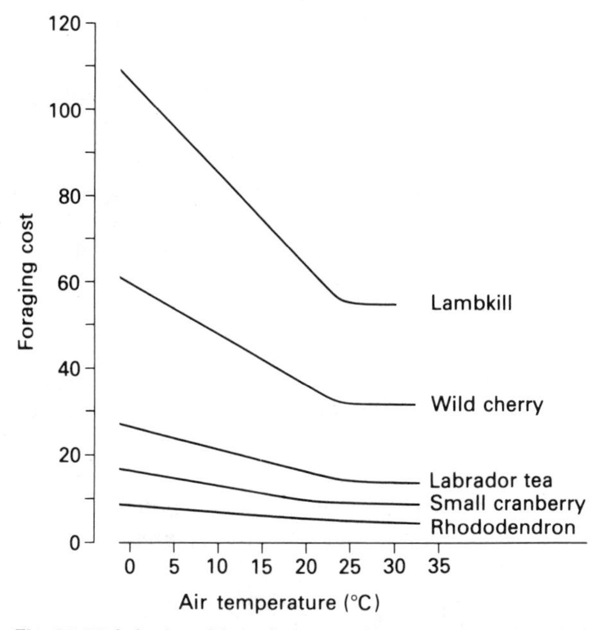

Fig. 24.25 Calculated foraging costs for queen bees (0.5 g) visiting different types of flower (After Heinrich, 1979)

The energy costs of foraging depend partly upon the environmental temperature. When foraging on fireweed the bees stop for only one or two seconds at each flower and remove tiny drops of nectar with a dab of the tongue. On the other types of flowers, however, they remain a number of minutes (Fig. 24.23). While perching on a flower, the bees do not allow their flight mechanism to cool, but keep it at flight temperature so they are ready to fly without delay. During prolonged perching they maintain the thorax at a temperature of about 32°C by shivering, but do not waste energy heating the abdomen, which is not involved in powered flight (Fig. 24.24). At environmental temperatures of about 25°C it is not necessary to heat the thorax during periods of inactivity; below this temperature the bee has to pay progressively more to maintain efficient foraging (Fig. 24.24). Some flower species produce more nectar than others, and it is possible to calculate the relative costs of foraging on different types of flowers at different temperatures (Fig. 24.25). Bumblebees forage on the profitable rhodendron flowers over a wide range of temperature, but they do not forage quickly enough lambkill and wild cherry at low temperatures because they cannot offset the extra energy required for thermoregulation (Heinrich, 1979).

Bumblebee foraging illustrates how time and energy constrain foraging efficiency. When the bee has to travel some distance to find productive flowers, then time becomes a limiting factor, and it is worthwhile for the bee to expend energy in order to save time. When

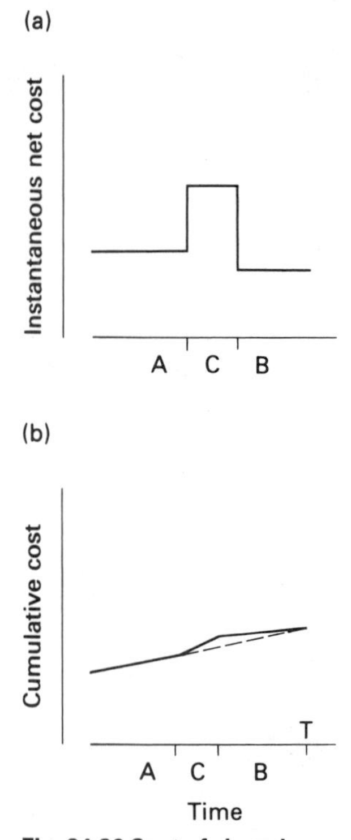

Fig. 24.26 Cost of changing from one activity to another. (*a*) The instantaneous cost of activity A is higher than that of B, but to change from A to B the animal must do the changing behavior C, which has an even higher instantaneous cost. (*b*) At time T the accumulated cost of the transition A-B-C is the same as it would have been if the animal had remained doing A (After Larkin and McFarland, 1978)

foraging on relatively unproductive flowers or when foraging at a low temperature, the bee may take more time in order to save energy.

Interactions between time and energy are important in many aspects of animal behavior. The animal has to budget for each particular type of activity and for the particular circumstances in which it occurs on a given occasion. In addition, it has to budget for the cost of changing from one activity to another. When an animal changes from one activity to another, there may be a cost involved. For example, a pigeon feeding in a cornfield may become thirsty as a result of eating dry seeds and have to fly a mile to obtain water. In addition to the physiological cost of the journey, the bird will spend time travelling from the feeding place to the drinking place, during which period it does not receive any of the benefits of feeding or drinking. Moreover, there may be risks involved in the journey, such as exposure to hawks or to farmers with guns. Thus, the cost of changing is the decrement of fitness that arises during the period in which the animal is changing from one activity to another and receiving benefits from neither. The cost may involve loss of valuable time, expenditure of energy, or risk from predators. It can be shown theoretically that that the *cost of changing* should be budgeted as if it were part of the cost of the activity about to be performed.

We can distinguish between the *instantaneous cost* that arises within each unit of the relevant period of time and the *cumulative cost* that occurs over the whole of the period under consideration (Fig. 24.26). The important thing is that a change in behavior is worth while if the cumulative cost a short time after changing is less than it would have been had the animal not changed behavior at all (Fig. 24.26). I showed by experiment that, when the cost of changing is high, doves change between feeding and drinking less often than they would otherwise do. Stephan Larkin and I obtained evidence that the birds do indeed allocate the cost of changing to the cost of the behavior about to be performed (Larkin and McFarland, 1978). They do not change their behavior until they have accounted for this cost. Larkin (1981) found that the patterns of feeding and drinking in doves were altered in a predictable manner when the cost of changing from one to the other was increased in terms of time or of energy expenditure.

In addition to considering minute to minute, animals take a more global account of their time and energy budgets. In general, we can expect behavior whose prime function is promoting the survival of the individual to take precedence over behavior that promotes other aspects of fitness, such as territorial mating and parental behavior. However, species vary considerably. Some aspects of behavior are essential, but others—like thermoregulatory behavior—may be important only when physiological mechanisms cannot cope. Thus, drinking is a daily necessity for some species, but other species can manage without drinking at all.

Each activity has value in terms of fitness, and animals need to

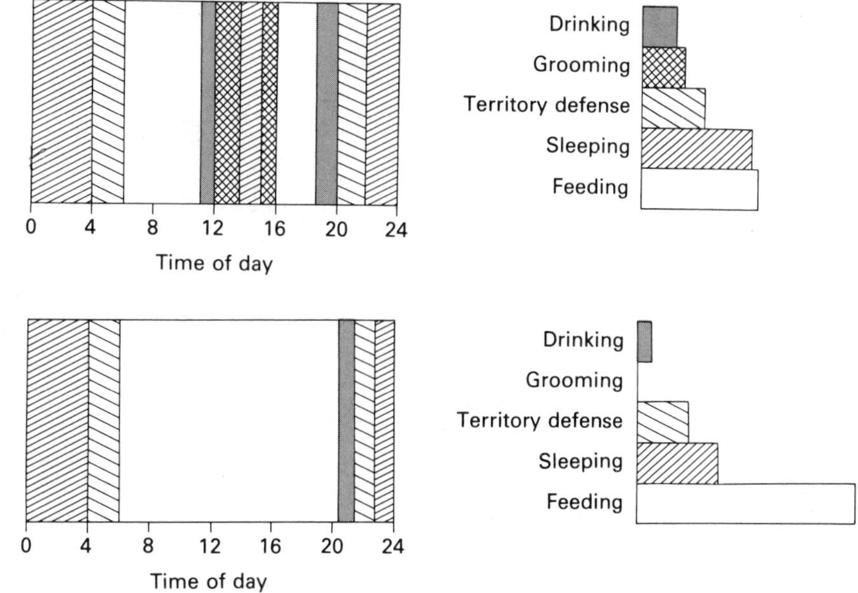

Fig. 24.27 Diagram showing how an animal might spend its time throughout the day

Fig. 24.28 Diagram showing how the animal in Figure 24.27 might adjust its daily routine when required to spend much more time to obtain the same amount of food

allocate priorities to activities in a general way, as well as from minute to minute. Alasdair Houston and I approached this problem by considering a measure of the cost to an animal of abstaining from each activity in its natural repertoire (Houston and McFarland, 1980). If an animal did no feeding, for instance, the cost would be high, but if it abstained from grooming, the cost might be relatively low. An animal with high motivation to feed and to groom but with not enough time for both would sacrifice less, in terms of fitness, if it devoted its available time to feeding.

Suppose an animal fills its typical day with useful activities, as illustrated in Figure 24.27. In an environment that was much the same from day to day, the animal would adjust its activities to the time available. Suppose, however, there is a change in the environment so it now takes very much longer to obtain the normal amount of food (Fig. 24.28). The animal can respond to the changed circumstances by spending the same amount of time feeding as before and settling for less food, or it can insist on the same amount of food as usual, or it can compromise between the two extremes. If the animal spent a long time obtaining the usual amount of food, then there would be less time available for all its other activities, which would have to be squashed into the remaining time. We found that the extent to which an activity resists squashing can be represented by a single parameter, which we called *resilience.* In the case where the animal feeds for the normal amount of time and ends up with a reduced food intake (Fig. 24.27), the resilience of feeding is relatively low because feeding has not compressed the other activities even though the animal's hunger is increased. In the case where the animal insists on the normal amount of food (Fig. 24.28), the resilience of feeding is relatively high because feeding ousts the other activities from the time available without itself being curtailed in any way.

Behavioral resilience is a measure of the extent to which any activity can be squashed in time by the animal's other activities. It also reflects the long-term importance of an activity: When time is a budget constraint, activities with low resilience will tend to be ignored. Indeed, if an activity completely disappears from an animal's repertoire when time is rationed, we might call it a luxury or leisure activity.

Behavioral resilience has been measured directly in few cases because of the practical difficulties involved—keeping a record of the behavior around the clock for a number of days in a situation where available time can be manipulated. Observations of this kind have been carried out by David Croft (1975) in studying how photoperiod length affects the activity and hormonal balance of female canaries (*Serinus canaria*). By manipulating photoperiods, Croft effectively altered the time available for activity, since canaries are inactive in the dark. Croft found that the birds spent the same amount of time feeding on short days as on long days but that they fed more efficiently on long days. During long days the birds expended more energy in various activities, so this result is not entirely unexpected. Birds kept on long days spent more daytime inactive and sleeping than birds on short days. Since the birds can sleep at night, it seems most likely that birds kept on long days filled spare time with sleep, a view supported by the fact that birds building a nest spend less time sleeping during the day.

Croft calculated the time available for nest building after time necessary for feeding, grooming, and traveling from place to place had been taken into account. As we can see from Table 24.3, there is adequate time for nest building on long days but barely enough time on short days. The interpretation of the observations is complicated, however, by the fact that long days stimulate hormone production so these birds are more motivated to build a nest than birds kept on short days. The nest-building behavior of birds kept on long days is more efficient than

Table 24.3 Estimated median time available for nest building by estrogen-implanted ovariectomized birds on a short (8 hr) and long (14 hr) photoperiod, compared to time required for nest building.

Measure	*Photoperiod* 8L:16D	*Photoperiod* 14L:10D
Time available for nest building	1.1 hr	7.5 hr
Time required for nest building	2.3 hr	2.3 hr

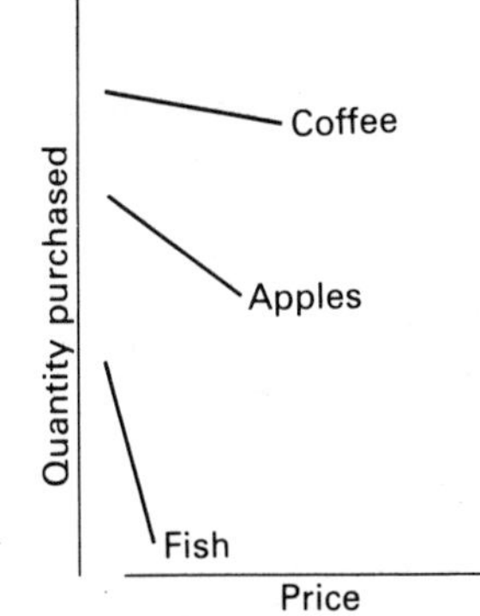

Fig. 24.29 Examples of economic demand curves. Demand for coffee is inelastic, demand for fish is elastic (both axes have log scales and arbitrary origins)

that of birds on short days, and less time is wasted on unnecessarily repetitious gathering and building movements. Thus, although *resilience* is a concept theoretically distinct from *motivation* (McFarland and Houston, 1981), it is difficult to isolate its effects in studies of time budgets.

However, resilience can be measured indirectly by means of *demand functions,* which are used by economists to express the relationship between the price and the consumption of a commodity. For example, when the price of coffee is increased in the supermarket, people continue to buy about the same amount as before, perhaps a little less (Fig. 24.29). As the price of fruit is increased, however, the demand for fruit falls off. When the price of fresh fish is increased, demand declines markedly. Presumably, people are willing to pay more to maintain their normal coffee-drinking habits. Demand for coffee is said to be *inelastic*. If the price of fresh fish increases, however, people tend to buy less and to switch to substitute foods such as meat or canned fish. Demand for fish is said to be *elastic*.

Exactly analogous phenomena occur in animal behavior. If an animal expends a certain amount of energy on a particular activity, then it usually does less of that activity if the energy requirement is increased (Fig. 24.30). Numerous studies have shown that demand functions in animals follow the same general pattern as those of humans (Kagel et al., 1980). For example, Steven Lea and Tim Roper (1977) found that demand for food was inelastic in rats required to work (press a lever) to obtain food rewards. As the work required (lever presses per reward) increased, the rats continued to work for about the same amount of food. However, these investigators also found that elasticity of demand

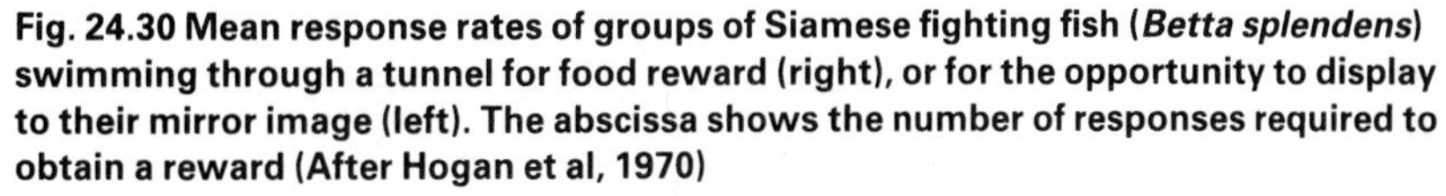

Fig. 24.30 Mean response rates of groups of Siamese fighting fish (*Betta splendens*) swimming through a tunnel for food reward (right), or for the opportunity to display to their mirror image (left). The abscissa shows the number of responses required to obtain a reward (After Hogan et al, 1970)

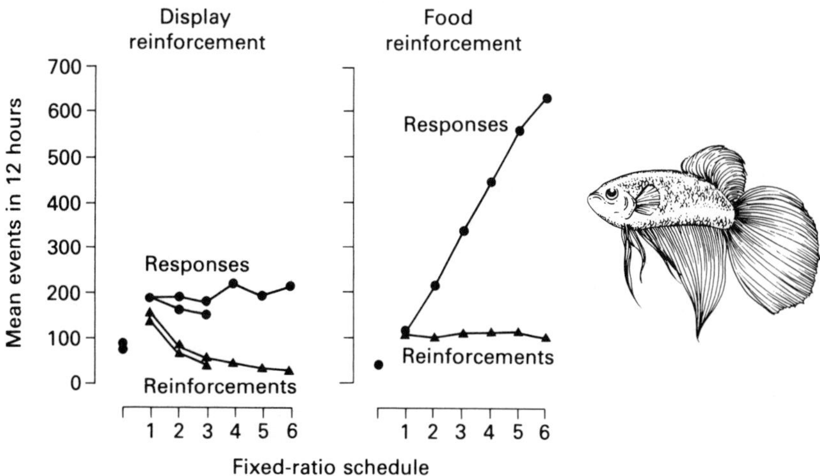

for food pellets increased when sucrose was available as a substitute.

The parallel between the demand phenomena of animals and people is a topic of considerable current interest (e.g. Allison, 1979, 1983; Lea, 1978; Rachlin, 1980). The elasticity of demand functions gives an indication of the relative importance of the commodities (or activities) on which the person (or animal) spends his or her money (or energy). There is a close relationship between elasticity of demand and resilience (Houston and McFarland, 1980). Thus, demand functions can be used as indirect measure of resilience. If activity A has higher resilience than activity B, then A will tend to show an inelastic demand function and activity B an elastic one (McFarland and Houston, 1981).

Points to remember

- Foraging strategies are evolutionary strategies designed to optimize some aspects of the costs and benefits of foraging.

- Most aspects of animal behavior involve some sort of trade-off with other types of behavior. This achieves a compromise between the costs and benefits of alternative courses of action.

- The sum of all possible costs and benefits can be expressed as a cost function. This can be defined as the instantaneous level of risk incurred by (and reproductive benefit available to) an animal in a particular internal state, engaged in a particular activity, in a particular environment.

- Animals can be assumed to make rational decisions in the sense that their choices are consistent and transitive.

- There are analogies between human consumer economics and the costs and benefits incurred by animals. Money is analogous to energy, and utility is analogous to benefit (increment in fitness). Time and energy budgets can be analyzed on this basis.

Further reading

McFarland, D. J. and Houston, A. (1981). *Quantitative Ethology: The State Space Approach*, Pitman, London.

Heinrich, B. (1979) *Bumblebee Economics*, Harvard University Press, Cambridge, Massachusetts.

25 Mechanisms of Decision-making

We saw in Chapter 24 that every aspect of animal behavior is likely to incur costs and benefits in contributing to or detracting from the animal's overall fitness. Moreover, every aspect of behavior is likely to involve some trade-off among various advantages and disadvantages. These disparate aspects of behavior can be evaluated in terms of a cost function, which provides a basis for choosing the optimal course of action in a given set of circumstances. We looked at animal decision-making from an economic point of view, and saw how we might account for rational decisions in terms of maximizing utility subject to certain constraints. However, such considerations are purely functional. They specify what animals ought to do to make the best decisions under particular circumstances, but they do not say anything about the mechanisms that animals might employ to attain these objectives. In this chapter we discuss the mechanisms of decision-making in animals.

25.1 Decision-making by rules

To say that an animal makes decisions implies no conscious intent but merely refers to the fact that the animal takes one behavioral alternative rather than another. Such a decision may be seen as a matter of chance or the outcome of some stochastic process (e.g., Dawkins and Dawkins, 1974). However, an animal that lives in a relatively stable and unchanging environment could make decisions on the basis of simple rules. In some species it appears that behavior occurs in a largely preprogrammed manner, making decisions being a matter of routine. For example, the lugworm *Arenicola* lives in a U-shaped burrow in the littoral zone of muddy seashores and sand flats (see Fig. 25.1). The worm feeds from sand sucked into the head end of the burrow, which passes through the gut to be deposited on the surface at the tail end of the burrow. Feeding behavior occurs in bursts that begin at regular seven-minute intervals and are separated by rest periods. The oxygen supply to the burrow is replenished by special irrigation behavior that recurs every fourth minute, even at low tide (though less vigorously), when oxygenated water is not available. The behavior of *Arenicola* is dominated by rhythms that persist over a wide range of conditions. After a prolonged

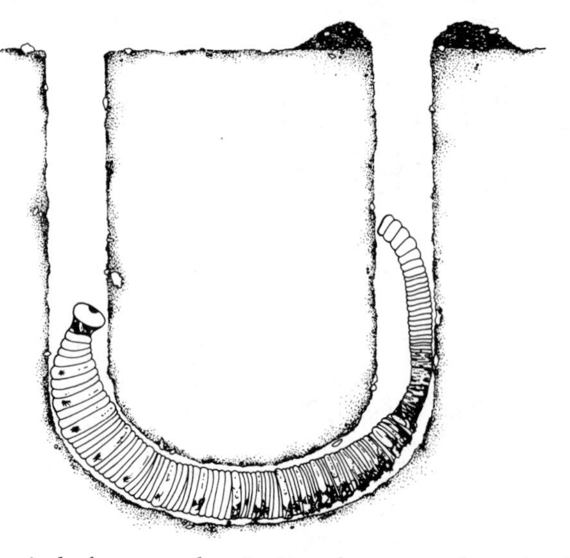

Fig. 25.1 The lugworm *Arenicola marina*

period of oxygen deprivation, however, the animal may break its rhythm and irrigate for a longer period than usual (G. P. Wells, 1966). Thus, it appears, the behavior of *Arenicola* is driven by a clocklike mechanism modified only in extreme circumstances. Such a rigid programmed decision strategy may be suitable for an animal living in a stable and predictable environment, but a more flexible approach is required in more changeable situations.

If an animal is going to depart from a strict routine, it immediately faces the problem of having to choose among behavioral alternatives. Normally, an animal will be able to pursue only one course of action at a time, either because incompatible movements are involved or because it can pay attention to only one set of stimuli at a time (see Chapter 12.5). Somehow, the animal has to achieve singleness of action in a situation where many types of actions may be called for simultaneously. One possibility is to have a strictly hierarchical rule like that found in the carnivorous gastropod *Pleurobranchus*. In this mollusk there is an order of precedence—egg laying, feeding, mating, other activities (Davis et al., 1974). Thus, the animal will abandon its mate, even during copulation, if offered food. During egg laying, however, there is hormonal suppression of feeding, which prevents the animal from eating its own eggs.

A more usual method of attaining singleness of action is for the animal somehow to gauge the relative strengths of the possible activities that could occur at a particular time. As we saw in Chapter 15, the strength of an animal's tendency to perform a particular activity depends upon both internal and external factors. Thus, the tendency to feed is determined by the degree of hunger and the external cues indicating food availability. An animal could have a simple rule that tells it to feed when the feeding tendency is stronger than any other tendency. This type of rule

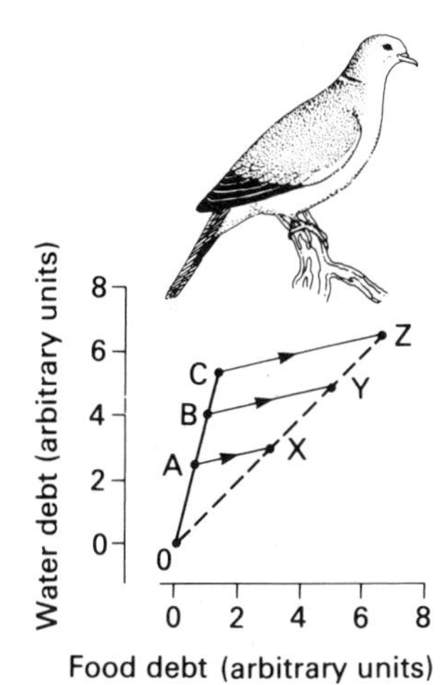

Fig. 25.2 Motivational state of doves deprived firstly of water, and then, at a specified time later (i.e. at points A, B, or C), of food also for a further specified period. Doves in states X, Y, and Z showed equivalent relative preferences for food and water (After McFarland, 1971)

was long assumed to be the mechanism by which animals made decisions. Competition among drives was thought to occur in such a way that the strongest drive was expressed in behavior, while the others were inhibited or prevented from influencing the animal's behavior. This may seem to be a straightforward theory, but it involves a number of difficulties. Quite apart from the problems associated with the drive concept (see Chapter 15), there are difficulties involved in evolving an animal based upon simple competition among motivational systems.

It is to be expected that the choice made by animals will depend upon motivational state. Some evidence supports this view. Thus, when doves are deprived of food and water in such a way that their hunger and thirst can be calculated precisely, they choose between food and water in a simple Y maze (or in a Skinner box) on a basis related systematically to the magnitude of their hunger and thirst, as illustrated in Figure 25.2. However, the doves do not seem to choose food or water according to the simple rule of "eat if hunger is greater than thirst and drink if thirst is greater than hunger," as can be seen from the results portrayed in Figure 25.3. This experiment was one of the first to raise doubts about simple competition as a mechanism of decision-making. Before abandoning the competition theory, however, we shall look into it more closely.

Fig. 25.3 Doves at various initial states of hunger–thirst (black dots) were given a food–water choice every 30 seconds. Their cumulative intake throughout the half-hour test period is shown by the white dots (one every 5 minutes). Note that the doves did not choose so as to equalize hunger and thirst as would be expected if simple motivational competition (Fig. 25.4) were operating (After McFarland, 1971)

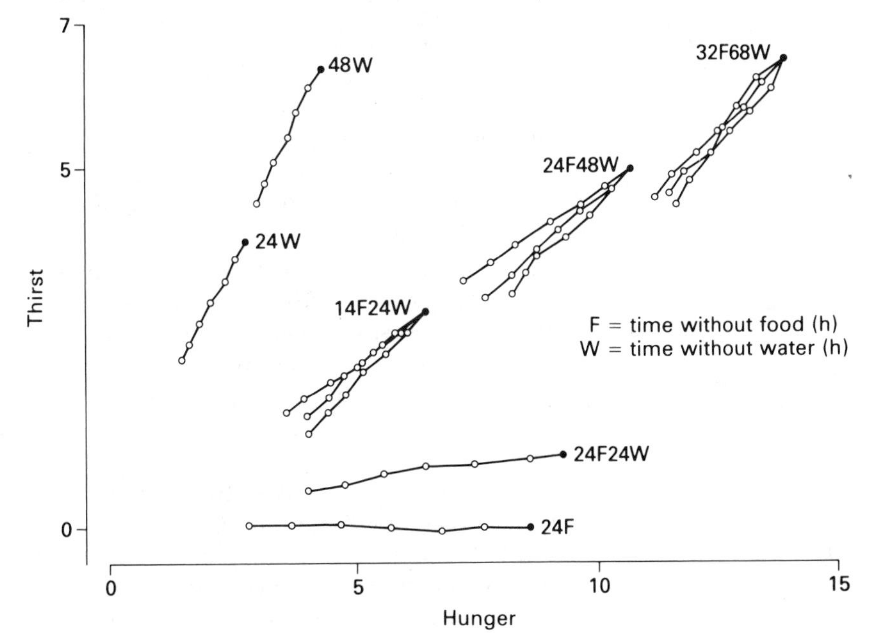

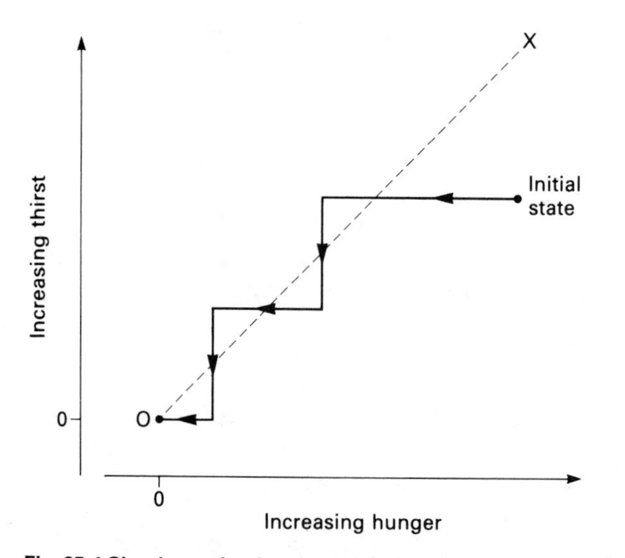

Fig. 25.4 Simple motivational competition. On the right of the dashed line hunger is greater than thirst and on the left thirst is greater than hunger. According to simple competition theory the animal reduces its initially higher motivation until it is equal to the other. It then alternates between the two (the size of the steps has been exaggerated in this diagram), until both are reduced to zero

25.2 Motivational competition

On a simple competition theory, we would expect a hungry and thirsty animal to behave in the manner shown in Figure 25.4; that is, eat whenever hunger is greater than thirst, and drink whenever thirst is greater than hunger. We would expect the animal to be rational in that its choices would be transitive. However, inspection of Figure 25.4 shows that it is very unlikely that animals will be designed this way. Along the line OX, where hunger and thirst have equal magnitude, the animal is continually dithering. If hunger is greater than thirst, one mouthful of food reduces hunger so that now thirst is greater than hunger. One sip of water reverses the situation again. Such dithering cannot be a very efficient way of proceeding, and a number of alternative mechanisms have been suggested. When engineers are faced with this type of problem in the design of some machine, they sometimes introduce *hysteresis* to prevent dithering. Hysteresis is a special mechanism that causes a delay between stimulus and response. It has been suggested (e.g., Toates and Oatley, 1970; Toates, 1980) as a component of the feeding- and drinking-control mechanisms of animals, such that the animal would not detect diminished hunger until well after it begins to eat. In the present context a hysteresis mechanism would have the

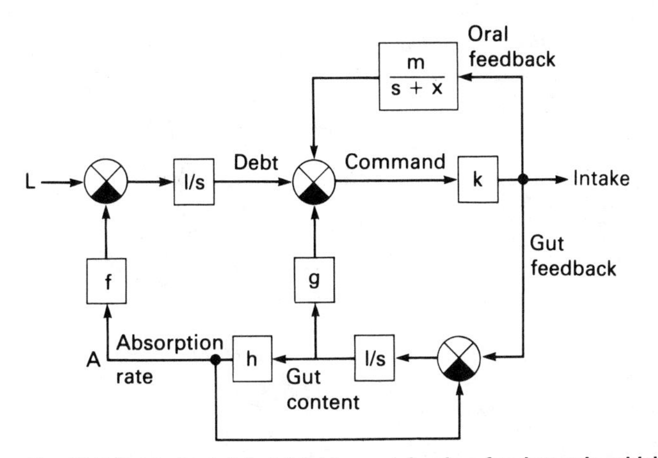

Fig. 25.5 Control model of drinking mechanism for doves, in which the oral factors provide positive feedback and the gut factors provide negative feedback (the circles containing X are summing points, in which a black quadrant changes the sign). L = rate of water loss, A = rate of water absorption into the blood. The other symbols are the parameters of the component mechanisms (After McFarland and McFarland, 1968)

effect of carrying the hunger-thirst trajectory beyond the hunger-equals-thirst dividing line.

Another possible mechanism that has much the same effect is *positive feedback*. There is some evidence that the tendency to feed (Wiepkema, 1971) or drink (McFarland and McFarland, 1968) temporarily increases as a result of the passage of food or water through the mouth. This positive feedback from the consequences of feeding and drinking occurs in parallel to the normal negative feedback that has satiating consequences (Fig. 25.5). In doves, for example, water delivered directly into the crop through a tube has the same satiating effect as water drunk normally, but has no reinforcing effect in an operant situation. The passage of water through the mouth acts as a positive reinforcer, but contributes nothing to satiation (McFarland, 1969b). The effect of positive feedback is to take the trajectory past the line where hunger equals thirst. It is evident, however, that the true situation is more complicated than this. I found that if doves were drinking and eating in a situation where they had to negotiate a barrier to change between eating and drinking and vice versa, then they changed less often when the barrier was large and more often when it was small (Fig. 25.6). In deciding when to change between feeding and drinking, therefore, it appears that doves take into account the difficulty or cost of changing (see Chapter 24).

We can now see that simple competition is unlikely to provide the mechanism by which animals decide to change from one activity to another. It remains possible, however, that some more sophisticated version of motivational competition could provide the necessary set of

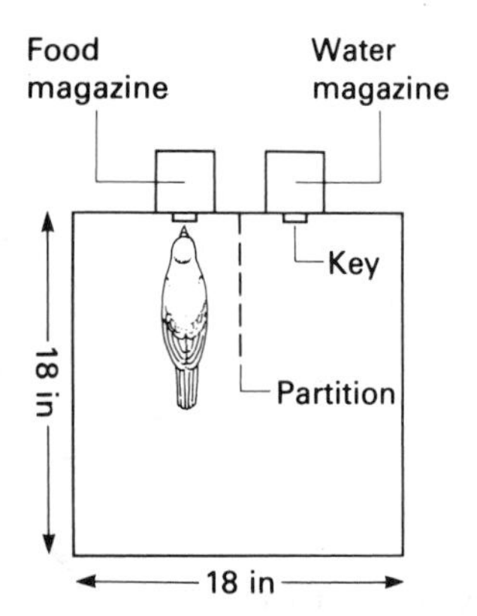

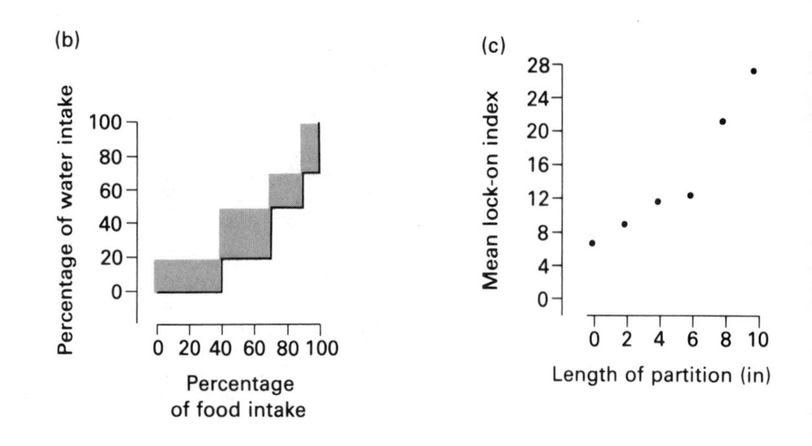

Fig. 25.6 (*a*) Plan view of Skinner box in which doves could obtain food and water by pecking at the appropriate key. The transparent partition between the keys could be varied in length. (*b*) Heavy line shows pattern of eating and drinking to satiation (100%). Note that the bird switches back and forth four times. By calculating the shaded area we generate a "lock-on index" inversely proportional to the number of alternations between the two activities. (*c*) Average lock-on index plotted as a function of partition length (After McFarland, 1971)

decision rules. In Chapter 15 we saw that an animal's motivational state may be made up of a variety of factors, including the animal's assessment of external stimuli (cue strength), its (primary) internal state, and, possibly, secondary motivational factors like the rate of change of state. These various factors combine to make up the animal's overall motivational state, which can be represented as a moving point (or trajectory) in a motivational state space. Different states can give rise to the same behavioral tendency, and these points may be joined by a motivational isocline. On this basis we may imagine that there is a simple competitive rule for deciding among tendencies but that each tendency is made up by a variety of factors, each carefully calibrated to reflect the animal's ultimate interests. For example, Richard Sibly (1975) proposed that doves operate on the basis of the following decision rule when choosing between feeding and drinking: Eat if food deficit times feeding incentive is greater than water deficit times drinking incentive; drink if food deficit times feeding incentive is less than water deficit times drinking incentive. Sibly defined *incentive* in terms of the rate at which food and water could be obtained. This formulation corresponds to competition between feeding and drinking tendencies, each based on hyperbolic motivational isoclines, as illustrated in Figure 25.7.

Incentive value of food rewards (g/s)

Food deficit (g)

Fig. 25.7 Hyperbolic isoclines, joining points of equal feeding tendency, in accordance with the hypothesis that feeding tendency = food deficit × incentive.

It is conceivable that some form of competition provides the basic decision rule employed by animals. Apparent violations to the rules can be explained by postulating suitable adjustments and calibrations to the various factors that combine to form the tendency to perform a particular type of behavior. This is a controversial topic, and the arguments involve complicated juggling of variables and parameters (e.g., Houston, 1982; McFarland, 1983). However, the student should keep in mind some general principles. First, a theory that claims that the strongest tendency always wins the competition, and so determines the animal's behavior, is in danger of creating a tautology. It implies that the observed behavior always reflects the strongest tendency. If such a theory effectively equates behavior and tendency, it is almost useless as an explanation of decision-making in animals. Second, the question arises as to whether a purely competitive mechanism of decision-making would make biological sense. Would not some low-priority aspects of behavior, like grooming, tend to be neglected? Would the animal be able to tailor desirable sequences of behavior to the prevailing circumstances, as seems suitable from an evolutionary viewpoint? Third, is there an alternative to a competition theory of decision-making? This is a difficult question. On the one hand, it might be claimed that competition is a necessary aspect of rational choice among alternatives that cannot be expressed simultaneously (e.g., Ludlow, 1980). On the other hand, it is possible to imagine non-rational decision-making like the cyclic routine of *Arenicola* discussed earlier that nevertheless suits the animal's circumstances.

25.3 The operational approach

An alternative to the model-making approach outlined previously is the *operational approach* to decision-making in animals. The operational approach defines phenomena purely in terms of their observable attributes, a valuable approach for making a start on the analysis of a complex process without any ad hoc preconceptions.

The idea that behavioral tendencies compete for overt expression implies that the different observed activities have equal status, since each is the winner in a competition. However, the phenomenon of displacement activities casts doubt on this assumption. The idea that displacement activities can occur as a result of disinhibition suggests that their status is somehow different than that of other activities. They do not assert themselves in competition with other potential activities but are permitted to occur during gaps in the predominant behavior. However, there are reasons for supposing that disinhibition occurs (see Chapter 21), not only in conflict and thwarting situations but also as part of ordinary behavior sequences. If this is true, then it contradicts the view that the activities observed in a sequence of behavior are necessarily the result of motivational competition and therefore have equal status. The problem is that it is not possible to tell if disinhibition is occurring by observation alone. There must be some experimental manipulation, and this must relate to the way in which competition and disinhibition are defined.

In Chapter 21, we saw that competition and disinhibition can be

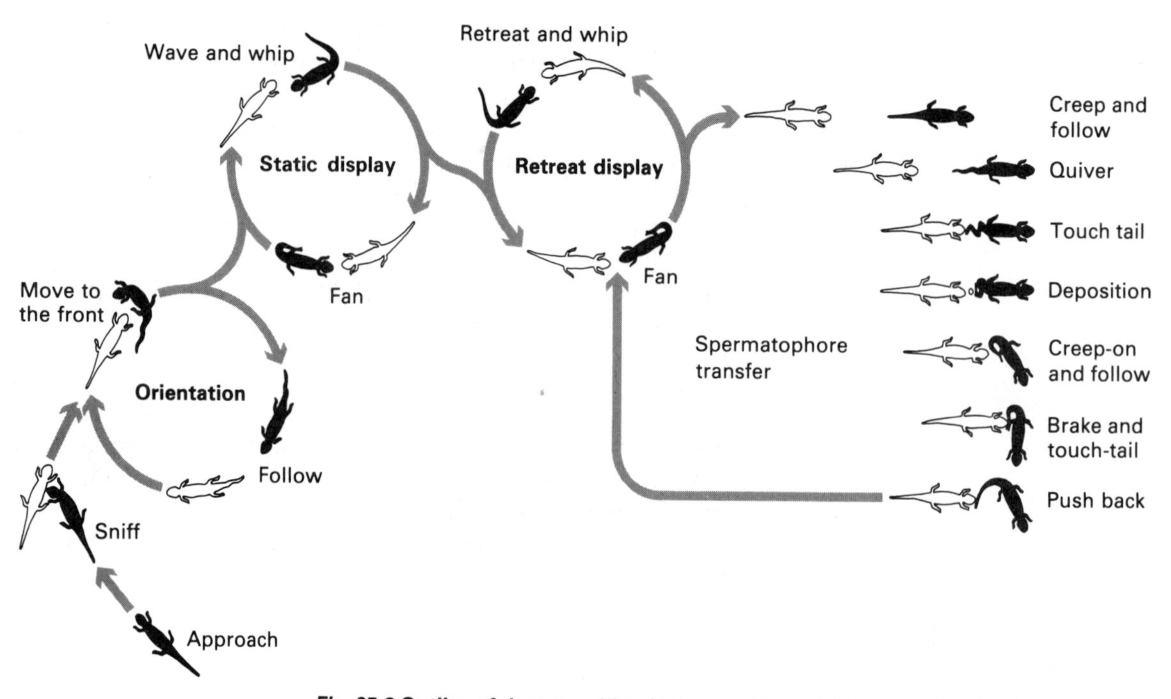

Fig. 25.8 Outline of the courtship of the smooth newt *Triturus vulgaris.* The male is shown in black (After Halliday, 1974)

operationally defined in terms of whether manipulation of a second-priority activity does or does not induce a change in the time at which the activity is observed to occur. We can show experimentally that the timing of a behavioral sequence A, B, A is not always affected by manipulation of factors relevant to B, even though the nature of B may be changed by such manipulation. Although this type of experiment is possible in the laboratory, it is difficult to carry out in the field or even in the seminatural conditions of animals in captivity. What is needed is some way of assessing the status of the activities that occur in normal behavior sequences.

In operational terms, the transition from one activity to another can occur in one of two ways. Either the level of causal factors for the second behavior is instrumental in inducing the occurrence of the behavior or it is not. In the former case we say that the behavior occurs as a result of motivational competition. In the latter case it must be due to disinhibition. Note that these labels are attached to transitions between one activity and another.

Sequences of behavior may be made up of one type of transition or both. For example, the courtship of the smooth newt (*Triturus vulgaris*) takes place on the bottom of a pond or aquarium. When a male encounters a female, he displays and initiates a courtship sequence made up of three distinct phases (see Fig. 25.8) that culminates in the deposition of a spermatophore by the male, followed by a stereotyped maneuver designed to induce the female to pick up the spermatophore with her cloaca (Halliday, 1974). During the first phase, the male attempts to orient himself in front of the female in order to display to her, and she initially moves away from him. If she is responsive, the female eventually remains stationary while the male displays to her with a variety of tail movements. Eventually, the female begins to advance toward the male, and he retreats before her while maintaining his display. The transition from static display to retreat display is initiated by the female. She provides the stimulus that induces the male to start his retreat display. His decision to respond is obviously due to motivational competition because its time of occurrence is the result of a change in his second-priority motivation (to perform retreat display) induced by the female's behavior.

Provided the female continues to advance, the male maintains his retreat display for about 30 seconds, after which he turns and moves slowly away from the female. The speed with which the male performs this part of the courtship seems to depend upon the number of spermatophores he is carrying (Halliday, 1976). Males with more than one spermatophore proceed quickly with the courtship sequence, deposit one spermatophore, and then return to retreat display and repeat the process again, as shown in Figure 25.8.

The male takes a more active role in the courtship than the female and probably uses up his available oxygen first. Although newts employ both cutaneous and pulmonary respiration, they are dependent upon oxygen obtained through the lungs to sustain muscular activity. Consequently, the courtship may be interrupted by one participant (usually the male) ascending to the surface of the water to breathe. When this happens in the natural situation, the male probably loses the courtship opportunity because the female wanders away or is approached by another male. Males are more likely to interrupt earlier parts of the courtship sequence to breathe and they never interrupt the spermatophore transfer phase. However, males often breathe immediately after the completion of spermatophore transfer, and the evidence suggests that this change in behavior is due to disinhibition (Halliday and Sweatman, 1976). It appears that the newts postpone breathing during spermatophore transfer and that breathing is disinhibited when spermatophore transfer no longer has overriding priority. Thus, the timing of breathing ascents seems to minimize the risk of wasting a spermatophore (Halliday, 1977a). Observation of newt courtship under atmospheres experimentally made deficient or rich in oxygen shows that oxygen availability has a marked effect on breathing ascents but only a small effect on the number of spermatophores deposited (Halliday, 1977b). Under reduced oxygen availability, the newts are able to compensate for the shorter available time by speeding up their courtship behavior. Enriched oxygen supply has the opposite effect. It seems that the male newts perform as much display behavior as they can manage to fit into the time available between breathing ascents. There is some evidence that the probability of successful fertilization of the female is related to the amount of display performed by the male (Halliday, 1974).

We can see from this example that, in some circumstances, motivational competition is the most appropriate and likely form of transition between two activities, particularly when a relevant external stimulus (like the female newt's approach behavior) changes the animal's motivational state. If a predator suddenly appeared on the scene, for example, it is most likely that whatever the animal was doing would be overruled by a sudden increase in escape motivation. In other situations, however, it may be important for the animal to finish the job (like the newt's spermatophore transfer behavior) and to inhibit other motivational tendencies until this is done. The subsequent behavior (e.g., breathing) is then disinhibited. We can also imagine that, in some circumstances, it may benefit the animal to interrupt its ongoing behavior to allow time to scan for predators or rivals.

25.4 Time-sharing

Behavior sequences can be made up of various combinations of competition and disinhibition, but one possibility, illustrated in Figure 25.9, has attracted particular interest. In this example, activity B occurs as a result of disinhibition by activity A. After a period of time T, activity A reestablishes itself by competition. If the onset and duration of activity B are under the control of the factors normally controlling activity A, then a *time-sharing* situation is said to exist (McFarland, 1974). The idea is that the system controlling activity A determines the pattern of A and that, in so doing, it periodically permits some other activity to occur for a short period of time T. The controlling system is said to be the *dominant* activity, while the activity that fills in the gaps in dominant behavior is

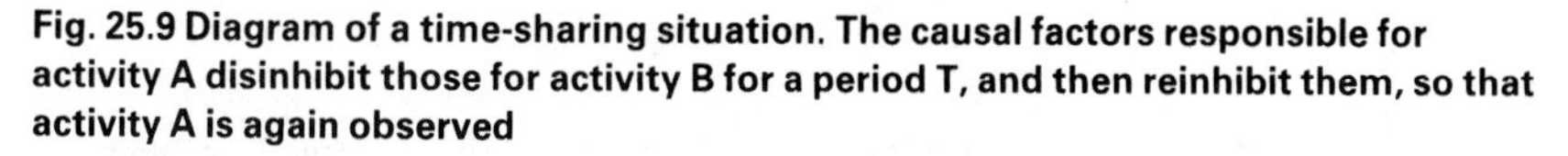

Fig. 25.9 Diagram of a time-sharing situation. The causal factors responsible for activity A disinhibit those for activity B for a period T, and then reinhibit them, so that activity A is again observed

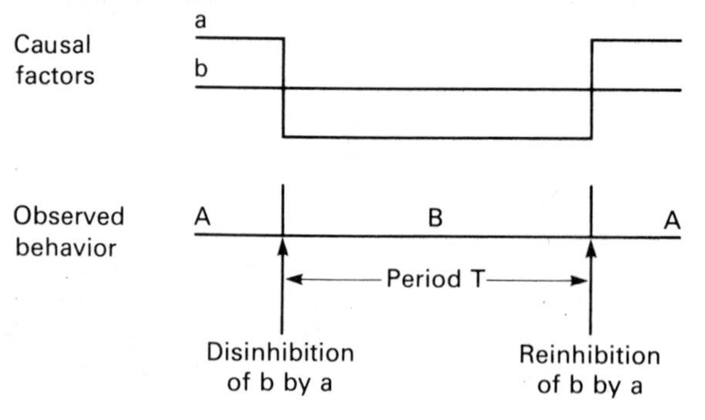

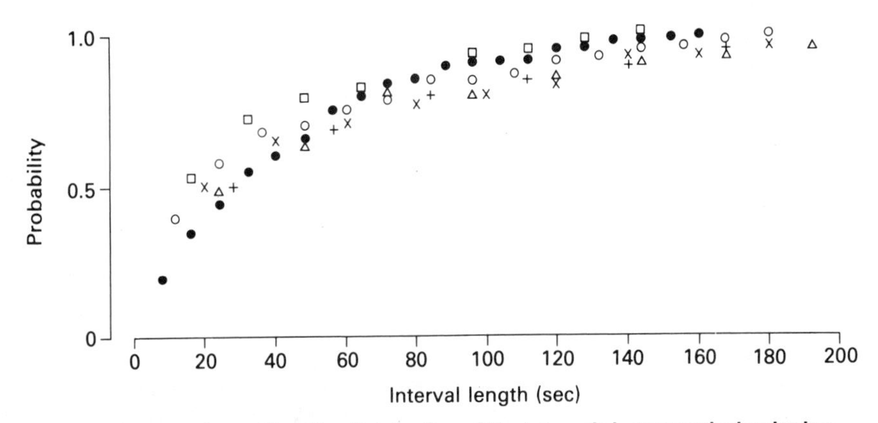

Fig. 25.10 Cumulative probability distribution of the intervals between the beginning of successive feeding bouts in doves able to work for both food and water. The different symbols indicate tests carried out at different reward rates (After McFarland and Lloyd, 1973)

called the *subdominant* activity (McFarland, 1974). It is important to realize that this terminology is purely descriptive and says nothing about the mechanisms involved. Various possible mechanisms for time-sharing have been suggested (e.g., Ludlow, 1980), but we do not consider them here. Various different types of evidence can be used to determine whether or not time-sharing is occurring in a particular situation.

Circumstantial evidence. It is possible for the time-sharing hypothesis to provide a plausible explanation of a set of behavioral observations, even though the hypothesis is not tested directly. For example, Ivor Lloyd and I, in a study of feeding and drinking in the Barbary dove (*Streptopelia risoria*), found that when both food and water were available in an operant situation, a hungry and thirsty dove will alternate between feeding and drinking (McFarland and Lloyd, 1973). The frequency of alternation, measured in terms of quantities ingested, depended largely upon the forced reward rates. However, the probability distribution of the intervals between the beginning of each feeding bout remained the same over a wide range of reward rates, even when the rate of ingestion was varied (see Fig. 25.10). This result rules out explanation of the time course of the behavior in terms of any type of temporary satiation of hunger or thirst, because any satiation process is dependent upon the quantity of food or water ingested as a function of time. Our result shows that the temporal organization of the behavior is not dependent upon quantities ingested but upon time itself. Clearly, the time-sharing hypothesis, eating being dominant and drinking subdominant, provides a possible explanation here.

Pattern of dominant behavior. When a behavior sequence is organized on a

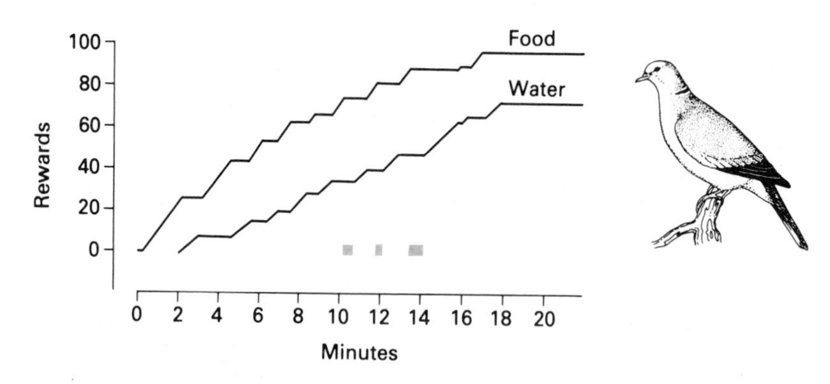

Fig. 25.11 Cumulative food and water intake in a primarily hungry dove working to obtain food and water in an operant situation. Shaded blocks indicate behavior other than feeding and drinking (After McFarland, 1974)

time-sharing basis, the opportunity and motivation for a particular type of subdominant behavior should make no difference to the pattern of dominant behavior. If a pattern of behavior can be shown to remain unaffected by manipulation of motivational factors relevant to alternative behavior, even though the alternative behavior is observed as an adjunct to the dominant behavior, then a time-sharing situation exists. There are a number of examples of this phenomenon. Richard Brown and I found that the temporal organization of copulatory behavior in male rats was not affected by 48 hours of food deprivation, even though food was present in the test situation, and the rats ate avidly during the pauses in copulatory behavior (Brown and McFarland, 1979). Thus, the pauses in copulatory behavior are preprogrammed and are filled in by whatever happens to be the second-priority activity at the time. This is a classic time-sharing situation. In both doves (McFarland, 1970b) and rats (McFarland, 1974), satiation curves for feeding have been shown to remain unaffected by availability of water or degree of thirst, even though drinking did occur during pauses in feeding behavior (see Fig. 25.11).

Masking experiments. The basis of masking experiments is to set up a situation in which ongoing behavior can be prevented from occurring by an experimentally controlled interruption. The assumption behind the method is that the temporal organization of the behavior in a time-sharing situation persists when the interruption occurs during subdominant behavior but not dominant behavior. The interruption must be motivationally neutral in not inducing alterations in motivational state, such as fear and aggression.

However, masking is not an infallible test for time-sharing, as Alasdair Houston (1982) has pointed out. In evaluating a possible time-sharing situation, it is necessary to take account of a variety of types of evidence. Figure 25.12 illustrates the basic time-sharing situa-

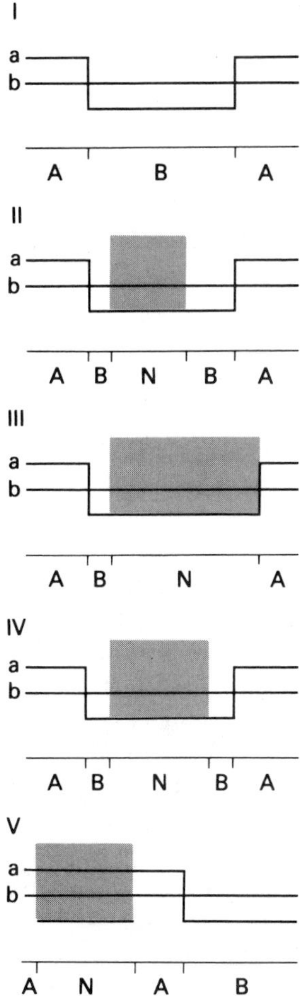

Fig. 25.12 Possible effects of interruption of behavior (shown by shaded blocks) in a time-sharing situation. Relative levels of causal factors shown by lines a and b, while A and B are the observed activities. N indicates that neither A or B is observed (After McFarland, 1974)

tion. The relative levels of causal factors relevant to two types of behaviors are indicated by the lines *a* and *b*. Initially, *a* has priority over *b*, so activity A is observed. The level of *a* then falls, thus disinhibiting *b* so that activity B is observed. After a period of time, *a* reinhibits *b*, and activity A is observed again. The problem is to show that the onset and duration of activity B is dependent on the system controlling A.

Possible effects of interruption are shown in Figure 25.12. In I there is no interruption. In II there is a short interruption of B, but B is resumed after the interruption because the interruption is terminated before B is reinhibited by A. In III there is a longer interruption of B, and now the point in time when B would have terminated occcurs during the interruption. At the end of the interruption, *a* has reasserted itself already and activity A is now observed. In IV the interruption is just about the right duration to mask most of the subdominant activity B. In V the same-sized interruption is applied to activity A, but this merely has the effect of postponing activity A and of shifting the disinhibition of *b* by *a* to a later point in time. The basic rationale illustrated in Figure 25.12 is that neutral interruptions postpone dominant activities and mask subdominant activities.

There have been numerous experimental tests of these ideas, and the masking technique has produced evidence for time-sharing in a variety of situations, including courtship in sticklebacks (Cohen and McFarland, 1979) and feeding and drinking in pigs (Sibly and McCleery, 1976) and in doves (McFarland, 1974). We end this chapter with a simple example of time-sharing that illustrates some of its basic features.

The fish of the genus *Amphiprion*, commonly known as anemone fish, are best known for their symbiotic association with sea anemones (see Mariscal, 1970a, for a review). The fish benefit from this association through protection from predation, reduced susceptibility to some diseases, and the ability to feed on the tissue, prey, and waste material of the anemone. The anemones benefit through reduced predation as a result of the territorial behavior of the fish as well as through the removal of parasites and waste material. In addition, some species of anemone fish, including *Amphiprion xanthurus*, often bring large pieces of food to their host anemone. One of the aspects of the association that has received considerable attention is the nature of the protection the fish develops from the stinging of the anemone's nematocysts (Davenport and Norris, 1958). Although there is considerable controversy concerning the degree to which the protection is a result of changes in the anemone or in the fish, scientists agree this protection is dependent upon a relatively continuous association. Separation of an hour has been shown to require some degree of re-acclimation, while separation of over 20 hours usually requires complete re-acclimation (Mariscal, 1970b).

Whether because of predation pressure, competition for territory, or

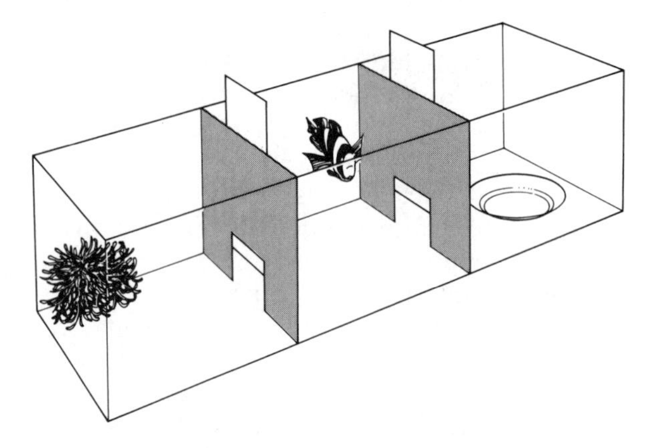

Fig. 25.13 Experimental tank for investigating time-sharing in anemone fish. An anemone is placed at one end and a food dish at the other. The end compartments are separated by two opaque partitions with sliding doors

the need to maintain the relationship with the anemone, anemone fish in the field are rarely seen far from their host (Mariscal, 1970a). If there is strong selection pressure for the anemone fish never to stray far from their host, one might expect them to have a proximal behavioral mechanism to achieve this end. One such possible mechanism is time-sharing. Shelly Cohen (1979) conducted a preliminary experiment designed to test the hypothesis that an anemone fish can use time-sharing as a mechanism to control the pattern of alternation between feeding behavior and anemone visits.

An experimental seawater aquarium 31 by 71 centimeters long was divided into three equal compartments by two opaque Plexiglass partitions, each with a remote-control sliding door in its center. One end compartment contained a sea anemone that the fish accepted as its host. The other end compartment contained a dish from which the fish was trained to receive all its food. The central compartment was divided further in half by another opaque partition, with an opening in the front portion of the tank (not shown in Fig. 25.13). This partition was to increase the amount of time needed to travel between the food and the anemone so that the fish could be trapped during transitions, and in addition, to make it impossible for the fish to see either the food or the anemone compartments while it was in the opposite compartment.

During the experimental session, 20 thin pieces of squid (approximately 5 by 5 millimeters) were dropped through the funnel in accordance with either one of two schedules. Under the ad lib schedule, one piece of food was dropped when the fish entered the feeding compartment, additional pieces being dropped at 10-second intervals if it remained in that compartment. Under the one-minute variable-interval schedule, a piece of food was dropped only when the fish was in the food compartment and the appropriate amount of time had elapsed

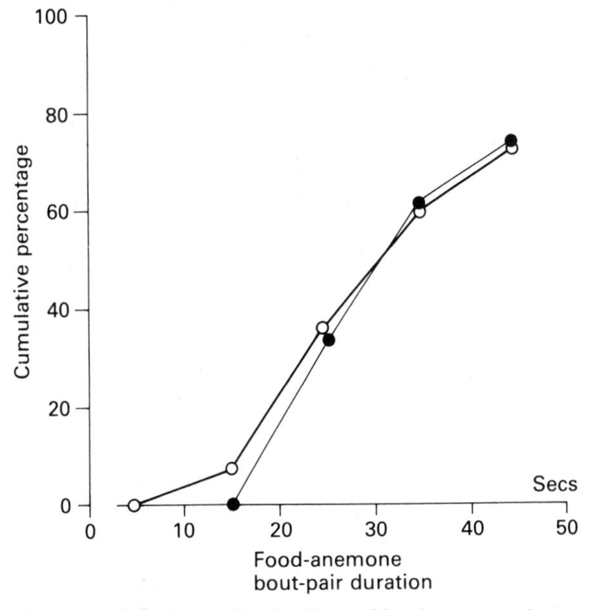

Fig. 25.14 Cumulative distributions of food-anemone bout pair durations in experiments involving ad libitum (white dots) and one-minute variable interval reward (black dots) schedules (After Cohen, 1979)

since the last piece of food was available. The intervals between the times when food was available randomly varied between 10 and 120 seconds and averaged one minute. The fish were fed only once a day, and only the data obtained from the first 10 rewards were used to minimize the effects of satiation.

During some of the experimental sessions, the fish was trapped in the central compartment for either 10 or 120 seconds during some of the transitions between the food and the anemone and between the anemone and the food. Since anemone fish are frightened easily, they required extensive training to habituate to the interruption procedure and to become familiar with the feeding regime.

The typical feeding behavior of the anemone fish was to alternate between foraging excursions and visits with the anemone. After receiving a food reward, it most often would dash quickly to the anemone, although on rare occasions it would wait for another reward. The rapid and predictable onset of the anemone visit following a food reward suggested that these transitions were occurring by disinhibition. The fish also would often return to the anemone compartment when a food reward was not received, but the timing of these transitions was much more variable. When the fish returned to the anemone with a reward, it often would spit the food into the center of the anemone. This behavior occurred with greater frequency toward the end of the sessions but was not always detectable and therefore was not recorded.

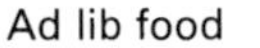

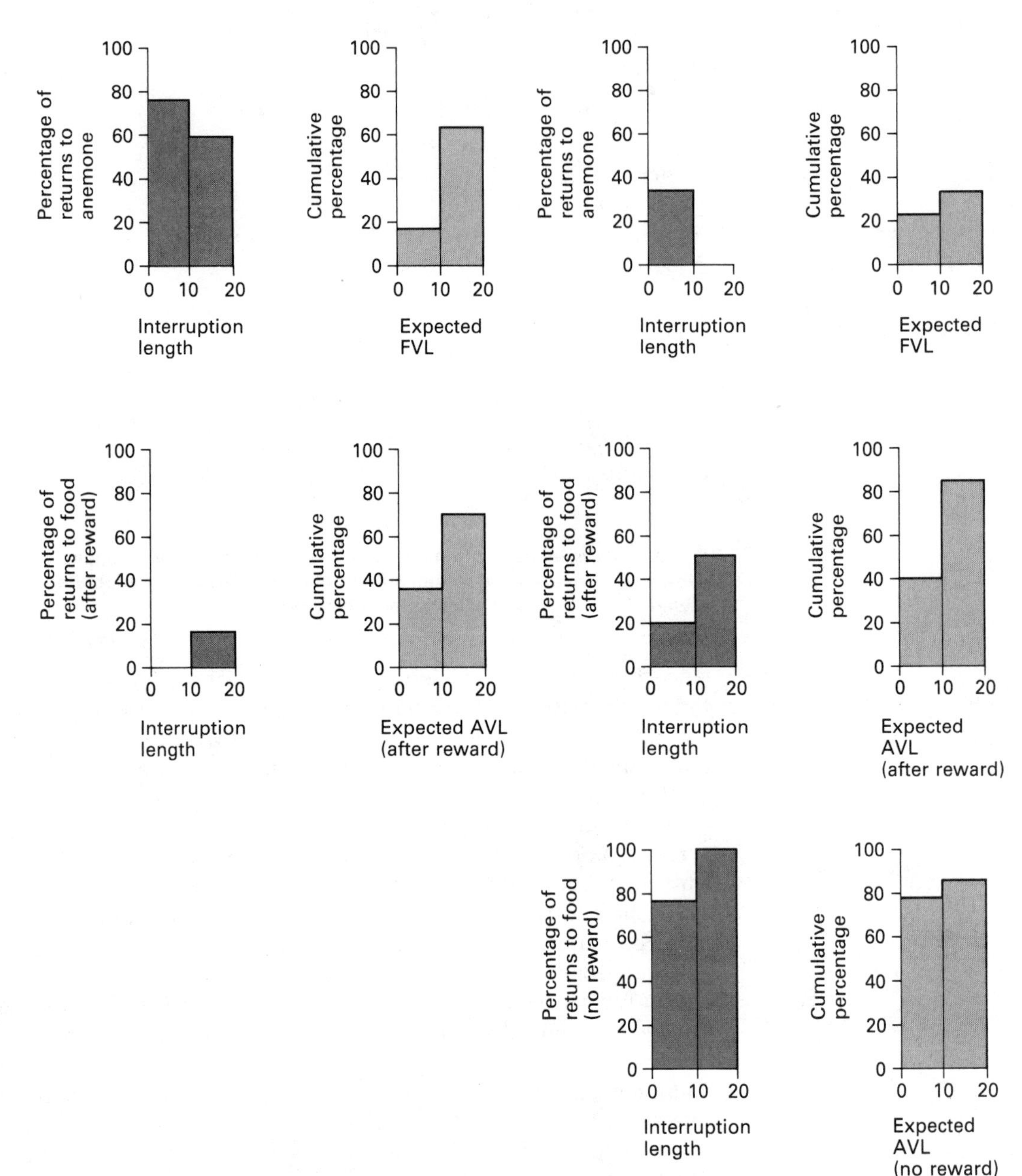

Fig. 25.15 The percentage returns to the anemone and to the food compartments after transition-dependent interruptions (dark) are compared to the expected behavior (pale) in the absence of the interruptions. AVL = anemone visit length, FVL = food visit length. The graphs show the percentage of returns after short (10 seconds) and long (20 seconds) interuptions under conditions of ad libitum food and variable interval feeding (VI: 1 minute)

The effects of the feeding schedule on the pattern of alternation between the food and the anemone are striking. Under the ad lib schedule, the anemone visits were significantly longer and the food visits were shorter, making the combined duration of each feeding-visit cycle unchanged (Fig. 25.14). Thus, the time between the initiation of one foraging excursion and the next is independent of the time it takes for food to be obtained or whether food is obtained at all. The amount of time spent with the anemone simply fills up the remaining time. This result is therefore compatible with the time-sharing hypothesis.

One prediction of the time-sharing hypothesis is that the behavior following a transition-dependent interruption should be dependent upon both the transition direction and the length of the interruption. If the animal is interrupted while going toward subdominant behavior, it should continue on to that behavior only if the interruption is shorter than the expected subdominant visit length. In contrast, if the animal is interrupted while going toward dominant behavior, it should always continue on to that behavior regardless of interruption length. The behavior following interruption, along with the expected behavior in the absence of interruption, is illustrated in Figure 25.15. The results of this experiment suggest that there is a temporal pattern of foraging behavior in anemone fish in which the initiation of foraging excursions occurs at regular intervals independent of the probability of success. In addition, results suggest that this temporal pattern of alternation between foraging and anemone visits might result from a time-sharing mechanism in which the tendency to visit the anemone is subdominant and disinhibited for a limited period of time between the dominant foraging attempts.

Note that the interruption tests reveal that the status of the foraging and anemone-visit activities is not the same. One is postponed by interruption, while the other is masked, a fundamental asymmetry, typical of time-sharing, that poses problems for the view that decision-making is basically a matter of competition among behavioral tendencies.

25.5 Optimal decision-making

So far in this chapter we have discussed mechanisms of decision-making in animals. Starting from simple rules of thumb, we considered the possibility that motivational competition is the basic decision rule employed by most animals. Conversely, the operational approach to decision-making suggests that a distinction be made between dominant and subdominant activities, and we discussed various experimental tests of this possibility. We now come to the question of whether animals make decisions that are most beneficial to them in terms of fitness. In other words, is their behavior optimal?

What does it mean to say that an animal is behaving optimally? This question can be answered in a number of different ways. First, an animal may or may not be behaving optimally with respect to natural selection. Considerations of optimal design apply both to the temporal organization of behavior and to the anatomical properties of animals; therefore, we might expect animals to spend their time in the manner most likely to maximize their fitness. However, it does not follow that we can expect the individual animal to behave optimally in its natural environment. Genetic variation among individuals, the patchy nature of the environment, and evolutionary lag (natural selection takes time to catch up with environmental changes) combine to make it unlikely that an animal ever can be adapted perfectly to its niche. Nevertheless, it is useful to imagine an animal that is adapted perfectly—that is, whose behavior represents the evolutionary best buy—so we then can specify what the animal would have to do to be adapted perfectly. This is an exercise in the *functional* explanation of behavior

Second, the animal may or may not be an optimizing machine in that its behavior conforms to a set of criteria embodied within the animal. A simple case would be an animal that had an internal representation of the goal it was trying to achieve. We could then ask to what extent the animal is taking the best route to achieve its goal. Clearly, this is a very different question from the first. An animal may attain its goal in the best possible manner, but its goal may not be the most appropriate goal. From an evolutionary viewpoint, the individual animal may be designed to optimize its behavior on the basis of certain internally represented criteria, but these criteria may not be the ones that maximize fitness. This second type of optimization is an exercise in the *causal* explanation of behavior. This chapter clarifies the distinction between functional and causal aspects of optimality.

25.6 Rules for optimal behavior

An apparently adaptive, or well-designed, behavior pattern may be the outcome of simple rules of thumb or of cognitive or intentional behavior (see Chapter 26.7). For example, a child may cross the road in accordance with strict rules of the highway code. If the child has been well trained, the road-crossing behavior will be automatic. An adult who has not had the benefit of the training, such as a foreigner, may think about crossing the road, estimate the speed and likely behavior of the oncoming traffic, etc. The road-crossing behavior of the child and the adult may be indistinguishable, yet one is achieved by rules of thumb, the other by cognition.

It is possible to achieve optimal behavior by means of a simple set of rules, an example of which is Richard Green's (1983) analysis of stopping rules for optimal foraging. In this analysis it is assumed that prey are distributed in patches that vary in quality, with higher capture rate in the better patches. The distributions of patch quality are different under different environmental circumstances. It is assumed that the predator is unable to distinguish between patch types except by assessing its success in each patch. Patches are not revisited and each is searched systematically until the predator decides to leave for another patch.

The optimal foraging strategy can be characterized by a stopping rule that determines when the predator should leave the patch. At any time in a patch the predator may decide to leave or to remain in the patch and continue searching. Green shows that the best stopping rule is based upon the number of prey found as a function of time spent searching in the patch. Alternative stopping rules include a *naive* strategy in which the predator relies upon knowledge of the average probability of finding prey in each patch, an *omniscient* strategy in which the predator can tell the quality of each patch without searching it and can thus avoid the poorer patches, and an *instantaneous-rate* strategy in which the patch is vacated when the rate of obtaining prey falls below a critical value. Green's best strategy involves assessing the quality of patches as they are searched: Its productivity is superior to that of the naive and instantaneous-rate strategies, and it is superior to the omniscient strategy in requiring less computing power from the individual animal. Green's strategy can be represented by a simple rule—remain in the patch as long as half the places searched have yielded prey; otherwise, leave—and can be implemented by means of a simple mechanism.

J. Waage (1979) suggested a simple decision mechanism for a parasitic wasp searching for suitable hosts in which to deposit eggs. Upon arrival at a patch, the wasp begins to search with a level of responsiveness (the tendency to turn back into a patch upon reaching the patch edge) set by a chemical stimulus provided by the hosts in the patch. The wasp's level of responsiveness decreases linearly with time until it reaches a threshold, or until a host is encountered. After each oviposition, the wasp's level of responsiveness increases by an amount that depends upon the time since the last oviposition, and the process continues until the responsiveness eventually reaches a threshold and the wasp leaves the patch.

The models of Green (1980; 1983) and Waage (1979) give similar results. It is important to remember, however, that Green's model is *functional*, specifying what the animal ought to do to achieve the best outcome. Waage's model is *mechanistic*, involving proximate causes of behavior.

One way of determining whether or not an animal is following fixed rules is to selectively interfere with its behavior. For example, in his study of the digger wasp *Ammophila campestris*, Gerard Baerends (1941) found that the female about to lay an egg digs a hole, kills or paralyzes a moth caterpillar, carries it to the hole, deposits an egg on the caterpillar,

Fig. 25.16 Diagram of the nesting activities of an individual male *Ammophila* wasp (After Baerends, 1941)

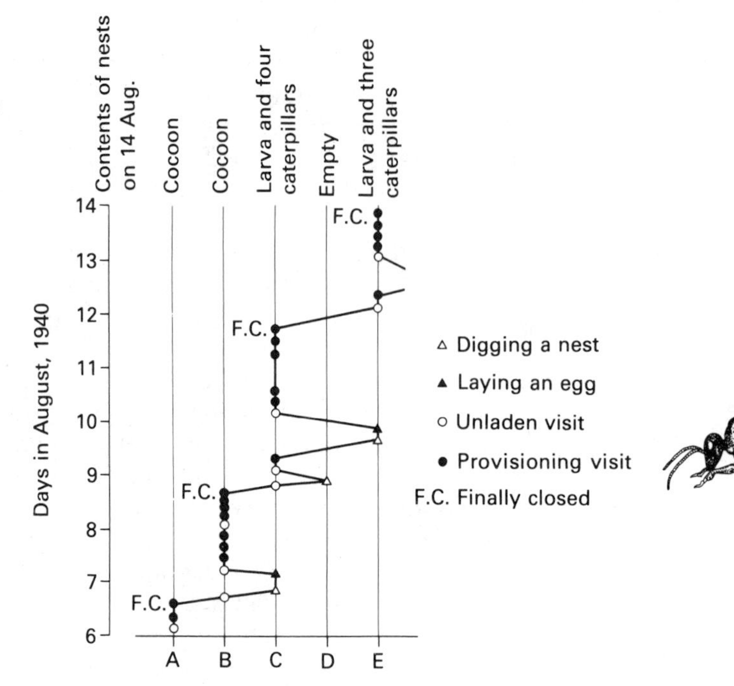

and stows it away in the hole. The female wasp then repeats this procedure with the second and subsequent eggs. Meanwhile, the first egg has hatched, and the larva has begun to consume the caterpillar. The wasp now returns to the first hole and provisions it with more caterpillars. She then may start another hole, or she may provision the second hole, depending upon the circumstances. In this way the female wasp can maintain up to five nests simultaneously, as shown in Figure 25.16.

Baerends found that the wasps inspect all the holes each morning before leaving for the hunting grounds. By robbing a hole Baerends could make the wasp bring more food than usual, and by adding caterpillars he could induce her to bring less food than usual. However, he could manipulate the wasp in this way only if he made changes to the nest before the wasp's first visit of the day. Changes made later in the day had no effect. The female wasp seems to operate by simple rules. There is a standard routine for laying an egg, which involves digging a hole and providing a caterpillar. There is a standard early morning inspection routine that usually determines which nest will be provisioned during the day. There is a standard stopping routine by which the wasp closes up the nest when sufficient caterpillars have been supplied. Although the wasp is capable of assessing the extent of provisions when she visits a nest, she does not always use this ability. Moreover, each routine, once started, is followed to its conclusion. Thus, a wasp will go on and on provisioning a nest if the caterpillars are removed systematically each time they are supplied. This example shows that complex behavior can be programmed on the basis of a set of rigid rules. The wasp behaves in an automatonlike way, although it may have some routines for extricating itself from difficulties, like removing an obstacle from the burrow.

As we have seen, interruption of an animal's behavior, under some circumstances, masks the behavior that would have happened had there been no interruption. Such time-sharing situations suggest that the animal is following certain rules that determine the organization and priority of its behavior patterns. Let us consider a particular example. When a hungry dove (*Streptopelia*) is feeding, either from a pile of grain or in a Skinner box, it typically pauses a few times, as shown in Figure 25.11. What the dove does during these pauses depends upon the circumstances. If the water is available, it may drink. Otherwise, it may preen or simply stand motionless. Experiments show that the timing of these pauses is not affected by manipulation of second-priority motivational factors like thirst. In one experiment a paper clip was fastened to each wing of doves that were primarily hungry. The doves ignored the paper clips while eating, but during the pauses they tried to remove them. However, the presence of paper clips did not affect the pattern of feeding behavior or alter the temporal position of the pauses (McFarland, 1970b). It seems as though the doves are preprogrammed to pause at certain times during feeding and that the rules governing feeding are unaffected by other motivational factors such as thirst or grooming, provided these tendencies do not become greater than the feeding tendency. This is typical of time-sharing.

If a primarily hungry dove is interrupted while feeding, it usually will continue feeding after the interruption; but if it is interrupted while drinking, the drinking usually will be masked, provided the interruption is long enough (McFarland and Lloyd, 1973). In an operant situation in which the dove pecks at illuminated keys to obtain food and water rewards, interruptions can be introduced simply by turning off the key lights. The birds soon learn not to peck when the keys are not illuminated. In a free feeding and drinking situation, interruptions can be introduced by plunging the room into darkness for about a minute. Comparisons show that the two types of interruptions have equivalent effects (Larkin and McFarland, 1978).

Time-shared feeding and drinking in doves has been the subject of numerous experiments aimed at discovering the rules by which the birds decide whether to eat or drink. The results show that either hunger or thirst can be dominant (McFarland and Lloyd, 1973; McFarland, 1974) and that the boundary between hunger dominance and thirst dominance does not change position (see Fig. 25.17) with repeated tests, with different initial conditions of hunger and thirst, or with alterations in the consequences of feeding and drinking (Sibly and McCleery, 1976). However, apparent rotations of this *dominance boundary*

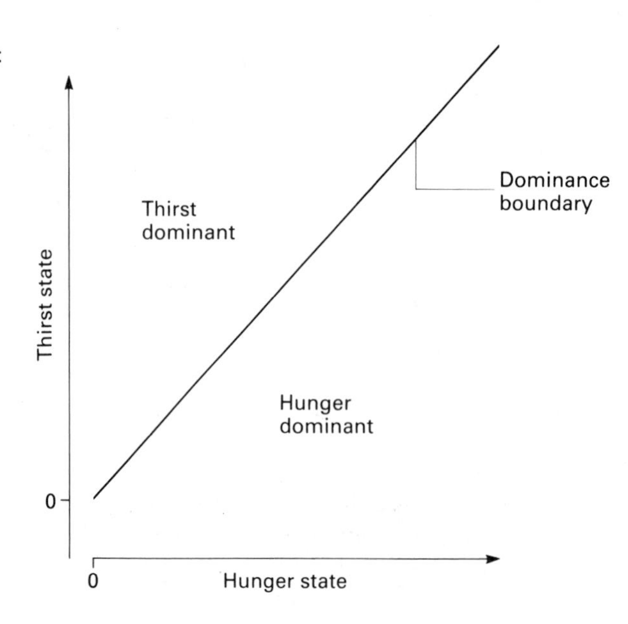

Fig. 25.17. The hunger-thirst dominance boundary

do occur when the animal's motivational state is altered during the course of the experimental session (see Fig. 25.18). Theoretical analysis of the situation shows there is no real change in the position of the dominance boundary but that a rotation is apparent within the experimenter's frame of reference because of the way in which the animal's motivational state is portrayed normally in two dimensions when other dimensions also are relevant (McFarland and Sibly, 1975). The magnitude of the apparent rotation of the dominance boundary has proved to be a useful tool in measuring motivational factors like the incentive value of food and water rewards (Sibly, 1975), the effectiveness of the external stimuli that signal the availability of food and water (McFarland and Sibly, 1975; Beardsley, 1983), and the cost (as gauged by the bird) of changing from feeding to drinking and vice versa (Larkin and McFarland, 1978). In general, it seems that both internal and external factors contribute to the feeding and drinking tendencies and that these tendencies compete for dominance (McFarland, 1974). The dominant system then periodically allows time for other (subdominant) activities. Why the behavior is organized in this way remains something of a mystery.

It is possible that the feeding behavior of doves is interspersed with pauses as part of a strategy for detecting predators. By feeding in flocks individual birds are able to spend more time feeding and less time watching for predators (Barnard, 1980; Bertram, 1980; Elgar and Catterall, 1981). Dennis Lendrem (1983) found that doves spend about 25 per cent of a two-minute feeding period looking around when feeding alone and about 20 per cent when in the presence of other birds. However, this difference was much more marked if the doves had seen a predator (a ferret) in the vicinity recently. Lone doves spend about 50 per cent of the time apparently looking around, whereas doves in the presence of two others spend only 25 per cent, and this time spent not feeding was reduced even further when more birds were in the flock. The doves obtain food faster as flock size increases and the overall rate of feeding is reduced when they have seen a predator recently. Detailed analysis of dove feeding behavior shows that doves lower their rate of feeding (as distinct from pausing between bouts of feeding) in risky situations. Thus, they feed more slowly in unfamiliar surroundings, when alone, and when they have seen a predator recently. In particular, the time spent raising the head after each peck is increased, and it is possible that this enhances the birds' ability to detect predators.

The rate of feeding is lowered also when doves have to discriminate between palatable and unpalatable foods. By administering lithium chloride in association with particular feeding situations, one can condition doves—like many other animals—to avoid wheat grains dyed a particular color (Lendrem and McFarland, 1985). The birds behave as if seeds of a certain color were noxious. Thus, some birds avoid yellow grains and others avoid red grains. When presented with a mixture of red and yellow grains, such birds have to discriminate between the two types of grain in order to avoid the color to which they are averse. Birds feeding from a mixture of noxious and innocuous seeds feed at a lower rate than birds presented with a mixture of innocuous seeds of different colors (Lendrem and McFarland, 1985).

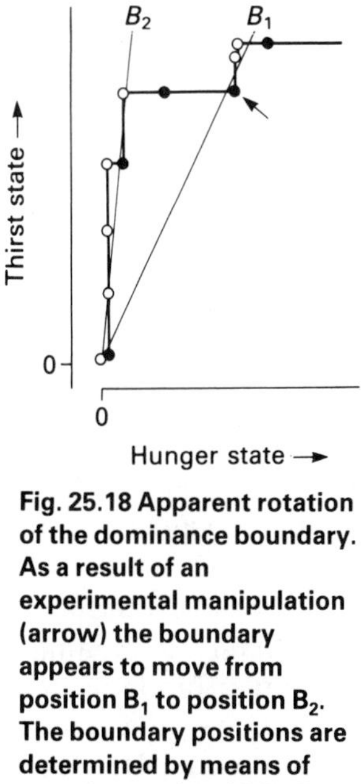

Fig. 25.18 Apparent rotation of the dominance boundary. As a result of an experimental manipulation (arrow) the boundary appears to move from position B_1 to position B_2. The boundary positions are determined by means of experiments in which all sub-dominant behavior is masked by interruption. Black dots indicate hunger dominance and white dots indicate thirst dominance (After McFarland, 1974)

If a dove is feeding more slowly than usual because it has to discriminate between noxious and innocuous seeds, it might be thought that it would have a reduced ability to detect predators, since it is paying more attention to its food. In fact, the speed at which doves respond to a model hawk flown overhead is enhanced if they are feeding from a mixture of noxious and innocuous seeds (Lendrem and McFarland, 1985). Birds previously exposed to a predator (and therefore with a lowered feeding rate) respond more quickly to a model hawk, as do birds that have to discriminate between noxious and innocuous food. Thus, it seems, feeding at a lowered rate, for whatever reason, enhances the birds' ability to detect predators. This finding is consistent with the idea that high rates of feeding (and of other behavior) carry high costs.

What then happens if we make the seed discrimination task more difficult by presenting the seeds on a cryptic background? As expected, we find a further lowering of the feeding rate (Fig. 25.19). This may be partly a result of the greater attention the birds have to devote to feeding, but it seems also to be an active tactic designed to maintain vigilance. Doves choosing between noxious and innocuous seeds against a cryptic background are quicker at detecting a model hawk than birds choosing between conspicuous seeds (Fig. 25.20) (Lendrem and McFarland, 1985). However, birds choosing seeds on a cryptic background make more errors (eat more noxious seeds) and pause less often

Fig. 25.19 Feeding rates of doves given a mixture of noxious and innocuous seeds in situations where the types of seeds were either difficult (cryptic condition) or easy (conspicuous condition) to discriminate (After Lendrem and McFarland, 1985)

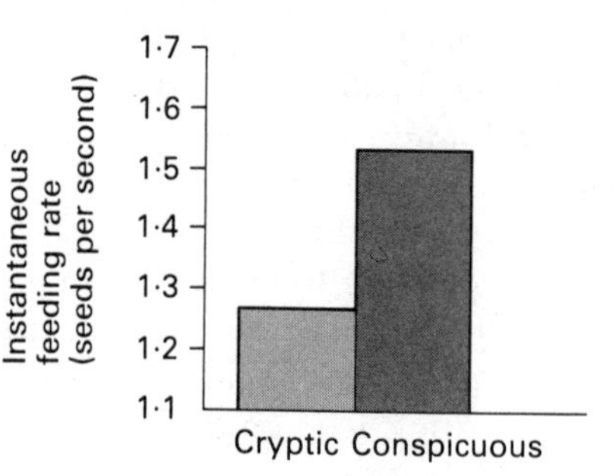

Fig. 25.20 Latency of response to a model hawk of doves feeding in the cryptic and conspicuous conditions of Figure 25.19. Note that although the doves feeding on cryptic seeds feed more slowly than those feeding on conspicuous seeds, their response to a potential predator is faster. This result suggests that the lower rate of feeding on cryptic seeds is not because the discrimination demands all the bird's attention, but rather because the situation is more risky (due to the higher probability of eating a noxious seed), and the doves pay more attention to the situation in general (After Lendrem and McFarland, 1985)

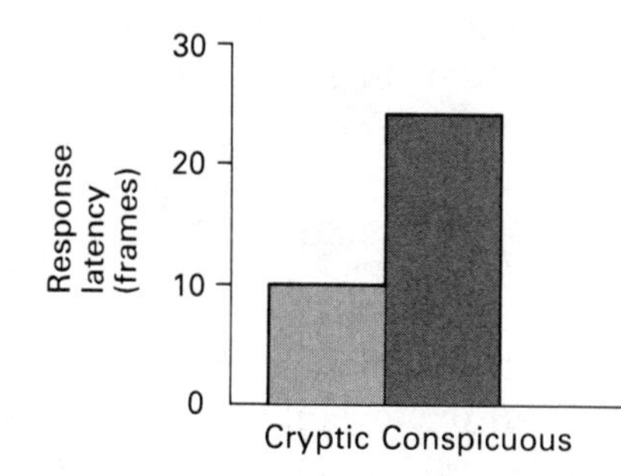

during feeding than those choosing between conspicuous seeds. Thus, there is obviously a trade-off between the demands of vigilance and those of feeding.

In conclusion, it appears that doves that feed quickly have a reduced chance of detecting predators. When they are wary—that is, in unfamiliar surroundings, when alone, or in a situation where a predator has been seen recently—doves feed at a lower rate. However, doves can lower their overall feeding rate in a number of different ways: They can pause more often, pause for a longer time, or reduce their rate of feeding. These methods can increase the chances of detecting a predator, and there is some evidence that the different methods compensate for each other (Lendrem and McFarland, 1985). It may be that doves rely on detecting extraneous movement during the recovery from each peck and pause to have a look around every so often. It may be that, having paused, the bird may spend a little time preening or drinking, thus accounting for the phenomenon called time-sharing. At the present time, we do not know enough about vision in birds to be able to substantiate these hypotheses. We also do not know if the birds are using a complex set of rules or if their behavior is cognitively controlled.

Points to remember

- Animals may make decisions on the basis of simple rules-of-thumb designed to fit particular environments.

- Changes in behavior due to motivational competition can be recognized when manipulation of the second-in-priority activity alters the timing of the change(s) in behavior. The alternative is that the change is due to disinhibition.

- Time-sharing occurs when the onset and duration of an activity is controlled by another activity.

- Optimal decision-making produces sequences of behavior that maximize some index of fitness under the prevailing circumstances. Any mismatch between the animal and its environment will mean that this is rarely achieved. However, animals may employ decision rules that closely approximate the optimal behavior.

Further reading

McFarland, D. J. (1977) 'Decision-making in animals,' *Nature* (London), **269,** 15–21.

Krebs, J. R. and McCleery, R. H. (1984) 'Optimisation in behavioural ecology.' In Krebs, J. R. and Davies, N. B. (eds) *Behavioural Ecology*, 2nd edn, Blackwell Scientific Publications, Oxford.

3.3 The Mentality of Animals

In this final group of three chapters we consider the controversial question of whether there is any affinity between the mental life of humans and other animals. Chapter 26 looks at language and mental representation, including questions such as whether apes are capable of language and whether animals have intentions. Chapter 27 looks at intelligence in animals, at tool use (sometimes taken as a sign of intelligence) and at cultural evolution. In Chapter 28 we discuss self-awareness and emotion in animals. We explore questions about consciousness and suffering in relation to animal welfare. Finally, we discuss the problem of animal suffering from an evolutionary perspective.

Edward Tolman (1886–1959)

Edward Tolman, an American psychologist, was a thorn in the flesh of his behaviorist contemporaries. His thinking was, in many respects, ahead of its time. He can be said to be the father of the modern cognitive approach to animal behavior.

Unlike other cognitive theorists of his time, such as George Romanes or Wolfgang Kohler, Tolman's view was not mentalistic. His system was purposive, but not anthropomorphic. He believed that animals behaved in a purposive manner, but did not assume they had mental images of their goals.

Tolman regarded himself as a behaviorist, but he espoused a molar rather than a molecular behaviorism. In the molar view, behavioral acts have distinctive properties of their own, which can be described independently of the particular physiological processes responsible for the behavior. Molecular behaviorism, on the other hand, is reductionist in that it seeks to account for behavior in terms of the underlying physics and physiology.

Tolman's major publication was *Purposive behavior in animals and men* (1932), but he subsequently published many papers criticizing contemporary views, both by argument and experiment, and refining his own cognitive perspective.

Tolman's (1932) cognitive view of Pavlovian conditioning has much in common with modern thinking (Rescorla, 1978). For instance, he believed that the animal learns about the reinforcer, and not simply because of the reinforcer. He attacked the stimulus-response theory prevalent in his time and was a pioneer of the view that the conditional stimulus is a sign that some other event is to follow.

Tolman also pioneered the idea of cognitive maps. His view was that animals acquire items of knowledge, or cognitions, which are organized so they can be utilized when needed. Challenging the prevalent view that animals "learn by doing", Tolman demonstrated by experiment that animals could learn general features of a room or maze without performing the relevant behavior. Tolman's evidence suggests that the animal acquires a cognitive map indicating how the relevant causal or spatial features of the environment relate to each other.

By courtesy of Regents, University of California

Tolman's theories were both hard-headed and sophisticated, and they were supported by experiments that provided a challenge other theories had difficulty in meeting. The criticism that Tolman's theories left the animal "buried in thought" (Guthrie, 1952) without predicting its behavior can be countered by the observation that the molecular behaviorist view leaves the animal "lost in action".

26 Language and Mental Representation

Do animals have a mental life similar to that of humans, or are they mindless automatons? For the past hundred years this question has troubled animal psychologists and ethologists. During this period the prevailing opinion has swung between the two extremes. There has been a vast increase in our knowledge of animal behavior and physiology, but the more we know, the more difficult the problem seems to become. In this chapter we draw together some lines of argument that have developed within different branches of animal psychology, and we attempt to arrive at a position from which the student can begin to form his or her own opinions about the mental life of animals.

There are many aspects of human behavior that seem to set us apart from other animals. It used to be thought that only humans could make and use tools, but we now know that many other species have this ability. As our knowledge of animal behavior has improved, so the difference between humans and other animals has appeared to diminish. Some human abilities, however, are difficult to substantiate in other species. One of these is language.

Language seems to us to be a uniquely human feature. Perhaps it is the only feature that distinguishes us from other animals. Whether or not this is true, we should not allow our anthropomorphic sentiments to cloud our judgment about the possibility of language in other animals. Unfortunately, there has long been a tendency to define language in a way that ensures it could only occur in humans. The most blatant of these attempts assert that language requires consciousness that only humans possess or that language depends upon speech of which only humans are capable. This is not an acceptable scientific procedure because it introduces an insurmountable bias into the investigation.

Unfortunately, language is not easy to define in objective terms because it has many necessary attributes. For example, we can agree that language is an aspect of communication but that not all communication is language. Human language usually involves speech, but it does not always require speech, as in the case of Morse code. Language is symbolic, but so are some aspects of communication in bees. Language is learned during a specifically sensitive period of development, but so is the song of some birds. Language can convey information about situations that are not immediate but are distant in space and time.

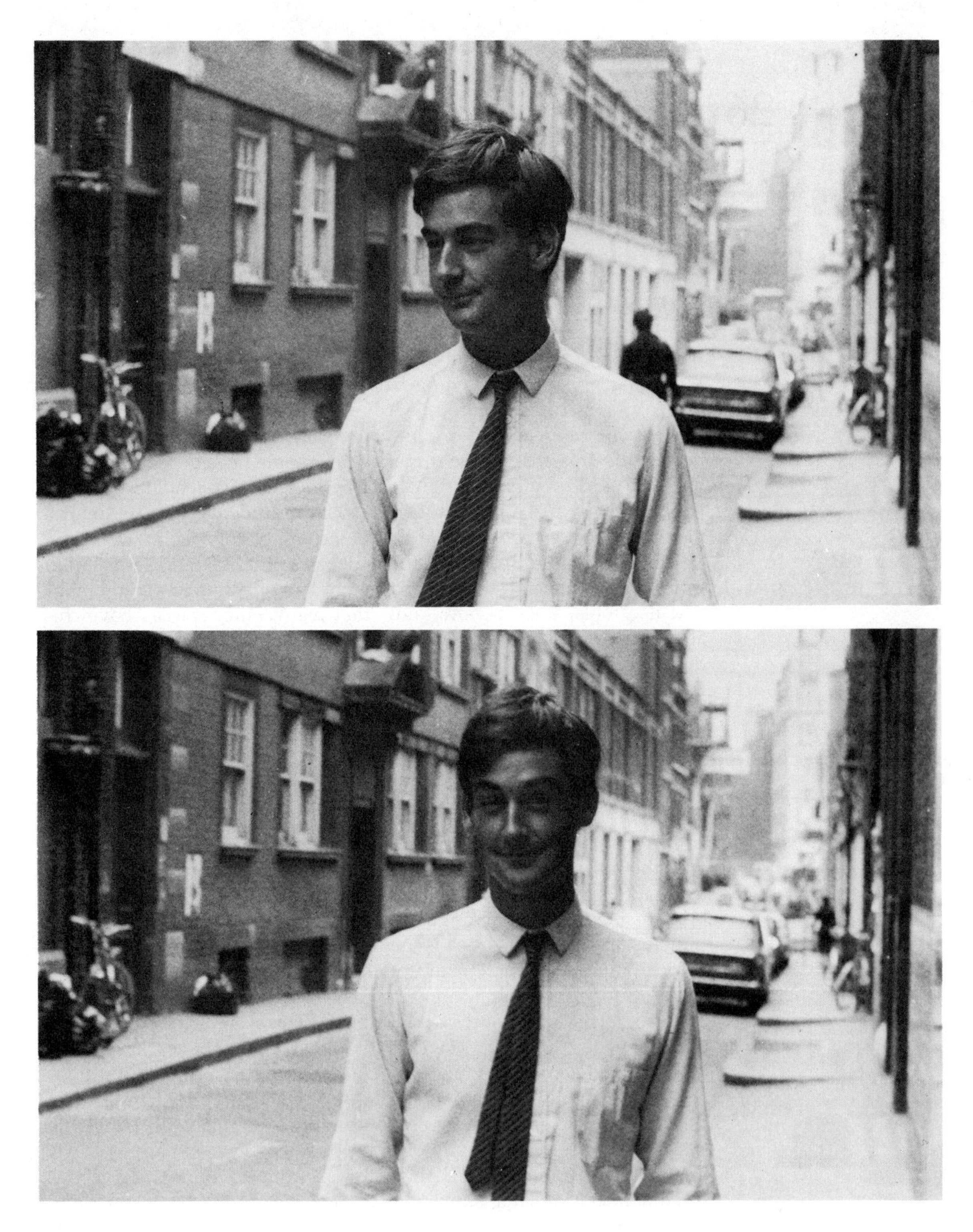

◁ **Fig. 26.1 Eyebrow flash: an example of facial expression in humans**

Some animal alarm calls have these attributes. Some aspects of language, like its grammatical rules, seem to separate it from other aspects of animal behavior, but even this is controversial. In exploring the occurrence of language in the animal kingdom, we have to tread carefully.

26.1 Non-verbal communication in humans

Darwin (1872) realized that there was much in common between the facial expressions of humans and those of other animals, especially the primates. Today, scientists widely recognize that the non-verbal aspect of communication in humans invites direct comparison with the displays of animals. However, we must recognize that some aspects of human non-verbal communication are related directly to language. An obvious example is the sign language used by deaf people. However, many simple gestures, like the thumb-up sign, probably are derived directly from language.

In Chapter 22, we saw that many of the displays and facial expressions of animals may be the result of ritualization. They may be derived evolutionarily from protective responses, intention movements, displacement activities, etc. Do equivalent stereotyped displays occur in humans?

One problem in studying human communication is that it may vary according to the cultural context. For example, a number of ways of signaling assent and denial involve head movements such as nodding and partial rotation (shaking) of the head. However, the meaning attached to these movements varies widely from culture to culture (Leach, 1972). In Greece and some other cultures, for instance, "no" is expressed by a strong backward jerk of the head. Cross-cultural studies do indicate nevertheless that some human facial expressions are universal and appear to have the same meaning in many cultures. An example is the eyebrow flash (see Fig. 26.1), in which both eyebrows are raised momentarily. This signal is usually given as a form of greeting at a distance (Eibl-Eibesfeldt, 1972). There are some small differences of usage between cultures. In Europe it is used as a greeting to good friends and relatives; in New Guinea it is used toward strangers; in Japan it is suppressed, being considered indecent. In general, the eyebrow flash is used as a friendly form of greeting or approval, but it may be omitted by people who are reserved or suspicious (Eibl-Eibesfeldt, 1972).

A number of basic expressions are universal, such as smiling, laughing, and crying. Not only are they found in different cultural groups, but also they occur in people born deaf and blind (Eibl-Eibesfeldt, 1970) and are prominent in young children (Blurton-Jones, 1972). These can be compared directly with the facial expressions of

Fig. 26.2 Open-mouth laughter of a human infant in a tickling game. This type of open-mouth laugh without retracted lips becomes rare in adulthood (From *The Oxford Companion to Animal Behaviour*, 1981)

other primates. Analysis of human facial expressions reveals a number of basic situations that give rise to fairly stereotyped responses (Hoof, 1972; 1976). These include *alertness*, which is evident in the relatively fixed gaze and a certain tension of the facial muscles. This expression is similar to that found in other primates, except that in some the ears also may be focused on the center of attraction. *Surprise* is characterized by prolonged eyebrow raising, open eyes, and often an open mouth, an expression that seems to have no counterpart in non-human primates. *Fear* is similar in a wide range of primates and is characterized by wide-open eyes and withdrawn lips. *Disgust* involves wrinkling of the nose and raising the upper lip, screwing up the eyes, and turning away the face, which are components derived from protective responses serving to exclude noxious stimuli. In humans the disgust expression is ritualized, but in other primates it is merely a collection of protective responses with no special role in communication. *Sadness* is accompanied by arched eyebrows, retracted corners of the mouth turned downward, and outward curling of the lips. Tears may be shed in intense cases. This expression appears to be ritualized in humans, and perhaps in chimpanzees. It elicits comfort behavior from other members of the group. *Anger* takes a variety of forms, usually involving withdrawn lips and bared teeth, staring eyes, and a frowning expression. *Joy* is expressed by smiling and laughter (Fig. 26.2), which often are associated with humor and therefore considered to be uniquely human. Some ethologists question this view however (e.g., Hoof, 1972; Eibl-Eibesfeldt, 1970).

Some monkeys and apes have a relaxed open-mouth display associated with playfulness (Fig. 26.3) that is superficially similar to human laughter, although there are a number of differences, particularly in the accompanying pattern of breathing. In both chimps and humans, the laughing-type display is elicited by an element of surprise in the situation. In chimpanzees and small children, this is of a purely physical nature, while in human adults it may be physical or purely intellectual. In both cases, however, the context is usually social. Human smiling is not simply a low-intensity form of laughter (though the two often are associated); smiling also occurs in situations involving mild fear or apprehension, such as social greeting and reassurance. In many primates a silent bared-teeth expression is typical of subordinate animals in acknowledging or appeasing dominant members of the group. In the chimpanzee, however, it is a reassuring and friendly expression, something akin to the human smile.

Fig. 26.3 Relaxed open-mouth face in a capuchin monkey being tickled by a human (From *The Oxford Companion to Animal Behaviour*, 1981)

The facial expressions of humans and of other mammals are usually regarded as expressions of emotion, or *affect* (Ekman, 1971). However, many other types of non-verbal communication can be classified in various ways. In addition to affect gestures, other classes of non-verbal communication can be distinguished (Ekman and Friesen, 1969), including *adaptors*, which serve both communication and non-communication

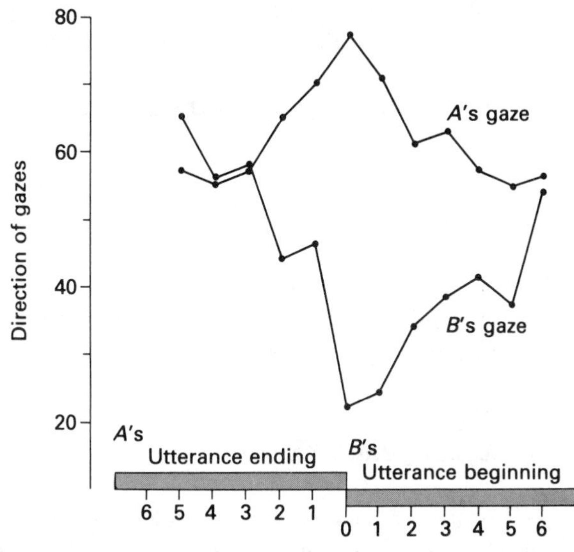

Fig. 26.4 Direction of gaze (ordinate) at the beginning and ending of long utterances is an important aspect of non-verbal communication between two conversing people. Same gaze direction indicates eye contact. Note that participants are furthest from eye contact at the switch from one speaker to the other. (After Kendon, 1967)

functions. Examples are grooming movements and intention movements. *Emblems* are non-verbal acts with a verbal counterpart and they include the sign languages used by deaf people, obscene gestures, and various signaling movements used at a distance, like beckoning. *Illustrators* are movements that illustrate points also being made verbally. They include gestures of emphasis, pointing, etc. *Regulators* are gestures used to control the flow of conversation between two people. Examples are head nodding, eye-contact movements, and various shifts of body posture.

Michael Argyle (1972) proposes a somewhat simpler classification based upon the consequences of the behavior. He recognizes three basic categories:

1. *Managing the immediate social situation*. This involves gestures and postures that convey the person's attitude and emotional state. Thus, a person may indicate feelings of superiority, dislike, sexual desire, or the like without expressing these attitudes in words.

2. *Sustaining verbal communication*. This includes regulators and illustrators and various gestures whose presence or absence may alter the meaning of the words being spoken. Eye contact has been found to be particularly important in this context (see Fig. 26.4).

3. *Replacing verbal communication*. Sign languages are found among many different cultures and may be employed whenever verbal communication is difficult or impossible. Communication among hunters at

moments when quiet is a necessity was probably one of the first situations in which sign language was found to be useful. A modern equivalent is the signaling used by workers in noisy surroundings. Sophisticated language-based signaling systems are used to communicate at a distance (e.g., semaphore) and with deaf people (e.g., American Sign Language, or Ameslan). Writing and Morse code perhaps should come into this category also.

Non-verbal communication can be classified into two basic types: (1) communication that is ancillary to the use of language or part of language and (2) communication that is independent of language, akin to the mode of communication employed by the majority of species.

26.2 Language

A number of the characteristic features of human language can be found in other animals. For example, the signals employed in human language are arbitrary in that they do not physically resemble the features of the world they represent. This abstract quality is found also in the communicative behavior of honeybees (*Apis mellifera*), the study of which was pioneered by Karl von Frisch (see Chapter 23).

The honeybee dance is symbolic in a number of respects: The rate of the waggle dance indicates the distance of the food source from the hive, the precise relationship between the rate of dancing and the distance being a matter of local convention. Different geographic races seem to have different dialects. Thus, one waggle indicates about 75 meters to a German honeybee, about 25 meters to an Italian bee and only 5 meters to an Egyptian bee. Provided all the bees in the colony agree on the convention, it does not matter what precise values are used. Some researchers argue (e.g., Hinde, 1974) that since the symbols (direction and speed dance) are physically related to the direction of the food source, they cannot be arbitrary. However, in any symbolic system that represents a range of values, there is bound to be some correspondence between the range of symbols and the reality.

Other ethologists (e.g., Gould and Gould, 1982) regard the honeybee dance as an example of arbitrary convention, arguing, for example, that the bees could take north as a reference point instead of the sun. The honeybee dance refers to situations remote from the communicating animal, a feature considered widely to be an important property of human language. The dance can refer not only to food sources remote in space (as far as 10 kilometers away) but also to those that may have been visited some hours previously. During the intervening period, the forager bee keeps mental track of the sun's movement and corrects the dance accordingly.

Another feature of human language is that it is an open system into

Fig. 26.5 Head and neck of adult human (*a*) and adult chimpanzee (*b*) (After Lieberman, 1975)

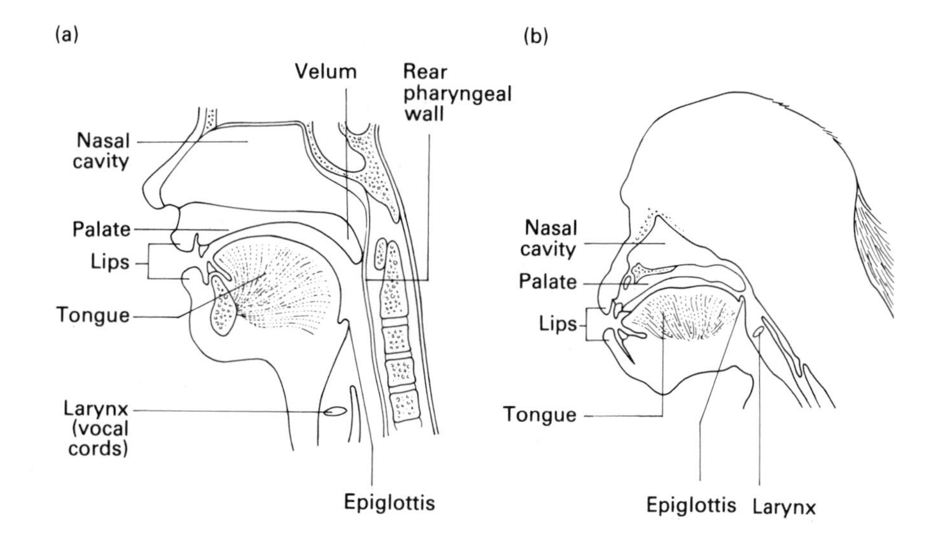

which new messages can be incorporated. The bee dance can refer to new sources of food, but this seems to be a rather restricted example of openness. However, the dance can be used also to direct bees to water, to *propolis* (a type of tree sap used to caulk the hive), and to possible new hive sites at swarming time (Gould, 1981).

Of course, some features of human language—like its acoustic quality—are not found in bee language. Some of these features may be seen in other aspects of animal communication, like bird song. Undoubtedly, human language is more sophisticated and complex than that of animals. But does this mean that there is a qualitative difference between human and animal communication, or is it merely a matter of degree? The question remains controversial.

If animals are capable of language, we might expect apes to be nearest to humans in this respect. The vocalizations and facial expressions of apes are subtle and complex, and it is possible that they converse among themselves, using a language we do not understand. There have been various attempts to discover whether or not apes are capable of language as we know it. The first of these were attempts to train chimpanzees to copy human speech. After several years of training, an orangutan was able to produce the words *papa* and *cup*, but no others. After prolonged training, the chimpanzee Viki managed the words *mamma, papa, cup,* and *up* (Hayes and Nissen, 1971). In both cases the words were enunciated poorly, and it became apparent that apes simply do not have the vocal apparatus necessary to reproduce human speech sounds. In the chimpanzee and in the human fetus, the larynx is positioned high in the vocal tract, whereas in humans it is in a low position (Fig. 26.5), an arrangement that makes it possible for humans to alter the shape of the pharyngeal cavity with the tongue and so produce a wide range of modulated sounds. Chimpanzees and other apes simply are incapable of producing these sounds (Jordan, 1971; Lieberman, 1975).

Even though apes cannot speak, other important questions about their communication remain: Do they communicate among themselves in a languagelike manner? Are they capable of understanding human speech sounds? Can they be trained to use human language through a medium other than speech?

Chimpanzees have a repertoire of about 13 sounds, and they can provide gradations between them. They use the sounds for communication both at a distance and at close quarters. They can recognize the voices of individuals they know, and they continually make use of sounds to keep contact with each other in undergrowth or among other obstacles to visual contact. From the intention movements, facial expressions, and scents and sounds produced by a chimpanzee, another member of the group usually can tell its identity, position, motivational state, and likely behavior. These are not, however, attributes of true language, so we must look for evidence of symbolic communication about the outside world.

A number of species have been reported to give alarm calls that differ according to the type of danger. Adult vervet monkeys (*Cercopithecus aethiops*) give different alarm calls when they sight a python, a leopard, or a martial eagle. When they hear the alarm, the other monkeys take action appropriate to the sighted predator. If it is a snake they look down, and if it is an eagle they look up. If they hear the leopard alarm, then they run into the trees (Seyfarth, Cheyney, and Marler, 1980). These observations suggest that the monkeys are able to communicate about external stimuli, but we can not be sure that this communication is not merely about different emotional states aroused by these stimuli.

Emil Menzel (1974; 1979) conducted experiments with chimpanzees to see if they could convey information about the location of food. He set up a series of tests for six chimpanzees kept in an area of field. Accompanied by one of the group, he hid food in the field and then released all six chimpanzees and allowed them to search for the food. Usually, the group ran enthusiastically directly to the food and found it very quickly. However, the chimp that had witnessed the food's being hidden did not necessarily lead the group. When Menzel hid a snake instead of food, the chimps approached it cautiously, with evident signs of fear. In one experiment, Menzel showed one cache of food to one chimpanzee and another cache to a different chimp. When all the chimps were released, they usually would make for the more desirable of the two sources of food. Thus, there might be more food in one location than in the other, or there might be fruit in one location and less desirable vegetables in the other.

It seems that chimpanzees can make deductions about their environment that are based upon the behavior of a companion. The chimp who knows the location of food indicates by his actions and emotions the desirability and direction of the goal. There is no direct communication about the location of the food, but the other chimpanzees are intelligent enough to draw their own conclusions (Menzel and Johnson, 1976). Some evidence indicates that chimps can learn to point to objects in the laboratory (Terrace, 1979; Woodruff and Premack, 1979) and that they can learn to use human pointing as a clue to the whereabouts of food (Menzel, 1979). However, they do not seem to use pointing or other indications of direction in communicating among themselves.

26.3 Teaching apes to converse

Although it seems that chimpanzees normally do not communicate about objects remote in time or space, it may be that they could be taught to do so. If we knew the extent to which apes were capable of handling various features of language, we perhaps could gain some understanding of our own abilities.

The Kellogs reared a chimpanzee named Gua in their home (Kellog and Kellog, 1933). They claimed she learned to understand 95 words and phrases in about eight months, about the same as their son Donald, who was three months older. Gua was tested by being given a card with four pictures on it. Another chimpanzee, named Ally, was also reared in a human house and gained some understanding of speech. She was taught the gestures that correspond to various words in Ameslan. She could make the correct sign when she heard the word spoken (Fouts et al., 1976). Other experiments, with a gorilla (Patterson, 1978) and a dog (Warden and Warner, 1928), suggest that these animals are able to associate sounds and visual cues.

Some experimenters have tried to investigate the extent to which primates have voluntary control over the sounds they make. In one experiment, rhesus monkeys were required to bark when a green light came on and to produce a coo to a red light. The monkeys learned to produce the correct sound to obtain food reward (Sutton, 1979). An orangutan has been trained to make three different sounds to obtain food, drink, or contact with the keeper (Laidler, 1978), and a chimpanzee has been trained to bark to induce a human to play (Randolph and Brooks, 1967). As a control, the chimpanzee was also trained to initiate play by touching its human companion when she was facing the chimp, and barking when she had her back to the chimp. These experiments suggest that apes and monkeys may have some limited voluntary control of the sounds they produce. Without special training, they do not seem to mimic the sounds they hear, even when they are living with a human family (Kellog, 1968). The fact that they readily imitate human actions, however, suggests that sound is not a medium that apes readily learn to use for communication beyond their normal, limited repertoire (Passingham, 1982).

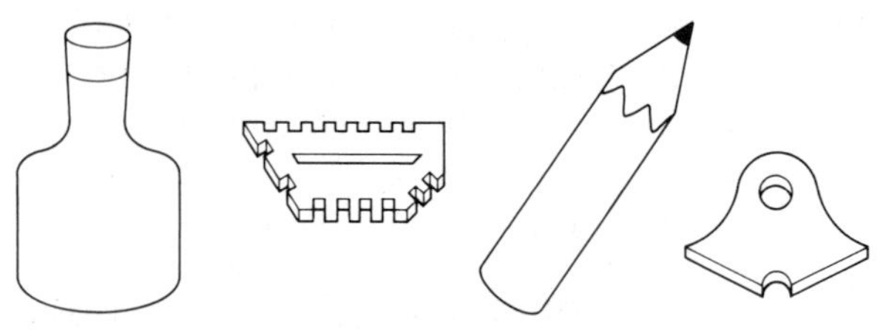

Question

"Is A (the bottle) the same as B (the pencil)?"

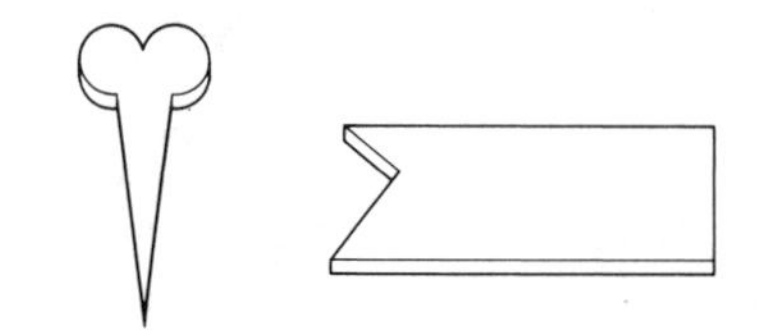

Alternatives:

"Yes" or "no"

Fig. 26.6 Example of a question asked of the chimpanzee Sarah by means of plastic shapes. Sarah could respond by choosing one of the alternatives below (After Premack, 1976)

Once it had become clear that speech was not a necessary aspect of language, and that the ability to produce or respond to sounds was not necessarily relevant to the question of whether or not animals were capable of language, then the way was open to investigate language by manipulating visual symbols (Gardner and Gardner, 1969; Premack, 1970). The Gardners started with a single chimpanzee named Washoe, to whom they taught Ameslan, in which words are represented by gestures of the hand and arm. Washoe was trained from the age of 11 months, and by the age of 5 years, she had mastered 132 signs (Gardner and Gardner, 1975). Washoe spontaneously learned to combine signs into strings of two to five words. Among her first combinations were "come open" and "gimme sweet" (Gardner and Gardner, 1971). The Gardners have since trained two chimpanzees from birth, and these progressed more quickly than Washoe. Others have been taught by Roger Fouts (1975), and a gorilla has been taught to use hand signs and to respond to spoken English commands (Patterson, 1978). Herbert Terrace (1979) and co-workers trained a chimpanzee named Nim Chimpsky to use Ameslan in a thorough study in which a transcript was kept of every sign Nim ever made and every combination he produced.

David Premack (1976; 1978) taught a chimpanzee named Sarah to read and write. He used colored plastic shapes to represent words. The shapes in no way resembled the things to which they referred (see Fig. 26.6). The shapes were presented on a vertical magnetic board, and Sarah could answer questions by placing appropriate shapes on the board. Sarah mastered 120 plastic symbols, although she was not pressed to attain a large vocabulary (Premack, 1976). She could carry out commands and answer questions using several symbols in combination. Other chimpanzees also have been trained in this method by Premack and his colleagues.

Another method was employed by Duane Rumbaugh (1977), who used an artificial grammar called *Yerkish* (Glasersfeld, 1977). A chimpanzee named Lana learned to operate a computer keyboard that displayed word symbols on a screen. The computer was programmed to recognize grammatical and incorrect usage of the symbols and to reward Lana accordingly. The advantage of this approach is that Lana was able to converse with the computer at any time of day instead of having to attend formal sessions. Other chimpanzees also have been trained to converse with each other by the computer-based method (Savage-Rumbaugh et al., 1978; 1980).

In assessing the results of these experiments, it is necessary to bear in mind the possibility that the chimps might cheat. They might make use of involuntary hints given by their trainers, or they might simply learn a succession of tricks, like a circus animal does.

The Gardners (1978) tested Washoe under conditions in which the trainer did not know the answer to the question. Washoe was required to name the object shown on a slide by making the appropriate sign to a nearby person who could not see the slide. A second person could see Washoe's gestures without being seen by Washoe and without being able to see the slides. Washoe was asked to name 32 items, each of which was shown to her four times. She gave the correct answer to 92 of 128 questions. Similar tests have been conducted on some of the other chimps used in these studies (Rumbaugh, 1977; Premack, 1976; Patterson, 1979).

It is possible that the chimps learn what to do when they perceive a certain cue, just as a circus animal learns what to do upon receipt of cues from its trainer. In order to discover whether or not the chimps understand the meaning of the signs and symbols they manipulate, it is necessary to conduct an experiment in which the chimp is required to name the object in a situation other than the learning situation. Various tests have been conducted with this point of view (e.g., Gardner and Gardner, 1978; Savage-Rumbaugh et al., 1980), and the results indicate that chimps can genuinely name objects. Sometimes, moreover, chimps do this spontaneously. Thus, Nim would give the sign for dog upon seeing a dog or a picture of a dog or upon hearing the bark of a dog (Terrace, 1979).

The evidence suggests that chimpanzees can come to understand the meaning of words; that is, they genuinely can name things. However, their abilities are not so evident in many other aspects of human language, some of which are of interest primary in assessing the cognitive abilities of chimpanzees. Of particular interest is the question of whether or not the chimp can incorporate new messages into its repertoire, an issue of some importance in assessing the bee dance (see Chapter 23).

It appears that chimpanzees sometimes coin novel phrases. Thus, Washoe is reported to have invented the word "candy drink" for watermelon and to have called a swan a "water bird." Such instances are difficult to interpret, however, because of the possibility that apparently novel usage is the result of simple generalization. For example, Washoe was introduced to the sign of a flower by being offered a real flower. She learned the sign but applied it not only to flowers but also to tobacco aroma and cooking smells. Apparently, Washoe associated the sign with the smell of the flower and generalized to other odors (Gardner and Gardner, 1969).

Another problem is that the chimps sometimes make novel word combinations that appear to make no sense. Nim's favorite food was banana, and he often combined this word with other words, such as drink, tickle, and toothbrush. Although it is possible that "banana toothbrush" is a request for a banana and for a toothbrush to clean the teeth after eating the banana, this seems unlikely because a banana and a toothbrush were never on view at the same time, and Nim never asked for objects he could not see (Ristau and Robbins, 1982). Perhaps the bizarre combinations of words are an example of word play similar to that found in children. Washoe was observed to sign to himself when playing alone, much as children talk to themselves.

Thus, we can say, the attempts to teach chimpanzees and other apes various types of human language have had limited success. The apes seem to be able to attain a standard that is equivalent to that of a young child. The difference between ape and child may simply be one of intelligence, but it is also possible that humans have an innate language acquisition device, as first suggested by Noam Chomsky (1972). In any event, the experiments with apes certainly have uncovered abilities not suspected previously, and they have given us considerable insight into apes' cognitive abilities.

26.4 The origins of human language

The origins of human language probably always will remain a mystery. The gap between the linguistic abilities of humans and other animals is so large that we can learn only a little by comparison of living species. Fossil evidence provides a few clues, but they are very difficult to interpret. The problem is that we do not know what to count as evidence of language ability (Passingham, 1982). If we believe that speech was a necessary precursor of language, then we can look for evidence relevant to the evolution of the vocal tract. Reconstructions based upon the bones of the skull and jaw suggest that the vocal tracts of Neanderthal man, australopithecines, and the chimpanzees were similar (Lieberman, 1975).

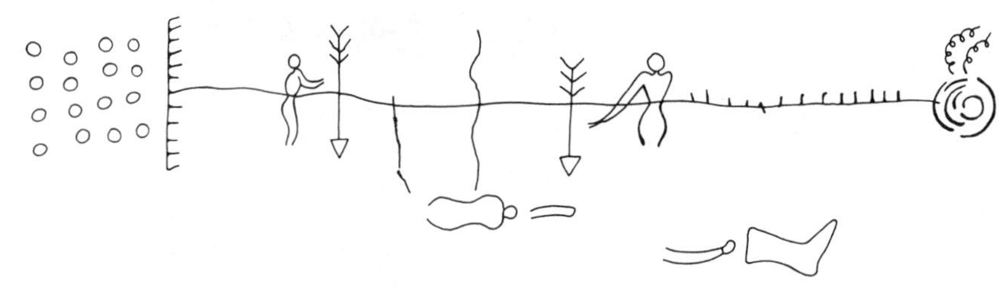

Fig. 26.7 Representation of a tribal feud by a Sioux Indian (After Klix, 1982)

Even if we ignore uncertainties about the reliability of such evidence, we cannot conclude that Neanderthal man was incapable of language. It may be that language evolved from the sophisticated use of gestures and that speech came later (Passingham, 1982).

Casts of the brains of fossil hominids can give us an indication of brain size and shape that can be compared with those of modern man. Although we can assume that a large brain is likely to be associated with language in primates, we do not know how large the brain needs to be. The Neanderthal brain is slightly larger than the human average, but does this mean that language was a feature of Neanderthal life? A large brain may be a sign of intelligence, but it may be that "possession of human language is associated with a specific type of mental organisation, not simply with a higher degree of intelligence" (Chomsky, 1972). However, the shape of the brain gives further clues. Asymmetry between the two hemispheres occurs in several hominids, the pattern being similar to that of modern man (LeMay, 1976). There is a connection between brain asymmetry and language in *Homo sapiens*, and there is a possibility that this is also the case in some early hominids (Passingham, 1982). There have also been attempts to identify in fossil brain casts some of the more detailed structures known to be involved in language in the brain of modern man (Holloway, 1976). With improved techniques (e.g., Holloway, 1981), it might be possible to take this type of analysis further, but at present the evidence justifies only extremely tentative suggestions.

Evidence of symbolic representation can be found in the various artifacts recovered from the sites of early man. Stone tools and other implements can be traced back as far as 2.5 million years (Lewin, 1981), but these are not necessarily evidence for the existence of language (Passingham, 1982). Decoration of artifacts is suggestive of a maker who possessed a symbolic language (Marshack, 1976), and such artifacts can be traced to about 300,000 years ago. However, representational designs seem to come only from *Homo sapiens* some 30,000 years ago. Written language appears to have developed from representational symbols with emotional or mystical significance toward a more informative representation of a particular situation (Fig. 26.7). However, we can recognize the properties of true language only when we find symbols for the sounds of speech rather than concepts (Klix, 1982).

26.5 Language and cognition

We have seen that chimpanzees and other apes can be taught to converse with humans by using sign language or by reading and writing with plastic shapes or symbols presented by a computer. The evidence suggests that chimpanzees can learn the meaning of words in the sense that they can name things. They can master a vocabulary of more than a hundred words, and it appears that they sometimes coin novel phrases.

A question of considerable interest is whether or not the various ape language projects reveal cognitive abilities, a question with various different aspects. One of these is the distinction between knowing *how* and knowing *that*. An ape may know how to ask for a reward in that it can learn to make the appropriate gesture. However, this is not the same as knowing that making a particular gesture will obtain a reward for it. To "know that" is to understand a relationship beyond the mere coupling of stimulus and response. In humans, "knowing how" may cover complex skills such as fast typing or golf, where the performer can achieve a good result without understanding or being able to describe the relationship between the objective of the behavior and its performance. In other cases, people clearly do know about the processes that led them to a particular objective and can describe them. The question of whether or not the results of ape language experiments allow us to make this distinction is discussed by Ristau and Robbins (1981; 1982).

Some of the results of the Lana project (Rumbaugh and Gill, 1977) are relevant here. Lana could use phases made up of plastic symbols to request specific rewards. She might write, "Please machine give banana," which can be described readily as knowing how to obtain a banana. However, attempts were also made to teach Lana to name two items, banana and M&M candy. She was presented with either a banana or an M&M on a tray and asked (via the experimenter's computer keyboard), "?What name-of this." The correct response (via a keyboard) was "Banana name-of this." Correct responses were rewarded. Lana required 1,600 trials to learn this task, even though she had requested these items previously, hundred of times, by using the standard phrase "Please machine give banana". This suggests that Lana did not know the meaning of the original symbols she had manipulated previously to request an M&M or banana. However, it is worth noting that Lana originally was not required to know the meaning and that requesting items in a routine knowing-how way could have become a habit difficult to break, an interpretation supported by the fact that subsequent training involving naming other items resulted in much more rapid success. The initial difficulties in labeling may have been the result of an inability to distinguish the requirements of the new situation from that in which the banana and M&M symbols were used originally.

In experiments with Sarah, Premack (1976) taught the concept "name-of" in the following way: The plastic symbol for apple was placed in

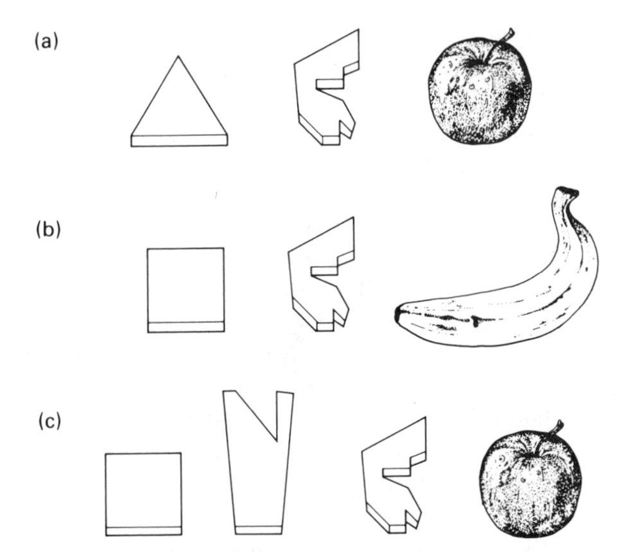

Fig. 26.8 Procedure used to teach Sarah the concept 'name of.' (*a*) The 'name of' symbol is placed between the symbol for apple and a real apple. (*b*) The 'name of' symbol is placed between the banana symbol and a real banana. (*c*) The 'name of' symbol, preceded by a 'not' symbol, is placed between the apple symbol and a banana (After Premack and Premack, 1972)

front of a real apple, with a gap between them. Sarah was required to place a new plastic chip, intended to mean "name-of," into this gap. She thus apparently created the sentence "'Apple' name-of object apple." "Not name-of" was formed by gluing the normal negative symbol onto the plastic symbol for "name-of." When the symbol for apple was placed in front of a banana, with a gap between, Sarah was required to select the correct symbol to fill the gap (Fig. 26.8). In this case, the choice of "not name-of" would earn Sarah a reward. She was able to use these labels correctly in both naming tests and sequences with other plastic symbols.

The ability to learn that abstract shapes are symbolic of objects in the real world suggests a kind of knowing-that that is similar to a declarative representation. However, it is difficult to devise a demonstration divorced of all procedural possibilities. Perhaps Sarah was merely learning that the two shapes "name-of" and "not name-of" indicated responses to be made when two things (i.e., the apple and its lexicon) were equivalent or not equivalent, which is a type of matching procedure (see Chapter 27.1).

26.6 Mental images

Issues of representation have been important in human cognitive psychology for more than a decade. A technique commonly used to obtain an objective measure of intangible phenomena like mental

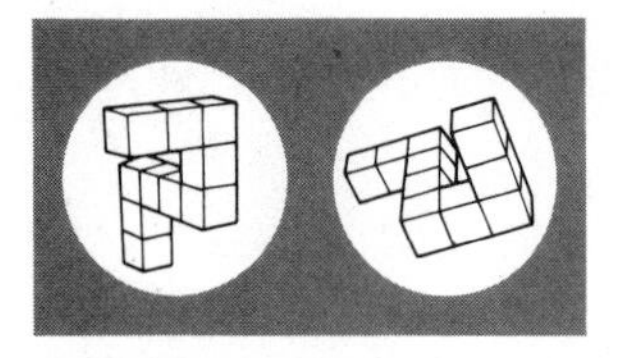

Fig. 26.9 Example of a pair of shapes used in mental chronometry. This is an identical pair differing by 80° rotation (After Shepard and Metzler, 1971)

representations is *mental chronometry*, which uses the time required to solve a spatial problem as an index of the processes involved (Posner, 1978). For example, a classic study (Shepard and Metzler, 1971) requires human subjects to view pairs of drawings of three-dimensional objects. On each trial the subject has to indicate whether two objects are the same in shape or are mirror images. As well as being different in shape, the objects can also differ in orientation, as shown in Figure 26.9. It was found that the time required to indicate whether or not the two objects are the same shape increases, in a regular manner, with the angular difference between the pairs of drawings presented to the subject (Fig. 26.10). The usual conclusion is that the subjects mentally rotated an internal representation of one object to make it line up with the other before they compared the shapes of the two representations. Although the precise nature of the representation used in this type of task is a matter of controversy (Kosslyn, 1981; Cooper, 1982), the most straightforward interpretation is that some form of mental imagery is involved and that the processes of rotation and comparison are carried out in series. The fact that reaction time is a function of presentation angle is taken to show that mental rotation takes time (about 30 milliseconds for every 20 degrees) (Cooper and Shepard, 1973).

Proficiency in the visual recognition of objects, regardless of their relative spatial orientation, is assessed by several intelligence and aptitude tests (Petrusic et al., 1978), and it would be of interest to compare species on this basis (see Chapter 27.1). So far, however, only one other species has been studied systematically, and the results are rather surprising.

Valerie Hollard and Juan Delius (1983) trained pigeons in a Skinner box to discriminate between shapes and their mirror images presented

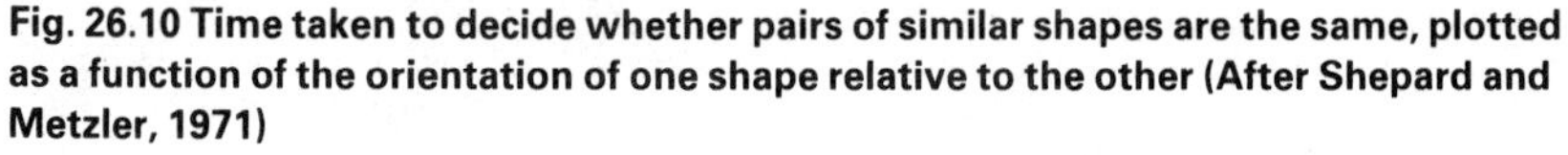

Fig. 26.10 Time taken to decide whether pairs of similar shapes are the same, plotted as a function of the orientation of one shape relative to the other (After Shepard and Metzler, 1971)

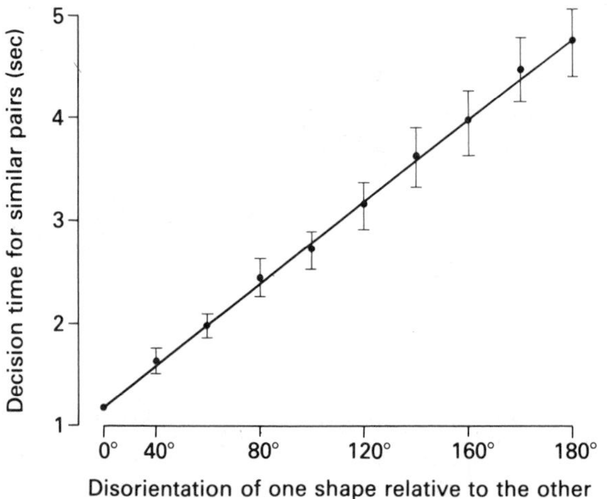

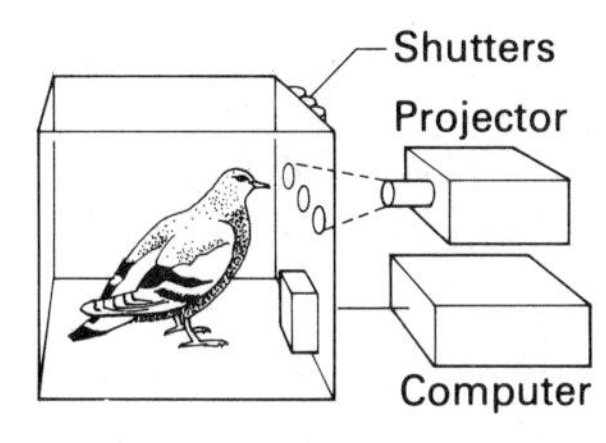

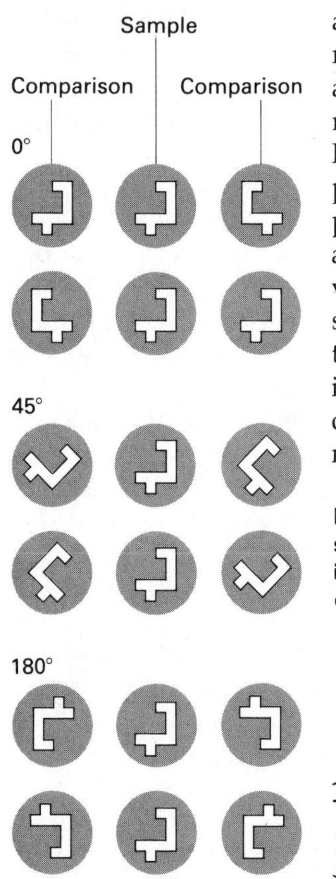

Fig. 26.11 Experimental apparatus (above) used to display symbols (below) to pigeons. The birds were trained to indicate which of the two comparison symbols the sample symbol most resembled. The comparison symbols were presented at 0°, 45°, and 180° rotation (After Hollard and Delius, 1983)

in various orientations, as shown in Figure 26.11. They then measured the pigeon's reaction times in tests for rotational invariance. When this part of the study was complete, the chamber was disassembled, and the test panel, containing the lights and keys, was presented to human subjects in a series of similar tests. In this way a direct comparison of the performance of pigeons and humans could be made on the basis of the same stimulus patterns.

The results showed that pigeons and humans were capable of similar accuracy, as judged by the errors made. However, whereas the human reaction time increased with the angular disparity between the sample and comparison forms, as shown by previous studies, the pigeon reaction time remained unaffected by the angular rotation, as shown in Figure 26.12. It appears that pigeons are able to solve this type of problem more efficiently than humans, presumably through some parallel form of processing. This result not only has implications for the assessment of intelligence in animals but also raises questions about the validity of studies of mental representations. Do the pigeon experiments suggest that mental representations are not necessarily involved in this type of task? If pigeons can solve such problems without using mental images, can we be sure that mental images are used in fact by humans, or is it merely that we find it easier to account for the results in terms of mental representation (see also Dennett, 1978).

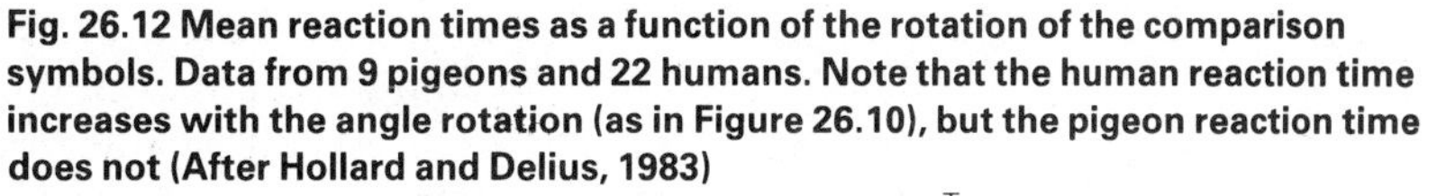

Fig. 26.12 Mean reaction times as a function of the rotation of the comparison symbols. Data from 9 pigeons and 22 humans. Note that the human reaction time increases with the angle rotation (as in Figure 26.10), but the pigeon reaction time does not (After Hollard and Delius, 1983)

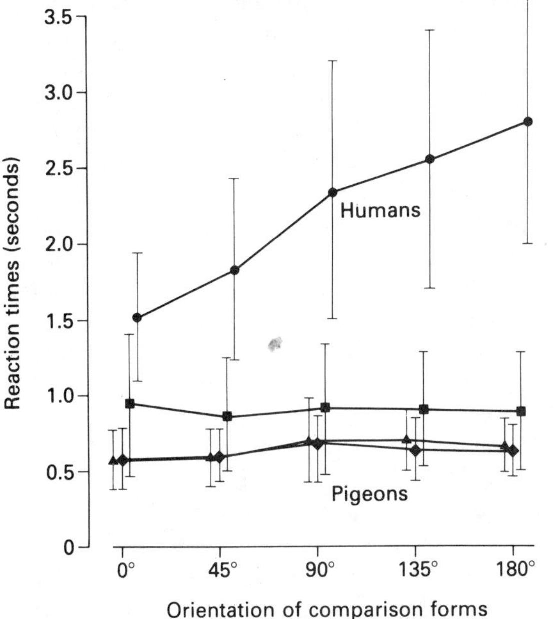

26.7 Intentional behavior

Human behavior can be said to be *intentional* when it involves some representation of a goal that is instrumental in guiding behavior. Thus, if I have a mental picture of the desirable arrangement of books on my shelf, and if this mental representation guides my behavior in placing books on the shelf, then I can be said to place them intentionally. If, however, I place books on the shelf haphazardly, or on the basis of simple habit rules, then my arrangement of the books may not be intentional. For my behavior to be intentional, my mental representation of the book arrangement does not have to be a conscious one. Although consciousness and intentionality are sometimes linked, it is better to treat them separately (Dennett, 1978).

Ethologists have long recognized intention movements (see Chapter 22.1) in animals as indications of what the animal was about to do. Both human observers and members of an animal's species can predict the future behavior of the animal from its intention movements, and it may seem silly to assume the animal cannot anticipate the next steps in its own behavior (Griffin, 1976). It is more likely, however, that intention movements are merely the initial stages of behavior patterns terminated prematurely, either because the animal is in a motivational conflict or because its attention is diverted to other possible aspects of behavior. Indeed, it is difficult to imagine how such incipient behavioral fragments could be avoided in an animal with a complex repertoire of activities. Although some so-called intention movements may have become ritualized during the course of evolution, this does not mean that they are intentional in the ordinary sense of the word.

Another aspect of animal behavior that has the appearance of intentionality is the injury-feigning distraction display of certain birds. When an incubating bird, like a sandpiper (*Ereunetes mauri*), is disturbed by a ground predator, it may leave the nest and act as though injured, trailing an apparently broken wing and luring the predator away from the nest. When the predator has been led a safe distance from the nest, the bird suddenly regains its normal behavior and flies away (Brown, 1963; Skutch, 1976). While most ethologists are content to account for this type of behavior in terms of ritualized display, some (e.g., Griffin, 1981) wish to keep open the possibility that the birds are behaving intentionally. Griffin notes that the possibility that animals have mental states or intentional behavior is often most vehemently denied in cases where the evidence is weakest. The problem is to know what does count as evidence. In Griffin's view, evidence is most likely to come from studies of animal communication, particularly that of primates.

Do the various studies of chimpanzee communication throw any light on this question? George Woodruff and David Premack (1979) explored the chimpanzee's ability in intentional communication by setting up a situation in which a human and a chimp could cooperate or compete in obtaining food. They communicated by means of non-verbal signals about the location of hidden food. When the human was cooperative and gave the chimpanzee all the food that was found, the chimpanzee could successfully send and receive behavioral signals about the location of the food. When the human and chimp were in competition with each other and the human kept the food that was found, the chimpanzee learned to mislead the human by withholding information and by discounting misleading behavioral cues given by the human. The chimpanzee's behavior suggests that they can infer purposes, or intentions, from the human behavior and that they can have some knowledge of the human's perception of their own behavior.

Hans Kummer (1982) reviews similar apparent abilities in free-ranging primates, and Donald Griffin (1981) comments on cases (Ruppell, 1969) in which a mother Arctic fox was competing for food with several of her nearly full-grown young. In order to reach the food first, a young fox might, for example, urinate in its mother's face. After several such tricks, the mother gave false alarm calls and seized the food for herself when the young ran off. Griffin notes that it is difficult to interpret such behavior without postulating short-term intentions and plans on the parts of both mother and young. However, we must ask if these animals are really inferring motives in others or simply learning effective means of obtaining food in various situations.

In the case of the chimpanzees, some observations suggest that the chimps really are practicing deception. One is the development of pointing, a new behavior that was used advantageously to indicate the correct or incorrect container. Chimpanzees rarely point in laboratory or field situations, though they do understand pointing by humans. A second observation concerns a chimpanzee Sadie, who pointed to an empty container when asked which contained food. When the human lifted the container and discovered no food, Sadie's head snapped in the direction of the other container, which she knew contained the food (Premack and Woodruff, 1978).

It appears at times that the argument between behaviorist and cognitive explanations of behavior will never end (see, e.g., the correspondence on Premack and Woodruff's paper in *Brain and behavioral sciences*, 1978). For any set of behavioral observations, it seems, each side can come up with an alternative explanation. However, Daniel Dennett (1978) notes that in considering the behavior of a complex system, we can take a number of different stances that are not necessarily contradictory. One is the *design* stance: If one knows exactly how a system is designed, one can predict its designed response to any particular situation. In ordinary language, we use this stance in predicting what will happen when we manipulate an object with a known function; hence, "Strike the match and it will light." In biology, we use the design stance in making predictions based upon the theory of natural selection. In fact, we have two stances at our disposal: the *maximize-inclusive-fitness* stance, and the *selfish gene-eye* view advocated by Dawkins (1976; 1982) (see Chapter 6). There is also the *physical* stance, which bases predictions on the physical state of the system. In the behavioral sciences, two alternative stances are in common use: the *physiological* and the *control-system* approaches; each has its advantages and disadvantages, and the two are complementary rather than contradictory (McFarland, 1971).

Dennett (1978) also advocates an *intentional* stance to the behavior of complex systems, which assumes that the system under investigation is an intentional system possessing certain information and beliefs that is directed by certain goals. In advocating this stance, Dennett is attempting not to refute behavioral or physiological explanations but to offer a higher level of explanation for the behavior of systems so complex that other stances become unmanageable:

> An intentional system is a system whose behavior can be (at least sometimes) explained and predicted by relying on ascriptions to the system of beliefs and desires (and other intentionally characterized features)—what I will call intentions here, meaning to include hopes, fears, intentions, perceptions, expectations, etc. There may, in every case be other ways of predicting and explaining the behavior of an intentional system—for instance, mechanistic or physical ways—but the intentional stance may be the handiest or most effective or in any case a successful stance to adopt, which suffices for the object to be an intentional system (Dennett, 1979).

Dennett (1983) maintains that ethologists and others studying animal behavior from a cognitive viewpoint are in need of a descriptive language and method that is open minded and capable of empirical verification. He proposes that inter.tional system theory can fulfill this role. He quotes the following example from Seyfarth et al. (1980):

> "Vervet monkeys give different alarm calls to different predators. Recordings of the alarms played back when predators were absent caused the monkeys to run into the trees for leopard alarms, look up for eagle alarms, and look down for snake alarms. Adults call primarily to leopards, martial eagles, and pythons, but infants give leopard alarms to various mammals, eagle alarms to many birds, and snake alarms to various snakelike objects. Predator classification improves with age and experience."

In adopting an intentional stance toward vervet monkeys, we regard the animal as an intentional system and attribute beliefs, desires, and rationality to it, according to the appropriate type or order of intentional system. A zero-order (or non-intentional) account of a particular monkey, called Tom, who gives a leopard alarm call in the presence of another vervet, might be as follows: Tom is prone to three types of anxieties—leopard anxiety, eagle anxiety and snake anxiety—each producing a characteristic vocalization that is produced automatically, without taking account of its effect upon other vervet monkeys.

A first-order intentional account might maintain that Tom wants to cause another monkey, Sam, to run into the trees. Tom uses a particular vocalization to stimulate this response in Sam. A second-order intentional account goes a step further in maintaining that Tom wants Sam to believe there is a leopard in the vicinity and that he should run into the trees. A third-order account might say that Tom wants Sam to believe that Tom wants Sam to run into the trees.

Dennett maintains that the question of which order of intentionality is appropriate is an empirical question. For example, suppose a lone male vervet monkey traveling among groups and out of hearing of other vervets silently seeks refuge in the trees upon seeing a leopard. We then might be inclined to dismiss the zero-order account. Vervet monkeys can recognize the individual voices of other members of the group, so if Tom's leopard alarm call was broadcast from a tape recorder in a situation where Sam could see Tom, then we might wish to abandon the third-order intentional explanation. If we observed that Sam did take to the trees under such circumstances, this behavior would not be rational, and rationality is an assumed property of an intentional system (Dennett, 1983).

The assumption that intentional systems are rational enables us to construct what Dennett (1983) calls the "Sherlock Holmes method":

> In *A Scandal in Bohemia*, Sherlock Holmes' opponent has hidden a very important photograph in a room and Holmes wants to find out where it is. Holmes has Watson throw a smoke bomb into the roof and yell "fire" when Holmes' opponent is in the next room, while Holmes watches. Then, as one would expect, the opponent runs into the room and takes the photograph from where it is hidden. Not everyone would have devised such an ingenious plan for manipulating an opponent's behavior, but once the conditions are described, it seems very easy to predict the opponent's actions (Cherniak, 1981).

The Sherlock Holmes method is not foolproof, however, as shown by the following story. As a graduate student, I used to experiment at frustrating pigeons by presenting food on some occasions and not others or by presenting food the bird could not obtain. One particular pigeon used to behave rather aggressively toward me, and I formed the impression that he did not like my treatment of him. One day I inadvertently performed a Sherlock Holmes experiment and allowed the pigeon to escape from its cage. It immediately marched over to the tangle of electrical wires that controlled the apparatus and started to pull them apart with his beak. Feeling rather shaken and guilty, I quit the room to obtain coffee. Upon returning I realized, from its behavior, that the pigeon regarded the wires as nest material and that its aggression toward me was typical of the early stages of courtship.

Nevertheless, such an approach might work in certain circumstances. For example, in the Premack and Woodruff (1978) experiments, Sadie points to the box containing food when the cooperative human enters the room and to the box containing no food when the uncooperative human enters. How could we test the theory that Sadie intended to deceive the uncooperative human? Dennett (1983) suggests that:

We introduce all the chimps in an entirely different context to transparent plastic boxes; they should come to know that since they—and anyone else—can see through them, anyone can see, and hence come to know, what is in them. Then on a one-trial, novel behavior test, we can introduce a clear plastic box and an opaque box one day, and place the food in the clear plastic box. The competitive trainer then enters, and lets Sadie see him looking right at the plastic box. If Sadie still points to the opaque box, she reveals, sadly, that she really doesn't have a grasp of the sophisticated ideas involved in deception. Of course this experiment is still imperfectly designed. For one thing, Sadie might point to the opaque box out of despair, seeing no better option. To improve the experiment, an option should be introduced that would appear better to her only if the first option was hopeless, as in this case. Moreover, shouldn't Sadie be puzzled by the competitive trainer's curious behavior? Shouldn't it bother her that the competitive trainer, on finding no food where she points, just sits in the corner and "sulks" instead of checking out the other box? Shouldn't she be puzzled to discover that her trick keeps working? She should wonder: can the competitive trainer be that stupid? Further, better-designed experiments with Sadie—and other creatures—are called for.

Points to remember

- Non-verbal communication occurs as part of everyday human behavior and has some features in common with animal communication.

- Language should not be defined as a uniquely human activity because there are many features of animal communication that are language-like.

- Apes cannot learn to speak, but they can learn to communicate with humans using symbols to represent words.

- According to Chomsky (1972), the possession of human language is associated with a specific type of mental organization, not simply with a higher degree of intelligence.

- The question of whether ape language involves cognition hinges upon distinctions such as 'knowing how' and 'knowing that'.

- Although mental images are thought to be important in human behavior, it seems that they are not necessary for the control of similar behavior in animals.

Further reading

Gardner, B. T. and Gardner, R. A. (1969) 'Teaching sign language to a chimpanzee,' *Science*, **165**, 664–72.

Griffin, D. R. (1982) *Animal Mind–Human Mind*, Springer–Verlag, Berlin.

Passingham, R. E. (1982) *The Human Primate*, Freeman, New York.

Rumbaugh, D. M. (1977) *Language Learning by a Chimpanzee*, Academic Press, New York.

Terrace, H. S. (1979) *Nim*, Eyre Methuen, London.

27 Intelligence, Tool Use, and Culture

Darwin believed in the evolutionary continuity of mental capabilities, and he opposed the widely held view that animals were merely automatons, far inferior to humans. In his *Descent of Man* (1871) Darwin argued that "animals possess some power of reasoning" and that "the difference in mind between man and higher animals, great as it is, certainly is one of degree and not kind." The intelligence of animals was exaggerated by Darwin's disciple George Romanes, whose publication of *Animal Intelligence* (1882) was the first attempt at a scientific analysis of animal intelligence. Much of Romanes's evidence was anecdotal, and his book was full of stories by respectable members of Victorian society.

Romanes defined intelligence as the capacity to adjust behavior in accordance with changing conditions. His uncritical assessment of the abilities of animals provoked a revolt by Conway Lloyd Morgan (1894) and the subsequent behaviorists, who attempted to pin down animal intelligence in terms of specific abilities. They were critical of the anecdotal approach and set up strict criteria for attributing mental capacities to animals. Lloyd Morgan inspired Edward Thorndike to investigate trial-and-error learning (see Chapter 7), and he thus indirectly founded the behaviorist school of animal psychology. In his book *Introduction to Comparative Psychology* (1894), Lloyd Morgan suggested that higher faculties evolved from lower ones, and he proposed a psychological scale of mental abilities.

Although the idea of a ladderlike evolutionary scale of abilities had a considerable influence upon animal psychology, it is not an acceptable view today. Studies of brain structure (e.g., Hodos, 1982) and of the abilities of various species (e.g., Macphail, 1982) make it abundantly clear that different species in different ecological circumstances exhibit a wide variety of types of intelligence. This makes intelligence difficult to define, but it emphasizes the importance of studying animal intelligence from a functional point of view as well as investigating the mechanisms involved.

In this chapter we look at some of the ways in which psychologists have tackled the problem of assessing the intelligence of animals. We look in some detail at tool use as an example of behavior most people would regard as indicating intelligence, and we discuss cultural transmission as a means by which animals acquire new forms of behavior.

27.1 Comparative aspects of intelligence

There are two main ways of assessing the intelligence of animals: One is to make a behavioral assessment and the other is to study the brain. In the past, both approaches have been dominated by the idea that there is a linear progression from lower, unintelligent animals with simple brains to higher, intelligent animals with complex brains. A survey of the animal kingdom as a whole tends to confirm this impression (see Chapter 11), but when we look closely at specific cases, we find many apparent anomalies. These are not exceptions to an overall rule but are due to the fact that evolution does not progress in a linear manner. It diverges among a multiplicity of routes, each involving adaptation to a different set of circumstances. This means that animals may exhibit considerable complexity in some respects but not others and that different species may reach equivalent degrees of complexity along different evolutionary routes.

In comparing the brains of animals of different species, we can expect to find a relationship between the relative size of a particular structure and the degree of sophistication of the behavior the structure controls. The more an animal makes use of a particular aspect of behavior in adapting to its environment, the larger the number of neurons and interconnections that will be present in the appropriate parts of the brain. This is easy to see when comparing specialized parts of the brain like those concerned with different sensory processes. It is less easy to interpret when we come to look at the more general-purpose parts, because they may have enlarged as a result of different selective pressures in different species (Jerison, 1973).

Many traditional ideas about the evolution of the vertebrate brain have been challenged. Thus, it has been claimed that, contrary to popular belief, there is no progressive increase in relative brain size in the sequence: fish, reptile, bird, mammal, or in the relative size of the forebrain in the sequence: lamprey, shark, bony fish, amphibian, reptile, bird, mammal (Jerison, 1973). Indeed, some sharks have forebrains equivalent to those of mammals in relative size (Northcutt, 1981). It was long thought that the telencephalon of sharks and bony fish is dominated by the olfactory sense, but it is now claimed that the representation of the olfactory sense in this region is no greater in non-mammalian vertebrates than in mammals (Hodos, 1982). The idea that an undifferentiated forebrain is characteristic of lower vertebrates also has been challenged (Hodos, 1982).

In reviewing our understanding of animal intelligence in the light of modern knowledge of neuroanatomy, William Hodos (1982) comes to the conclusion that:

> If we are to find signs of intelligence in the animal kingdom and relate them to developments in neural structures, we must abandon the unilinear, hierarchical

models that have dominated both searches. We must accept a more general definition of intelligence than one closely tied to human needs and values. We must accept the fact that divergence and nonlinearities characterize evolutionary history, and we must not expect to find smooth progressions from one major taxon to another. Finally, we must not allow ourselves to be biased by our knowledge of the mammalian central nervous system in our search for neural correlates of intelligence in other vertebrate classes. Without such changes in our thinking, we would appear to have little hope of progressing any further than we have in our attempt to understand the relationships between the human mind and the animal mind and their respective neural substrates.

We now turn to the question of how to assess animal intelligence behaviorally. Since intelligence tests for people were introduced by Albert Binet in 1905, considerable progress has been made in improving and refining them. This has been possible largely because different tests can be evaluated by checking on the subsequent educational progress of individuals. Modern I.Q. tests are reasonably accurate in predicting how well a person will progress in intellectual achievement. However, difficulties remain, especially in attempting to compare the general intelligence of people from differing cultural backgrounds. The difficulties in assessing the intelligence of animals are much more severe because there is no way of checking on the validity of a test and because animals of different species differ greatly in their performance in particular respects.

Until recently, attempts to assess animal intelligence concentrated upon abilities that normally would be taken as a sign of intelligence in humans. A modern I.Q. test includes various subtests designed to assess a person's memory, arithmetic and reasoning power, language ability, and ability to form concepts. As we have seen, pigeons appear to have a prodigious ability to form concepts such as water, tree, and human beings. Are we to take this as a sign of great intelligence? In discussing language abilities in animals, we came to the conclusion that human abilities, in this respect, far exceed those of any other animal, however well trained. Does this show that humans have greater general intelligence or merely that human intelligence is highly specialized in the use of language?

In comparing the intelligence of different species, it is difficult to devise a test that is not biased in one way or another. Many of the early tests of animal problem-solving ability were unreliable (Warren, 1973). Sometimes the same test, with the same species, gave different results according to the type of apparatus employed.

Various attempts have been made to discover whether or not animals can master problems that require the learning of a general rule. Animals can be trained to choose from an array one item that matches a sample. Primates learn to solve this type of problem quickly, but pigeons require a large number of trials. Harry Harlow (1949) devised a test to measure the ability of animals to follow rules and to make valid inferences.

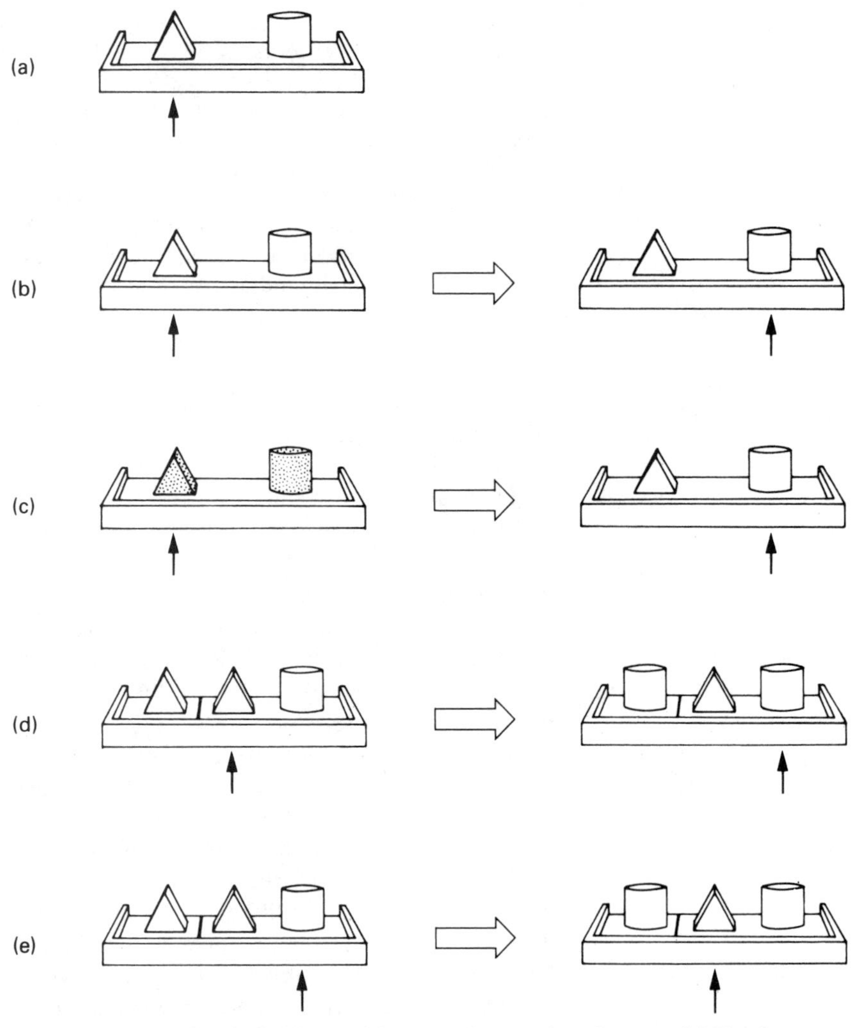

Fig. 27.1 Series of discrimination problems used to test learning sets. (*a*) Simple discrimination (the arrow indicates the correct choice). (*b*) Reversal task (the animal has to reverse its originally correct choice). (*c*) Conditional task (one object is correct when both are grey, the other when both are white). (*d*) Matching task (the animal has to match the sample at left on tray). (*e*) Oddity task (the animal has to choose the odd man out) (After Passingham, 1981)

Instead of testing monkeys on a single visual discrimination like that shown in Figure 27.1, Harlow presented them with a series of tests, all of which require the same rule. Thus, the animal might be given a series of discrimination problems of the type shown in the figure. Although the objects are changed from problem to problem the rule is that the food reward is to be found always under the same object, trial after trial, irrespective of the position of the object. If the animal improves over a series of such problems, it is said to have acquired a *learning set*.

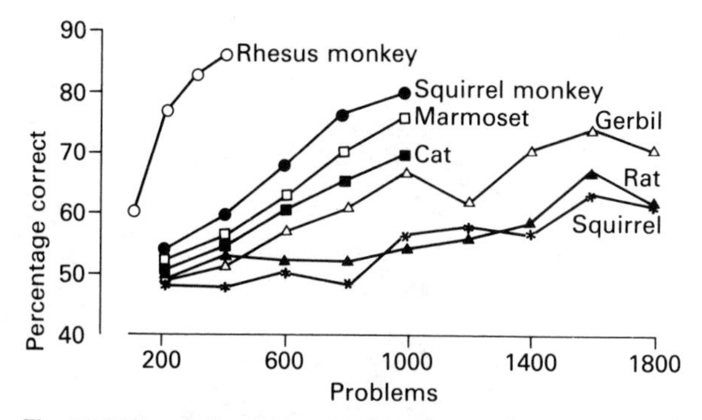

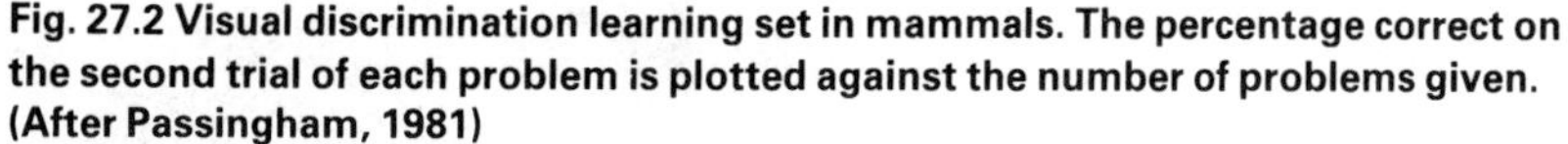
Fig. 27.2 Visual discrimination learning set in mammals. The percentage correct on the second trial of each problem is plotted against the number of problems given. (After Passingham, 1981)

As shown in Figure 27.1, various types of problems can be used to test an animal's ability to learn that a general rule is common to a set of problems and that the correct choice is governed by a single principle. Despite the criticism that the ability of different species to master learning sets depends largely upon the way the tests are set up (Hodos, 1970), there do seem to be genuine differences among species (see Fig. 27.2) once these criticisms have been taken into account (Passingham, 1981). When these animals are ranked in terms of their rate of improvement over a series of problems, their rank can be predicted from an index of brain development (Ridell, 1979; Passingham, 1982), which is an estimate of the number of nerve cells in the brain that are additional to those needed for the control of bodily functions (Jerison, 1973). Thus, it seems, tests of animal intelligence can be devised that are similar to those used in human intelligence tests and that differentiate members of different species.

The claim that such tests are genuine measures of intelligence is strengthened by the finding that performance on such tests is correlated with an index of brain size. Similar results are obtained with the other types of tests illustrated in Figure 27.1. Thus, rhesus monkeys and chimpanzees, but not cats, improve more rapidly on a series of object-discrimination problems if they have prior experience of *reversal* problems (in which the correct solution is periodically reversed) (Warren, 1974). Common principles apply to the two types of problems that these primates are able to make use of, while cats do not seem to have this ability. Similar differences between cats and monkeys occur with series of *oddity* problems (in which the animal has to select the odd symbol from a group) (Warren, 1965). Critics of these types of tests argue that the test inevitably is presented in a way that makes it easy to animals of some species but not for others (Macphail, 1982). Even if the demonstrated differences are taken at face value, they represent only one aspect of intelligence, and it is not surprising that monkeys and apes

Fig. 27.3 A vulture about to hurl a stone at an ostrich egg (*Photograph: J. M. Pearson (Biofotos)*)

perform well on tests derived from human I.Q. tests because all are primates.

27.2 Tool use in animals

The ability to use tools has long been regarded as an aspect of intelligence, and the ability to make tools once was regarded as a characteristic that set humans apart from other animals. Now that we know much more about tool using in animals, the issue is not so clear-cut, although the manufacture of tools is regarded widely as an important influence upon human evolution.

Tool use can be defined as the use of an external object as a functional extension of the body to attain an immediate goal (Lawick-Goodall, 1970). This definition excludes some cases of manipulation of objects by animals and includes others. For example, we saw that crows will take whelks up into the air and drop them on a rock to break them open. Other birds exhibit similar behavior. Thus, song thrushes (*Turdus philomelos*) hold snails in their beak and smash them against a rock anvil, while ravens (*Corvus corax*) and some vultures (*Gypaëtus barbatus*) drop bones in order to crack them open and feed on the marrow. The use of a stone anvil to break open food items, however, does not count as tool use because the anvil is not an extension of the animal's body.

The Egyptian vulture (*Neophron percnopterus*) is known to break ostrich eggs by throwing them against a stone anvil. This does not count as tool use. However, the vultures also may carry a stone into the air and drop it onto an ostrich nest or pick up a stone in the beak and throw it at an egg (Fig. 27.3). These uses of a stone do count as tool use because the stone can be regarded as an extension of the vulture's body.

An animal that scratches or rubs itself against a tree is not using the tree as a tool, but an elephant or horse that picks up a stick to scratch itself is using the stick as an extension of its body for a short-term purpose. However, a bird that carries twigs to build a nest is using the twig as material and not as an extension of its body. A nest normally is not regarded as a tool for raising the young because it achieves a long-term rather than a short-term objective.

The Galapagos woodpecker finch (*Cactospiza pallida*) probes for insects in crevices in the bark of trees by holding a cactus spine in its beak (Fig. 27.4). This clearly counts as tool use by the earlier definition, but should it count as a sign of intelligence? From the functional point of view, the use of a cactus spine to probe for food is an intelligent solution to a particular problem. A human that hit upon this solution to the problem normally would be regarded as showing signs of intelligence. Those who wish to judge intelligence purely on the basis of appropriate responses to circumstances have to allow that the finch is behaving intelligently.

If the behavior of the woodpecker finch were largely innate, then we might not want to count its probing behavior as intelligent. Observations of a juvenile woodpecker finch taken from the nest as a fledgling (Eibl-Eibesfeldt, 1967) showed that the bird manipulated twigs from an early age. If the hungry bird was presented with an insect in a hole, it would drop the twig and try to obtain the insect with its beak. Gradually, the bird began to probe for insects with a twig, and it seems likely that learning plays some part in the development of the behavior. Does this make us more inclined to regard the behavior as an indication of intelligence? We have to be very careful here. Even if it is true that learning plays a part in the development of the behavior, it seems likely that woodpecker finches are predisposed genetically to learn this particular type of manipulation in much the same way that some birds are predisposed to learn particular types of song. Conversely, the probing behavior is the functional equivalent of intelligent behavior, and we must resist the temptation to say that the bird is not really intelligent just because it is a bird and not a mammal. We must be careful to use the same criteria in judging the bird's behavior as we would in judging similar behavior in a chimpanzee.

Chimpanzees in the wild have been observed to make use of sticks, twigs, and grass stems to probe for food items. Grass stems may be used

Fig. 27.4 A Galapagos finch using a stick to probe for prey (*Photograph: Alan Root, Survival Anglia, Oxford Scientific Films Limited*)

to probe for termites, as illustrated in Figure 27.5. These are chosen with care and can be modified to make them more suitable for their purpose. For example, if the end becomes bent, the chimps may bite it off (Lawick-Goodall, 1970). Juvenile chimps may manipulate grass stems during play, but they do not use them for probing for food until they are about three years old. Even then they are often clumsy and may select tools that are inappropriate for the task. The skill required in probing for termites does not appear to be learned easily.

Wild chimpanzees have also been observed to use sticks to obtain honey from bees' nests and to dig up plants with edible roots. They may use leaves as a sponge to obtain drinking water from a hole in a tree or to clean various parts of the body. Although tool use has been studied most intensively in wild chimpanzees, it also has been observed in other wild primates. Thus, baboons may use stones to squash scorpions, and twigs to probe for insects (Kortland and Kooij, 1963).

The ability to use tools in a natural situation probably develops in the individual through a mixture of imitative and instrumental learning. In these respects tool use in primates is difficult to separate from the

Fig. 27.5 A chimpanzee using a twig to probe for termites (*Photograph: Peter Davey: by courtesy of Bruce Coleman Ltd*)

development of probing behavior in the woodpecker finch. Some biologists, while admitting that tool use is not in itself a sign of intelligence, argue that it sets the stage for truly intelligent behavior involving innovation.

That innovations do occur among chimpanzees probing for termites is suggested by comparison of the methods used by different chimpanzee populations. Chimpanzees at Gombe, in East Africa, use twigs without previously peeling off the bark. They may use each end of the twig to probe in turn. Chimps at Okorobiko, in Central Africa, usually peel the bark off a twig before using it as a probe, and they use only one end. Chimpanzees from Mount Assirik in Senegal, West Africa, do not use twigs as a probe but instead use relatively large sticks to make holes in the termite mound through which they pick out the termites by hand (McGrew et al., 1979). These differences suggest that a certain amount of variation exists within a population that may lead to innovations that suit local circumstances. The technique of fishing for termites is learned by imitation and is passed through the population by cultural tradition.

That some aspects of behavior typical of a certain population are maintained by cultural means is well documented (Bonner, 1980; Mundinger, 1980). In some few cases, inventions have been observed and their spread through the population recorded. A celebrated case concerns the Japanese macaques (*Macaca fuscata*) of Koshima Island (Kawamura, 1963). To bring the monkeys into the open where they could be observed more readily, scientists supplemented their diet by scattering sweet potatoes on the beach. A 16-month-old female, called Imo, was observed to wash the sand off her potatoes in a stream. She continued this practice on a regular basis and soon was imitated by other monkeys, particularly those of her own age. Within 10 years, the habit had been acquired by the majority of the population, with the exception of adults more than 12 years old and infants of less than a year. Two years later Imo invented another food-cleaning procedure. Scientists had been scattering grain on the beach, and the monkeys picked them up one at a time. Imo gathered handfuls of mixed sand and grain and threw them into the sea. The sand sank and the grain was scooped easily from the water surface. The new procedure spread through the population in a manner similar to that of potato washing. The new behavior was adopted first by monkeys of Imo's own age. Mothers learned from juveniles and adult males were the last to catch on.

Was Imo an especially intelligent monkey? On the one hand, we might argue that she learned in a manner similar to the so-called insight learning of Köhler's apes, discussed in Chapter 19. Imo made her discoveries by chance and learned to exploit them. She did not necessarily have any special insight into the situation. On the other hand, many human inventions came about in a similar way. If the invention leads to a genuine improvement in the monkeys' circumstances, it counts as a form of adaptive behavior arrived at through the efforts of a single individual. As judged by the results, Imo appears to be highly intelligent, but if we require that intelligence involves mechanisms like reasoning, then we need to know more about Imo's thought processes before we make a judgment.

27.3 Cultural aspects of behavior

Evolution occurs as a result of natural selection, and inheritance of acquired characteristics is not normally possible. However much an individual animal adapts to its environment, whether by learning or by physiological adaptation, the acquired adaptations cannot be passed to the offspring by genetic means. So much is widely accepted amongst biologists. However, as we have seen, information can be passed from parent to offspring by imitation and by imprinting. In general, the passage of information from one generation to the next by non-genetic means is known as *cultural exchange*.

We have seen (Chapter 20) that sensitive periods of learning occur in the early life of many animals. During such periods they often learn from their parents. For example, the white-crowned sparrow will remember its parents' song provided it hears it during the sensitive period between 10 and 50 days old. Individuals prevented from hearing the song of their own species during this period never produce a proper white-crowned sparrow song during later life. Whereas the juvenile white-crowned sparrow will not learn songs that are much different from the song of its own species, other birds, such as the bullfinch, will learn the song of a completely different species. Thus, a juvenile bullfinch fostered by a canary will adopt canary song. The tendency to copy the song of the parents leads to regional variations even among species that will learn only songs similar to that of their own species. In the region of San Francisco, for example, populations of white-crowned sparrows separated by only a few miles have distinct dialects (Marler and Tamura, 1964).

Animal dialects represent an elementary form of tradition. Juvenile white-crowned sparrows are inevitably exposed to the song dialect characteristic of the locality in which they are born, because their sensitive period of learning occurs before they are mobile. Other forms of traditional behavior include the migration routes of some mammals and birds: Geese, ducks, and swans migrate in flocks composed of mixed juveniles and adults. The juveniles learn the route characteristic of their population, stopping at traditional rest places, and breeding and overwintering localities (Schmidt-Koenig, 1979). There are many other examples. Reindeer also show fidelity to traditional migration routes and calving grounds. Migratory salmon hatch in freshwater streams and migrate to the sea. The juveniles become imprinted on the odor of their native stream and return to the same stream as adults to spawn in the traditional places. Some game trails of deer and other mammals are known to have been used for centuries. In parts of Britain and Germany, where roads have been built across the traditional migratory routes of toads, conservationists stand guard during the breeding season to prevent the toads from being run over by motorists.

Simple forms of traditional behavior do not require any special learning abilities, or any special teaching. They arise as an inevitable consequence of the circumstances in which the young are raised and of the tendency of juvenile animals to become imprinted upon their habitat, their parents, and their peers. In some animals, however, more complex forms of learning are involved. We have seen how tool use in primates can spread through a population, probably as a result of imitative learning.

Imitation is not necessarily a sign of high intelligence, however. Animals may copy each other as a result of a simple social facilitation. Many animals eat more when fed in groups than when fed alone. This has been demonstrated experimentally in chickens, puppies, fish, and opossums. If a chicken is allowed to eat until completely satiated, and is then introduced to others that are still feeding, the satiated bird will resume eating. Domestic chicks also have a tendency to peck when others peck, even if there is no food available. Chicks tend to peck at the same type of foodlike particle as the mother hen. The tendency to concentrate on the type of food being eaten by others has also been demonstrated in sparrows and chaffinches (Davis, 1973).

Avoidance of noxious food can also be socially facilitated. Attempts to kill large flocks of the common crow in the United States, by providing poisoned bait, were not successful because the majority avoided the bait after a few individuals had been poisoned. Similarly, the reaction of just one rat to novel food may be sufficient to determine the reactions of other rats in the group. If the one rat eats the food, then others will join in, but if the rat sniffs the food and rejects it, the others will reject it. Sometimes, the pioneer rat urinates on the bait, thus warning others to reject it (Rozin, 1976b).

Social attraction to, and avoidance of, food often result from a tendency to investigate places where other members of the species have been observed, rather than direct copying of the behavior of others. This is probably the case with the tradition of stealing milk among blue tits in England. For many years milk has been delivered to houses in England and left on the doorstep early in the morning. Blue tits, and sometimes great tits, peck through the foil bottle tops and help themselves to the rich cream floating on top (Fig. 27.6). Hinde and Fisher (1951) reported that this practice first appeared in particular localities and gradually spread, suggesting that the birds were learning from each other.

Another interesting example of traditional feeding behavior is seen in Mike Norton-Griffiths' study of young oystercatchers (Norton-Griffiths, 1967; 1969), a shorebird that feeds on mussels, a bivalve mollusk with a hard shell. The adults use one of two methods to obtain the flesh. They may hammer very hard with their beaks at the weakest point of the shell, or they may insert their bill into the open siphon when the mussel is underwater and cut the adductor muscle, which holds the two halves of the shell together. Adult birds are either hammerers or stabbers, and the two techniques are not used by the same bird. Moreover, both members of a mated pair use the same method. By switching the eggs of hammerers and stabbers, Norton-Griffiths showed that the young learn the method of opening mussels from their parents. The appropriate movements for both hammering and stabbing are seen in all birds at a young age, but they do not seem to practice these movements. It is possible that both methods are to a certain extent innate. How the juvenile oystercatchers acquire the necessary skills is not fully understood. They stay with their parents for 18–26 weeks, whereas the young of oystercatchers that feed mainly on other prey stay with their parents for only 6–7 weeks. Those born into the mussel-feeding tradition have to be fed by their parents while they serve their apprenticeship. Perhaps

Fig. 27.6 A blue tit opening a milk bottle to obtain food (*Photograph: John Markham: by courtesy of Bruce Coleman Ltd*)

they learn by observation, or perhaps the necessary coordination requires a long time to develop.

From these examples it is obvious that culture and tradition do not require great intelligence on the part of individuals. Although highly developed culture and highly developed intelligence are both primarily human traits, they are not necessarily causally related. There are many cultural practices found in primitive human societies that are biologically adaptive but not obviously rational. For example, most of the traditional methods of cooking maize (corn) by the indigenous people of the New World involve some kind of alkali treatment. Often the maize is boiled for about 40 minutes in water containing wood ash, lye, or dissolved lime, and then eaten directly, or converted into dough, tortillas, etc. The alkali treatment is practiced for purely traditional reasons, although it is said by some to make the food more palatable. However, the alkali treatment has important nutritional consequences. The local maize has low levels of available lysine, a nutritionally essential amino acid. Most of the lysine in maize occurs as part of an indigestible protein called glutelin. The alkali treatment breaks up the glutelin, releases the lysine, and greatly improves the nutritional quality of the food. Some kind of alkali treatment is found in all indigenous cultures that rely on maize, and it is probable that natural selection, acting through malnutrition, has eliminated those people who did not follow the traditional practice.

Thus, we see that although culture can short-circuit biological heredity, and lead to very rapid evolution, the products of cultural evolution are still subject to natural selection. Although some learning ability is essential for cultural change, no great intelligence is necessary, and those people who follow traditional practices are often unable to give a rational explanation of their behavior.

Points to remember

- It is difficult to compare intelligence among different species because many have special skills and particular disabilities. Should intelligence be judged by the type of problem solved or by the way it is solved?

- The ability to use tools is widespread in the animal kingdom, and some species can even make their own tools. Should this ability be regarded as an indication of intelligence?

- Culture and tradition, involving the transfer of information from one generation to the next by non-genetic means, occurs in many species. It may result from imprinting or from imitation.

Further reading

Bonner, J. T. (1980) *The Evolution of Culture in Animals*, Princeton University Press, Princeton, New Jersey.

Jerison, H. J. (1973) *Evolution of the Brain and Intelligence*, Academic Press, New York.
Macphail, E. M. (1982) *Brain and Intelligence in Vertebrates*, Clarendon Press, Oxford.

28 Animal Awareness and Emotion

Emotion is an important part of human experience. At the common-sense level we can all agree what we mean by emotion, and we usually agree on the attributes of the different emotions. At the scientific level, however, emotion poses considerable problems.

Emotion has subjective, physiological, and behavioral manifestations that are difficult to reconcile with each other. At the subjective level, emotion is an essentially private experience. There is no way we can know what the emotional experiences of another person are like. We tend to assume they are the same as our own experiences, but we have no logical way of verifying this. When it comes to the subjective emotional experience of animals, we are in even more difficulty. We tend to assume that animals similar to ourselves, like primates, have similar emotional experiences and that dissimilar animals, like insects, probably have rather different experiences, if any. However, this is a common-sense and not a scientific view. In scientific terms, we cannot assume that animals have particular subjective feelings any more than we are entitled logically to make such assumptions about other people.

In physiological terms, emotional states in humans typically are accompanied by autonomic changes, but these are not a reliable guide to particular emotional states. In animals especially, while physiological indexes such as increased heart rate or changes in hormonal balance do provide evidence of emotional arousal—e.g., fear, aggression, or sex—the emotions are poorly differentiated. In other words, most animals react physiologically the same way, whether their emotional response is one of fear, of aggression, or of a sexual nature. Thus, although some insight can be gained from physiological investigations, interpretation of the physiological responses is difficult.

Charles Darwin (1872) stressed the communicative aspect of emotion. As we saw in Chapter 22, he postulated that facial expressions and other behavioral signs of emotion had evolved from protective responses and other utilitarian aspects of behavior. Although Darwin's evolutionary thesis, as developed and expanded by the early ethologists, is accepted widely, his ideas about emotion were naive. Darwin and his disciple George Romanes did not hesitate in labeling animal emotions in human terms. Thus, the expression of a dog that had done wrong was taken to indicate "shame" (Darwin, 1872); fish experienced "jealousy" and parrots had a sense of "pride" in their utterances (Romanes, 1882). This anthropomorphic approach led to a revolt among psychologists like Conway Lloyd Morgan (1894), who advocated an approach devoid of speculation about the private thoughts and feelings of animals.

In this chapter, we explore the inner life of animals from the subjective, physiological, and behavioral viewpoints and discuss the implications of this research for the welfare of animals and their treatment by human beings.

28.1 Self-awareness in animals

The behaviorist attitude that the private mental experiences of animals cannot be a subject of scientific investigations was dominant for the first three-quarters of this century. During this period scientists like Edward Tolman (1932) dissented, but they did not have much influence (see Griffin, 1976, for a review). The behaviorist position seems unassailable on logical grounds, but it can be circumvented in various ways. One argument is that, although we cannot prove that animals have subjective experiences, it may be true nevertheless. What difference would it make if it were true? Another approach is to argue that it is unlikely, on evolutionary grounds, that there is a marked discontinuity between humans and other animals in this respect.

Donald Griffin (1976), who was one of the first to mount a concerted attack on the behaviorist position, uses both arguments. In his opinion, the study of animal communication is most likely to provide evidence that "they have mental experiences and communicate with conscious intent." However, the study of animal language in more recent years has not fulfilled its early promise. The interpretation of the behavior of chimpanzees trained in some aspect of human language remains controversial, and there is doubt that it will ever reveal much about the private experiences of these animals (Terrace, 1979; Ristau and Robbins, 1982). There have been various attempts to investigate self-awareness in animals by other means, and it is to these that we now turn.

Are animals aware of themselves in the sense that they know what posture they are adopting and what action they are taking? Sensory information from the joints and muscles is available to the brain, so it seems that animals ought to be aware of their behavior. In an experiment designed to test this, rats were trained to press one of four levers, depending upon which of four activities the animal was engaged in when a buzzer sounded (Beninger et al., 1974). If a signal occurred when the rat was grooming, for example, then it would require pressing the grooming lever to obtain a food reward. The rats learned to press a different lever depending on whether they were grooming, walking, rearing up, or remaining still when the buzzer sounded. The results of similar studies (Morgan and Nicholas, 1979) show that rats can base

Fig. 28.1 Viki imitating a photograph of herself (Drawn from a photograph)

operant behavior on signals emanating from their behavior as well as from the external environment. In a sense, the rats must be aware of their actions, but this does not necessarily mean they are conscious of them; they may be aware of their actions in the same way that they are aware of external stimuli.

Many animals respond to a mirror as if they were seeing another member of their species. However, there is some evidence that chimpanzees and orangutans can recognize themselves in a mirror. Young wild-born chimpanzees will use mirrors to groom parts of the body they cannot otherwise see. Gallup (1977; 1979) painted small patches of red dye on the eyebrow and opposite ear of chimps while they were under a light anesthetic. When they awoke from the anesthetic, he confirmed that they did not touch those parts of their body more than normal. Then he provided the chimps with a mirror. The chimps examined their reflections and repeatedly touched their own dyed eyebrows and ears.

Does the ability to respond to parts of one's body seen in a mirror indicate self-awareness? The question is not far removed from a wider question. Does the ability to imitate the actions of others indicate self-awareness? Chimpanzees are extraordinarily proficient at imitating each other and human beings. Although true imitation has to be separated carefully from other forms of social learning (Davis, 1973), there is little doubt about its occurrence among the primates. For example, the chimpanzee Viki was fostered by the Hayes family. They demonstrated a series of 70 acts that Viki was encouraged to copy. Many of them she had never seen before, and she copied 10 of these the first time she saw them. Viki learned to produce 55 acts in response to the appropriate demonstration (Fig. 28.1). She also learned fairly complicated household tasks such as washing dishes and dusting furniture (Hayes and Hayes, 1952). Many of these actions were imitated spontaneously, without prompting. However, chimpanzees do not match up to the imitative abilities of human children. Imitation has been claimed for in human infants as young as 12 to 21 days.

Although the ability to imitate sometimes is taken as a sign of intelligence, its occurrence in very young babies and in a wide variety of non-mammalian species must cast some doubt on this view. The study of bird song shows that some form of imitation occurs in the song learning of many species and that some species are especially proficient at imitating sounds. Parrots and mynah birds can reproduce human sounds with extraordinary fidelity (Nottebohm, 1976).

To be able to imitate, an animal must perceive the external auditory or visual example and match it with a set of motor instructions of its own. For example, a baby that imitates an adult sticking out its tongue must somehow associate the sight of the tongue with its own motor instructions for sticking out its tongue. The baby does not have to be aware of the fact that it has a tongue, it merely has to connect a particular perception with a particular set of motor commands. How this is done is a mystery, but the question of whether imitation necessarily involves self-awareness is debatable.

Part of the problem is that we need to be clear about what we mean by self-awareness. As Griffin (1982) notes, many philosophers distinguish between awareness and consciousness. Awareness is a form of perception, while consciousness involves a special kind of self-awareness, which is not simple awareness of parts of one's body or of processes occurring within the brain. Consciousness, in this view, involves a propositional awareness that it is *I* who am feeling or thinking, *I* am the animal aware of the circumstances. We have reviewed some of the evidence that animals are aware in the perceptual sense. However, the ability to report on one's actions, to imitate the actions of others or to recognize one's image in a mirror need not necessarily require consciousness as defined here.

A dissociation between conscious and unconscious perception can occur in people with brain damage. Some people with damage to certain areas of the brain used in visual processing report that they are partially blind. They cannot name objects presented to them in certain parts of the visual field. They claim they cannot see such objects, yet when asked to point, they often can do so accurately (Weiskrantz, 1980). One patient could guess accurately whether he was being shown a horizontal or diagonal line, yet he was unaware of seeing anything (Weiskrantz et al., 1974). This phenomenon, called *blind sight*, is due to damage to parts of the brain responsible for visual awareness, which leaves intact other parts of the brain involved in visual processing. These other parts enable patients to make correct judgments even though they are not aware of what they see.

28.2 Physiological aspects of emotion

Emotional states are accompanied by an increase in the activity of the *autonomic nervous system*, the division of the vertebrate nervous system that serves the internal organs, such as the blood vessels, heart, intestines, lungs, and certain glands. The autonomic nervous system is controlled by the brain and provides two types of innervations that have antagonistic effects upon internal organs. The *sympathetic* nerve pathways become active under conditions of stress or exertion and have an emergency function. They have the effect of increasing the blood supply to the muscles, brain, heart, and lungs; of increasing the heart rate; and of reducing the blood supply to the intestine and peripheral parts of the body. The *parasympathetic* nervous system is anatomically distinct from the sympathetic, and it serves a recuperative function, restoring the blood supply to normal and counteracting other effects of sympathetic activity. Emotional arousal is accompanied by sympathetic activity. In humans this includes increased heart rate, sweating, and changes in peripheral blood circulation so that the face becomes pale or flushed.

The psychologist William James proposed that the subjective feelings of emotion are generated by the sensory receptors involved in the emotional reaction. For example, a frightening stimulus would elicit certain behavioral and physiological changes, and these would give rise to the subjective experience of fear. As James (1890) put it, when we meet a bear we run. We do not run because we are afraid; we are afraid because we run. Carl Lange, a contemporary of James, proposed a similar explanation of emotional experience, and this view is now known as the *James-Lange theory*.

Walter Cannon (1927) pointed out that sympathetic arousal was much the same whatever emotion was being experienced. He showed that cats exhibit emotional behavior such as hissing and spitting even after all autonomic sensory feedback has been eliminated. Cannon concluded that autonomic activity is not necessary for emotional experience independent of its effects on autonomic activity. When human subjects are injected with the hormone adrenaline, a wide range of sympathetic activity occurs, including sweating, increased heartbeat rate, etc. However, the subjects do not report emotional experiences. Some report the physical symptoms and others say that they feel a cold emotion, as if they were angry or afraid (Landis and Hunt, 1932).

Autonomic arousal may provide a basis for emotional experience, but it provides no differentiation among the various emotional situations. However, the situation, as perceived by the subject, may lead to an interpretation of the autonomic arousal in terms of the appropriate emotional experience. This is the hypothesis proposed by Stanley Schachter and J. Singer (1962). They injected human subjects with adrenaline, pretending it was a vitamin supplement. Some subjects were told that the injection would cause an increase in heart rate, flushing, etc. (i.e., the real effects of adrenaline), while others were told it might have side effects such as itching or numbness (i.e., they were misinformed). All subjects then were asked to sit in a waitingroom prior to a vision test. In this waitingroom they were observed by a hidden experimenter and were accompanied by a confederate of the experimenter. In one test the confederate behaved in a sullen and angry manner, while in another test he was frivolous and playful. When the subjects were asked to rate their emotional experience following their stay in the waitingroom, marked differences were found between the two test conditions: Those that had been misinformed about the injection felt happier than the correctly informed subjects when in the presence of the playful confederate. In the presence of the sullen confederate, they felt angrier and more resentful than the controls. Although this experiment has been the subject of some criticism (e.g., Plutchik and Ax, 1970), it does suggest that both our perceived physiological state and our assessment of the external situation contribute to our emotional feelings.

In evaluating the subjective experiences of animals from a physiological viewpoint, it is possible to make direct comparisons with humans, but none of these is entirely satisfactory. For example, do animals experience pain? When subjected to stimuli painful to humans, many animals have similar physiological responses. Pain-killing drugs, which alleviate subjective, physiological and behavioral reactions to pain in humans, produce corresponding effects on standard behavioral indexes of pain in animals. However, the interpretation of such findings rests largely on the behavioral index used. Thus, a simple withdrawal reflex probably would not be a good indication of pain (as subjectively experienced) because such reflexes are widespread in the animal kingdom and occur in very primitive animals. Even the criterion of crying out in pain is difficult to evaluate. While a dog or monkey would scream with pain if severely injured, an antelope torn to pieces by a predator remains relatively silent. Screams may serve to elicit help from other members of the group, but they may also endanger those attracted by the calls.

Comparative aspects of brain anatomy sometimes are used to assess the likelihood that animals have subjective experiences similar to those of humans. Usually, a fairly simple criterion such as the size of the brain or the cortex is used as an index in an assumed progression from simple to more intelligent animals. However, thorough anatomical studies reveal that some brain structures show marked differences when different animal groups are compared. Hodos (1982) suggests that some of these specialized areas may contribute to a type of mental activity normally associated with the cortex in humans. Moreover, in assessing the intelligence of animals in relation to their brain structure, it is often difficult to provide a fair test devoid of anthropomorphic assumptions (Macphail, 1982).

28.3 Consciousness and suffering

The question of consciousness in animals is fraught with difficulties. The spectrum of scientific opinion is very wide. There are those who believe that consciousness does not occur in animals, and there are those who maintain that it occurs in most animals. There are those who believe that consciousness is not a suitable subject for scientific study, and there are those who think it is a neglected topic. The situation is confounded by the difficulty of arriving at an acceptable definition of consciousness.

Donald Griffin (1976) defines consciousness as the presence of mental images and their use by the animal to regulate its behavior. This is not far removed from the definition given in the Oxford English Dictionary. To be conscious is to be "aware of what one is doing or intending to do, having a purpose and intention in one's actions" (see Griffin, 1982). According to Griffin (1976), "an intention involves mental images of future events in which the intender pictures himself as a participant and makes a choice as to which image he will try to bring to reality." Although Griffin and others (e.g., Thorpe, 1974) see intention and consciousness as part and parcel of the same phenomenon, this is not a universal view. As we saw in Chapter 26, it is possible to imagine intentional behavior that does not involve consciousness.

As we saw earlier, consciousness is thought by many researchers to involve more than simple perceptual awareness. For example, Humphrey (1978) sees consciousness as self-knowledge used to predict the behavior of other individuals, and Hubbard (1975) suggests that it involves knowledge of oneself as distinct from other selves. Such knowledge could be used as a basis for communication, but this does not mean that consciousness necessarily involves language. We can agree with Passingham (1982) that "spoken language has revolutionised thought. The use of language for thought vastly amplifies the level of intelligence that can be achieved. Animals think, but people can think in a totally new way, with a completely different code." Undoubtedly, the advent of language has altered the way we think about ourselves. It is hard for us to imagine consciousness without language. However, this does not entitle us to assume that animals that have no language, or only primitive language, do not have consciousness. We saw previously that some sort of self-awareness can be demonstrated in animals with no language equivalent to ours. Therefore, we should not equate language and consciousness.

Do animals have to be conscious to suffer? At the common-sense level, we are inclined to suppose that they do. When we are unconscious we do not suffer pain or mental anguish because parts of our brain are deactivated. However, we do not know whether these parts are involved only in consciousness or in consciousness plus other aspects of brain activity. Thus, we cannot use the fact that we do not experience pain when unconscious to infer that consciousness and suffering go hand in hand. It may be that whatever makes us unconscious also stops pain but that the two are not causally connected.

We have no conception of what the conscious experiences of animals might involve, if they exist. Therefore, we can draw no conclusions about the relationship between consciousness and suffering in animals. In our ignorance, it would be wrong to assume that suffering in animals is confined to those that are intelligent, that use language, or that show evidence of conscious experience.

There are both dangers and virtues in using ourselves as models of what animals might feel (Dawkins, 1980). The scientific basis for the analogy between the mental experiences of ourselves and other animals is weak. It would not be good scientific practice to come to conclusions about the mental experiences of animals on the basis of such evidence. In contrast, we do come to conclusions about the mental experiences of other people on the basis of analogy with ourselves. When we see another person suffering or hear their cries of pain, we do not ignore them because we cannot prove their mental experiences are the same as ours. We give them the benefit of the doubt and go to their aid. Perhaps we should also give members of other species the benefit of the doubt.

28.4 The evolutionary perspective

It is natural for us to offer help to other people in distress. Although there may be some differences among cultures, our sympathy probably has an innate basis. Even when we are being cruel to another person, we recognize that we are causing them to suffer. We automatically assume their mental experiences will be similar to those we would experience in the same situation. It would appear that we are designed by natural selection to assume that other people have similar mental experiences to ourselves. Why?

Some researchers have argued (e.g., Humphrey, 1979; Crook, 1980) that the evolution of close-knit societies made recognition of others advantageous. By "recognition of others" we mean some awareness of others as beings with feelings similar to our own. This recognition might facilitate an ability to respond to the individual attributes of others and perhaps to the development of a language capable of expressing this sympathy. Griffin (1981) suggests that this may have occurred in other species and "that animals that are consciously aware of their sociobiological goals can achieve them more effectively than would otherwise be the case." Certainly, the social and political life of some primates seems complicated enough to warrant such a development. The problem is that we can always invent a plausible adaptive advantage for an observed or supposed trait, and such speculation does not lead very far.

We may make more progress by concentrating on the possible evolutionary origins of more widespread and simple attributes. Why, for example, might animals evolve to feel pain? If we imagine a population of animals equipped with simple avoidance behavior devoid of pain or suffering, we can postulate a mutant that had some primitive awareness of pain. Could such a mutant successfully invade the population? Obviously, for the new trait to have a selective advantage, it would have to make some difference to the animal's behavior. If the mental experience were entirely private and ineffectual, it is hard to see how natural selection could act upon it. Perhaps the trait would improve the animal's ability to communicate with its kin, or perhaps it somehow could enable it to learn more effectively about the dangers of the environment. We do not know, but we can at least improve the cutting edge of our speculations by conforming to the recognized procedures of evolutionary biology. Could the proposed new trait successfully invade the population? Would there be an evolutionarily stable strategy (see Chapter 7)? And so on.

In evaluating animal suffering from an evolutionary viewpoint, we should give the animal the benefit of the doubt and assume that it suffers in situations that it normally would take steps to avoid. If animals do suffer, they presumably do so in response to circumstances that are functionally disadvantageous. In general, we can expect animals to be designed to choose situations that lead to increased fitness and to avoid those likely to result in decreased fitness. Perhaps, therefore, we can use the choice behavior of animals as a guide to their welfare.

In 1880 Herbert Spencer suggested that the subjective experience of pleasure and pain evolved to help animals choose suitable habitats and living conditions. While it is well known that wild animals have marked habitat preferences (Lack, 1937; Hilden, 1965; Partridge, 1978), this does not mean that subjective feelings are necessary for them to exercise a choice. However, we can expect that animal habitat preferences are related to their welfare, in the evolutionary sense. In the case of domestic or laboratory breeds of animals, the connection between preference and fitness is not so clear-cut. Nevertheless, the choices made by such animals probably offer a reasonable rule of thumb to their welfare (Dawkins, 1980).

An illustration of the value of this approach comes from attempts to specify the conditions under which battery hens should be kept. In the United Kingdom, the government set up a committee to inquire into the welfare of animals kept under intensive livestock husbandry systems. In its report, the committee recommended that the floors of battery cages should not be made of fine-gauge hexagonal wire because they thought

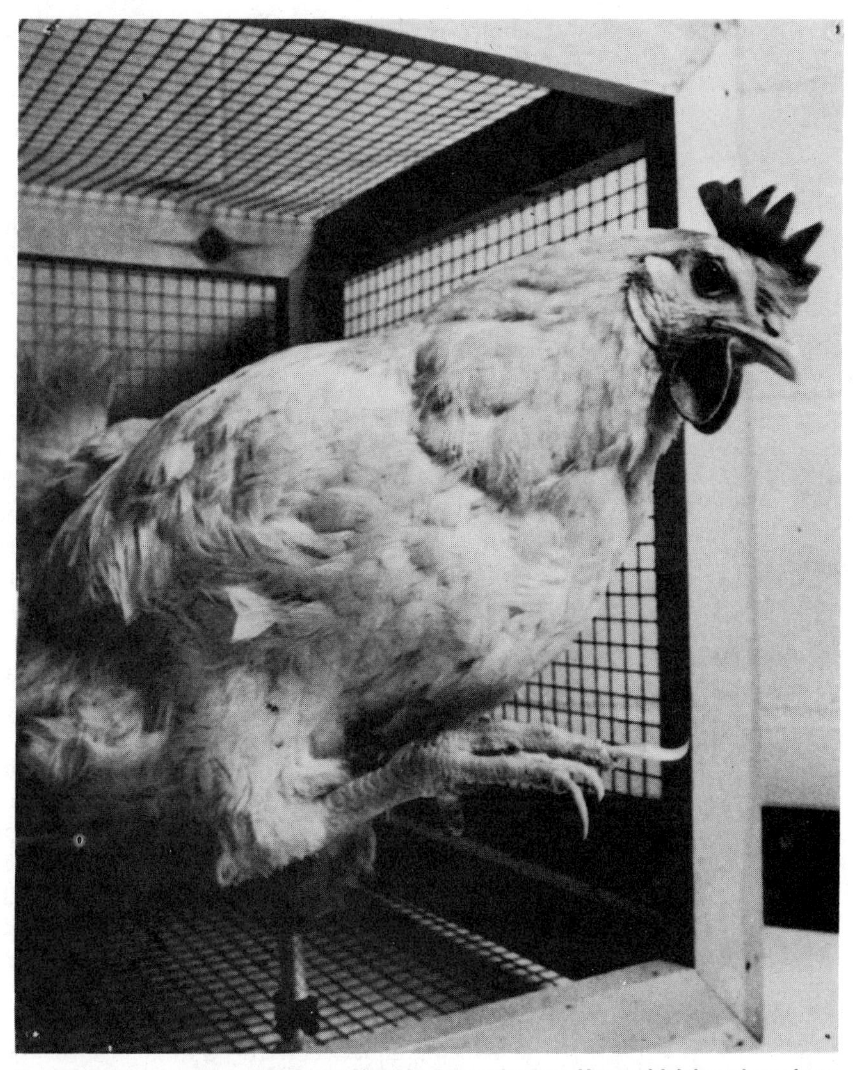

Fig. 28.2 A hen from one of Marian Dawkins' experiments. Hens which have been kept in a battery for a number of weeks may initially select the familiar battery cage in preference tests between battery and outdoor cage (*Photograph: Tony Allen*)

that this was uncomfortable for the hens to stand on (Brambell, 1965). However, when battery hens were given preference tests between this type of floor and a floor made of the heavy rectangular metal mesh recommended by the government committee, the researchers found that the birds preferred the hexagonal wire floors (Hughes and Black, 1973). Photographs taken from below the floors showed that their feet gained more support from the fine mesh.

Extensive investigations of habitat preference in domestic fowl have been carried out by Marian Dawkins. In one study she offered hens a choice between a battery cage and an outside run in the garden. When the birds arrived at a T junction in a choice corridor, they could see the outside run on one side and the battery house environment on the other. The hens were free to enter both environments, and before being tested again, they were confined for five minutes in the one they chose. Hens that just previously had been living in an outside run chose the outside run from the very first choice test. Hens that had been living in battery cages initially chose the battery cage environment (Fig. 28.2). However, after repeated choice tests they preferred the outside run. Thus, even a short recent experience of the outside run was sufficient to alter their preferences (Dawkins, 1976; 1977).

Choice tests can be criticized on the grounds that the preferences of animals reflect a number of different influences. Thus, habitat preferences may be influenced by genetic factors, imprinting, familiarity, or recent experience. Genetic differences can be investigated by testing different genetic strains of domestic animals (Dawkins, 1980). An animal's early experience may have long-term effects upon its habitat preferences, but these can be controlled in investigations on domestic animals.

In conducting choice tests, care must be taken to ensure that the animals are familiar with the alternatives. Many animals avoid novel or unfamiliar situations, and it is obviously possible that they might prefer a particular alternative once they become familiar with it. Farmers sometimes say their animals must like the conditions in which they are kept because they return there after being freed. This is similar to the argument that some people prefer to be in prison because they find it difficult to adjust to life outside after a prolonged sentence. The problem has to do with the interaction between familiarity and recent experience. The initial choice of an animal may be its most recently familiar surroundings, but this preference can wear off quickly, as with the battery hens just discussed.

Preference tests have been criticized on the grounds that they do not discriminate between the animal's short-term and long-term preferences. Ian Duncan (1977; 1978) showed that hens will chose to enter a trap nest to lay their eggs, even though this results in confinement for several hours without food. The preference for the nest can be so strong that the hens will enter it day after day. However, such phenomena must be judged in relation to the alternatives available to the animal. How would an animal budget its time if it were able to choose among different daily routines? In the wild we might expect animals to arrange their daily routines in an optimal manner (see Chapter 16), but adjustment between one daily routine and another might be slow. Animals have both short-term and long-term mechanisms of adjustment to environmental changes, and these affect behavior in complex ways. Dawkins (1980; 1982) has suggested that measures of behavioral resilience (see Chapter 24.6) might go some way to resolving these difficulties. It can be argued that preferences exhibited under pressure are more likely to be reliable than those that result from idle whim. Domestic animals are not normally subject to time and energy constraints in the way that wild animals are, so perhaps we should provide this kind of pressure when conducting preference tests. Economists know that the apparent preferences of consumers may change radically with changes in prices. The equivalent demand function in animals (see Chapter 24.6) would give us a more coherent picture of animal preferences.

If the evidence were available, then we might be able to argue on evolutionary grounds that departures from the natural way of life that the animal was willing to pay to restore would be a good index of welfare. In what other way could suffering have evolved?

Points to remember

- Animals may be aware of their own body-state and behavior, but this does not necessarily involve consciousness.

- Physiological manifestations of emotion occur in many animals, but this does not mean that animals experience emotion in the way that humans do.

- In judging whether animals are suffering, we have to make assumptions about their mental state which are not testable scientifically.

- Evolutionary consideration lead to the suggestion that animals usually know what is best for them and make choices on this basis.

Further reading

Dawkins, M. S. (1980) *Animal Suffering*, Chapman and Hall, London.

Gallup, G. G., Jr. (1979) 'Self-awareness in primates', *American Scientist*, **67**, 417–21.

Griffin, D. R. (1976, 1981) *The Question of Animal Awareness*, The Rockefeller University Press, New York.

Indexes

References and Author Index

The numbers in square brackets refer to the page or pages in the text where mention of a given work or person is made.

Adler, H. E. (1971) Orientation: sensory basis. *Ann. N.Y. Acad. Sci.* **188,** 1–408. [243]

Adler, N. T. (1969) The effect of male's copulatory behaviour on successful pregnancy of the female rat. *J. comp. physiol. Psychol.* **69,** 613. [148]

Adolph, E. F. (1972) Some general concepts of physiological adaptations. In *Physiological Adaptations,* eds M. K. Yousef, S. M. Horvath and R. W. Bullard, 1–7, Academic Press, New York. [288]

Allison, J. (1979) Demand economics and experimental psychology. *Behavioral Science* **24,** 403–415. [454]

Allison, J. (1983) Behavioral substitutes and complements. In *Animal Cognition and Behavior,* ed. R. L. Malgren, North-Holland, Amsterdam. [454]

Amoore, J. E. (1963) Stereochemical theory of olfaction. *Nature, London* **198,** 271–272. [189, 190]

Andrew, R. J. (1956a) Some remarks on behaviour in conflict situations, with special reference to *Emberiza* spp. *Brit. J. anim. Behav.* **4,** 41–45. [401]

Andrew, R. J. (1956b) Normal and irrelevant toilet behaviour in *Emberiza* spp. *Brit. J. anim. Behav.* **4,** 85–91. [384, 386, 388]

Andrew, R. J. (1956c) Intention movements of flight in certain passerines, and their use in systematics. *Behaviour* **10,** 179–204. [399]

Andrews, R. J. (1963) The origins and evolution of the cells and facial expressions of the primates. *Behaviour* **20,** 1–109. [398]

Anthoney, T. R. (1968) The ontogeny of greeting, grooming and sexual motor patterns in captive baboons. *Behaviour* **31,** 358–372. [31]

Argyle, M. (1972) Non-verbal communication in human social interaction. In *Non-Verbal Communication,* ed. R. A. Hinde, Cambridge University Press, Cambridge. [487]

Armstrong, E. A. (1947) *Courtship and Display amongst Birds.* Lindsay Drummond, London. [382]

Aschoff, J. (1960) Exogenous and endogenous components in circadian rhythms. *Cold Spring Harbor Symp. quant. Biol.* **25,** 11–28. [291]

Aschoff, J. (1965) Circadian rhythms in man. *Science* **148,** 1427–1432. [292]

Assem, J. van den (1967) Territory in the three-spined stickleback, *Gasterosteus aculeatus. Behaviour Suppl. 16,* 1–164. [93]

Avery, O. T., Macleod, C. M. and McCarty, M. (1944) Studies on the chemical nature of the substance inducing transformation of pneumococcal types. *J. exp. Med.* **79,** 137–158. [23]

Baerends, G. P. (1941) Fortpflanzungsverhalten und Orientierung der Grabwespe *Ammophilia campestris* Jur. *Tijdschr. Ent.* **84,** 68–275. [473, 474]

Baerends, G. P. (1950) Specialisations in organs and movements with a releasing function. *Symp. soc. exp. Biol. IV,* 337–360. [365, 418]

Baerends, G. P. (1975) An evaluation of the conflict hypothesis as an explanatory principle for the evolution of displays. In *Function and Evolution in Behaviour,* eds G. P. Baerends, C. Beer and A. Manning, Clarendon Press, Oxford. [401]

Baerends, G. P., Brouwer, R. and Waterbolk, H. Tj. (1955) Ethological studies on *Lebistes reticulatus* (Peters), I: An analysis of the male courtship pattern. *Behaviour* **8,** 249–334. [280]

Baerends, G. P. and Drent, R. H. (1970) The herring gull and its egg, Part I. *Behaviour Suppl. XVIII,* 265–310. [212]

Baerends, G. P. and Drent, R. H. (1982) The herring gull and its egg, Part II. *Behaviour* **82,** 1–416. [5]

Baerends, G. P. and Kruijt, J. P. (1973) Stimulus selection. In *Constraints on learning: Limitations and Predispositions,* eds R. A. Hinde and J. Stevenson-Hinde, 23–50, Academic Press, London. [216–219]

Baker, A. G. and Mackintosh, N. J. (1977) Excitatory and inhibitory conditioning following uncorrelated presentations of the CS and US. *Animal Learning and Behaviour* **5,** 315–319. [351]

Baker, P. T. and Weiner, J. (1966) *Biology of Human Adaptability*. Oxford University Press, New York. [123]

Baker, R. R. (1981) Man and other vertebrates: a common perspective to migration and navigation. In *Animal Migration*, ed. D. J. Aidley, Cambridge University Press, Cambridge. [226, 251]

Baldwin, E. (1948) *An Introduction to Comparative Biochemistry* (3rd edition). Cambridge University Press, Cambridge. [284]

Barnard, C. J. (1980) Flock feeding and time budgets in the house sparrow *Passer domesticus* L. *Anim. Behav.* **28,** 295–309. [431, 432, 476]

Barnard, C. J. and Sibly, R. M. (1981) Producers and scroungers: a general model and its application to captive flocks of house sparrows. *Anim. Behav.* **29,** 543–550. [428]

Barrington, E. J. W. (1968) *The Chemical Basis of Physiological Regulation*. Scott, Foresman and Company, Glenview, Illinois. [284]

Bastock, M. (1956) A gene mutation which changes a behaviour pattern. *Evolution* **10,** 421–439. [39]

Bastock, M., Morris, D. and Moynihan, M. (1953). Some comments on conflict and thwarting in animals. *Behaviour* **6,** 66–84. [384–387]

Bateson, P. P. G. (1964) An effect of imprinting on the perceptual development of domestic chicks. *Nature, London* **202,** 421–422. [367, 386, 375]

Bateson, P. P. G. (1966) The characteristics and context of imprinting. *Biol. Rev.* **41,** 177–220. [370, 374]

Bateson, P. P. G. (1973) Internal influences on early learning in birds. In *Constraints on Learning*, eds R. A. Hinde and J. Stevenson-Hinde, 101–116, Academic Press, London. [376]

Bateson, P. P. G. (1978) Early experience and sexual preferences. In *Biological Determinants of Sexual Behaviour*, ed. J. B. Hutchison, 29–53, Wiley, London. [370]

Bateson, P. P. G. (1979) How do sensitive periods arise and what are they for? *Anim. Behav.* **27,** 470–486. [328, 369, 370, 372, 377, 378]

Bateson, P. P. G. (1980) Optimal outbreeding and the development of sexual preferences in Japanese quail. *Z. Tierpsychol.* **53,** 231–44. [378]

Bateson, P. P. G., Lotwick, W. and Scott, D. K. (1980) Similarities between the faces of parents and offspring in Bewick's swans and the differences between mates. *J. Zool. Lond.* **191,** 61–74. [378, 379]

Bateson, P. P. G. and Reese, E. P. (1969) The reinforcing properties of conspicuous stimuli in the imprinting situation. *Anim. Behav.* **17,** 692–699. [375]

Bateson, P. P. G. and Wainwright, A. A. P. (1972) The effects of prior exposure to light on the imprinting process in domestic chicks. *Behaviour* **42,** 279–290. [375, 376]

Beer, C. G. (1970) Individual recognition of voice in the social behavior of birds. In *Advances in the Study of Behavior*, eds D. S. Lehrman, R. Hinde and E. Shaw, Academic Press, New York. [377]

Békésy, G. von (1952) Mechanics of hearing. *Nature, London* **169,** 241–242. [196]

Békésy, G. von (1960) *Experiments on Hearing*. McGraw-Hill, New York. [196]

Bekhterev, V. M. (1913) *La psychologie objective*. Alcan, Paris. [323]

Beninger, R. J., Kendall, S. B. and Vanderwof, C. H. (1974) The ability of rats to discriminate their own behaviors. *Can. J. Psychol.* **28,** 79–91. [521]

Bentley, D. R. (1971) Genetic control of an insect network. *Science* **174,** 1139–1141. [45]

Bentley, D. R. and R. R. Hoy (1972) Genetic control of the neuronal network generating cricket (*Teleogryllus gryllus*) song patterns. *Anim. Behav.* **20,** 478–492. [44–46]

Bergman, G. (1965) Der sexuelle Grössendimorphismus der Anatiden als Anpassung an das Höhlenbrüten. *Commentationes Biologicae, Soc. Sci. Fenn.* **28,** 1–10. [122]

Bernard, C. (1859) *Leçons sur les propriétés physiologiques et les altérations pathologiques des liquides de l'organisme*. Bailliere, Paris. [259]

Bernstein, J. I. (1970) Anatomy and physiology of the central nervous system. In *Fish Physiology*, eds W. S. Hoar and D. J. Randall, Vol. 4, 1–90, Academic Press, New York. [242]

Berthold, P. (1973) Relationship between migratory restlessness and migration distance in six *Sylvia* species. *Zool. Jb. Syst.* **115,** 594–599. [301]

Berthold, P. (1974) Circannual rhythms in birds with different migratory habits. In *Circannual Clocks*, ed. E. T. Pengelley, Academic Press, New York. [298, 299]

Bertram, B. C. R. (1979) Ostriches recognise their own eggs and discard others. *Nature, London* **279,** 233–234.

Bertram, B. C. R. (1980) Vigilance and group size in ostriches. *Anim. Behav.* **28,** 278–286. [476]

Biederman, G. B., D'Amato, M. R. and Keller, D. M. (1964) Facilitation of discriminated avoidance learning by dissociation of CS and manipulandum. *Psychon. Sci.* **1,** 229–230. [332]

Bischof, N. (1975) Comparative ethology of incest avoidance. In *Biosocial Anthropology*, ed. R. Fox, 37–67, Malaby Press, New York. [377]

Black, R. (1971) Hatching success in the three-spined stickleback (*Gasterosteus aculeatus*) in relation to changes in behaviour during the parental phase. *Anim. Behav.* **19,** 532–541. [93]

Blakemore, R. P. (1975) Magnetic bacteria. *Science* **190,** 377–379. [226]

Blest, A. D. (1957) The evolution of protective displays in the Saturnioidae and Sphingidae (Lepidoptera). *Behaviour* **11,** 257–309. [405, 406]

Blest, A. D. (1961) The concept of ritualization. In *Current Problems in Animal Behaviour*, eds W. H. Thorpe and O. L. Zangwill, Cambridge University Press, London. [409]

Bligh, J. (1976) Reproduction. In *Environmental Physiology of Animals* by J. Bligh, J. L. Cloudsley-Thompson and A. G. Macdonald, Blackwell Scientific Publications, Oxford. [292, 293]

Blough, D. S. (1955) Method for tracing dark adaptation in the pigeon. *Science* **121,** 703–704. [210]

Blurton Jones, N. G. (1968) Observations and experiments on the causation of threat displays of the great tit (*Parus major*). *Anim. Behav. Monogr.* **1,** 2. [403]

Blurton Jones, N. G. (ed.) (1972) *Ethological Studies of Child Behaviour*. Cambridge University Press, New York. [485]

Blurton Jones, N. G. and Sibly, R. M. (1978) Testing adaptiveness of culturally determined behaviour: do Bushman women maximize their reproductive success by spacing births widely and foraging seldom? In *Human Behaviour and Adaptation*, eds V. Reynolds and N. Blurton Jones, Taylor and Francis, London. [152]

Bodmer, W. F. and Cavalli-Sforza, L. L. (1976) *Genetics, Evolution, and Man*. W. H. Freeman, New York. [53, 65]

Boeuf Le, B. J. (1974) Male–male competition and reproductive success in elephant seals. *Amer. Zool.* **14,** 163–176. [149]

Bolles, R. C. (1967) *Theory of Motivation*. Harper and Row, New York. [275, 278]

Bolles, R. C. (1970) Species-specific defense reactions and avoidance learning. *Psychol. Rev.* **77,** 32–48. [332]

Bonner, J. T. (1980) *The Evolution of Culture in Animals*. Princeton University Press, Princeton, New Jersey. [514]

Bookman, M. A. (1978) Sensitivity of the homing pigeon to an earth-strength magnetic field. In *Animal Migration, Navigation and Homing*, eds K. Schmidt-Koenig and W. T. Keeton, 127–134, Proceedings in Life Sciences, Springer Verlag, Berlin. [254]

Booth, D. A. (ed.) (1978) *Hunger Models: Computable Theory of Feeding Control*. Academic Press, London. [274, 277]

Brambell, F. W. R. (1965) (Chairman) *Report of the Technical Committee to Enquire into the Welfare of Animals kept under Intensive Livestock Husbandry Systems*. Cmnd 2836, H.M.S.O., London. [529]

Braun, H. W. and Geiselhart, R. (1959) Age differences in the acquisition and extinction of the conditioned eyelid response. *J. exp. Psychol.* **57,** 386–388. [314]

Breland, K. and Breland, M. (1961) The misbehavior of organisms. *Amer. Psychol.* **16,** 661–664. [325, 329]

Brockmann, H. Jane and Barnard, C. J. (1979) Kleptoparasitism in birds. *Anim. Behav.* **27,** 487–514. [428]

Brockmann, H. J. and Dawkins, R. (1979) Joint nesting in a digger wasp as an evolutionarily stable preadaptation to social life. *Behaviour* **71,** 203–45. [106]

Brockmann, H. J., Grafen, A. and Dawkins, R. (1979) Evolutionarily stable nesting strategy in a digger wasp. *J. theor. Biol.* **77,** 473–496. [107]

Brogden, W. J. (1939) Sensory pre-conditioning. *J. exp. Psychol.* **25,** 323–332. [322]

Brogden, W. J., Lipman, E. A. and Cullen, E. (1938) The role of incentive in conditioning and extinction. *Amer. J. Psychol.* **51,** 109–117. [322, 333]

Brower, J. V. (1958) Experimental studies of mimicry in some North American butterflies, I: The Monarch, *Danaus plexippus*, and Viceroy, *Limenitis archippus archippus*. *Evolution* **12,** 32–47. [98]

Brower, L. P. (1969) Ecological chemistry. *Scientific American* **220** (Feb.), 22–29. Offprint 1133. [99, 337]

Brown, J. L. (1970) Cooperative breeding and altruistic behaviour in the Mexican jay, *Aphelocoma ultramarina*. *Anim. Behav.* **18,** 366–378. [138]

Brown, J. L. (1974) Alternative routes to sociality in jays—with a theory for the evolution of altruism and communical breeding. *Amer. Zool.* **14,** 61–78. [129, 138]

Brown, J. L. (1975) *The Evolution of Behavior*. W. W. Norton, New York. [29, 129, 159]

Brown, P. L. and Jenkins, H. M. (1968) Auto-shaping of the pigeon's key-peck. *J. exp. anim. Behav.* **11,** 1–8. [331]

Brown, R. E. and McFarland, D. J. (1979) Interaction of hunger and sexual motivation in the male rat: a time-sharing approach. *Anim. Behav.* **27,** 887–896. [466]

Brown, R. G. B. (1962) The aggressive and distraction behaviour of the western sandpiper *Ereunetes mauri*. *Ibis*, **104,** 1–12. [500]

Brun, R. (1914) Die Raumorientierung der Ameisen und das Orientierungsproblem im Allgemeinen. Jena, 242 pp. [246]

Budgell, P. (1970a) Modulation of drinking by ambient temperature changes. *Anim. Behav.* **18,** 753–757. [260, 276]

Budgell, P. (1970b) The effect of changes in ambient temperature on water intake and evaporative water loss. *Psychonomic Sci.* **20,** 275–276.

Bullock, T. H. (1977) *Introduction to Nervous Systems*. W. H. Freeman, New York. [174–178]

Bunning, E. (1967) *The Physiological Clock*. Springer Verlag, New York. [292]

Burghardt, G. M. (1970) Chemical perception in reptiles. In *Communication by Chemical Signals*, eds J. W. Johnson, D. G. Moulton and A. Turk, Appleton-Century-Crofts, New York. [212]

Burghardt, G. M. (1975) Chemical preference polymorphism in newborn garter snakes *Thamnopphis sirtalis*. *Behaviour* **52,** 202–225. [212]

Burrows, M. and Horridge, G. A. (1974) The organisation of inputs to motoneurons of the locust metathoracic leg. *Phil. Trans. R. Soc. B.* **269,** 49–94. [241]

Bykov, K. M. (1957) *The Cerebral Cortex and the Internal Organs*. Chemical Publishing Co, New York. [344]

Cade, T. J. and Dybas, J. A. (1962) Water economy of the budgerigar. *Auk.* **79,** 345–364. [260]

Cade, W. (1979) The evolution of alternative male reproductive strategies in field crickets. In *Sexual Selection and Reproductive Competition in Insects*, eds M. Blum and N. A. Blum, 343–379, Academic Press, London. [149]

Cain, A. J. (1964) The perfection of animals. In *Viewpoints in Biology*, Vol. 3, eds J. D. Carthy and C. L. Duddington, 36–63, Butterworth, London. [427]

Campbell, C. (1969) Development of specific preferences in thiamine-deficient rats: evidence against mediation by aftertastes. Unpublished master's thesis, Library of the University of Illinois at Chicago Circle. [336]

Cannon, W. B. (1927) The James–Lange theory of emotions: a critical examination and an alternative theory. *Amer. J. Psychol.* **39,** 106–124. [524]

Cannon, W. B. (1932) *The Wisdom of the Body*. Kegan Paul, London. [259, 273]

Capretta, P. J. (1961) An experimental modification of food preferences in chickens. *J. comp. physiol. Psychol.* **54,** 238–242. [337]

Caspari, E. (1972) Sexual Selection in Human Evolution. In *Sexual Selection and the Descent of Man*, ed. B. Campbell, Heinemann, London. [125]

Cavalli-Sforza, L. L. (1967) Genetic drift in an Italian population. *Scientific American*, August 1967. [64]

Cavalli-Sforza, L. L. and Bodmer, W. F. (1971) *The Genetics of Human Populations*. W. H. Freeman, New York. [54]

Cavalli-Sforza, L. L. and Feldman, M. W. (1974) Cultural versus biological inheritance: phenotypic transmission from parent to children (a theory of the effect of parental phenotypes on children's phenotype). *Amer. J. Human Genetics* **25,** 618–637. [54]

Chagnon, N. A., Neel, J. V., Weitkamp, L., Gershowitz, H. and Ayres, M. (1970) The influence of cultural factors on the demography and pattern of gene flow from the Makiritare to the Yanomama Indians. *Amer. J. phys. Anthrop.* **32,** 339–350. [124]

Chance, M. R. A. (1960) Köhler's chimpanzees—how did they perform? *Man* **60,** 130–135. [349]

Cherniak, C. (1981) Minimal rationality. *Mind* **XC,** 161–177. [503]

Chitty, D. (1954) *Control of Rats and Mice, Vol. 2: Rats.* Oxford University Press, Oxford. [335]
Chomsky, N. (1972) *Language and Mind.* Harcourt Brace Jovanovich, New York. [494]
Church, R. M. (1963) The varied effects of punishment on behavior. *Psychol. Rev.* **70,** 369–402. [323]
Church, R. M. (1969) Response suppression. In *Punishment and Aversive Behaviour*, eds B. A. Campbell and R. M. Church, 111–156, Appleton-Century-Crofts, New York. [323]
Church, R. M. (1978) The internal clock. In *Cognitive Processes in Animal Behavior*, eds S. H. Hulse, H. Fowler and W. K. Honig, Lawrence Erlbaum Associates, Hillsdale, New Jersey. [344]
Cloudsley-Thompson, J. L. (1980) *Biological Clocks.* Weidenfeld and Nicolson, London. [303]
Clutton-Brock, T. H. and Albon, S. D. (1979) The roaring of red deer and the evolution of honest advertisement. *Behaviour* **69,** 145–170. [113]
Clutton-Brock, T. H. and Harvey, P. H. (1977) Primate ecology and social organisation. *J. Zool. Lond.* **183,** 1–39. [75, 158]
Clutton-Brock, T. H. and Harvey, P. H. (1979) Comparison and adaptation. *Proc. R. Soc. Lond. B.* **205,** 547–565. [75]
Cohen, S. (1979) The temporal organisation of courtship behaviour in the three-spined stickleback, *Gasterosteus aculeatus.* D. Phil. thesis, University of Oxford. [460–470]
Cohen, S. and McFarland, D. J. (1979) Time-sharing as a mechanism for the control of behaviour sequences during the courtship of the three-spined stickleback (*Gasterosteus aculeatus*). *Anim. Behav.* **27,** 270–283. [467]
Coon, C. S. (1962) *The Origin of Races.* Alfred A. Knopf, New York. [123, 182]
Cooper, L. A. (1982) Internal representation. In *Animal Mind—Human Mind*, ed. D. R. Griffin, Springer Verlag, Berlin. [498]
Cooper, L. A. and Shepard, R. N. (1973) The time required to prepare for a rotated stimulus. *Mem. Cognit.* **1,** 246–250. [498]
Cooper, R. M. and Zubek, J. P. (1958) Effects of enriched and restricted early environments on the learning ability of bright and dull rats. *Can. J. Psychol.* **12,** 159–164. [52, 53]
Coppinger, R. P. (1969) The effect of experience and novelty on avian feeding behaviour with reference to the evolution of warning coloration in butterflies, Part I: reactions of wild caught adult blue jays to novel insects. *Behaviour* **35,** 45–60. [405]
Coppinger, R. P. (1970) The effect of experience and novelty on avian feeding behavior with reference to the evolution of warning coloration in butterflies, II: Reactions of naive birds to novel insects. *Amer. Natur.* **104,** 323–335. [405]
Cox, C. R. and Le Boeuf, B. J. (1977) Female incitation of male competition: a mechanism of mate selection. *Amer. Natur.* **111,** 317–335. [117]
Croft, D. B. (1975) The effect of photoperiod length on the diurnal activity of canaries. Unpublished Ph.D. thesis, University of Cambridge. [452]
Croghan, P. C. (1976) Ionic and osmotic regulation of aquatic animals. In *Environmental Physiology of Animals*, eds J. Bligh, J. L. Cloudsley-Thompson and A. G. Macdonald, Blackwell Scientific Publications, Oxford. [284]
Crook, J. H. (1964) The evolution of social organization and visual communication in the weaverbirds (Ploceidae). *Behaviour Suppl. 10,* 1–178. [74]
Crook, J. H. (1972) Sexual selection, dimorphism, and social organization in the primates. In *Sexual Selection and the Descent of Man 1871–1971*, ed. B. Campbell, 231–281, Aldine, Chicago. [124, 125]
Crook, J. H. (1980) *The Evolution of Human Consciousness.* Oxford University Press, New York. [527]
Crook, J. H. and Gartlan, J. S. (1966) Evolution of primate societies. *Nature, London* **210,** 1200–1203. [75]
Croze, H. J. (1970) Searching image in carrion crows. *Z. Tierpsychol. Beiheft* **5,** 86. [213]
Cullen, E. (1957) Adaptations in the kittiwake to cliff-nesting. *J. comp. Psychol.* **99,** 275–302. [71, 73]
Curio, E. (1975) The functional organization of anti-predator behaviour in the pied flycatcher: a study of avian visual perception. *Anim. Behav.* **23,** 1–115. [220]
Daan, S. (1981) Adaptive daily strategies in behavior. In *Handbook of Behavioral Neurobiology, Vol. 4: Biological Rhythms*, ed. J. Aschoff, Plenum Press, New York. [305, 306]
Daanje, A. (1950) On locomotory movements in birds and the intention movements derived from them. *Behaviour* **3,** 48–98. [399]

Daly, M. and Wilson, M. (1978) *Sex, Evolution and Behavior*, Duxbury Press, North Scituate, Massachusetts. [153]

Dampier, W. (1929) *History of Science*. Cambridge University Press, London. [85]

Darwin, C. (1841) *The Zoology of the Voyage of H.M.S. 'Beagle', Pt. III: Birds, by J. Gould. Notice of their habits and ranges, by Charles Darwin*, 98–106. [62]

Darwin, C. (1859) *On the Origin of Species by Natural Selection or the Preservation of Favoured Races in the Struggle for Life*. John Murray, London. [62, 113, 363]

Darwin, C. (1868) *The Variation of Animals and Plants under Domestication*. London. [15]

Darwin, C. (1871) *The Descent of Man and Selection in Relation to Sex*. John Murray, London. [7, 61, 113, 119, 122, 150, 505]

Darwin, C. (1872) *The Expression of the Emotions in Man and the Animals*. John Murray, London. [7, 394, 396, 398, 485, 520]

Davenport, D. and Norris, K. S. (1958) Observations on the symbiosis of the sea anemone *Stoichactis* and the pomacentrid fish, *Amphiprion percula*. *Biol. Bull. Woods Hole* **115**(3), 397–410. [410]

Davies, N. B. (1978) Ecological questions about territorial behaviour. In *Behavioural Ecology*, eds J. R. Krebs and N. B. Davies, Blackwell Scientific Publications, Oxford. [92, 103, 104]

Davies, N. B. and Halliday, T. R. (1977) Optimal mate selection in the toad *Bufo bufo*. *Nature, London* **269,** 56–58. [90, 91]

Davis, J. D. (1980) Homeostasis and behaviour. In *Analysis of Motivational Processes*, eds F. M. Toates and T. R. Halliday, Academic Press, London. [277]

Davis, J. M. (1973) Imitation: a review and critique. In *Perspectives in Ethology*, eds P. P. G. Bateson and P. H. Klopfer, Plenum Press, New York. [516, 522]

Davis, J. M. (1975) Socially induced flight reactions in pigeons. *Anim. Behav.* **23,** 597–601. [398]

Davis, W. J., Mpitsos, E. J. and Pineo, J. M. (1974) The behavioral hierarchy of the mollusk *Pleurobranchaea*: I. The dominant position of feeding behavior, *J. comp. Physiol.* **90,** 207–224. II. Hormonal suppression of feeding associated with egg laying. *Ibid*. 225–243. [456]

Dawkins, M. (1971a) Perceptual changes in chicks: another look at the "search image" concept. *Anim. Behav.* **19,** 566–574. [213–216]

Dawkins, M. (1971b) Shifts of "attention" in chicks during feeding. *Anim. Behav.* **19,** 575–582. [214–216]

Dawkins, M. (1976) Towards an objective method of assessing welfare in domestic fowl. *Applied animal Ethology* **2,** 245. [530]

Dawkins, M. (1977) Do hens suffer in battery cages? Environmental preference and welfare. *Anim. Behav.* **25,** 1034. [530]

Dawkins, M. S. (1980) *Animal Suffering*. Chapman and Hall, London. [527, 528, 530]

Dawkins, M. and Dawkins, R. (1974) Some descriptive and explanatory stochastic models of decision making. In *Motivational Control Systems Analysis*, ed. D. J. McFarland, Academic Press, London. [455]

Dawkins, R. (1976) *The Selfish Gene*. Oxford University Press, Oxford. [36, 90, 127, 128, 130, 145, 146, 502]

Dawkins, R. (1978) Replicator selection and the extended phenotype. *Z. Tierpsychol.* **47,** 61–76. [80]

Dawkins, R. (1979) Twelve misunderstandings of kin selection. *Z. Tierpsychol.* **51,** 184–200. [89]

Dawkins, R. (1980) Good strategy or evolutionarily stable strategy? In *Sociobiology: Beyond Nature/Nurture?* eds G. W. Barlow and J. Silverberg, 331–367, Westview Press, Boulder, Colorado. [527, 528, 530]

Dawkins, R. (1982) *The Extended Phenotype*. W. H. Freeman, Oxford. [502]

Dawkins, R. and Brockmann, H. J. (1980) Do digger wasps commit the Concorde fallacy? *Anim. Behav.* **28,** 892–896. [107, 108]

Dawkins, R. and Carlisle, T. R. (1976) Parental investment, mate desertion and a fallacy. *Nature, London* **262,** 131–133. [108, 130]

Dawkins, R. and Krebs, J. R. (1978) Animal signals: information or manipulation? In *Behavioural Ecology*, eds J. R. Krebs and N. B. Davies, 282–309, Blackwell Scientific Publications, Oxford. [409]

Delius, J. D. and Emmerton, J. (1978) Sensory mechanisms related to homing in pigeons. In *Animal Migration, Navigation and Homing*, eds K. Schmidt-Koenig and W. T. Keeton, 35–41, (Proceedings in Life Sciences) Springer Verlag, Berlin. [254]

Delius, J. D., Perchard, R. J. and Emmerton, J. (1976) Polarized light discrimination by pigeons and an electroretinographic correlate. *J. comp. physiol. Psychol.* **90,** 560–571. [258]

Delius, J. D. and Vollrath, F. W. (1973) Rotation compensation reflexes independent of the labyrinth: neurosensory correlates in pigeons. *J. comp. Physiol.* **83,** 123–134. [328]

Denenberg, V. H. (1962) An attempt to isolate critical periods of development in the rat. *J. comp. physiol. Psychol.* **55,** 813–815. [372]

Dennett, D. C. (1978) *Brain Storms: Philosophical Essays on Mind and Psychology*. Bradford Books, Montgomery, Vermont; (1979) Harvester Press, Brighton, UK. [354, 499–503]

Dennett, D. C. (1983) Intentional systems in cognitive ethology: the "Panglossian paradigm" defended. *The Behavioral and Brain Sciences* **6,** 343–390. [502, 503]

Denton, E. J. and Warren, F. J. (1957) The photosensitive pigments in the retinae of deep-sea fish. *J. mar. biol. Assoc. U.K.* **36,** 651–662. [223]

Dickinson, A. (1980) *Contemporary Animal Learning Theory*. Cambridge University Press, Cambridge. [340, 350–356]

Dobbing, J. (1976) Vulnerable period in brain growth and somatic growth. In *The Biology of Human Growth*, eds D. F. Roberts and A. M. Thompson, Taylor and Francis, London. [372]

Dobzhansky, Th. (1964) *The Heredity and the Nature of Man*. Harcourt, Brace and World, New York. [51]

Dobzhansky, T. (1972) Genetics and the races of Man. In *Sexual Selection and the Descent of Man*, ed. B. Campbell, Heinemann, London. [123]

Domjan, M. and Wilson, N. E. (1972) Contribution of ingestive behaviours to taste-aversion learning in the rat. *J. comp. physiol. Psychol.* **80,** 403–412. [353]

Dorst, J. and Dandelot, P. (1970) *A Field Guide to the Large Mammals of Africa*. Houghton Mifflin, Boston. [159]

Doty, R. W. (1968) Neural organization of deglutition. In *Handbook of Physiology—Alimentary canal, IV*, ed. W. Heidel, American Physiological Society, Washington. [240]

Dowling, J. E. and Boycott, B. B. (1966) Organisation of the primate retina: electron microscopy. *Proc. R. Soc. Lond. B.* **166,** 80–111. [200]

Drent, R. H. (1970) Functional aspects of incubation in the herring gull. In "The herring gull and its egg", Part I, eds G. P. Baerends and R. H. Drent, Behaviour, **82,** 1–132. [77, 436–438]

Duke-Elder, S. (1958) System of ophthalmology I: The eye in evolution. Henry Kimpton, London. [204]

Duncan, I. J. H. (1977) Behavioural wisdom lost. *Applied animal Ethology* **3,** 193. [530]

Duncan, I. J. H. (1978) The interpretation of preference tests in animal behaviour. *Applied animal Ethology* **4,** 197. [530]

Edmunds, M. (1974) *Defence in Animals*. Longman, Harlow, Essex. [4, 94, 97, 99, 407]

Edwards, W. (1954) The theory of decision making. *Psychol. Bull.* Vol. 51, No. 4, 380–417. [440, 441]

Edwards, W. (1961) Behavioural decision theory. *Ann. Rev. Psychol.* **12,** 473–498. [441]

Ehrman, L. and Parsons, P. A. (1976) *The Genetics of Behavior*. Sinauer Associates, Sunderland, Massachusetts. [43, 44]

Eibl-Eibesfeldt, I. (1967) Concepts of ethology and their significance in the study of human behaviour. In *Early Behavior: Comparative and Developmental Approaches*, ed. H. W. Stevenson, Wiley, New York. [511]

Eibl-Eibesfeldt, I. (1970) *Ethology: The Biology of Behavior*. Holt Rinehard and Winston, New York. [485, 486]

Eibl-Eibesfeldt, I. (1972) Similarities and differences between cultures in expressive movements. In *Non-Verbal Communication*, ed. R. A. Hinde, Cambridge University Press, London. [485]

Ekman, P. (1971) Universals and cultural differences in facial expressions of emotion. *Nebr. Symp. Motiv.* 207–284. [486]

Ekman, P. and Friesen, W. V. (1969) The repertoire of nonverbal behavior: categories, origins, usage and coding. *Semiotica* **1,** 49–98. [486]

Elgar, M. A. and Catterall, C. P. (1981) Flocking and predator surveillance in house sparrows: test of a hypothesis. *Anim. Behav.* **29,** 868–872. [476]

Emlen, S. T. (1967) Migratory orientation in the indigo bunting, *Passerina cyanea*. Part I: The evidence for use of celestial cues. *Auk.* **84,** 309–342; Part II: Mechanisms of celestial orientation. *Auk.* **84,** 463–489. [257]

Emlen, S. T. (1972) The ontogenetic development of orientation capabilities. In *Animal Orientation and Navigation*, eds S. R. Galler, K. Schmidt-Koenig, G. J. Jacobs and R. E. Belleville, 191–210, NASA SP-262 US Govt. Printing Office, Washington, D.C. [257]

Emlen, S. T. (1978) The evolution of cooperative breeding in birds. In *Behavioural Ecology: An Evolutionary Approach*, eds J. R. Krebs and N. B.

Davies, Blackwell Scientific Publications, Oxford. [137–138]
Emlen, S. T. and Emlen, J. T. (1966) A technique for recording migratory orientation of captive birds. *Auk.* **83,** 361–367. [257]
Estes, W. K. (1944) An experimental study of punishment. *Psychol. Monogr. 57,* No. 263, 40, 98, 109–112. [323]
Etienne, A. S. (1969) Analyse der schlagauslösenden Bewegungsparameter einer punktförmigen Beuteattrappe bei der Aeschnalarve. *Z. vergl. Physiol.* **64,** 71–110. [232]
Evans, S. M. (1966) Non-associative avoidance learning in Nereid polychaetes. *Anim. Behav.* **14,** 102–106. [343]
Evarts, E. V. J. (1968) Motorneuron firing in pyramidal tract correlated with movement. *Neurophysiol.* **31,** 14–27. [243]
Ewert, J. P. (1980) *Neuroethology.* Springer Verlag, Heidelberg. [228, 231, 233–235]
Ewert, J. P. and Burghagen, H. (1979) Ontogenetic aspects of visual 'size-constancy' phenomena in the midwife toad (*Alytes obstetricans* (Laur) *Brain Behaviour and Evolution,* **16,** 99–112. [231]
Ewert, J. P. and Hock, F. (1972) Movement sensitive neurons in the toad's retina. *Expl. Brain Res.* **16,** 41–59. [234]
Falconer, D. S. (1960) *Introduction to Quantitative Genetics.* Oliver and Boyd, Edinburgh. [62]
Fincke, O. M. (1982) Lifetime mating success in a natural population of the damsel fly, *Enallagma hageni* (Walsh) (Odonata: Coenagrionidae). *Behavioural Ecology and Sociobiology* **10,** 293–302. [91]
Fischer, H. (1965) Das Triumphgeschrei der Graugans (*Anser anser*). *Z. Tierpsychol.* **22,** 247–304. [116]
Fisher, R. A. (1918) The correlation between relatives on the supposition of Mendelian inheritance. *Trans. R. Soc. Edin.* **52,** 399–433. [51]
Fisher, R. A. (1930) The genetical theory of natural selection (2nd edition, 1958). Oxford University Press, Oxford. [85, 89, 118, 121]
Fitzsimons, J. T. (1971) The physiology of thirst: a review of the extraneural aspects of the mechanisms of drinking. In *Progress in Physiological Psychology, Vol. 4,* eds E. Stellar and J. M. Sprague, 119–201, Academic Press, New York. [269]
Fitzsimons, J. T. (1976) The physiological basis of thirst. *Kidney International* **10,** 3–11. [269]
Fitzsimons, J. T. and Le Magnen, J. (1969) Eating as a regulatory control of drinking in the rat. *J. comp. physiol. Psychol.* **67,** 273–283. [276]
Follett, B. K. (1973) Circadian rhythms and photoperiodic time measurement in birds. *J. reprod. Fert., Suppl.* **19,** 5–18. [293]
Ford, E. B. and Huxley, J. S. (1927) Mendelian genes and rates of development in *Gammarus chevreuxi. Brit. J. exp. Biol.* **5,** 112–133. [28]
Foree, D. D. and LoLordo, V. M. (1973) Attention in the pigeon: the differential effects of food-getting versus shock-avoidance procedures. *J. comp. physiol. Psychol.* **85,** 551–558. [338]
Fouts, R. S. (1975) Capacities for language in great apes. In *Socioecology and Psychology of Primates,* ed. R. H. Tuttle, 371–390, Mouton, The Hague. [492]
Fouts, R. S., Chown, B. and Goodwin, L. (1976) Transfer of signed responses in American sign language from vocal English stimuli to physical object stimuli by a chimpanzee (*Pantroglodytes*), *Learn. and Motiv.* **7,** 458–475. [491]
Fox, R. (1972) Alliance and constraint: sexual selection in the evolution of human kinship systems. In *Sexual Selection and the Descent of Man 1871–1971,* ed. B. Campbell, Heinemann, London. [125]
Fraenkel, G. S. and Gunn, D. L. (1940) *The Orientation of Animals.* Clarendon Press, Oxford (Dover Books, New York, 1961). [243–245]
Freud, S. (1915) Instincts and their vicissitudes. *Coll. Papers Vol. IV,* Hogarth Press, London. [362]
Frisch, K. von (1967) *The Dance Language and Orientation of Bees* (translated by L. E. Chadwick). Harvard University Press, Cambridge, Massachusetts. [257]
Frisch, K. von and Lindauer, M. (1954) Himmel und Erde in Konkurrenz bei der Orientierung der Bienen. *Naturwissenschaften* **41,** 245–253. [419]
Fry, F. E. J. and Hochachka, P. W. (1970) Fish. In *Comparative Physiology of Thermoregulation, Vol. 1,* ed. G. C. Whittow, 79–134, Academic Press, New York. [288]
Fuller, J. L. and Thompson, W. R. (1960) *Behaviour genetics.* Wiley, New York. [39]
Futuyma, D. J. (1979) *Evolutionary Biology.* Sinauer Associates, Sunderland, Massachusetts. [63, 65]
Gallagher, J. E. (1977) Sexual imprinting: a sensitive period in Japanese quail (*Coturnix coturnix japonica*). *J. comp. physiol. Psychol.* **91,** 72–78. [371, 378]

Gallup, G. G., Jr. (1977) Self-recognition in primates. A comparative approach to the bidirectional properties of consciousness. *Amer. Psychol.* **32,** 329–338. [522]
Gallup, G. G., Jr. (1979) Self-awareness in primates. *Amer. Sci.* **67,** 417–421. [522]
Galton, F. (1869) *Hereditary Genius: An Inquiry into its Laws and Consequences*. Macmillan, London. [38]
Galton, F. (1883) *Inquiries into Human Faculty and its Development*. Macmillan, London. [39]
Gandolfi, G., Mainardi, D., Rossi, A. C. (1968) La reazione di paura e lo svantaggio individuale dei pesci allarmisti (esperimenti con modelli). *Zoologia* **102,** 8–14. [232]
Garcia, J., Ervin, F. R., Yorke, C. H. and Koelling, R. A. (1967) Conditioning with delayed vitamin injection. *Science* **155,** 716–718. [336]
Garcia, J., Kimmeldorf, D. J. and Hunt, E. L. (1961) The use of ionizing radiation as a motivating stimulus. *Psychol. Rev.* **68,** 383–385. [334]
Garcia, J., Kimmeldorf, D. J. and Koelling, R. A. (1955) Conditioned aversion to saccharin resulting from exposure to gamma radiation. *Science* **122,** 157–158. [334]
Garcia, J. and Koelling, R. A. (1966) Relation of cue to consequence in avoidance learning. *Psychonomic Sci.* **4,** 123–124. [337]
Garcia, J., Kovner, R. and Green, K. F. (1970) Cue properties versus palatability of flavours in avoidance learning. *Psychonomic Sci.* **20,** 313–314. [337]
Garcia, J., McGowan, B. K., Ervin, F. R. and Koelling, R. A. (1968) Cues: their effectiveness as a function of the reinforcer. *Science* **160,** 794–795. [337]
Gardner, B. T. and Gardner, R. A. (1969) Teaching sign language to a chimpanzee. *Science* **165,** 664–672. [492, 494]
Gardner, B. T. and Gardner, R. A. (1971) Two-way communication with an infant chimpanzee. In *Behavior of Non-Human Primates*, eds A. M. Schrier and F. Stollnitz, Vol. IV, Chapter 3, Academic Press, New York. [492]
Gardner, B. T. and Gardner, R. A. (1975) Evidence for sentence constituents in the early utterances of child and chimpanzee. *J. exp. Psychol.* **104,** 244–267. [492]
Gardner, R. A. and Gardner, B. T. (1978) Comparative psychology and language acquisition. *Ann. N.Y. Acad. Sci.* **309,** 37–76. [493]
Gaston, A. J. (1976) Factors affecting the evolution of group territories of babblers (Turdoides) and long-tailed tits. Unpublished dissertation, Oxford University. [139]
Geist, V. (1971) *Mountain Sheep: A Study in Behaviour and Evolution*. University of Chicago Press, Chicago. [115]
Gibson, E. J. and Walk, R. D. (1956) The effect of prolonged exposure to visually presented patterns on learning to discriminate them. *J. comp. physiol. Psychol.* **49,** 239–242. [375]
Gibson, E. J., Walk, R. D. and Tighe, T. J. (1959) Enhancement and deprivation of visual stimulation during rearing as factors in visual discrimination learning. *J. comp. physiol. Psychol.* **52,** 74–81. [375]
Glaserfeld, E. von (1977) Linguistic communication: theory and definition. In *Language Learning by a Chimpanzee*, ed. D. Rumbaugh, 55–71, Academic Press, New York. [493]
Goldsmith, T. H. (1972) The natural history of invertebrate visual pigments. In *Handbook of Sensory Physiology*, ed. H. J. A. Dartnall, 685–719, Springer Verlag, Berlin. [224]
Goodhart, C. B. (1964) A biological view of toplessness. *New Scientist* **23,** 558–560. [125]
Gormezano, I. (1966) Classical conditioning. In *Experimental Methods and Instrumentation in Psychology*, ed. J. B. Sidowski, 385–420, McGraw-Hill, New York. [342]
Goss-Custard, J. D. (1977a) Optimal foraging and size-selection of worms by redshank *Tringa totanus*. *Anim. Behav.* **25,** 10–29. [429, 430]
Goss-Custard, J. D. (1977b) Predator responses and prey mortality in the redshank *Tringa totanus* (L.) and a preferred prey *Corophium volutator* (Pallas). *J. anim. Ecol.* **46,** 21–36. [430]
Gottlieb, G. (1961) The following-response and imprinting in wild and domestic ducklings of the same species (*Anas platyrhynchos*). *Behaviour* **18,** 205–228. [369]
Gottlieb, G. (1963) A naturalistic study of imprinting in wood ducklings (*Aix sponsa*). *J. comp. physiol. Psychol.* **56,** 86–91. [367]
Gottlieb, G. (1970) Ontogenesis of sensory function in birds and mammals. In *Biopsychology of Development*, ed. E. Tobach, Academic Press, New York. [32]
Gottlieb, G. (1971) *Development of Species Identification in Birds*. University of Chicago Press, Chicago. [32, 33, 369, 376]

Gould, J. L. (1976) The dance language controversy. *Q. Rev. Biol.* **57,** 211–244. [424]
Gould, J. L. (1980) The case for magnetic-field sensitivity in birds and bees (such as it is). *American Scientist* **68,** 256–267. [254, 421]
Gould, J. L. (1981) Language. In *The Oxford Companion to Animal Behaviour,* ed. D. J. McFarland, Oxford University Press, Oxford. [489]
Gould, J. L. (1982) *Ethology.* W. W. Norton, New York. [194, 423]
Gould, J. L. and Gould, C. G. (1982) The insect mind: physics or metaphysics? In *Animal Mind—Human Mind,* ed. D. R. Griffin, Dahlem Konferenzen, Springer Verlag, Berlin. [426, 488]
Gould, S. J. (1978) *Ever Since Darwin.* Burnett Books, London. [114]
Gould, S. J. (1980) Is a new and general theory of evolution emerging? *Paleobiology* **6,** 119–130. [427]
Gould, S. J. and Eldredge, N. (1977) Punctuated equilibria: the tempo and mode of evolution reconsidered. *Paleobiology* **3,** 115–151. [427]
Grafen, A. (1982) How not to measure inclusive fitness. *Nature, London* **298,** 425–426. [89, 90]
Grafen, A. (1984) Natural selection, kin selection and group selection. In *Behavioural Ecology,* 2nd edition, eds J. R. Krebs and N. B. Davies, Blackwell Scientific Publications, Oxford. [89, 128]
Gray, J. (1950) The role of peripheral sense organs during locomotion in the vertebrates. *Symp. Soc. exp. Biol.* **4,** 112–126. [242]
Green, R. F. (1980) Bayesian birds: a simple example of Oaten's stochastic model of optimal foraging. *Theoretical Population Biology* **18,** 244–256. [473]
Green, R. F. (1983) Stopping rules for optimal foragers. (Unpublished manuscript). [472, 473]
Grether, W. F. (1938) Pseudo-conditioning without paired stimulation encountered in attempted backward conditioning. *J. comp. Psychol.* **25,** 91–96. [343]
Griffin, D. R. (1958) *Listening in the Dark.* Yale University Press, New Haven, Connecticut. [228]
Griffin, D. R. (1976) *The Question of Animal Awareness.* The Rockefeller University Press, New York. [500, 526]
Griffin, D. R. (1981) *The Question of Animal Awareness* (2nd edition). Rockefeller University Press, New York. [500, 527]
Griffin, D. R. (1982) *Animal Mind—Human Mind.* Springer Verlag, Berlin. [523, 526]
Grinnell, J. (1917) The niche relationships of the California thrasher. *Auk.* **21,** 364–382. [67]
Grohmann, J. (1939) Modifikation oder Funktionsreifung? Ein Beitrag zur Klärung der wechselseitigen Beziehungen zwischen Instinkthandlung und Erfahrung. *Z. Tierpsychol.* **25,** 132–144. [30]
Gross, M. R. and Charnov, E. L. (1980) Alternative male life histories in bluegill sunfish. *Proc. nat. Acad. Sci. U.S.A.* **77,** 6937–6940. [149, 150]
Grossen, N. E. and Kelley, M. J. (1972) Species-specific behaviour and acquisition of avoidance behaviour in rats. *J. comp. Physiol. Psychol.* **81,** 307–310. [333]
Guiton, P. (1959) Socialisation and imprinting in brown Leghorn chicks. *Anim. Behav.* **7,** 26–34. [369]
Gumma, N. R., South, F. E. and Allen, J. N. (1967) Temperature preference in golden hamsters. *Anim. Behav.* **15,** 534–537. [290]
Guthrie, D. M. (1980) *Neuroethology: An Introduction.* Blackwell Scientific Publications, Oxford. [169, 228]
Gwinner, E. (1971) A comparative study of circannual rhythms in warblers. In *Biochronometry,* ed. M. Menaker, 405–427, National Academy of Science, Washington D.C. [298]
Gwinner, E. (1972) Endogenous timing factors in birds' migration. In *Animal Orientation and Navigation,* eds S. R. Galler, K. Schmidt-Koenig, G. J. Jacobs and R. E. Belleville, 321–338, NASA, Washington, D.C. [300, 301]
Gwinner, E. and Wiltschko, W. (1978) Endogenously controlled changes in migratory direction of the garden warbler, *Sylvia borin. J. comp. Physiol.* **125,** 267–273. [301, 302]
Hailman, J. P. (1965) Cliff-nesting adaptations of the Galapagos swallow-tailed gull. *Wilson Bulletin* **77,** 346–362. [73, 75]
Haldane, J. B. S. (1954) *Introducing Douglas Spalding. Brit. J. anim. Behav.* **2,** 1. [366]
Haldane, J. B. S. (1955) Population genetics. *New Biology* **18,** 34–51. [89]
Halliday, T. R. (1974) The sexual behaviour of the smooth newt, *Triturus vulgaris* (Urodela, Salamandridae). *J. Herpetol.* **8,** 277–292. [144, 462–464]
Halliday, T. R. (1976) The libidinous newt. An analysis of variations in the sexual behaviour of the smooth newt, *Triturus vulgaris. Anim. Behav.* **24,** 398–414. [144, 463]

Halliday, T. R. (1977a) The effects of experimental manipulation of breathing behaviour on the sexual behaviour of the smooth newt, *Triturus vulgaris*. *Anim. Behav.* **25,** 39–45. [144, 463]
Halliday, T. R. (1977b) The courtship of European newts: an evolutionary perspective. In *The Reproductive Biology of Amphibians*, eds D. H. Taylor and S. I. Guttman, 85–232, Plenum Press, New York. [463]
Halliday, T. R. (1978) Sexual Selection and Mate Choice. In *Behavioural Ecology: An Evolutionary Approach*, eds J. R. Krebs and N. B. Davies, Blackwell Scientific Publications, Oxford. [118]
Halliday, T. R. (1980) *Sexual Strategy*. Oxford University Press, Oxford. [118, 120, 154]
Halliday, T. R. and Sweatman, H. P. A. (1976) To breathe or not to breathe; the newt's problem. *Anim. Behav.* **24,** 551–561. [390, 463]
Ham, R. G. and Veomett, M. J. (1980) *Mechanisms of Development*. Mosby, St. Louis, Missouri. [28, 31]
Hamburger, V. (1963) Some aspects of the embryology of behaviour. *Quart. Rev. Biol.* **38,** 342–365. [32]
Hamilton, W. D. (1964) The genetical theory of social behaviour (I and II). *J. theor. Biol.* **7,** 1–16 and 17–32. [89, 139]
Hamilton, W. D. (1979) Wingless and fighting males in fig wasps and other insects. In *Several Selection and Reproductive Competition in Insects*, eds M. S. Blum and N. A. Blum, Academic Press, London. [105]
Harcourt, A. H. (1979) Social relationships between adult male and female gorillas in the wild. *Anim. Behav.* **27,** 325–342. [151]
Harlow, H. F. (1949) The formation of learning sets. *Psychol. Rev.* **56,** 51–65. [507]
Harris, G. W., Michael, R. P. and Scott, P. P. (1958) Neurological site of action of stilboestrol in eliciting sexual behaviour. *Ciba Foundation Symposium on the Neurological Basis of Behaviour*. Churchill, London. [184]
Harris, L. J., Clay, J., Hargreaves, F. and Ward, A. (1933) Appetite and choice of diet. The ability of the vitamin B deficient rat to discriminate between diets containing and lacking the vitamin. *Proc. Roy. Soc. Lond. B.* **113,** 161–190. [334]
Harrison, C. J. O. (1969) Helpers at the nest in Australian passerine birds. *Emu* **69,** 30–40. [129]
Hart, J. S. (1964) Insulative and metabolic adaptation to cold in vertebrates. *Symp. soc. exp. Biol.* **18,** 31–48. [155]
Haskell, P. T. (1961) *Insect Sounds*. Witherby, London. [193]
Hasler, A. D. (1960) Homing orientation in migrating fishes. *Ergebnisse der Biologie* **23,** 94–115. [252]
Hasler, A. D. and Schwassmann, H. O. (1960) Sun orientation of fish at different latitudes. *Cold Spring Harbour Symp. quant. Biol.* **25,** 411–429. [291]
Hayes, K. J. and Hayes, C. (1952) Imitation in a home-raised chimpanzee. *J. comp. physiol. Psychol.* **45,** 450–459. [523]
Hayes, K. J. and Nissen, C. H. (1971) Higher mental functions of a home-raised chimpanzee. In *Behaviour of Nonhuman Primates*, eds A. M. Schrier and F. Stollnitz, Vol. 4, 59–115, Academic Press, New York. [489]
Hebb, D. O. (1953) Heredity and environment in mammalian behaviour. *Brit. J. anim. Behav.* **1,** 43–47. [29]
Hediger, H. (1964) *Wild Animals in Captivity: an Outline of the Biology of Zoological Gardens*. Dover Publications, New York. [404]
Heider, K. G. (1969) Attributes and categories in the study of material culture: New Guinea Dani attire. *Man* **4,** 379–391. [126]
Heiligenberg, W. (1976) The interaction of stimulus patterns controlling aggressiveness in the cichlid fish *Haplochromis burtoni*. *Anim. Behav.* **24,** 452–458. [220]
Heiligenberg, W., Kramer, U. and Schulz, V. (1972) The angular orientation of the black eye-bar in *Haplochromis burtoni* (Cichlidae: Pisces) and its relevance to aggressivity. *Z. vergl. Physiol.* **76,** 168–176. [220]
Heinrich, B. (1979) *Bumblebee Economics*. Harvard University Press, Cambridge, Massachusetts. [446–449]
Heinroth, O. (1910) Beiträge zur Biologie, insbesondere Psychologie und Ethologie der Anatiden. *Verhandlungen des internationalen Ornithologenkongresses*, Berlin. [366, 367]
Helmholtz, H. von (1867) *Handbuch der physiologischen Optik*. Voss, Leipzig. [196, 249]
Henton, W. W., Smith, J. C. and Tucker, D. (1966) Odour discrimination in pigeons. *Science* **153,** 1138–1139. [254]
Herrnstein, R. J., Loveland, D. H. and Cable, C. (1976) Natural concepts in pigeons. *J. exp.*

Psychol., Animal Behaviour Processes **2,** 285–302. [345]

Hess, E. H. (1958) Imprinting in animals. *Scientific American* **198,** 81–90. [369]

Hess, E. H. (1959a) Imprinting. *Science* **130,** 133–141. [369, 370]

Hess, E. H. (1959b) The conditions limiting critical age of imprinting. *J. comp. physiol. Psychol.* **52,** 515–518. [369–371]

Hilden, O. (1965) Habit selection in birds. *Ann. Zool. Fenn.* **2,** 53–75. [528]

Hill, K. G., Loftus-Hills, J. J. and Gartside, D. F. (1972) Premating isolation between the Australian field crickets, *Teleogryllus commodus* and *T. oceanicus*. *Aust. J. Zool.* **20,** 153–163. [44, 46]

Hinde, R. A. (1952) The behaviour of the great tit (*Parus major*) and some other related species. *Behaviour Suppl. 2*. [403, 407]

Hinde, R. A. (1959) Unitary drives. *Anim. Behav.* **7,** 130–141. [278]

Hinde, R. A. (1960) Energy models of motivation. *Sym. Soc. exp. Biol.* **14,** 199–213. [278]

Hinde, R. A. (1966) *Animal Behavior* (1st edition). McGraw-Hill, New York. [366, 389, 401]

Hinde, R. A. (1970) *Animal Behavior* (2nd edition). McGraw-Hill, New York. [34, 243, 244, 245, 275, 276, 374, 375, 389]

Hinde, R. A. (1974) *Biological Bases of Human Social Behavior*. McGraw-Hill, New York. [372, 375, 488]

Hinde, R. A. and Fisher, J. (1951) Further observations on the opening of milk bottles by birds. *Brit. Birds* **44,** 393–396. [516]

Hirsch, J. (1963) Behavior genetics and individuality understood. *Science* **142,** 1436–1442. [39, 43]

Hirsch, J. (ed.) (1967) *Behavior-Genetic Analysis*. McGraw-Hill, New York. [39]

Hirsch, J. and Erlenmeyer-Kimling, L. (1962) Individual differences in behavior and their genetic basis. In *Roots of Behavior*, ed. E. L. Bliss, Harper and Row, New York. [42, 43]

Hodgkin, A. L. and Huxley, A. F. (1945) *J. Physiol.* **104,** 176–195. [168]

Hodos, W. (1970) Evolutionary interpretation of neural and behavioral studies of living vertebrates. In *The Neuro-Sciences; Second Study Program*, ed. F. O. Schmitt, 26–39, Rockefeller University Press, New York. [509]

Hodos, W. (1982) Some perspectives on the evolution of intelligence and the brain. In *Animal Mind—Human Mind*, ed. D. R. Griffin, Springer Verlag, Berlin. [182, 505, 506, 525]

Hodos, W. and Campbell, C. B. G. (1969) Scala Naturae: Why there is no theory in comparative psychology. *Psychol. Rev.* **76,** 337–350. [182]

Hoffman, H. S., Searle, J. L., Toffrey, S. and Kozma, F., Jr. (1966) Behavioral control by an imprinted stimulus. *J. exp. Anal. Behav.* **9,** 177–189. [375]

Hoffman, K. (1954) Versuche zu der im Richtungsfinden der Vögel enthaltenen Zeitschätzung. *Z. Tierpsychol.* **11,** 453–475. [254]

Hogan, J. A. (1980) Homeostasis and Behaviour. In *Analysis of Motivational Processes*, eds F. M. Toates and T. R. Halliday, Academic Press, London. [277]

Hogan, J. A., Kkeist, S. and Hutchings, C. S. L. (1970) Display and food as reinforcers in the Siamese fighting fish (*Betta splendens*). *J. comp. physiol. Psychol.* **70,** 351–357. [453]

Holland, P. C. (1977) Conditioned stimulus as a determinant of the form of the Pavlovian conditioned response. *J. exp. Psychol: Animal Behaviour Processes* **3,** 77–104. [355]

Holland, P. C. and Straub, J. J. (1979) Differential effects of two ways of devaluing the unconditioned stimulus after Pavlovian appetitive conditioning. *J. exp. Psychol.: Animal Behaviour Processes* **5,** 65–78. [355]

Hollard, V. D. and Delius, J. D. (1983) Rotational invariance in visual pattern recognition by pigeons and humans. *Science* **218,** 804–806. [499]

Holloway, R. L. (1976) Paleoneurological evidence for language origins. *Ann. N.Y. Acad. Sci.* **280,** 330–348. [495]

Holloway, R. L. (1981) Exploring the dorsal surface of hominoid brain endocasts by stereoplotter and discriminant analysis. *Phil. Trans. R. Soc. Lond. B.* **292,** 155–166. [495]

Holst, E. von (1939) Entwurf eines Systems der Lokomotorischen Periodenbildungen bei Fischen. *Z. vergl. Physiol.* **26,** 481–528. [242]

Holst, E. von (1954) Relations between the central nervous system and the peripheral organs. *Brit. J. anim. Behav.* **2,** 89–94. [250]

Holst, E. von (1973) *The Behavioural Physiology of Animals and Man*. Methuen, London. [242]

Holst, E. von and Mittelstaedt, H. (1950) Das Reafferenzprinzip. *Naturwissenschaften* **37,** 464–476. [250]

Hooff, J. A. R. A. M. van (1972) A comparative

approach to the phylogeny of laughter and smiling. In *Non-Verbal Communication*, ed. R. A. Hinde, 209–238, Cambridge University Press, Cambridge. [486]

Hooff, J. A. R. A. M. van (1976) The comparison of facial expression in man and higher primates. In *Methods of Inference from Animal to Human Behaviour*, ed. M. von Cranach, 165–196, Mouton, The Hague. [486]

Hoogland, R., Morris, D. and Tinbergen, N. (1957) The spines of sticklebacks (*Gasterosteus* and *Pygosteus*) as means of defence against predators (*Perca* and *Essox*). *Behaviour* **10,** 205–236. [80–82]

Houston, A. I. (1982) Transitions and time-sharing. *Anim. Behav.* **30,** 615–625. [390, 461, 466]

Houston, A. I. and McFarland (1976) On the measurement of motivational variables. *Anim. Behav.* **24,** 459–475. [221, 281]

Houston, A. I. and McFarland (1980) Behavioral resilience and its relation to demand functions. In *Limits to Action: The Allocation of Individual Behaviour*, ed. J. E. R. Staddon, 177–203, Academic Press, New York. [451, 454]

Howard, I. P. (1982) *Human Visual Orientation*. Wiley, New York. [335]

Howard, I. P. and Templeton, W. B. (1966) *Human Spatial Orientation*. Wiley, London. [172, 249]

Howard, R. D. (1978) The evolution of mating strategies in bullfrogs *Rana catesbiana*. *Evolution* **32,** 850–871. [148]

Howard, R. D. (1979) Big bullfrogs in a little pond. *Nat. Hist. Mag.* **88,** 30–36. [149]

Howard, R. R. and Brodie, E. D. (1971) Experimental study of mimicry in salamanders involving *Notophthalmus viridescens viridescens* and *Pseudotriton ruber schencki*. *Nature, London* **233,** 277. [97]

Hoy, R. R. (1974) Genetic control of acoustic behavior in crickets. *Amer. Zool.* **14,** 1067. [47]

Hoy, R. R. and Paul, R. L. (1973) Genetic control of song specificity in crickets. *Science* **180,** 82–83. [46]

Hubbard, J. I. (1975) *The Biological Bases of Mental Activity*. Addison-Wesley, Reading, Massachusetts. [526]

Hubel, D. H. and Wiesel, T. N. (1965) Receptive fields and functional architecture in two nonstriate visual areas (18 and 19) of the cat. *J. Neurophysiol.* **28,** 229–289. [32]

Hughes, A. (1977) The topography of vision in mammals. In *Handbook of Sensory Physiology*, VII/5, ed. F. Crescitelli, 613–756, Springer Verlag, Berlin. [205]

Hughes, B. O. and Black, A. J. (1973) The preference of domestic hens for different types of battery cage floor. *British Poultry Science* **14,** 615. [529]

Humphrey, N. K. (1978) Nature's psychologists. *New Scientist* **78,** 900. [526]

Humphrey, N. K. (1979) Nature's psychologists. In *Consciousness and the Physical World*, eds B. Josephson and B. S. Ramachandra, Pergamon, New York. Reprinted in E. Sunderland and M. T. Smith (1980) *The Exercise of Intelligence*, Garland STPM Press, New York. [527]

Huxley, J. S. (1914) The courtship habits of the great crested grebe (*Podiceps cristatus*) with an addition to the theory of sexual selection. *Proc. zool. Soc. Lond.* **2,** 491–562. [381]

Huxley, J. S. (1923) Courtship activities of the red-throated diver (*Columbus stellatus Pontopp*) together with a discussion on the evolution of courtship in birds. *J. Linn. Soc. Lond.* **25,** 253–292. [392]

Iersel, J. J. A. van, and Bol, A. C. A. (1958) Preening in two tern species: a study on displacement activities. *Behaviour* **13,** 1–88. [385, 386, 388]

Immelmann, K. (1969) Über den Einfluss frühkindlicher Erjahrungen auf die geschichtliche Objektfixierung bei Estrildiden. *Z. Tierpsychol.* **26,** 677–691. [376]

Immelmann, K. (1972) Sexual and other long-term aspects of imprinting in birds and other species. In *Advances in the Study of Behavior*, Vol. 4, eds D. S. Lehrman, R. A. Hinde and E. Shaw, 147–174, Academic Press, New York. [373, 374]

Isaac, G. Ll. (1978) The food-sharing behavior of protohuman hominids. *Scientific American* **238,** (April) 90–100. [160]

James, W. (1890) *The Principles of Psychology*. Holt, New York. [524]

Jander, R. (1957) Die optische Richtungsorientierung der roten Waldameise (*Formica rufa* L.). *Z. vergl. Physiol.* **40,** 162–238. [246]

Jarman, P. J. (1974) The social organisation of antelopes in relation to their ecology. *Behaviour* **48,** 215–267. [75]

Jenkins, H. M. and Harrison, R. H. (1958) Auditory generalisation in the pigeon. Air research and development command. TN 58-443, Astia document 158248. [318]

Jenni, D. A. and Collier, G. (1972) Polyandry in the American jacana (*Jacana spinosa*). *Auk.* **89,** 743–765. [155]

Jerison, H. J. (1973) *Evolution of the Brain and Intelligence*. Academic Press, New York. [182, 506]

Jones, F. R. H. (1955) Photo-kinesis in the ammocoete larva of the brook lamprey. *J. exp. Biol.* **32,** 492–503. [244]

Jordan, J. (1971) Studies of the structure of the organ of voice and vocalization in the chimpanzee: I. *Folia morphol.* **30,** 97–126; III. *Folia morphol.* **30,** 322–340. [490]

Kaas, J. H., Guillery, R. W. and Allman, J. M. (1972) Some principles of organisation in the dorsal lateral geniculate nucleus. *Brain Behav. Evol.* **6,** 253–299. [204]

Kacelnik, A. (1979) The foraging efficiency of great tits (*Parus major*) in relation to light intensity. *Anim. Behav.* **27,** 237–241. [434]

Kagel, J. H., Battalio, R. C., Green, L. and Rachlin, H. (1980) Consumer demand theory applied to choice behaviour of rats. In *Limits to Action*, ed. J. E. R. Staddon, Academic Press, New York. [453]

Kaissling, K. E. and Priesner, E. (1970) Die Riechschwelle des Seidenspinners. *Naturwissenschaften* **57,** 23–28. [189]

Kalat, J. W. and Rozin, P. (1971) Role of interference in taste-aversion learning. *J. comp. Physiol.* **77,** 53–58. [339]

Kamin, L. J. (1969) Predictability, surprise, attention and conditioning. In *Punishment and Aversive Behaviour*, eds B. A. Campbell and R. M. Church, 279–296, Appleton-Century-Crofts, New York. [352]

Kawamura, S. (1963) The process of sub-culture propagation among Japanese macaques. In *Primate Social Behaviour*, ed. C. H. Southwick, 82–90, Van Nostrand, New York. [514]

Keeton, W. T. (1972) *Biological Science* (2nd edition). W. W. Norton, New York. [172]

Kellogg, W. N. (1968) Chimpanzees in experimental homes. *Psychol. Rec.* **18,** 489–498. [491]

Kellogg, W. N. and Kellogg, I. A. (1933) *The Ape and the Child*. McGraw-Hill, New York. [491]

Kendon, A. (1967) Some functions of gaze direction in social interaction. *Acta psychol.* **26,** 1–47. [487]

Kennedy, D. (1976) Neural elements in relation to network function. In *Simpler Networks and Behavior*, ed. J. C. Fentress, Sinauer Associates, Sunderland, Massachusetts. [241]

Kennedy, J. S. (1945) Classification and nomenclature of animal behaviour. *Nature, London* **156,** 754. [243]

Kettlewell, H. B. D. (1955) Selection experiments on industrial melanism in the Lepidoptera. *Heredity* **9,** 323–342. [59]

Kettlewell, H. B. D. (1956) Further selection experiments on industrial melanism in the Lepidoptera. *Nature, London* **175,** 934. [59]

Kettlewell, H. B. D. (1973) *The Evolution of Melanism*. Oxford University Press, Oxford. [60]

Kling, J. W. and Riggs, L. A. (1971) *Experimental Psychology*. Methuen, London. [207]

Klix, F. (1982) On the evolution of cognitive processes and performances. In *Animal Mind—Human Mind*, ed. D. R. Griffin, Springer Verlag, Berlin. [495]

Klopfer, P. H. and Gamble, J. (1966) Maternal "imprinting" in goats: the role of chemical senses. *Z. Tierpsychol.* **23,** 588–592. [377]

Knudsen (1981) The hearing of the barn owl. *Scientific American*, **245** (Dec), 83–91. [228]

Köhler, W. (1925) *The Mentality of Apes*. Harcourt Brace, New York. [346, 348]

Konishi, M. (1965) The role of auditory feedback in the control of vocalization in the white-crowned sparrow. *Z. Tierpsychol.* **22,** 770–783. [35]

Korringa, P. (1947) Relations between the moon and periodicity in the breeding of marine animals. *Ecol. Monogr.* **17,** 347–381. [303]

Kortlandt, A. (1940) Eine Übersicht der angeborenen Verhaltensweisen des Mitteleuropäischen Kormorans (*Phalacrocorax carbo sinensis*). *Arch. néérl. Zool.* **14,** 401–442. [381, 383]

Kortlandt, A. and Kooij, M. (1963) Protohominid behaviour in primates (preliminary communication). *Symp. zool. Soc. Lond.* **10,** 61–88. [512]

Kosslyn, M. (1981) The medium and the message in mental imagery; a theory. *Psychol. Rev.* **88,** 46–66. [498]

Kramer, G. (1951) Eine neue Methode zur Erforschung der Zugorientierung und die bisher damit erzielten Ergebnisse. *Proc. X int. ornithol. Congr. Uppsala, 1950,* 269–280. [254]

Krebs, J. R. (1971) Territory and breeding density in the great tit, *Parus major* L. *Ecology* **52,** 2–22. [434]

Krebs, J. R. (1973) Behavioural aspects of predation. In *Perspectives in Ethology*, eds P. P. G. Bateson and P. H. Klopfer, 73–111, Plenum Press, New York. [216]

Krebs, J. R. (1978) Optimal foraging: decision rules for predators. In *Behavioural Ecology: an Evolutionary Approach*, eds J. R. Krebs and N. B. Davies, Blackwell Scientific Publications, Oxford. [409]

Krebs, J. R. and Davies, N. B. (1981) *An Introduction to Behavioural Ecology*. Blackwell Scientific Publications, Oxford. [75, 121]

Kreithen, M. L. (1978) Sensory mechanisms for animal orientation—can any new ones be discovered? In *Animal Migration, Navigation and Homing*, eds K. Schmidt-Koenig and W. T. Keeton, 25–34, Proceedings in Life Sciences, Springer Verlag, Berlin. [254]

Kreithen, M. L. and Keeton, W. T. (1974) Detection of changes in atmospheric pressure by the homing pigeon, *Columba livia*. *J. comp. Physiol.* **89,** 73–82. [254, 258]

Krieckhaus, E. E. (1970) 'Innate recognition' aids rats in sodium regulation. *J. comp. physiol. Psychol.* **73,** 117–122. [272]

Krog, J., Folkow, B., Fox, R. H. and Anderson, K. L. (1960) Hand circulation in the cold of Lapps and North Norwegian fishermen. *JAP* **15,** 4: 654–658. [123]

Kruijt, J. P. (1964) Ontogeny of social behaviour in Burmese red junglefowl (*Gallus gallus spadiceus* Bonnaterre). *Behaviour Suppl. 12*. [31]

Kruuk, H. (1964) Predators and anti-predator behaviour of the black-headed gull (*Larus ridibundus* L.). *Behaviour Suppl. 11*, 1–130. [82, 83, 84]

Kummer, H. (1968) *Social Organization of Hamadryas Baboons*. University of Chicago Press, Chicago. [151]

Kummer, H. (1982) Social knowledge in free-ranging primates. In *Animal Mind—Human Mind*, ed. D. R. Griffin, Springer Verlag, Berlin. [354]

Kuo, Z. Y. (1932) Ontogeny of embryonic behaviour in Aves, IV: The influence of embryonic movements upon the behavior after hatching. *J. comp. Psychol.* **14,** 109–122. [31]

Kuo, Z. Y. (1967) *The Dynamics of Behavior Development*. Random House, New York. [31]

Lack, D. (1937) The psychological factor in bird distribution. *British Birds* **31,** 130. [528]

Lack, D. (1943) *The Life of the Robin*. London. [211]

Lack, D. (1954) *The Natural Regulation of Animal Numbers*. Oxford University Press, Oxford. [87]

Lack, D. (1966) *Population Studies of Birds*. Clarendon Press, Oxford. [87]

Lack, D. (1968) *Ecological Adaptations for Breeding in Birds*. Methuen, London. [75, 292, 293]

Laidler, K. (1978) Language in the Orang-utan. In *Action, Gesture and Symbol*, ed. A. Lock, 133–155, Academic Press, New York. [491]

Laming, P. R. (1981) *Brain Mechanisms of Behaviour in Lower Vertebrates*. Cambridge University Press, Cambridge. [180]

Landis, C. and Hunt, W. A. (1932) Adrenalin and emotion. *Psychol. Rev.* **39,** 467–485. [524]

Landsberg, J. W. (1976) Posthatch age and developmental age as a baseline for determination of the sensitive period for imprinting. *J. comp. physiol. Psychol.* **90,** 47–52. [369]

Larkin, S. (1981) Time and energy in decision-making. Unpublished D.Phil. thesis, University of Oxford. [450]

Larkin, S. and McFarland, D. J. (1978) The cost of changing from one activity to another. *Anim. Behav.* **26,** 1237–1246. [450, 475, 476]

Lawick-Goodall, J. van (1970) Tool-using in primates and other vertebrates. In *Advances in The Study of Behavior*, Vol. 3, eds D. S. Lehrman, R. A. Hinde and E. Shaw, Academic Press, New York. [510, 512]

Lea, S. E. G. (1978) The psychology and economics of demand. *Psychol. Bull.* **85,** 441–466. [454]

Lea, S. E. G. and Roper, T. J. (1977) Demand for food on fixed-ratio schedules as a function of the quality of concurrently available reinforcement. *J. exp. Anal. Behav.* **27,** 371–380. [453]

Leach, E. (1972) The influence of cultural context on non-verbal communication in man. In *Non-Verbal Communication*, ed. R. A. Hinde, Cambridge University Press, Cambridge. [485]

Lee, R. B. (1972) The Kung Bushmen of Botswana. In *Hunters and Gatherers Today*, ed. M. G. Bicchieri, Holt Rinehart and Winston, New York. [151]

Lee, R. B. and DeVore, I. (eds) (1968) *Man the Hunter*. Aldine, Chicago. [160]

Lehrman, D. S. (1953) A critique of Konrad Lorenz's theory of instinctive behaviour. *Quart. Rev. Biol.* **28,** 337–363. [32, 36]

Lehrman, D. S. (1955) The physiological basis of parental feeding behaviour in the ring dove (*Streptopelia risoria*). *Behaviour* **7,** 241–286. [184]

Lehrman, D. S. (1970) Semantic and conceptual issues in the nature-nurture problem. In *Development and Evolution of Behavior*, eds L. R. Aronson, E. Tobach, D. S. Lehrman and J. S. Rosenblatt, W. H. Freeman, New York. [28]

LeMay, M. (1976) Morphological cerebral asymmetries of modern man, fossil man, and nonhuman primates. *Ann. N.Y. Acad. Sci.* **280,** 349–366. [495]

Lendrem, D. W. (1983) Vigilance in birds. Unpublished D.Phil. thesis, University of Oxford. [476]

Lendrem, D. W. and McFarland, D. (1985) Selective attention and vigilance. In *Attention and Vigilance*, eds D. W. Lendrem and D. McFarland, Pitman, London. [477, 478]

Leong, C. Y. (1969) The quantitative effect of releasers on the attack readiness of the fish *Haplochromis burtoni* (Cichlidae: Pisces). *Z. vergl. Physiol.* **65,** 29–50. [219, 220]

Leroy, Y. (1964) Transmission du parametre fréquence dans le signal acoustique des hybrides F_1 et P × P_1, de deux grillons: *Teleogryllus commodus* Walker et *F. oceanicus* Le Guillon (Orthoptères, ensifères). *C.R. Acad. Sci.* **259,** 892–895. [44]

Lettvin, J. W., Maturana, H. R., McCulloch, W. S. and Pitts, W. H. (1959) What the frog's eye tells the frog's brain. *Proc. I.R.E.* **47,** 1940–1951. [211, 234]

Lewin, R. (1981) Ethiopian stone tools are world's oldest. *Science* **211,** 806–807. [495]

Lewis, R. A. (1975) Social influences on marital choice. In *Adolescence in the Life Cycle*, eds S. E. Dragastin and G. H. Elder, 211--225, Wiley, New York. [379]

Lieberman, P. (1975) *On the Origins of Language*. Macmillan, New York. [489, 490, 494]

Liley, N. R. (1966) Ethological isolating mechanisms in four sympatric species of poeciliid fishes. *Behaviour Suppl. 13*. [143, 144]

Lill, A. (1966) Some observations of social organization and non-random mating in captive Burmese red jungle fowl (*Gallus gallus spadiceus*). *Behaviour* **26,** 228–242. [143]

Lill, A. (1968) An analysis of sexual isolation in the domestic fowl, I: The basis of homogamy in males. *Behaviour* **30,** 8–126; II: The basis of homogamy in females. *Behaviour* **30,** 127–145. [143]

Lindauer, M. (1960) Time-compensated sun orientation in bees. *Cold Spring Harbour Symp. quant. Biol.* **25,** 371–377. [291]

Lindauer, M. (1961) *Communication among Social Bees*. Harvard University Press, Cambridge, Massachusetts. [420]

Lindauer, M. (1963) Kompassorientierung. *Ergeb. Biol.* **26,** 158–181. [419]

Linker, E., Moore, M. E. and Galanter, E. (1964) Taste thresholds, detection models, and disparate results. *J. exp. Psychol.* **67,** 59–66. [209]

Logan, F. A. and Boice, R. (1969) Aggressive behaviours of paired rodents in an avoidance context. *Behaviour* **34,** 161–83. [332]

Lorenz, K. (1932) Betrachtungen über das Erkennen der arteigenen Triebhandlungen der Vögel. *J. Ornithol.* **80,** 50–98. [394]

Lorenz, K. (1935) Der Kumpan in der Umwelt des Vogels. *J. Ornithol.* **83,** 137–213. [364, 367, 369, 374, 376, 394]

Lorenz, K. (1937) Über die Bildung des Instinktbegriffes. *Naturwissenschaften* **25,** 289, 300, 307–318, 324–331. [363]

Lorenz, K. (1939) Vergleichende Verhaltens-forschung. *Zoo. Anz. Suppl. Bd. 12*, 69–102. [36]

Lorenz, K. (1941) Vergleichende Bewegungsstudien an Anatiden. *Suppl. J. Ornith.* **89,** 194–294. [393]

Lorenz, K. (1950) The comparative method in studying innate behaviour patterns. *Sym. Soc. exp. Biol.* **4,** 221–268. [275, 363, 364]

Lorenz, K. (1965) *Evolution and Modification of Behavior*. University of Chicago Press, Chicago. [28, 29, 32, 327]

Loschiavo, S. R. (1968) Effect of oviposition on egg production and longevity in *Trogoderma parabile* (Coleoptera: Dermestidae). *Canad. Ent.* **100,** 86–89. [86]

Louw, G. N. and Holm, E. (1972) Physiological, morphological and behavioural adaptations of the ultrapsammophilous Namib Desert lizard *Aporosaura anchietae* (Bocage). *Madogua* **1,** 67–85. [265]

Lovejoy, C. O. (1981) The origin of man. *Science* **211,** 341–350. [154]

Ludlow, A. R. (1980) The evolution and simulation of a decision maker. In *Analysis of Motivational*

Processes, eds F. M. Toates and T. R. Halliday, Academic Press, London. [461, 465]

Lush, J. L. (1940) Intra-sire correlations or regressions of offspring on dam as a method of estimating heritability of characteristics. *Thirty-third Annual Proceedings of the American Society of Animal Production*, 293–301. [52]

Lythgoe, J. N. (1979) *The Ecology of Vision*. Clarendon Press, Oxford. [205, 223, 225]

McClearn, G. E. (1963) The inheritance of behavior. In *Psychology in the Making*, ed. L. J. Postman, Alfred A. Knopf, New York. [22]

McClearn, G. E. and DeFries, J. C. (1973) *Introduction to Behavioral Genetics*. W. H. Freeman, New York. [18, 44]

McDonald, D. L. (1972) Some aspects of the use of visual cues in directional training of homing pigeons. In *Animal Orientation and Navigation*, eds S. R. Galler, K. Schmidt-Koenig, G. J. Jacobs and R. E. Belleville, 293–304, NASA SP-262, US Govt Printing Office, Washington DC. [255]

McDonald, D. L. (1973) The role of shadows in directional training and homing of pigeons, *Columba livia*. *J. exp. Zool.* **183,** 267–280. [255]

McDougall, W. (1908) *An Introduction to Social Psychology*. Methuen, London. [275, 362]

McFarland, D. J. (1965) Hunger, thirst and displacement pecking in the Barbary dove. *Anim. Behav.* **13,** 292–300. [384, 386, 388]

McFarland, D. J. (1966a) The role of attention in the disinhibition of displacement activity. *Quart. J. exp. Psychol.* **18,** 19–30. [388]

McFarland, D. J. (1966b) On the causal and functional significance of displacement activities. *Z. Tierpsychol.* **23,** 217–235. [386, 387]

McFarland, D. J. (1969a) Mechanisms of behavioural disinhibition. *Anim. Behav.* **17,** 238–242. [389]

McFarland, D. J. (1969b) Separation of satiating and rewarding consequences of drinking. *Physiol. Behav.* **4,** 987–989. [459]

McFarland, D. J. (1970a) Recent developments in the study of feeding and drinking in animals. *J. psychosom. Res.* **14,** 229–237. [276]

McFarland, D. J. (1970b) Adjunctive behaviour in feeding and drinking situations. *Rev. Comp. Anim.* **4,** 64–73. [390, 466, 475]

McFarland, D. J. (1971) *Feedback Mechanisms in Animal Behaviour*. Academic Press, London. [250, 275–278, 457, 460, 502]

McFarland, D. J. (1973) Stimulus relevance and homeostasis. In *Constraints on Learning*, eds R. A. Hinde and J. Stevenson-Hinde, Academic Press, London. [335, 336]

McFarland, D. J. (1974) Time-sharing as a behavioral phenomenon. In *Advances in The Study of Behavior*, Vol. 4, eds D. S. Lehrman, J. S. Rosenblatt, R. A. Hinde and E. Shaw, Academic Press, New York. [277, 278, 464–467, 475–477]

McFarland, D. J. (1976) Form and function in the temporal organisation of behaviour. In *Growing Points in Ethology*, eds P. P. G. Bateson and R. A. Hinde, 55–93, Cambridge University Press, Cambridge. [280, 435]

McFarland, D. J. (1977) Decision-making in animals. *Nature, London* **269,** 15–21. [435, 436]

McFarland, D. J. (1983) Behavioural transitions: a reply to Roper and Crossland (1982): Time-sharing: a reply to Houston (1982). *Anim. Behav.* **31,** 305–308. [390, 461]

McFarland, D. J. and L'Angellier, A. B. (1966) Disinhibition of drinking during satiation of feeding behaviour in the Barbary dove. *Anim. Behav.* **14,** 463–467. [390]

McFarland, D. J. and Baher, E. (1968) Factors affecting feather posture in the Barbary dove. *Anim. Behav.* **16,** 171–177. [402]

McFarland, D. J. and Budgell, P. (1970) The thermoregulatory role of feather movements in the Barbary dove (*Streptopelia risoria*). *Physiol. Behav.* **5,** 763–771. [263]

McFarland, D. J. and Houston, A. (1981) Quantitative ethology: the state space approach, Pitman, London. [220, 221, 281, 290, 328, 442–446, 453, 454]

McFarland, D. J. and Lloyd, I. (1973) Time-shared feeding and drinking. *Quart. J. exp. Psychol.* **25,** 48–61. [465, 475]

McFarland, D. J. and McFarland, F. J. (1968) Dynamic analysis of an avian drinking response. *Med. biol. Engng.* **6,** 659–668. [459]

McFarland, D. J. and McGonigle, B. (1967) Frustration tolerance and incidental learning as determinants of extinction. *Nature, London* **215,** 786–787. [388]

McFarland, D. J. and Nunez, A. T. (1978) Systems analysis and sexual behaviour. In *Biological Determinants of Sexual Behaviour*, ed. J. B. Hutchinson, 615–652, Wiley, Chichester. [277]

McFarland, D. J. and Sibly, R. M. (1972) "Unitary

drives'' revisited. *Anim. Behav.* **20,** 548–563. [278, 279]

McFarland, D. J. and Sibly, R. M. (1975) The behavioural final common path. *Phil. Trans. R. Soc. B.* **270,** 265–293. [476]

McFarland, D. J. and Wright, P. (1969) Water conservation by inhibition of food intake. *Physiol. and Behav.* **4,** 95–99. [267]

McGonigle, B., McFarland, D. J. and Collier, P. (1967) Rapid extinction following drug-inhibited incidental learning. *Nature, London* **214,** 531–532. [388]

McGraw, M. B. (1945) *The Neuromuscular Maturation of the Human Infant*. Columbia University Press, New York. [152]

McGrew, W. C., Tutin, C. E. G. and Baldwin, P. J. (1979) Chimpanzees, tools and termites: cross-cultural comparison of Senegal, Tanzania and Rio Muni. *Man* **14,** 185–214. [160, 513]

Mackintosh, N. J. (1973) Stimulus selection: learning to ignore stimuli that predict no change in reinforcement. In *Constraints on Learning*, eds R. A. Hinde and J. Stevenson-Hinde, 75–96, Academic Press, London. [351]

Mackintosh, N. J. (1974) *The Psychology of Animal Learning*. Academic Press, London. [314–316, 322, 332–338, 343, 352]

Mackintosh, N. J. (1976) Overshadowing and stimulus intensity. *Anim. Learn. and Behav.* **4,** 186–192. [351, 352]

Mackintosh, N. J. (1983) *Conditioning and Associative Learning*. Clarendon Press, Oxford. [314]

McLeese, D. W. (1956) Effects of temperature, salinity and oxygen on the survival of the American lobster. *J. Fish. Res. Bd. Canada* **13,** 247–272. [287]

Macphail, E. M. (1982) *Brain and Intelligence in Vertebrates*. Clarendon Press, Oxford. [505, 509, 526]

Maier, S. F. and Seligman, M. E. P. (1976) Learned helplessness. Theory and evidence. *J. exp. Psychol. General* **105,** 3–46. [351]

Makkink, G. F. (1936) An attempt at an ethogram of the European avocet (*Recurvirostra avosetta* L.) with ethological and psychological remarks. *Ardea* **25,** 1–60. [381]

Malott, R. W. and Siddall, J. W. (1972) Acquisition of the people concept in pigeons. *Psychol. Rep.* **31,** 3–13. [345]

Manning, A. (1956) The effect of honey-guides. *Behaviour* **9,** 114–139. [418]

Mariscal, R. N. (1970a) An experimental analysis of the protection of *Amphiprion xanthurus* Cuv. and Val. and some other anemone fishes from sea anemones. *J. exp. mar. Biol. Ecol.* **4,** 134–149. [467]

Mariscal, R. N. (1970b) A field and laboratory study of the symbiotic behaviour of fishes and sea anemones from the tropical Indo-Pacific. *Calif. Pub. in Zoology* **91** (33 pages, 16 figs). [467]

Marler, P. and Mundinger, P. (1971) Vocal learning in birds. In *The Ontogeny of Vertebrate Behavior*, ed. H. Moltz, Academic Press, New York. [34]

Marler, P. and Tamura, M. (1964) Culturally transmitted patterns of vocal behavior in sparrows. *Science* **146,** 1483–1486. [515]

Marshack, A. (1976) Implications of the Paleolithic symbolic evidence for the origins of language. *American Scientist* **64,** 136–140. [495]

Marshall, A. J. (1970) Environmental factors other than light involved in the control of sexual cycles in birds and mammals. In *La Photorégulation de la Reproduction chez les Oiseaux et les Mammifères*, eds J. Benoit and I. Assenmacher, C.N.R.S. Editeur, Paris. [294]

Martin, G. R. (1977) Absolute visual threshold and scotopic spectral sensitivity in the tawny owl, *Strix aluco*. *Nature, London* **268,** 636–628. [223]

Mast, S. O. (1911) *Light and the Behavior of Organisms*. Wiley, London. [244]

Masterson, F. A. (1970) Is termination of a warning signal an effective reward for a rat? *J. comp. physiol. Psychol.* **72,** 471–477. [333]

Matthews, G. V. T. (1955) *Bird Navigation*. Cambridge University Press, Cambridge. [255]

Matthews, G. V. T. (1968) *Bird Navigation* (2nd edition). Cambridge University Press, Cambridge. [255]

Maynard Smith, J. (1958) Sexual selection. In *A Century of Darwin*, ed. S. A. Barnett, 231–244, Harvard University Press, Cambridge, Massachusetts. [124]

Maynard Smith, J. (1964) Group selection and kin selection. *Nature, London* **201,** 1145–1147. [127, 128]

Maynard Smith, J. (1976) Sexual selection and the handicap principle. *J. theor. Biol.* **57,** 239–242. [102, 103, 118]

Maynard Smith, J. (1977) Parental investment—a prospective analysis. *Anim. Behav.* **25,** 1–9. [130–132]

Maynard Smith, J. (1978a) *The Evolution of Sex*. Cambridge University Press, Cambridge. [130]

Maynard Smith, J. (1978b) The ecology of sex. In *Behavioural Ecology: an Evolutionary Approach*, eds J. R. Krebs and N. B. Davies, Blackwell Scientific Publications, Oxford. [131, 132, 377]

Maynard Smith, J. (1982) *Evolution and the Theory of Games*, Cambridge University Press, Cambridge. [101]

Meddis, R. (1965) On the function of sleep. *Anim. Behav.* **23,** 676–691. [305]

Menzel, E. W. (1974) A group of young chimpanzees in a one-acre field. In *Behavior of Nonhuman Primates*, eds A. M. Schrier and F. Stollnitz, Vol. 3, 83–153, Academic Press, New York. [490]

Menzel, E. W. (1978) Cognitive mapping in chimpanzees. In *Cognitive Processes in Animal Behavior*, eds S. H. Hulse, H. Fowler and W. K. Honig, 375–422, Lawrence Erlbaum Associates, Hillsdale, New Jersey. [346, 490]

Menzel, E. W. (1979) Communication of object-locations in a group of young chimpanzees. In *The Great Apes*, eds D. A. Hamburg and E. R. McGown, 359–371, Benjamin/Cummings, Menlo Park, California. [490, 491]

Menzel, E. W. and Johnson, M. K. (1976) Communication and cognitive organization in human and other animals. *Ann. N.Y. Acad. Sci.* **280,** 131–142. [491]

Menzel, R. (1979) Behavioral access to short-term memory in bees. *Nature,* **281,** 331–337. [425]

Menzel, R., Erber, J. and Masuhr, T. (1974) Learning and memory in the honey bee. In *Experimental Analysis of Insect Behaviour*, ed. L. Barton Browne, Springer Verlag, New York. [425]

Merkel, F. W. and Wiltschko, W. (1965) Magnetismus und Richtungsfinden zugunruhiger Rotkelchen (*Erithacus rubecula*). *Vogelwarte* **23,** 71–77. [254]

Merton, P. A. (1964) Absence of conscious position sense in the human eyes. In *The Oculomotor System*, ed. M. B. Bender, Harper and Row, New York. [249]

Messenger, J. B. (1968) The visual attack of the cuttlefish *Sepia officinalis. Anim. Behav.* **16,** 342–369. [232, 241]

Meyer, D. B. (1977) The avian eye and its adaptation. In *Handbook of Sensory Physiology*, Vol. III, ed. F. Crescitelli, 549–611, Springer Verlag, Berlin. [205]

Meyerriecks, A. J. (1960) Comparative breeding behavior of four species of North American herons. *Nuttal Ornithological Club Publication* **2,** 1–158. [399]

Miller, N. E. (1948) Studies of fear as an acquirable drive. *J. exp. Psychol.* **38,** 89–101. [333]

Miller, R. S. (1967) Pattern and process in competition. In *Advances in Ecological Research*, Vol. 4, ed. J. B. Cragg, Academic Press, New York. [70]

Milsum, J. H. and Roberge, F. A. (1973) Physiological regulation and control. In *Foundations of Mathematical Biology*, Vol. 3, ed. R. Rosen, 1–95, Academic Press, London. [432]

Mittelstaedt, M. (1964) Basic control patterns of orientational homeostasis. *Symp. Soc. exp. Biol.* **18,** 365–385. [328]

Moore, B. R. (1973) The role of directed Pavlovian reactions in simple instrumental learning in the pigeon. In *Constraints on Learning*, eds R. A. Hinde and J. Stevenson-Hinde, Academic Press, London. [329–331]

Moore, M. J. and Capretta, P. J. (1968) Changes in colored or flavored food preferences in chicks as a function of shock. *Psychonomic. Sci.* **12,** 195–196. [338]

Morgan, C. L. (1894) *Introduction to Comparative Psychology*. Scott, London. [8, 320, 505, 521]

Morgan, C. L. (1900) *Animal Behaviour*. Scott, London. [320]

Morgan, M. and Nicholas, D. J. (1979) Discrimination between reinforced action patterns in the rat. *Learn. and Motiv.* **10,** 1–22. [521]

Morris, D. (1957) "Typical intensity" and its relation to the problem of ritualisation. *Behaviour* **11,** 1–12. [392]

Morris, D. (1967) *The Naked Ape*. Cape, London. [125]

Morse, D. H. (1971) The insectivorous bird as an adaptive strategy. *Ann. Rev. ecol. Syst.* **2,** 177–200. [70]

Moynihan, M. (1955) Some aspects of reproductive behaviour in the black-headed gull (*Larus ridibundus ridibundus* L.) and related species. *Behaviour Suppl. 4,* 1–201. [401, 402]

Mundinger, P. C. (1980) Animal cultures and a general theory of cultural evolution. *Ethol. Sociobiol.* **1,** 183–223. [514]

Muntz, W. R. A. (1981) Colour vision. In *The Oxford Companion to Animal Behaviour*, ed D. McFarland, Oxford University Press, Oxford. [201]

Munz, F. W. (1958) Photosensitive pigments from the retinae of certain deep sea fishes. *J. Physiol.* **140,** 220–225. [223]

Nachman, M. (1970) Learned taste and temperature aversions due to lithium chloride sickness after temporal delays. *J. comp. physiol. Psychol.* **73,** 22–30. [338]

Nethersole Thompson, C. and D. (1942) Eggshell disposal by birds. *British Birds* **35,** 162–169, 190–200, 214–224, 241–250. [78]

Nicol, J. A. C. (1965) Migration of choroidal tapetal pigment in the spur dog, *Squalus acanthias. J. mar. biol. Ass. UK* **45,** 405–427. [224]

Niebuhr, V. (1981) An investigation of courtship feeding in herring gulls *Larus argentatus. Ibis* **123,** 218–223. [144]

Nisbet, I. C. T. (1973) Courtship feeding, egg-size and breeding success in common terns. *Nature, London* **241,** 141–142. [144]

Nisbet, I. C. T. (1977) Courtship feeding and clutch size in common terns *Sterna hirundo*. In *Evolutionary Ecology*, eds B. Stonehouse and C. M. Perrins, Macmillan, London. [144]

Norton-Griffiths, M. N. (1967) Some ecological aspects of the feeding behaviour of the oystercatcher *Haematopus ostralegus* on the edible mussel *Mytilus edulis. Ibis* **109,** 412–424. [516]

Norton-Griffiths, M. N. (1969) The organisation, control and development of parental feeding in the oystercatcher (*Haematopus ostralegus*). *Behaviour* **34,** 55–114. [516]

Nottebohm, F. (1976) Vocal tract and brain: a search for evolutionary bottlenecks. *Ann. N.Y. Acad. Sci.* **280,** 643–649. [523]

O'Brien, W. J., Slade, N. A. and Vinyard, G. L. (1976) Apparent size as the determinant of prey selection by Bluegill sunfish (*Lepomis machrochirus*). *Ecology* **57,** 1304–1311. [429]

Orians, G. H. and Willson, M. F. (1964) Interspecific territories of birds. *Ecology* **45,** 736–745. [71]

Packer, C. (1977) Reciprocal altruism in *Papio anubis. Nature, London* **265,** 441–443. [133]

Palmer, J. D. (1973) Biological clocks of the tidal zone. *Scientific American*, 70–79, [304]

Papi, F. (1960) "Orientation by night: the moon". *Cold Spring Harbour Symp. quant. Biol.* **25,** 475–80. [291]

Parker, G. A. (1978) Searching for mates. In *Behavioural Ecology: An Evolutionary Approach*, eds J. R. Krebs and N. B. Davies, 214–244. Blackwell Scientific Publications, Oxford. [148]

Parker, G. A. (1979) Sexual selection and sexual conflict. In *Sexual Selection and Reproductive Competition in Insects*, eds M. S. Blum and N. A. Blum, 123–166, Academic Press, New York. [142]

Parker, G. A., Baker, R. R. and V. G. F. Smith (1972) The origin and evolution of gamete dimorphism and the male–female phenomenon. *J. theor. Biol.* **36,** 529–553. [121]

Parsons, P. A. (1967) *The Genetic Analysis of Behaviour*. Methuen, London. [124]

Partridge, L. (1978) Habitat selection. In *Behavioural Ecology: An Evolutionary Approach*, eds J. R. Krebs and N. B. Davies, Blackwell Scientific Publications, Oxford. [93, 528]

Passingham, R. E. (1975) Changes in the size and organization of the brain in man and his ancestors. *Brain Behav. Evol.* **11,** 73–90. [152]

Passingham, R. E. (1981) Primate specializations in brain and intelligence. *Symp. zool. Soc. Lond.* **46,** 361–388. [508, 509]

Passingham, R. E. (1982) *The Human Primate*. W. H. Freeman, New York. [151, 154, 158, 160, 183, 346, 491, 494, 495, 509, 526]

Patterson, F. G. (1978) The gestures of a gorilla: language acquisition in another pongid. *Brain and Lang.* **5,** 72–97. [491–493]

Patterson, F. G. (1979) Linguistic capabilities of a young lowland gorilla. Unpublished dissertation, Stanford University. [493]

Patterson, I. J. (1965) Timing and spacing of broods in the black-headed gull *Larus ridibundus. Ibis* **107,** 433–459. [83, 84]

Pavlov, I. P. (1927) *Conditioned Reflexes*. Oxford University Press, London. [311, 316–319]

Payne, R. S. (1962) How the barn owl locates prey by hearing. *Living Bird*, **1,** 151–159. [227]

Payne, T. L. (1974) Pheromone perception. In *Pheromones*, ed. M. C. Birch, North Holland, Amsterdam. [189]

Pearson, K. G. and Iles, J. F. (1970) Central programming and reflex control of walking in the cockroach. *J. exp. Biol.* **56,** 173–193. [241]

Pengelley, E. T. (ed.) (1974) *Circannual Clocks—Annual Biological Rhythms*. Academic Press, New York. [292]

Pengelley, E. T. and Asmundson, S. J. (1974) Circannual rhythmicity in hibernating mammals. In *Circannual Clocks*, ed. E. T. Pengelley, Academic Press, New York. [296]

Pennycuick, C. J. (1960) The physical basis of astronavigation in birds: theoretical considerations. *J. exp. Biol.* **37,** 573–593. [255]

Perdeck, A. C. (1958) Two types of orientation in migrating starlings *Sturnus vulgaris* L., and chaffinches, *Fringilla coelebs* L., as revealed by displacement experiments. *Ardea* **46,** 1–37. [251, 252]

Perdeck, A. C. (1967) Orientation of starlings after displacement to Spain. *Ardea* **55,** 194–202. [251]

Perril, S. A., Gerhardt, H. C. and Daniel, R. (1978) Sexual parasitism in the green tree frog, *Hyla cinerea. Science* **200,** 1179–1180. [149]

Perrins, C. M. (1965) Population fluctuation and clutch-size in the great tit (*Parus major* L.). *J. anim. Ecol.* **34,** 601–647. [87]

Peters, R. S. (1958) *The Concept of Motivation.* Routledge and Kegan Paul, London. [275]

Petrusic, W. A., Varro, L. and Jamieson, D. G. (1978) Mental rotation validation of two spatial ability tests. *Psychol. Res.* **40,** 139–148. [498]

Pitcher, T. J., Partridge, B. L. and Wardle, C. S. (1976) A blind fish can school. *Science* **194,** 963–965. [225]

Plutchik, R. and Ax, A. F. (1970) A critique of determinants of emotional state by Schachter and Singer (1962). In *Feelings and Emotions: the Loyola Symposium,* ed. M. B. Arnold, Academic Press, New York. [525]

Posner, M. I. (1978) *Chronometric Explorations of Mind.* Lawrence Erlbaum Associates, Hillsdale, New Jersey. [498]

Powell, G. V. N. (1974) Experimental analysis of the social value of flocking by starlings (*Sturnus vulgaris*) in relation to predation and foraging. *Anim. Behav.* **22,** 501–505. [432]

Premack, D. (1970) A functional analysis of language. *J. exp. Anal. Behav.* **14,** 107–125. [492]

Premack, D. (1976) *Intelligence in Ape and Man.* Lawrence Erlbaum Associates, Hillsdale, New Jersey. [346, 492–496]

Premack, D. (1978) Chimpanzee theory of mind, Part II: The evidence for symbols in chimpanzee. *Behav. Brain Sci.* **1,** 625–629. [345, 492]

Premack, A. J. and Premack, D. (1972) Teaching language to an ape. *Scientific American* **227,** (Oct.) 92–99. [497]

Premack, D. and Woodruff, G. (1978) Does the chimpanzee have a theory of mind? *Behav. Brain Sci.* **1,** 515–526. [503]

Prosser, C. L. (1973) *Comparative Animal Physiology* (3rd edition). W. B. Saunders, Philadelphia. [173, 174, 191, 198, 203, 243]

Prout, T. (1971) The relation between fitness components and population prediction in *Drosophila.* I: The estimation of fitness components. *Genetics* **68,** 127–149. [88]

Provine, R. R. (1981) Wing-flapping development in chickens made flightless by feather mutations. *Developmental Psychobiology* **14,** 481–486. [31]

Pulliam, H. R. (1973) On the advantages of flocking. *J. theor. Biol.* **38,** 419–422. [432]

Raber, H. (1948) Analyse des Balzverhaltens eines domestizierten Truthahns (*Meleagris*). *Behaviour* **1,** 237–266. [384]

Rachlin, H. (1980) Economics and behavioral psychology. In *Limits to Action,* ed. J. E. R. Staddon, Academic Press, New York. [454]

Randolph, M. C. and Brooks, B. A. (1967) Conditioning of a vocal response in a chimpanzee through social reinforcement. *Folia primat.* **5,** 70–79. [401]

Reinberg, A. (1974) Aspects of circannual rhythms in Man. In *Circannual Clocks,* ed. E. T. Pengelley, Academic Press, New York. [290]

Rescorla, R. A. (1971) Variations in the effectiveness of reinforcement and nonreinforcement following prior inhibitory conditioning. *Learn. and Motiv.* **2,** 113–123. [352, 353]

Rescorla, R. A. (1978) Some implications of a cognitive perspective on Pavlovian conditioning. In *Cognitive Processes in Animal Behavior,* eds S. H. Hulse, H. Fowler and W. K. Honig, Lawrence Erlbaum Associates, Hillsdale, New Jersey. [342, 344]

Rescorla, R. A. and Cunningham, C. L. (1979) Spatial contiguity facilitates Pavlovian second-order conditioning. *J. exp. Psychol.: Animal Behaviour Processes* **5,** 152–161. [353]

Rescorla, R. A. and Furrow, D. R. (1977) Stimulus similarity as a determinant of Pavlovian conditioning. *J. exp. Psychol.: Animal Behaviour Processes* **3,** 203–215. [353]

Revusky, S. H. (1967) Hunger level during food consumption: effects on subsequent preference. *Psychonomic. Sci.* **7,** 109–110. [336]

Revusky, S. H. (1971) The role of interference in association over delay. In *Animal Memory,* eds W. K. Honig and P. H. R. James, 55–213, Academic Press, New York. [339, 353]

Revusky, S. H. (1977) Learning as a general process with an emphasis on data from feeding experiments. In *Food Aversion Learning*, eds N. W. Milgram, L. Krames and T. M. Alloway, Plenum Press, New York. [337, 339]

Revusky, S. H. and Garcia, J. (1970) Learned associations over long delays. In *Psychology of Learning and Motivation*, Vol. 4, ed. G. H. Bower, 1–83, Academic Press, New York. [260, 335–338]

Richelle, M. and Lejeune, H. (1980) *Time in Animal Behaviour*. Pergamon Press, Oxford. [344]

Richter, C. P. (1943) Total self-regulatory functions in animals and human beings. *Harvey Lect.* **38,** 63–103. [260, 274, 276]

Richter, C. P. (1955) Self-regulatory functions during gestation and lactation. *Trans. Conf. Gestation.* Princeton, New Jersey. [274]

Richter, C. P., Holt, L. E. and Barelare, B. Jr. (1937) Vitamin B_1 craving in rats. *Science* **86,** 354–355. [334]

Riddell, W. I. (1979) Cerebral indices and behavioral differences. In *Development and Evolution of Brain Size*, eds M. E. Hahn, C. Jensen and B. C. Dudek, 89–109, Academic Press, New York. [509]

Ridley, M. and Rechten C. (1981) Female sticklebacks prefer to spawn with males whose nests contain eggs. *Behaviour* **76,** 1–2. [119]

Riggs, S. K. and Sargent, F. (1964) Physiological regulation in moist heat by young American negro and white males. *Human Biol.* **36,** 339–353. [123]

Ristau, C. A. and Robbins, D. (1981) Language in the great apes: a critical review. In *Advances in the Study of Behavior*, Vol. 12, eds J. Rosenblatt, R. A. Hinde, C. Beer and M. C. Busnel, Academic Press, New York. [494]

Ristau, C. A. and Robbins, D. (1982) Cognitive aspects of ape language experiments. In *Animal Mind—Human Mind*, ed. D. R. Griffin, Springer Verlag, Berlin. [496, 521]

Roberts, S. (1981) Isolation of an internal clock. *J. exp. Psychol.: Animal Behavior Processes* **7,** 242–268. [344]

Robinson, M. H. (1970) Insect anti-predator adaptations and the behaviour of predatory primates. *Congr. Latin. Zool.* **II,** 811–836. [233]

Rodgers, W. and Rozin, P. (1966) Novel food preferences in thiamine-deficient rats. *J. comp. physiol. Psychol.* **61,** 1–4. [336]

Roeder, K. D. (1963) *Nerve Cells and Insect Behaviour*. Harvard University Press, Cambridge, Massachusetts. [191, 193]

Roeder, K. D. (1970) Episodes in insect brains. *American Scientist* **58,** 378–389. [191, 193]

Rohwer, S. and Rohwer, F. C. (1978) Status signalling in Harris sparrows: experimental deceptions achieved. *Anim. Behav.* **26,** 1012–1022. [410]

Rolls, B. J. and Rolls, E. T. (1982) *Thirst*. Cambridge University Press, Cambridge. [274, 277]

Romanes, G. J. (1882) *Animal Intelligence*. Kegan, Paul, Trench, London. [8, 505, 521]

Romer, A. S. (1958) *Vertebrate Paleontology*. Chicago University Press, Chicago. [61]

Root, R. B. (1967) The niche exploitation pattern of the blue-gray gnatcatcher. *Ecol. Monogr.* **37,** 317–350. [68–70]

Roper, T. J. and Crossland, G. (1982) Mechanisms underlying eating-drinking transitions in rats. *Anim. Behav.* **30,** 602–614. [390]

Rothenbuhler, N. (1964) Behavior genetics of nest cleaning in honey bees. Responses of F_1 and backcross generations to disease-killed brood. *Amer. Zool.* **4,** 111–123. [39, 40]

Rowell, C. H. F. (1961) Displacement grooming in the chaffinch. *Anim. Behav.* **9,** 38–63. [385, 386, 388]

Rowell, C. H. F. and Horn, G. (1968) Dishabituation and arousal in the response of single nerve cells in an insect brain. *J. exp. Biol.* **49,** 171–183. [317]

Rowley, I. (1965) White-winged choughs. *Aust. Nat. Hist.* **15,** 81–85. [139]

Rozin, P. (1967) Specific aversions as a component of specific hungers. *J. comp. physiol. Psychol.* **64,** 237–242. [273]

Rozin, P. (1968) Specific aversions and neophobia as a consequence of vitamin deficiency and/or poisoning in half-wild and domestic rats. *J. comp. physiol. Psychol.* **66,** 82–88. [336]

Rozin, P. (1969) Adaptive food sampling in vitamin deficient rats. *J. comp. physiol. Psychol.* **69,** 126–132. [338]

Rozin, P. (1976a) The selection of foods by rats, humans and other animals. In *Advances in the Study of Behavior*, eds. J. S. Rosenblatt, R. A. Hinde, E. Shaw and C. Bear, Vol. 6, 21–76, Academic Press, New York. [273, 274, 336]

Rozin, P. (1976b) The evolution of intelligence and access to the cognitive unconscious. In *Progress in Psychobiology and Physiological Psychology*, Vol. 6, eds J. M. Sprague and A. N. Epstein, 245–276, Academic Press, New York. [516]

Rozin, P. and Kalat, J. (1971) Specific hungers and poison avoidance as adaptive specializations of learning. *Psychol. Rev.* **78,** 459–486. [260, 273, 274, 329, 336–338]

Rozin, P. and Kalat, J. (1972) Learning as a situation-specific adaptation. In *Biological Boundaries of Learning,* eds M. E. P. Seligman and J. Hager, 66–97, Appleton, New York. [339]

Rozin, P. and Mayer, J. (1961) Thermal reinforcement and thermoregulatory behavior in the goldfish. *Science* **134,** 942–943. [288]

Rumbaugh, D. M. (1977) *Language Learning by a Chimpanzee.* Academic Press, New York. [493]

Rumbaugh, D. M. and Gill, T. V. (1977) Lana's acquisition of language skills. In *Language Learning by a Chimpanzee,* ed. D. M. Rumbaugh, 165–192, Academic Press, New York. [496]

Ruppel, G. (1969) Eine "Lüge" als gerichtete Mitteilung beim Eisfuchs (*Alopex lagopus* L.). *Z. Tierpsychol.* **26,** 371–374. [501]

Rusak, B. (1981) Vertebrate behavioral rhythms. In *Handbook of Behavioral Neurobiology, Vol. 4: Biological Rhythms,* ed. J. Aschoff. Plenum Press, New York. [305]

Rutledge, J. T. (1974) Circannual rhythm of reproduction in male European starlings (*Sturnus vulgaris*). In *Circannual Clocks,* ed. E. T. Pengelley, Academic Press, New York. [299]

Rzoska, J. (1953) Bait shyness, a study in rat behaviour. *Brit. J. anim. Behav.* **1,** 128–135. [335]

Sadoglu, P. (1975) Genetic paths leading to blindness in *Astyanax mexicanus.* In *Vision in Fishes,* ed. M. A. Ali, 419–426, Plenum Press, New York. [225]

Salzano, F. M., Neel, J. V. and Maybury-Lewis, D. (1967) Further studies on the Xavante Indians. *Amer. J. human Genetics* **19,** 463–489. [124]

Sauer, F. and Sauer, E. (1955) Zur Frage der nächtlichen Zugorientierung von Grasmücken. *Rev. suisse Zool.* **62,** 250–259. [256]

Saunders, D. S. (1976) *Insect Clocks.* Pergamon Press, Oxford. [246, 291]

Savage-Rumbaugh, E. S., Rumbaugh, D. M. and Boysen, S. (1978) Symbolic communication between two chimpanzees (*Pan troglodytes*). *Science* **201,** 641–644. [493]

Savage-Rumbaugh, E. S., Rumbaugh, D. M. and Boysen, S. (1980) Do apes use language? *American Scientist* **68,** 49–61. [493]

Schachter, S. and Singer, J. (1962) Cognitive, social and physiological determinants of emotional state. *Psychol. Rev.* **69,** 379–399. [525]

Scharrer, E. (1964) Photo-neuro-endocrine systems: general concepts. *Ann. N.Y. Acad. Sci.* **117,** 13–22. [292]

Scheller, R. H. and Axel, R. (1984) How genes control behavior. *Scientific American* **250,** 44–52. [27]

Schiller, P. (1952) Innate constituents of complex responses in primates. *Psychol. Rev.* **59,** 177–191. [349]

Schmidt-Koenig, K. (1958) Experimentelle Einflussnahme auf die 24-Stunden-Periodik bei Brieftauben und deren Auswirkungen unter besonderer Berücksichtigung des Heimfindevermögens. *Z. Tierpsychol.* **15,** 301–331. [254]

Schmidt-Koenig, K. (1960) Internal clocks and homing. *Cold Spring Harbour Symp. quant. Biol.* **25,** 389–393. [254, 255]

Schmidt-Koenig, K. (1961) Die Sonne als Kompass im Heim-Orientierungssystem der Brieftauben. *Z. Tierpsychol.* **68,** 221–244. [254]

Schmidt-Koenig, K. (1979) *Avian Orientation and Navigation.* Academic Press, London. [252–258, 291, 299, 302, 515]

Schmidt-Nielsen, K. (1964) *Desert Animals: Physiological Problems of Heat and Water.* Clarendon Press, Oxford. [264, 268, 294]

Schneider, D. (1969) Insect olfaction: deciphering system for chemical messages. *Science* **163,** 1031–1036. [188]

Schneidermann, N., Fuentes, I. and Gormezano, I. (1962) Acquisition and extinction of the classically conditioned eyelid response in the albino rabbit. *Science* **136,** 650–652. [314, 315]

Schneirla, T. C. (1965) Aspects of stimulation and organization in approach/withdrawal processes underlying vertebrate behavioral development. In *Advances in the Study of Behavior,* eds D. S. Lehrman, R. A. Hinde and E. Shaw, Vol. 1, 1–74, Academic Press, New York. [32]

Schutz, F. (1965) Sexuelle Prägung bei Anatiden. *Z. Tierpsychol.* **22,** 50–103. [371, 373, 376, 377]

Schutz, F. (1971) Prägung des Sexualverhaltens von Enten und Gänsen durch Sozialeindrücke während der Jugendphase. *J. Neuro-visc. Rel. Suppl. 10,* 339–357. [373, 374]

Schüz, E. (1963) On the northwestern migration

divide of the white stork. *Proc. int. ornithol. Congr.* **13,** 475–480. [300]

Schüz, E. (1971) *Grundriss der Vogelzugskunde*. Parey Verlag, Berlin. [300]

Schwartz, E. (1974) Lateral-line mechanoreceptors in fishes and amphibians. In *Handbook of Sensory Physiology*, ed. A. Fessard, III/3, 257–278, Springer Verlag, New York. [225]

Scott, E. M. and Verney, E. L. (1947) Self-selection of diet, VI: The nature of appetites for B vitamins. *J. Nutr.* **34,** 471–480. [334]

Scott, J. P. and Fuller, J. L. (1965) *Dog Behavior: The Genetic Basis*. University of Chicago Press, Chicago. [45, 51]

Seitz, A. (1940) Die Paarbildung bei einigen Cichliden I. *Z. Tierpsychol.* **4,** 40–84. [218]

Selander, R. K. (1964) Speciation in wrens of the genus *Campylorhynchus*. *Univ. Calif. Publ. Zool.* **74,** 1–224. [137]

Selander, R. K. (1972) Sexual selection and dimorphism in birds. In *Sexual Selection and the Descent of Man*, ed. B. Campbell, Heinemann, London. [122]

Seligman, M. E. P. (1970) On the generality of the laws of learning. *Psychol. Rev.* **77,** 406–418. [327]

Senturia, J. B. and Johansson, B. W. (1974) Physiological and biochemical reflections of circannual rhythmicity in the European hedgehog and man. In *Circannual Clocks*, ed. E. T. Pengelley, Academic Press, New York. [290]

Sevenster, P. (1961) A causal analysis of a displacement activity (fanning in *Gasterosteus aculeatus* L.). *Behaviour Suppl. 9*, 1–170. [385, 386]

Sevenster, P. (1968) Motivation and learning in sticklebacks. In *The Central Nervous System and Fish Behavior*, ed. D. Ingle, 233–245, University of Chicago Press, Chicago. [325, 388]

Sevenster, P. (1973) Incompatibility of response and reward. In *Constraints on Learning: Limitations and Predispositions*, eds R. A. Hinde and J. Stevenson-Hinde, Academic Press, London. [325]

Seyfarth, R. M., Cheney, D. L. and Marler, P. (1980) Monkey responses to three different alarm calls: evidence of predator classification and semantic communication. *Science* **210,** 801–803. [490, 502]

Sheffield, F. D. (1965) Relation between classical conditioning and instrumental learning. In *Classical Conditioning: A Symposium*, ed. W. F. Prokasy, Appleton-Century-Crofts, New York. [319]

Shepard, R. N. and Metzler, J. (1971) Mental rotation of three-dimensional objects. *Science* **171,** 701–703. [498]

Shepher, J. (1971) Self-imposed incest-avoidance and exogamy in second generation Kibbutz adults. Unpublished doctoral dissertation, Rutgers University, New Brunswick, New Jersey. [379]

Sheppard, P. M. (1961) Some contributions to population genetics resulting from the study of Lepidoptera. In *Advances in Genetics*, Vol. 10, eds E. W. Caspari and J. M. Thoday, 165–216, Academic Press, New York. [41]

Sherrington, C. S. (1918) Observations on the sensual role of the proprioceptive nerve-supply of the extrinsic ocular muscles. *Brain* **41,** 332–343. [249]

Sherry, D. F., Mrosovsky, N. and Hogan, J. A. (1980) Weight loss and anorexia during incubation in birds. *J. comp. physiol. Psychol.* **94,** 89–98. [271]

Shettleworth, S. J. (1972) Constraints on learning. In *Advances in the Study of Behavior*, Vol. 4, eds D. S. Lehrman, R. Hinde and E. Shaw, Academic Press, New York. [325, 329]

Shumake, S. A., Smith, J. C. and Tucker, D. (1969) Olfactory intensity-difference thresholds in the pigeon. *J. comp. physiol. Psychol.* **67,** 64–69. [254]

Sibly, R. M. (1975) How incentive and deficit determine feeding tendency. *Anim. Behav.* **23,** 437–446. [460, 476]

Sibly, R. M. and McCleery, R. H. (1976) The dominance boundary method of determining motivational state. *Anim. Behav.* **24,** 108–124. [467, 475]

Sibly, R. M. and McCleery, R. H. (1985) Optimal decision rules for gulls. *Anim. Behav.* (in press). [438, 439]

Sibly, R. M. and McFarland, D. J. (1974) A state-space approach to motivation. In *Motivational Control Systems Analysis*, ed. D. J. McFarland, 213–250, Academic Press, London. [289, 290]

Siegel, R. G. and Honig, W. K. (1970) Pigeon concept formation: successive and simultaneous acquisition. *J. exp. Analysis Behav.* **13,** 385–390. [345]

Siegfried, W. R. and Underhill, L. G. (1975) Flocking as an anti-predator strategy in doves. *Anim. Behav.* **23,** 504–508. [432]

Simmons, J. A. (1971) Echolocation in bats: signal processing of echoes for target range. *Science* **171,** 925–928. [231]

Sinclair, A. R. E. (1977) *The African Buffalo*. University of Chicago Press, Chicago. [115]
Skinner, B. F. (1937) Two types of conditioned reflex: a reply to Konorski and Miller. *J. Gen. Psychol.* **16,** 272–279. [323]
Skinner, B. F. (1938) *The Behavior of Organisms*. Appleton-Century-Crofts, New York. [322, 323, 326]
Skinner, B. F. (1953) *Science and Human Behavior*. Macmillan, New York. [323]
Skinner, B. F. (1958) Reinforcement today. *Amer. Psychol.* **13,** 94–99. [325]
Skinner, B. F. (1975) *About Behaviourism*, Cape, London. [325]
Skutch, A. F. (1969) *Life Histories of Central American Birds, Vol. III: The Golden naped Woodpecker*. Pacific Coast Avifauna, 35, 479–517. [139]
Skutch, A. F. (1976) *Parent Birds and Their Young*. University of Texas Press, Austin, Texas. [500]
Sluckin, W. (1964) *Imprinting and Early Learning*. Methuen, London. [369]
Smart, J. L. (1977) Early life malnutrition and late learning ability: a critical analysis. In *Genetics and Intelligence*, ed. A. Oliveira, 215–235, North Holland, Amsterdam. [372]
Smith, N. G. (1966) Evolution of some arctic gulls (*Larus*): an experimental study of isolating mechanisms. *Ornithol. Monogr.* **4,** 1–99. [142, 143]
Sparks, J. (1982) *The Discovery of Animal Behaviour*. Collins, London. [364]
Sutherland, N. S. (1964) Discrimination learning: non-additivity of cues. *Nature, London* **201,** 528–530. [388]
Sutherland, N. S. and Mackintosh, N. J. (1971) *Mechanisms of Animal Discrimination Learning*. Academic Press, New York. [215, 216]
Sutton, D. (1979) Mechanisms underlying vocal control in nonhuman primates. In *Neurobiology of Social Communication*, eds H. D. Steklis and M. J. Raleigh, 45–67, Academic Press, New York. [491]
Sweeney, B. M. (1969) *Rhythmic Phenomena in Plants*. Academic Press, New York. [292]
Swets, J. A., Tanner, W. P. and Birdsall, T. G. (1961) Decision processes in perception. *Psychol. Rev.* **68,** 301–340. [208]
Tansley, K. (1965) *Vision in Vertebrates*. Methuen, London. [224]
Terrace, H. S. (1979) *Nim*. Eyre Methuen, London. [491–493]
Testa, T. J. (1975) Effects of similarity of location and temporal intensity pattern of conditioned and unconditioned stimuli on acquisition of conditioned suppression in rats. *J. exp. Psychol.: Animal Behavior Processes* **1,** 114–121. [353]
Thoday, J. M. (1953) Components of fitness. *Symp. Soc. exp. Biol.* **7,** 96–113. [85]
Thompson, R. F. (1965) *Foundations of Physiological Psychology*. Harper International, New York. [318]
Thorndike, E. L. (1898) Animal intelligence: an experimental study of the associative processes in animals. *Psychol. Rev. Monogr. Suppl. 2* **8,** 1, 16. [320]
Thorndike, E. L. (1911) *Animal Intelligence*. Macmillan, New York. [321, 327]
Thorndike, E. L. (1913) *The Psychology of Learning* (Educational Psychology II). Teachers College, New York. [320, 323]
Thorndike, E. L. (1932) Reward and punishment in animal learning. *Comp. Psychol. Monogr. 8* **39,** 26, 27, 47. [323]
Thornhill, R. (1980) Rape in Panorpa scorpion flies and a general rape hypothesis. *Anim. Behav.* **28,** 52–59. [142]
Thorpe, W. H. (1956) *Learning and Instinct in Animals*. Methuen, London. [275]
Thorpe, W. H. (1963) Ethology and the coding problem in germ cell and brain. *Z. Tierpsychol.* **20,** 529–551. [32]
Thorpe, W. H. (1974) *Animal Nature and Human Nature*. Methuen, London. [526]
Thorpe, W. H. (1979) *The Origins and Rise of Ethology*. Heinemann, London. [366]
Tinbergen, L. (1960) The natural control of insects in pinewoods, I: factors influencing the intensity of predation by song birds. *Arch. Néérl. Zool.* **13,** 265–343. [216]
Tinbergen, N. (1940) Die Übersprungbewegung. *Z. Tierpsychol.* **4,** 1–10. [381]
Tinbergen, N. (1942) An objective study of the innate behaviour of animals. *Biblioth. Biother,* **1,** 39–98. [36, 384]
Tinbergen, N. (1949) De functie van de rode vlek op de snavel van de zilvermeeuw. *Bijdr. tot de Dierk.* **28,** 453–465. *Sym. Soc. exp. Biol.* **4,** 305–312. [395]
Tinbergen, N. (1950) The hierarchical organization of nervous mechanisms underlying instinctive behaviour. [364]
Tinbergen, N. (1951) *The Study of Instinct*. Oxford

University Press, Oxford. [233, 363, 365, 366, 382–384, 392, 394, 395]

Tinbergen, N. (1952) Derived activities: their causation, biological significance, origin and emancipation during evolution. *Quart. Rev. Biol.* **27,** 1–32. [382, 384, 392]

Tinbergen, N. (1953) *The Herring Gull's World.* Collins, London. [364, 394, 396, 399]

Tinbergen, N. (1959) Comparative studies of the behaviour of gulls (Laridae): a progress report. *Behaviour* **15,** 1–70. [401, 402]

Tinbergen, N. (1962) The evolution of animal communication—a critical examination of methods. *Symp. zool. Soc. Lond.* **8,** 1–6. [78, 401]

Tinbergen, N. (1963) On aims and methods in ethology. *Z. Tierpsychol.* **20,** 410–433. [360]

Tinbergen, N., Broekhuysen, G. J., Feekes, F., Houghton, J. C. W., Kruuk, H. and Szulc, E. (1962). Eggshell removal by the black-headed gull, *Larus ridibundus* L.: a behaviour component of camouflage. *Behaviour* **19,** 74–118. [78–80]

Tinbergen, N. and Perdeck, A. C. (1950) On the stimulus situation releasing the begging response in the newly hatched herring gull chick (*Larus a. argentatus* Pont.). *Behaviour* **3,** 1–38. [395]

Tinbergen, N. and Iersel, J. J. A. van (1947) "Displacement reactions" in the three-spined stickleback. *Behaviour* **1,** 56–63. [382]

Tinkle, D. W. (1969) The concept of reproductive effort in its relation to the evolution of life histories of lizards. *Amer. Natur.* **103,** 501–516. [84, 86]

Toates, F. M. (1975) *Control Theory in Biology and Experimental Psychology.* Hutchinson, London. [277]

Toates, F. M. (1980) *Animal Behaviour—A Systems Approach.* Wiley, Chichester. [271, 273, 276, 277, 458]

Toates, F. M. and Oatley, K. (1970) Computer simulation of thirst and water balance. *Med. biol. Engng.* **8,** 71–87. [458]

Tolman, E. C. (1932) Purposive behavior in animals and men. Appleton-Century-Crofts, New York. (Reprinted University of California Press, 1949.) [521]

Tolman, E. C. (1938) The determiners of behavior at a choice point. *Psychol. Rev.* **45,** 1–41. [320, 344]

Treisman, M. (1977) Motion sickness: an evolutionary hypothesis. *Science* **197,** 493–495. [335]

Trivers, R. L. (1971) The evolution of reciprocal altruism. *Quart. Rev. Biol.* **46,** 35–57. [127, 134]

Trivers, R. L. (1972) Parental investment and sexual selection. In *Sexual Selection and the Descent of Man*, ed. B. Campbell, Aldine, Chicago. [121]

Tryon, R. C. (1942) Individual differences. In *Comparative Psychology* (2nd edition), ed. F. A. Moss, Prentice-Hall, Englewood Cliffs, New Jersey. [52]

Turnbull, C. M. (1966) *Wayward Servants: The Two Worlds of the African Pygmies.* Eyre & Spottiswoode, London. [160]

Uexkull, J. von (1934) *Streifzüge durch die Umwelten von Tieren und Menschen.* Springer Verlag, Berlin. Translated in *Instinctive Behaviour*, ed. C. H. Schiller, Methuen, London. [213, 364]

Ullyott, P. (1936) The behaviour of *Dendrocoelum lacteum*, I and II. *J. exp. Biol.* **13,** 253–264, 265–278. [244]

Ulrich, R. E. and Azrin, N. H. (1962) Reflexive fighting in response to aversive stimulation. *J. exp. Analysis Behavior* **5,** 511–520. [338]

Verney, E. G. (1947) The antidiuretic hormone and the factors which determine its release. *Proc. R. Soc. B.* **135,** 25–106. [334]

Vidal, J. M. (1976) L'empreinte chez les animaux. *La Recherche* **63,** 24–35. [371, 378]

Vince, M. A. (1969) Embryonic communication, respiration and the synchronization of hatching. In *Bird Vocalizations*, ed. Hinde, 233–260, Cambridge University Press, London. [32]

Waage, J. K. (1979) Foraging for patchily-distributed hosts by the parasitoid, *Nemeritis canescens. J. anim. Ecol.* **48,** 353–371. [473]

Wahlsten, D. L. and Cole, M. (1972) Classical and avoidance training of leg flexion in the dog. In *Classical Conditioning II: Current Research and Theory*, eds A. H. Black and W. F. Prokasy, Appleton-Century-Crofts, New York. [333]

Wallraff, H. G. (1969) Über das Orientierung-vermögen von Vögeln unter natürlichen und künstlichen Sternmustern. Dressurversuche mit Stockenten. *Ver. deut. zool. Ges.* Innsbruck, 348–357. [257]

Warden, C. J. and Warner, L. H. (1928) The sensory capacities and intelligence of dogs, with a report on the ability of the noted dog "Fellow" to respond to verbal stimuli. *Quart. Rev. Biol.* **3,** 1–28. [491]

Ware, D. M. (1972) Predation by rainbow trout (*Salmo gairdneri*): the influence of hunger, prey density, and prey size. *J. Fish. Res. Bd. Can.* **29,** 1193–1201. [232]

Warren, J. M. (1965) Primate learning in comparative perspective. In *Behavior of Nonhuman Primates*, Vol. 1, eds A. M. Schrier, H. F. Harlow and F. Stollnitz, 249–281, Academic Press, New York. [509]

Warren, J. M. (1973) Learning in vertebrates. In *Comparative Psychology: A Modern Survey*, eds D. A. Dewsbury and D. A. Rethlingshafer, 471–509, McGraw-Hill, New York. [507]

Warren, J. M. (1974) Possibly unique characteristics of learning by primates. *J. hum. Evol.* **3,** 445–454. [509]

Warriner, C. C., Lemmon, W. B. and Ray, T. S. (1963) Early experience as a variable in mate selection. *Anim. Behav.* **11,** 221–224. [372, 373]

Wasserman, E. A., Franklin, S. and Hearst, E. (1974) Pavlovian appetitive contingencies and approach versus withdrawal to conditioned stimuli in pigeons. *J. comp. physiol. Psychol.* **86,** 616–627. [350]

Watson, J. B. (1907) Kinesthetic and organic sensations: their role in the reactions of the white rat to the maze. *Psychol. Monogr. 8,* 33–49. [311]

Watson, J. B. (1913) Psychology as the behaviorist views it.' *Psychol. Rev.* **20,** 158–177. [8, 311]

Watson, J. B. (1914) *Behavior: An Introduction to Comparative Psychology*. Holt, New York. [311]

Watson, J. B. (1916) The place of the conditioned reflex in psychology. *Psychol. Rev.* **23,** 89–116. [311]

Watson, J. B. (1930) *Behaviorism*. W. W. Norton, New York. [39]

Watson, J. D. and Crick, F. H. (1953) Molecular structure of nucleic acids: a structure for deoxyribose nucleic acid. *Nature, London* **171,** 737–738. [23]

Weihaupt, J. G. (1964) Geophysical biology. *Bioscience* **14,** 18–24. [291]

Weiskrantz, L. (1980) Varieties of residual experience. *Quart. J. exp. Psychol.* **32,** 365–386. [523]

Weiskrantz, L., Warrington, E. K., Sanders, M. D. and Marshall, J. (1974) Visual capacity of the hemianopic field following a restricted occipital ablation. *Brain* **97,** 709–728. [523]

Wells, G. P. (1966) The lugworm (*Arenicola*): a study in adaptation. *Netherlands J. Sea Res.* **3,** 294–313. [456]

Wells, M. J. (1966) *Brain and Behaviour in Cephalopods*. Heinemann, London. [180]

Wells, M. J. (1968) *Lower Animals*. Weidenfeld and Nicolson, London. [343]

Wendler, G. (1966) Coordination of walking movements in arthropods. *Symp. Soc. exp. Biol.* **20,** 229–249. [241]

Wenner, A. M. (1962) Sound production during the waggle dance of the honey bee. *Anim. Behav.* **10,** 79–95. [423]

Wenner, A. M. (1964) Sound communication in honey bees. *Scientific American,* **210**(4), 116–124. [423]

Werner, E. E. and Hall, D. J. (1974) Optimal foraging and the size selection of prey by the bluegill sunfish (*Lepomis macrochirus*). *Ecology* **55,** 1216–1232. [429]

West, G. C. and Norton, D. W. (1975) Metabolic adaptations in tundra birds. In *Physiological Adaptations to the Environment*, ed. F. J. Fernberg, Intext Educational Publ., New York. [292]

Westermark, E. (1891) *The History of Human Marriage*. Macmillan, London. [379]

Whiten, A. (1972) Operant study of sun altitude and pigeon navigation. *Nature, London* **237,** 405–406. [255]

Whitfield, M. (1976) The evolution of the oceans and the atmosphere. In *Environmental Physiology of Animals*, eds J. Bligh, J. L. Cloudsley-Thompson and A. G. Macdonald, Blackwell Scientific Publications, Oxford. [284]

Wickler, W. (1966) Ursprung und biologische Deutung des Genitalpräsentierens männlicher Primaten. *Z. Tierpsychol.* **23,** 422–437. [124–126]

Wickler, W. (1967) Socio-sexual signals and their intra-specific imitation among primates. In *Primate Ethology*, ed. D. Morris, Weidenfeld and Nicolson, London. [125]

Wickler, W. (1968) *Mimicry in Plants and Animals*. Weidenfeld and Nicolson, London. [407]

Wiepkema, P. R. (1971) Positive feedbacks at work during feeding. *Behaviour* **39,** 2–4. [281, 459]

Wiepkema, P. R., Alingh Prins, A. J. and Steffens, A. B. (1972) Gastrointestinal food transport in relation to meal occurrence in rats, 1. *Physiol. Behav.* **9,** 759–763. [273]

Wilcoxon, H. C., Dragoin, W. B. and Kral, P. A.

(1971) Illness-induced aversions in rat and quail: relative salience of visual and gustatory cues. *Science* **171,** 826–828. [338]

Wilkinson, P. F. and Shank, C. C. (1977) Rutting-fight among musk oxen on Banks Island, Northwest Territories, Canada. *Anim. Behav.* **24,** 756–758. [105]

Williams, D. R. and Williams, H. (1969) Auto-maintenance in the pigeon: sustained pecking despite contingent non-reinforcement. *J. exp. Anal. Behav.* **12,** 511–520. [331]

Williams, G. C. (1966) *Adaptation and Natural Selection.* Princeton University Press, Princeton, New Jersey. [142]

Wilson, D. M. (1968) Inherent asymmetry and reflex modulation of the locust flight pattern. *J. exp. Biol.* **48,** 631–641. [241]

Wilson, E. O. (1975) *Sociobiology: The New Synthesis.* The Belknap Press of Harvard University Press, Cambridge, Massachusetts. [128, 148]

Wilson, J. A. (1979) *Principles of Animal Physiology* (2nd edition). Macmillan, New York. [268]

Wolf, A. P. (1966) Childhood association, sexual attraction and the incest taboo: a Chinese case. *Amer. Anthrop.* **68,** 883–898. [379]

Wolf, A. P. (1970) Childhood association and sexual attraction: a further test of the Westermark hypothesis. *Amer. Anthrop.* **72,** 503–515. [379]

Woodward, W. T. and Bitterman, M. E. (1973) Pavlovian analysis of avoidance conditioning in the goldfish (*Carassius auratus*). *J. comp. Physiol.* **82,** 123–129. [333]

Woodfield, A. (1976) *Teleology.* Cambridge University Press, Cambridge. [427]

Woodruff, G. and Premack, D. (1979) Intentional communication in the chimpanzee: the development of deception. *Cognition* **7,** 333–362. [491, 500]

Woodworth, R. S. (1918) *Dynamic Psychology.* Columbia University Press, New York. [275]

Woolfenden, G. E. (1973) Nesting and survival in a population of Florida scrub jays. *Living Bird* **12,** 25–49. [138]

Wootton, R. J. (1976) *The Biology of the Sticklebacks.* Academic Press, London. [93, 119]

Wright, P. and McFarland, D. J. (1969) A functional analysis of hypothalamic polydipsia in the Barbary dove (*Streptopelia risoria*). *Physiol. Behav.* **4,** 877–883. [268]

Wright, S. (1921) Systems of mating. *Genetics* **6,** 111–178. [63]

Wurtman, R. J., Axelrod, J. and Kelly, D. E. (1968) *The Pineal.* Academic Press, New York. [292]

Yodlowski, M. L., Kreithen, M. L. and Keeton, W. T. (1977) Detection of atmospheric infrasound by homing pigeons. *Nature, London* **265,** 725–726. [254]

Young, J. Z. (1960) The failures of discrimination learning following removal of the vertical lobes in Octopus. *Proc. R. Soc. Lond. B.* **153,** 18–46. [343]

Young, P. T. (1961) *Motivation and Emotion: A Survey of the Determinants of Human and Animal Activity.* Wiley, New York. [334]

Zach, R. (1979) Shell dropping: decision making and optimal foraging in Northwestern crows. *Behaviour* **68,** 106–117. [432]

Zahavi, A. (1974) Communal nesting by the Arabian babbler: a case of individual selection. *Ibis* **116,** 84–87. [138, 140]

Zahavi, A. (1975) Mate selection—a selection for a handicap. *J. theor. Biol.* **53,** 205–214. [118]

Zahavi, A. (1976) Cooperative nesting in Eurasian birds. *Proc. XVI int. orn. Congr. (Canberra, Australia)* 685–693. [140]

Zahavi, A. (1977) The cost of honesty. (Further remarks on the handicap principle). *J. theor. Biol.* **67,** 603–605. [118]

Zahorik, D. M. and Maier, S. F. (1969) Appetite conditioning with recovery from thiamine deficiency as the unconditioned stimulus. *Psychol. Sci.* **17,** 309–310. [336]

Zahorik, D. M., Maier, S. F. and Pies, R. W. (1974) Preferences for tastes paired with recovery from thiamine deficiency in rats. Appetite conditioning or learned safety? *J. comp. physiol. Psychol.* **87,** 1083–1091. [336]

Zimmerman, J. L. (1971) The territory and its density dependent effect in *Spiza americana. Auk* **88,** 591–612. [92]

Subject Index